Advanced Wireless Transmission Technologies

Elucidating fundamental design principles by means of accurate trade-off analysis of relevant design options, this is the first book to provide a coherent treatment of transmission technologies essential to current and future wireless systems. Develop in-depth knowledge of the capabilities and limitations of wireless transmission technologies in supporting high-quality wireless transmission services, and foster a thorough understanding of various design trade-offs, to make informed decisions for your own application requirements. Key technologies such as advanced diversity combining, multiuser scheduling, multiuser multi-antenna transmission, relay transmission, and cognitive radio are examined, making this book an indispensable reference for senior graduate students, researchers, and engineers working in wireless communications.

Hong-Chuan Yang is Professor in the Department of Electrical and Computer Engineering at the University of Victoria. He co-authored *Order Statistics in Wireless Communications* (Cambridge, 2011) with Mohamed-Slim Alouini.

Mohamed-Slim Alouini is Professor of Electrical Engineering at KAUST, and a Fellow of the IEEE.

Advanced Wireless Transmission Technologies

Analysis and Design

HONG-CHUAN YANG
University of Victoria

MOHAMED-SLIM ALOUINI
King Abdullah University of Science and Technology

CAMBRIDGE
UNIVERSITY PRESS

University Printing House, Cambridge CB2 8BS, United Kingdom

One Liberty Plaza, 20th Floor, New York, NY 10006, USA

477 Williamstown Road, Port Melbourne, VIC 3207, Australia

314–321, 3rd Floor, Plot 3, Splendor Forum, Jasola District Centre, New Delhi – 110025, India

79 Anson Road, #06–04/06, Singapore 079906

Cambridge University Press is part of the University of Cambridge.

It furthers the University's mission by disseminating knowledge in the pursuit of education, learning, and research at the highest international levels of excellence.

www.cambridge.org
Information on this title: www.cambridge.org/9781108420198
DOI: 10.1017/9781108332835

© Cambridge University Press 2020

This publication is in copyright. Subject to statutory exception and to the provisions of relevant collective licensing agreements, no reproduction of any part may take place without the written permission of Cambridge University Press.

First published 2020

Printed in the United Kingdom by TJ International Ltd, Padstow Cornwall

A catalogue record for this publication is available from the British Library.

Library of Congress Cataloging-in-Publication Data
Names: Yang, Hong-Chuan, 1973– author. | Alouini, Mohamed-Slim, author.
Title: Advanced wireless transmission technologies : analysis and design / Hong-Chuan Yang and Mohamed-Slim Alouini.
Description: First edition. | New York : Cambridge University Press, 2019. | Includes bibliographical references and index.
Identifiers: LCCN 2019028529 (print) | LCCN 2019028530 (ebook) | ISBN 9781108420198 (hardback) | ISBN 9781108332835 (epub)
Subjects: LCSH: Wireless communication systems.
Classification: LCC TK5103.2 .Y3385 2019 (print) | LCC TK5103.2 (ebook) | DDC 621.384–dc23
LC record available at https://lccn.loc.gov/2019028529
LC ebook record available at https://lccn.loc.gov/2019028530

ISBN 978-1-108-42019-8 Hardback

Cambridge University Press has no responsibility for the persistence or accuracy of URLs for external or third-party internet websites referred to in this publication and does not guarantee that any content on such websites is, or will remain, accurate or appropriate.

**To
our families**

Contents

	Preface		*page* xi
1	**Digital Wireless Communications**		1
	1.1	Wireless Channel Modeling	1
	1.2	Digital Communications over Fading Channels	9
	1.3	Effect of Frequency-Selective Fading	16
	1.4	Summary	19
	1.5	Further Reading	19
	References		19
2	**Transmission Technologies for Flat Fading Channels**		20
	2.1	Diversity Combining Techniques	20
	2.2	Channel Adaptive Transmission	31
	2.3	MIMO Transmission	36
	2.4	Summary	39
	2.5	Further Reading	39
	References		40
3	**Transmission Technologies for Selective Fading Channels**		41
	3.1	Equalization	41
	3.2	Multicarrier Transmission/OFDM	44
	3.3	Spread Spectrum Transmission	48
	3.4	Summary	53
	3.5	Further Reading	53
	References		54
4	**Advanced Diversity Techniques**		55
	4.1	Generalized Selection Combining (GSC)	55
	4.2	GSC with Threshold Test per Branch (T-GSC)	59
	4.3	Generalized Switch and Examine Combining	66
	4.4	GSEC with Post-Examining Selection (GSECps)	72
	4.5	Summary	81
	4.6	Further Reading	82
	References		82

5 Adaptive Diversity Combining — 84
- 5.1 Output-Threshold Maximum Ratio Combining — 84
- 5.2 Minimum Selection GSC — 91
- 5.3 Output-Threshold GSC — 102
- 5.4 Adaptive Transmit Diversity — 113
- 5.5 RAKE Finger Management during Soft Handover — 120
- 5.6 Joint Adaptive Modulation and Diversity Combining — 128
- 5.7 Summary — 143
- 5.8 Further Reading — 143
- References — 143

6 Multiuser Scheduling — 146
- 6.1 Multiuser Selection Diversity — 146
- 6.2 Performance Analysis of Multiuser Selection Diversity — 148
- 6.3 Multiuser Diversity with Limited Feedback — 154
- 6.4 Multiuser Parallel Scheduling — 160
- 6.5 Power Allocation for Parallel Scheduling — 169
- 6.6 Summary — 174
- 6.7 Further Reading — 176
- References — 176

7 Multiuser MIMO Transmissions — 179
- 7.1 Introduction — 179
- 7.2 Zero-forcing Beamforming Transmission — 182
- 7.3 Random Unitary Beamforming (RUB) Transmission — 189
- 7.4 RUB with Conditional Best Beam Index Feedback — 203
- 7.5 RUB Transmission with Receiver Linear Combining — 211
- 7.6 Summary — 222
- 7.7 Further Reading — 222
- References — 223

8 Relay Transmission — 226
- 8.1 Basic Relaying Strategies — 226
- 8.2 Opportunistic Nonregenerative Relaying — 230
- 8.3 Cooperative Opportunistic Regenerative Relaying — 240
- 8.4 Incremental Opportunistic Regenerative Relaying — 246
- 8.5 Summary — 254
- 8.6 Further Reading — 254
- References — 254

9 Cognitive Transmission — 256
- 9.1 Introduction to Cognitive Radio — 256
- 9.2 Temporal Spectrum Opportunity Characterization — 258
- 9.3 Extended Delivery Time Analysis — 263

	9.4	Spectrum Sharing: Single Antenna Transmitter	281
	9.5	Spectrum Sharing: Transmit Antenna Selection	289
	9.6	Summary	299
	9.7	Further Reading	299
		References	299
10	**Application: Hybrid FSO/RF Transmission**		303
	10.1	Switching-Based Hybrid FSO/RF Transmission	303
	10.2	Hybrid FSO/RF Transmission with Adaptive Combining	319
	10.3	Joint Adaptive Modulation and Combining for Hybrid FSO/RF Transmission	323
	10.4	Summary	330
	10.5	Further Reading	331
		References	332
11	**Application: Sensor Transmission with RF Energy Harvesting**		335
	11.1	Cognitive Transmission with Harvested RF Energy	335
	11.2	Cooperative Beam Selection for RF Energy Harvesting	351
	11.3	Summary	363
	11.4	Further Reading	364
		References	365
12	**Application: Massive MIMO Transmission**		367
	12.1	Antenna Subset Selection for Massive MIMO	367
	12.2	Hybrid Precoding for Massive MIMO	377
	12.3	Summary	388
	12.4	Further Reading	389
		References	389
Appendix	**Order Statistics**		391
	A.1	Basic Distribution Functions	391
	A.2	Distribution of Partial Sum of the Largest Order Statistics	393
	A.3	Joint Distributions of Partial Sums	397
	A.4	Limiting Distributions of Extreme Order Statistics	404
	A.5	Summary	405
	A.6	Further Reading	405
		References	405
		Index	407

Preface

Motivation and Goal

The wireless communication industry continues its exciting era of rapid development. To meet the ever growing demand for ubiquitous wireless connectivity and high-data-rate wireless services, various advanced transmission technologies have been developed and investigated. In particular, channel-adaptive transmission technology was proposed and has been widely adopted to improve spectrum utilization efficiency while satisfying a certain reliability requirement. Multiple antenna transmission and reception, commonly referred to as MIMO, technologies have demonstrated as an effective solution to enhance transmission reliability, efficiency, and capacity by exploring the spatial degree of freedom. Relay transmission was introduced as a viable option to extend the high-data-rate coverage of wireless communication systems. Last but not least, cognitive radio transmission offers an attractive solution to address the spectrum scarcity problem by exploiting underutilized spectra. These technologies will remain the essential building blocks of future wireless communication systems.

Emerging application scenarios, such as vehicular networks, Internet of Things, big data applications, and smart grid communications, bring new technical challenges to wireless communications. Wireless communication systems will continue to evolve to satisfy the diverse quality of service requirements of existing and future applications. Meanwhile, after several decades of intensive research and development in the field, physical layer wireless transmission over fading channels has reached a new level of maturity. The wireless community has developed a sufficient understanding of those advanced wireless transmission technologies listed above and readily applied them to various modern wireless systems. It is anticipated that future wireless system designers will need to make suitable design choices, from these technologies and their various implementation options, for their specific target application scenarios. As such, they need to possess an in-depth understanding of these transmission technologies and their capabilities/limitations in delivering high-quality wireless transmission services.

Several books on these advanced transmission technologies have been published, in addition to the rich literature on them in the form of published journal articles and conference papers. In particular, multiple books have been dedicated to MIMO transmission technologies [1–5]. Cognitive radio received extensive attention in early 2000s, which led to a few books on the subject of cognitive wireless communications and networks

[6–9]. There are also several edited and authored volumes on cooperative relay transmission [10, 11]. On the other hand, there does not exist a single-volume book that systematically presents these advanced transmission technologies in a unified fashion. The treatment of other advanced transmission technologies, such as user scheduling and advanced diversity techniques, are scattered across various journal and conference publications. Researchers, engineers, and graduate students interested in wireless communications need to read several different books and many journal papers to obtain a good understanding of these important technologies.

The primary reason for writing this book is to provide a coherent treatment of various advanced wireless transmission technologies. The goal is to establish a solid understanding of the fundamental design principles of these technologies and their application potentials in various scenarios among readers with basic digital wireless communication knowledge [12–14]. Such a single-volume book will serve as a valuable reference for senior graduate students, faculty members, researchers, and engineers in the field, and greatly benefit the wireless communication community. Through this book, graduate students interested in wireless communications can obtain a general but in-depth knowledge of these technologies before diving into specific research topics. Researchers in the field can attain a new perspective toward these key enablers for highly spectrum- and energy-efficient wireless services. Researchers in related areas, such as wireless networking and component design, can efficiently learn the essential ideas behind these physical-layer transmission solutions. System engineers can develop a comprehensive understanding of different transmission technologies before further exploring individual technologies for their specific design needs.

Approach and Features

Wireless communications is a broad field. There are many technical aspects associated with each individual transmission technology, ranging from information theory to system engineering, from channel modeling to performance analysis, from signal processing algorithms to standard incorporation. Even a dedicated book may not be able to thoroughly cover all of these perspectives with sufficient depth/details. As such, it is a serious challenge to systematically present various advanced transmission technologies in a single volume, especially when a certain theoretical depth is expected. We would have to settle at a certain compromise between technical breadth and theoretical depth. It is also critical to maintain a logical connection between different topics for coherence and readability of the book.

In this book, following a rather theoretical approach, we concentrate on the fundamental design principles behind each transmission technology and their demonstration with representative designs. To foster a deep understanding, we address their exact performance and complexity characterization in a fading environment. Building upon a common analytical framework of digital wireless transmission over fading channels, different design options of each transmission technology are systematically analyzed and compared. Through such solid analytical exposure, readers can thoroughly appreciate the design principles as well as the essential design trade-offs involved. We briefly explain the implementation challenges associated with each technology, but leave the hardware implementation and standard incorporation to readers' future exploration.

The efficient application of advanced wireless transmission technologies in real-world systems relies heavily on the accurate prediction of their performance over general wireless fading channels and the corresponding system complexity. Theoretical performance and complexity analysis are invaluable in this process, because they can help circumvent the time-consuming computer simulation and expensive field-test campaigns. Whenever feasible, elegant closed-form solutions for important performance metrics are obtained with related discussion. These analytical results will bring important insights into the dependence of the performance, as well as complexity, on system design parameters and, as such, facilitate the determination of the most suitable design choice in the face of practical implementation constraints.

Mathematical and statistical tools play a critical role in the performance and complexity analysis of digital wireless communication systems over fading channels. In fact, the proper utilization of these tools can help either simplify the existing results, which do not allow efficient numerical evaluation, or provide new analytical solutions that were previously deemed infeasible. To analyze these advanced transmission technologies, we employ various mathematical tools, including moment generation function (MGF), Markov chains, and order statistics. Through this book, readers will gain an extensive exposure to these diverse mathematical tools and their applications in wireless system analysis.

We strive to achieve the ideal balance between theory and practice. Special emphasis is placed on the important trade-off of performance versus complexity throughout the book. Whenever feasible, the associated complexity measures are quantified and plotted together with the performance curves for clear trade-off illustration. The essential results on order statistics are summarized in the Appendix so that the main body can focus on the practical design insights of different transmission technologies. Such an arrangement allows easy reading and convenient referencing.

Organization

This book provides a systematic treatment of various advanced wireless transmission technologies. First, we summarize the basics of digital wireless communications over fading channels. Fundamental transmission technologies for both flat and selective fading channels are presented together with their analytical framework, which serves as the necessary background for the rest of the book. Building upon the fundamentals, we then present various advanced transmission technologies, including advanced diversity techniques, channel-adaptive transmission and reception techniques, multiuser scheduling techniques, multiuser MIMO techniques, cooperative relaying transmission, and cognitive radio transmission. Besides presenting various design options for each transmission technologies, special emphasis is placed on their performance and complexity analysis.

The accurate mathematical formulation will help the readers obtain a thorough and accurate understanding of the various design options and their associated design trade-offs. In Chapters 10–12, as contemporary applications of these advanced transmission technologies, we present several representative designs for hybrid FSO/RF transmission, sensor transmission with RF energy harvesting, and massive MIMO transmission. Finally, the key results of order statistics are summarized in the Appendix.

The content of each chapter is summarized here.

Chapter 1 presents the basics of digital wireless communications, including channel modeling, digital bandpass modulation, and performance analysis over fading channels. Furthermore, the key performance metrics and the general approaches to evaluating them are introduced.

Chapter 2 reviews fundamental transmission technologies for frequency flat fading channels, including diversity combining, transmit diversity, adaptive transmission, and MIMO transmission. We also present a trade-off analysis on different threshold combining schemes.

Chapter 3 summarizes fundamental transmission technologies for frequency-selective fading channels, including equalization, multicarrier transmission, and spread spectrum transmission. The design principle and practical challenges of orthogonal frequency division multiplexing (OFDM) the discrete implementation of multicarrier transmission, are also presented in detail.

Chapter 4 presents four advanced diversity combining schemes for diversity-rich environment, including generalized selection combining (GSC), GSC with threshold test per branch (T-GSC), generalized switch and examine combining (GSEC), and GSEC with post-examining selection (GSECps). The performance versus complexity trade-off of these schemes is accurately quantified after deriving the exact statistics of the combiner output signal-to-noise ratio (SNR).

Chapter 5 introduces adaptive diversity combining techniques for system complexity reduction and processing power savings. Three sample adaptive combining schemes, namely output threshold MRC (OT-MRC), minimum selection GSC (MS-GSC), and output threshold GSC (OT-GSC), are discussed. Then, we apply the general idea of adaptive combining to transmit diversity systems, RAKE receiver design over soft handover region, and joint design with rate-adaptive transmission.

Chapter 6 studies multiuser scheduling techniques. Both single-user scheduling and multiuser parallel scheduling schemes are investigated. We focus mainly on those low-complexity schemes that are readily applicable in practical multiuser wireless systems. Through accurate mathematical analysis, we quantify their unique trade-offs of performance versus fairness, feedback load, and system complexity.

Chapter 7 studies essential transmission technologies for multiuser MIMO systems. We focus on low-complexity linear transmission techniques, including zero-forcing beamforming (ZFBF) and random unitary beamforming (RUB), and study their performance and complexity through accurate mathematical analysis. Here, we present the most comprehensive analysis on various RUB schemes.

Chapter 8 discusses several essential relay transmission strategies. After presenting fundamental amplify-and-forward (AF) and decode-and-forward (DF) relaying schemes, we introduce the concepts of opportunistic relaying, cooperative relaying, and incremental relaying. These concepts are illustrated through accurate performance quantification of sample designs in terms of outage probability and average end-to-end error rate.

Chapter 9 presents essential analysis and design problems of cognitive transmission. After introducing the basic idea of cognitive radio and its key enabling provisions, we first characterize temporal transmission opportunities and secondary transmission delay for opportunistic spectrum access implementation. For spectrum-sharing implementation, we study the performance of secondary transmission with power adaptation, diversity reception, and transmit antenna selection under primary interference constraints.

Chapter 10 considers hybrid FSO/RF transmission systems. Several practical transmission schemes are developed for hybrid FSO/RF links with the application of switched diversity combining, adaptive diversity combining, and joint adaptive modulation and combining. Special emphasis is placed on the candidate solutions to key challenges of hybrid FSO/RF transmission.

Chapter 11 investigates secondary sensor transmission with harvested RF energy from existing wireless systems. We develop and apply the exact statistics of harvested RF energy over fading environment to the analysis and design of secondary packet transmission in the presence of primary interference. We also study energy harvesting performance improvement with cooperative beamforming.

Chapter 12 studies several low-complexity transmission schemes for massive MIMO systems. Following the antenna subset selection approach, we present a trace-based sequential selection strategy for massive MIMO transmission. We also investigate a hybrid procoding solution for massive MIMO systems for partially connected beamforming structure.

The Appendix presents a summary of the key order statistics results used in the analysis of various wireless transmission technologies. Specifically, we introduce several new results on order statistics, including the distribution and joint distribution of the partial sums of ordered random variables. We also summarize the limiting distributions of extreme statistics.

Usage

This book is intended for senior graduate students, researchers, and practicing engineers in the field of wireless and mobile communications. The reader should have a basic knowledge of digital wireless communications [12–14]. The material in this book has been used in a one-term graduate-level course on advanced wireless communications at the University of Victoria, Canada, and an intensive three-week short course at the Tsinghua University, China and King Abdullah University of Science and Technology, Saudi Arabia. It is an ideal avenue for students to enhance their analytical skills as well as to expand their knowledge of advanced wireless transmission technologies.

Acknowledgments

We would like to acknowledge our current and past students, post-docs, and collaborators for their contributions in the works that have been included in this book. Specifically, the authors would like to thank Professor Seyeong Choi, Professor David Gesbert, Professor Mazen Hasna, Professor Young-chai Ko, Professor Geir Oien, Professor Khalid A. Qaraqe, Dr. Peng Lu, Dr. Sung-Sik Nam, Dr. Ki-Hong Park, Dr. Lin Yang, Dr. Muhamed Hanif, Dr. Kamel Tourki, Professor Tamer Rakia, Dr. Tianqing Wu, Dr. Wenjing Wang, Mr. Muneer Usman, Mr. Seung-Sik Eom, Mr. Noureddine Hamdi, Mr. Bengt Holter, and Mr. Le Yang. The authors want to thank Dr. Julie Lancashire at Cambridge University Press, whose remarkable talent identified the potential of this project. Finally, the valuable support from Ms. Sarah Strange, Ms. Heather Brolly, Annie Toynbee, and Samuel Fearnley at Cambridge University Press during the preparation of the book is much appreciated.

References

[1] E. Biglieri, R. Calderband, A. Constantinides, A. Goldsmith, A. Paulraj, and H. V. Poor, *MIMO Wireless Communications*, Cambridge University Press, 2007.
[2] B. Clerckx and C. Oestges, *MIMO Wireless Networks*, Academic Press, 2013.
[3] T. Marzetta, E. G. Larsson, H. Yang, and H. Q. Ngo, *Fundamentals of Massive MIMO*, Cambridge University Press, 2016.
[4] R. Kshetrimayum, *Fundamentals of MIMO Wireless Communications*, Cambridge University Press, 2017.
[5] R. W. Heath and A. Lozano, *Foundations of MIMO Communication*, Cambridge University Press, 2018.
[6] K.-C. Chen and R. Prasad, *Cognitive Radio Networks*, Wiley, 2009.
[7] E. Hossain, D. Niyato, and Z. Han, *Dynamic Spectrum Access and Management in Cognitive Radio Networks*, Cambridge University Press, 2009.
[8] R. C. Qiu, Z. Hu, H. Li, and M. C. Wicks, *Cognitive Radio Communications and Networking*, Wiley, 2012
[9] E. Biglieri, A. Goldsmith, L. J. Greenstein, N. B. Mandayam, and H. V. Poor, *Principles of Cognitive Radio*, Cambridge University Press, 2007.
[10] M. Dohler and Y. Li, *Cooperative Communications: Hardware, Channel and PHY*, Wiley, 2010.
[11] Y.-W. P. Hong, W.-J. Huang, and C.-C. J. Kuo, *Cooperative Communications and Networking*, Springer, 2010.
[12] A. J. Goldsmith, *Wireless Communications*, Cambridge University Press, 2005.
[13] D. Tse and P. Viswanath, *Fundamentals of Wireless Communication*, Cambridge University Press 2005.
[14] H.-C. Yang, *Introduction to Digital Wireless Communications*, IET Press, 2017.

1 Digital Wireless Communications

Wireless communications is a fascinating field with a very broad scope. This book primarily focuses on physical wireless transmission over point-to-point links. To facilitate the presentation of advanced wireless transmission technologies in later chapters, we present a brief overview of digital wireless communications in the first three chapters. The discussion here is by no means meant to be comprehensive. The main objective is to introduce the essential background for advanced wireless technologies covered in later chapters. For a thorough treatment of these subjects, the reader may refer to [1–3].

In this chapter, we first present the basic models for three main effects of wireless channels, i.e., path loss, shadowing, and fading. We then introduce the statistical fading channel models commonly used in wireless system analysis. After that, we discuss the digital modulation schemes and their performance analysis over fading channels, including the well-known moment generating function (MGF) based approach [3]. Finally, the effect of frequency-selective fading is discussed.

1.1 Wireless Channel Modeling

Wireless channels rely on the physical phenomenon of electromagnetic wave propagation due to the pioneering discoveries of Maxwell and Hertz. Radio waves propagate through several mechanisms, including direct line of sight (LOS), reflection, diffraction, and scattering. As such, there usually exist multiple propagation paths between the transmitters and the receivers, as illustrated in Fig. 1.1. In general, when the LOS path exists, as in microwave systems and indoor applications, the transmitted radio signal experiences less attenuation. If the LOS path does not exist, the radio signal can still reach the receiver through other mechanisms, but with severe attenuation.

The complicated propagation environment and the unpredictable nature of the propagation process make the modeling of wireless channels very challenging, especially considering the mobility of the transmitter and/or the receiver. To avoid the modeling complications associated with the propagation details, the wireless channel is usually characterized by three major effects: (1) path loss, for the general trend of power dissipation as propagation distance increases; (2) shadowing, due to large objects, such as buildings and trees, along the propagation path; and (3) fading, as a result of the random superposition of signals from different propagation paths at the receiver. The received

Digital Wireless Communications

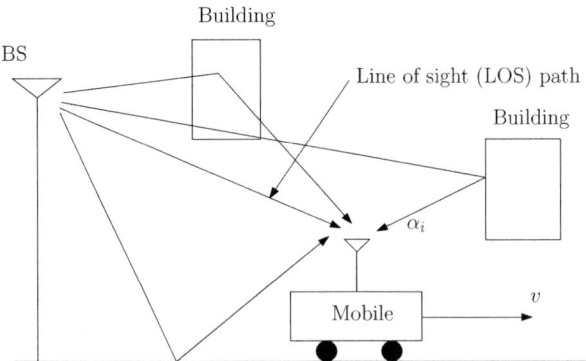

Figure 1.1 Multipath propagation.

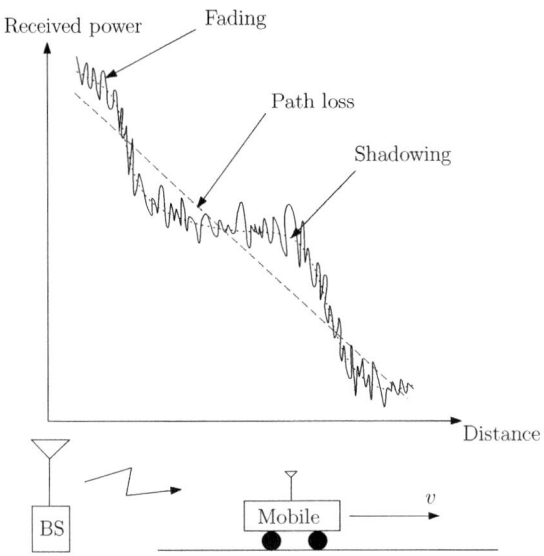

Figure 1.2 Received signal power attenuation as propagation distance increases.

signal power variation due to these three effects is demonstrated in Fig. 1.2. In general, path loss and shadowing are called large-scale propagation effects, whereas fading is referred to as a small-scale effect, since the fading effect manifests at much smaller temporal/spatial scales.

1.1.1 Path Loss and Shadowing

Path loss characterizes the general trend of power attenuation as the propagation distance increases. The linear path loss (PL) is defined as the ratio of the transmitted signal power P_t versus the received signal power P_r, i.e., $PL = P_t/P_r$. The log-distance model is the most convenient path loss model for high-level system analysis among all available path

1.1 Wireless Channel Modeling

loss models [2]. Specifically, under the log-distance model, path loss at a distance d in the dB scale is predicted as

$$PL(d)\text{dB} = PL(d_0)\text{dB} + 10\beta \log_{10}(d/d_0), \quad (1.1)$$

where d_0 is the reference distance, usually set to 1–10 m for indoor applications and 1 km for outdoor applications, $PL(d_0)$dB is the path loss at d_0, and β is the path loss exponent, which can be estimated by minimizing the mean square error (MSE) between the measurement data and the model. Note that the path loss model only captures the general trend of power dissipation, ignoring the effect of surrounding objects or multipath.

Shadowing characterizes the blocking effect of large objects in the propagation environment. As the size, position, and properties of such objects are in general unknown, the shadowing effect has to be described in the statistical sense. The most popular shadowing model is the log-normal model, which has been empirically confirmed [4]. With the log-normal model, the path loss in dB scale at distance d, denoted by ψ_{dB}, is modeled as a Gaussian random variable with mean value $PL(d)$ dB, from appropriate path loss model, and variance σ_{dB}^2, which varies with the environment. As such, the probability density function (PDF) of ψ_{dB} is given by

$$p_{\psi_{\text{dB}}}(x) = \frac{1}{\sqrt{2\pi}\sigma_{\text{dB}}} \exp\left(-\frac{(x - PL(d)\text{dB})^2}{2\sigma_{\text{dB}}^2}\right). \quad (1.2)$$

The log-normal model is so named as the path loss in linear scale ψ is a log-normal random variable, with PDF given by

$$p_{\psi}(x) = \frac{10/\ln(10)}{\sqrt{2\pi}x\sigma_{\text{dB}}} \exp\left(-\frac{(10\log_{10}x - PL(d)\text{dB})^2}{2\sigma_{\text{dB}}^2}\right). \quad (1.3)$$

With the path loss and shadowing model, we can address some interesting system design problem, e.g., for a given transmitting power and target service area, what is the percentage of coverage after considering the path loss and shadowing effects? Assuming a location is covered if the received signal power considering path loss and shadowing effects is above the threshold P_{min}, the percentage of coverage can be calculated by averaging the probability that the received signal power at a distance r is less than P_{min} over the target area. Mathematically, we have

$$C = \frac{1}{\pi R^2} \int_0^{2\pi} \int_0^R \Pr[P_r(r) > P_{\text{min}}] r \, dr \, d\theta, \quad (1.4)$$

where R is the radius of the cell. The coverage probability at distance r under the combined log-distance path loss model and log-normal shadowing model is given by

$$\Pr[P_r(r) > P_{\text{min}}] = \int_{-\infty}^{P_t - P_{\text{min}}} p_{\psi_{\text{dB}}}(x) dx, \quad (1.5)$$

where $p_{\psi_{\text{dB}}}(\cdot)$ was given in (1.2) together with (1.1). After carrying out integration, the coverage percentage can be calculated as

$$C = Q(a) + \exp\left(\frac{2 - 2ab}{b^2}\right) Q\left(\frac{2 - ab}{b}\right), \quad (1.6)$$

where

$$a = \frac{P_{\min} - P_t + PL(d_0)\text{dB} + 10\gamma \log_{10}(R/d_0)}{\sigma_{\text{dB}}}, b = \frac{10\gamma \log_{10}(e)}{\sigma_{\text{dB}}},$$

and $Q(\cdot)$ is the Gaussian Q-function,[1] defined as

$$Q(x) = \frac{1}{\sqrt{2\pi}} \int_x^{\infty} e^{-\frac{t^2}{2}} \, dt. \tag{1.7}$$

1.1.2 Multipath Fading

Fading characterizes the effect of the random superposition of signal copies that arrive at the receiver from different propagation paths. These signal replicas may add together constructively or destructively, which leads to a large variation in received signal strength. Let us assume the following bandpass signal is transmitted over a wireless channel:

$$s(t) = \text{Re}\{u(t)e^{j2\pi f_c t}\}, \tag{1.8}$$

where $u(t)$ is a complex baseband envelope. The received signal after multipath propagation becomes

$$r(t) = \text{Re}\left\{\sum_{n=0}^{N(t)} \alpha_n(t) u(t - \tau_n(t)) e^{j2\pi f_c(t - \tau_n(t)) + \phi_{D_n}(t)}\right\}, \tag{1.9}$$

where $N(t)$ is the number of paths, $\tau_n(t)$ is the delay, $\alpha_n(t)$ is the amplitude, and $\phi_{D_n}(t)$ is the phase shift, all for the nth path at time t. Note that the phase shift $\phi_{D_n}(t)$ is related to the Doppler frequency shift as $\phi_{D_n}(t) = \int_t 2\pi f_{D_n}(t) \, dt$. After some manipulation, while focusing on the complex baseband input and output relationship, the complex baseband impulse response of the wireless channel can be obtained as

$$c(\tau, t) = \sum_{n=0}^{N(t)} \alpha_n(t) e^{-j\phi_n(t)} \delta(\tau - \tau_n(t)),$$

where $\phi_n(t) = 2\pi f_c \tau_n(t) - \phi_{D_n}(t)$. As such, the multipath channel is modeled as a linear time-variant system. Fig. 1.3 illustrates the impulse response of the multipath fading channel. We can see that the wireless channels introduce time-varying gains and delay spreads to the transmitted signal. Based on the relative severity of these effects with respect to the transmitted signal, the multipath fading channels can be classified into slow/fast fading and frequency flat/selective fading.

The root mean square (RMS) delay spread, denoted by σ_T, is a commonly used metric to quantify power spread of the received signal along the delay axis due to multipath propagation. The definition of σ_T is based on the power delay profile, which describes

---

[1] The Gaussian Q-function is related to the complementary error function erfc($\cdot$) by $Q(x) = \frac{1}{2}\text{erfc}\left(\frac{x}{\sqrt{2}}\right)$.

1.1 Wireless Channel Modeling

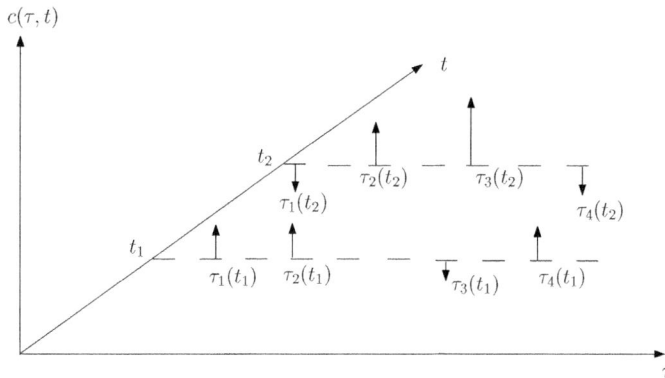

Figure 1.3 Impulse response of linear time-variant channel.

the average signal power distribution along the delay axis and is mathematically calculated as $A_c(\tau) = \mathbf{E}_t[c^2(\tau,t)]$, where $\mathbf{E}_t[\cdot]$ denotes the temporal averaging operation. Typically, $A_c(\tau)$ is a discrete function of τ and can be specified by the mapping from specific delay values τ_n to the corresponding channel power gain α_n^2, for $n = 0, 1, 2, \ldots$. Specifically, RMS delay spread σ_T can be calculated as

$$\sigma_T = \sqrt{\frac{\sum_{n=0}^{N} \alpha_n^2 (\tau_n - \mu_T)^2}{\sum_{n=0}^{N} \alpha_n^2}}, \tag{1.10}$$

where μ_T is the average delay spread, given by

$$\mu_T = \frac{\sum_{n=0}^{N} \alpha_n^2 \tau_n}{\sum_{n=0}^{N} \alpha_n^2}. \tag{1.11}$$

If σ_T is large compared to the transmit symbol period, then the delay spread will lead to significant intersymbol interference (ISI). Noting that time-domain delay spread translates to frequency selectiveness in the frequency domain, the channel coherence bandwidth serves as an alternative metric for quantifying power spread along the delay axis. By definition, channel coherence bandwidth, denoted by B_c, is the bandwidth over which the channel frequency response remains highly correlated. In particular, let $A_C(\Delta f)$ stand for the autocorrelation function of the channel frequency response, i.e., $A_C(\Delta f) = \mathbf{E}[C^*(f_1,t)C(f_2,t)]$. Then, B_c should satisfy $A_C(\Delta f) \approx 1$ for all $\Delta f = |f_1 - f_2| < B_c$. We can intuitively expect that $B_c \propto 1/\sigma_T$. Correspondingly, the wireless channel is considered frequency flat if $\sigma_T \ll T_s$, or equivalently, $B_c \gg B_s$, where B_s is the signal bandwidth. Otherwise, the wireless channel introduces selective fading.

Wireless systems will use wideband channels to support high-data-rate services. As such, most advanced systems will be operating in a frequency-selective fading environment. The most popular model for selective fading is the tapped delay line model, which essentially models the wireless channel as a discrete-time filter. Under this model, the channel impulse response is given by

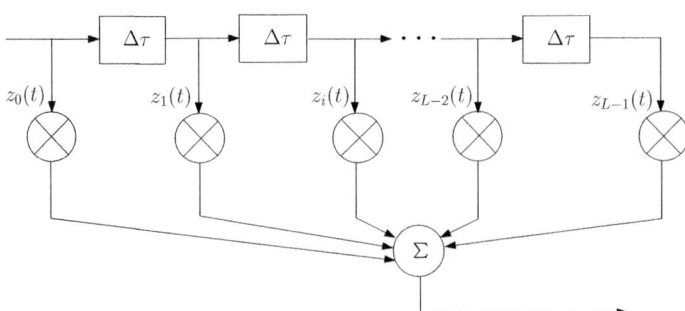

Figure 1.4 Tapped delay line channel model for selective fading.

$$c(\tau, t) = \sum_{i=0}^{L-1} z_i(t)\delta(\tau - i\Delta\tau),$$

where L is the total number of delay bins, also known as the channel length, $\Delta\tau$ is the width of each delay bin, and $z_i(t)$ is the composite time-varying gain of all paths in the ith bin. Fig. 1.4 illustrates the tapped delay line model for selective-fading channels. Meanwhile, most advanced wireless technologies, such as diversity combining, multiuser scheduling, and multi-antenna transmission, are still designed and analyzed under the frequency flat fading channel models. The basic premise of such an approach is that the wideband frequency-selective channel can be converted into multiple parallel frequency flat fading channels with the well-known multicarrier/orthogonal frequency division multiplexing (OFDM) transmission technique [1]. We follow the same approach and focus on the statistical channel model for the flat fading environment in the following subsections. The effect of selective fading will be discussed in later sections.

1.1.3 Frequency Flat Fading

A wireless channel introduces frequency flat fading when the RMS delay spread is small with respect to the symbol period, i.e., $\sigma_T \ll T_s$. In this case, the signal copies from multiple paths can be deemed to arrive at the receiver at the same time. The channel impulse response simplifies to

$$c(\tau, t) = \left(\sum_{n=0}^{N(t)} \alpha_n(t) e^{-j\phi_n(t)}\right) \delta(\tau) = z(t)\delta(\tau), \tag{1.12}$$

which means that the channel introduces a time-varying complex gain of $z(t) = \sum_{n=0}^{N(t)} \alpha_n(t) e^{-j\phi_n(t)}$. Correspondingly, the complex baseband input/output relation of the channel becomes

$$r(t) = z(t)u(t) + n(t), \tag{1.13}$$

where $u(t)$ is the complex baseband envelope of transmitted signal and $n(t)$ is the additive Gaussian noise. To develop statistical models for the complex channel gain $z(t)$, we rewrite $z(t)$ as $z_I(t) + jz_Q(t)$, where

1.1 Wireless Channel Modeling

$$z_I(t) = \sum_{n=0}^{N(t)} \alpha_n(t) \cos \phi_n(t), \quad z_Q(t) = \sum_{n=0}^{N(t)} \alpha_n(t) \sin \phi_n(t). \tag{1.14}$$

with the application of central limit theorem (CLT), $z(t)$ can be modeled as a complex Gaussian random process when $N(t)$ is sufficiently large, Starting from this basic result, we can arrive at the instantaneous statistics of the channel gain, depending on whether an LOS propagation path exists or not.

When there is no LOS component, the random process $z(t)$ can be assumed to have zero mean. With the additional assumption that $\phi_n(t)$ follows a uniform distribution over $[-\pi, \pi]$ and is independent of $\alpha_n(t)$, we can show that $z_I(t)$ and $z_Q(t)$ are independently Gaussian random variables with zero mean and common variance, denoted by σ^2. It follows that the channel amplitude $|z(t)| = \sqrt{z_I(t)^2 + z_Q(t)^2}$ is Rayleigh distributed with the distribution function

$$p_{|z|}(x) = \frac{x}{\sigma^2} \exp\left[-\frac{x^2}{2\sigma^2}\right], \quad x \geq 0. \tag{1.15}$$

The channel phase $\theta(t) = \arctan(z_Q(t)/z_I(t))$ is uniformly distributed over $[0, 2\pi]$. We can further show that the channel power gain $|z(t)|^2$, which is proportional to the instantaneous received signal power, follows an exponential distribution with PDF given by

$$p_{|z|^2}(x) = \frac{1}{2\sigma^2} \exp\left[-\frac{x}{2\sigma^2}\right], x \geq 0,$$

where $2\sigma^2$ is the average channel power gain, depending upon the path loss/shadowing effects.

When the LOS component exists, $z_I(t)$ and $z_Q(t)$ are modeled as Gaussian random process with nonzero mean. In this case, the instantaneous channel amplitude $|z| = \sqrt{z_I^2 + z_Q^2}$ follows a Rician distribution with the distribution function

$$p_{|z|}(x) = \frac{x}{\sigma^2} \exp\left[-\frac{x^2 + s^2}{2\sigma^2}\right] I_0\left(\frac{xs}{\sigma^2}\right), \quad x \geq 0, \tag{1.16}$$

where $s^2 = \alpha_0^2$ is the power gain of the LOS path, $2\sigma^2$ is the average power gain of all non-LOS components, and $I_0(\cdot)$ is the modified Bessel function of zeroth order. Alternatively, the Rician distribution function is given in terms of the so-called Rician fading parameter $K = s^2/2\sigma^2$ and the total average power gain $\Omega = s^2 + 2\sigma^2$. We can show that the channel power gain $|z(t)|^2$ follows a noncentral χ^2 distribution with the distribution function

$$p_{|z|^2}(x) = \frac{K+1}{\Omega} \exp\left[-K - \frac{(K+1)x}{\Omega}\right] I_0\left(2\sqrt{\frac{K(K+1)x}{\Omega}}\right). \tag{1.17}$$

The Nakagami model is another statistical fading channel model for the LOS scenario and was developed from experimental measurements. With the Nakagami model, the channel amplitude is modeled as a random variable with the distribution function

$$p_{|z|}(x) = \frac{2m^m x^{2m-1}}{\Gamma(m)\Omega} \exp\left[-\frac{mx^2}{\Omega}\right],$$

where $\Gamma(\cdot)$ is the gamma function and $m \geq 1/2$ is the Nakagami fading parameter. It follows that the distribution function of the channel power gain under the Nakagami model is given by

$$p_{|z|^2}(x) = \left(\frac{m}{\Omega}\right)^m \frac{x^{m-1}}{\Gamma(m)} \exp\left[-\frac{mx}{\Omega}\right]. \tag{1.18}$$

By setting the Nakagami parameter m to the proper values, the Nakagami model can be applied to many fading scenarios. Specifically, when $m = 1$ (or $K = 0$ for the Rician fading model), we have Rayleigh fading. If m approaches ∞ (or K approaches ∞ for the Rician model), then the model corresponds to the no-fading environment. The Nakagami model can well approximate Rician fading with $m = \frac{(K+1)^2}{2K+1}$. Finally, when $m < 1$, the Nakagami model applies a fading scenario that is more severe than Rayleigh fading.

An immediate application of these statistical models would be outage performance analysis. Outage occurs when the instantaneous received signal power is too low for reliable information transmission. Since the received signal power over fading channels is proportional to the channel power gain, the outage probability can be calculated by evaluating the cumulative distribution function (CDF) of the channel power gain at a certain outage threshold, i.e., $P_{\text{out}} = \Pr[|z(t)|^2 < P_0] = F_{|z|^2}(P_0)$, where P_0 is the normalized power outage threshold. For example, the outage probability of a point-to-point link under the Rician fading model is given by

$$P_{\text{out}} = 1 - Q_1\left(\sqrt{2K}, \sqrt{\frac{2(1+K)}{\Omega}P_0}\right), \tag{1.19}$$

where $Q_1(\cdot, \cdot)$ is the Marcum Q-function, defined as

$$Q_1(\alpha, \beta) = \int_\beta^\infty x \exp\left(-\frac{x^2 + \alpha^2}{2}\right) I_0(\alpha x) dx. \tag{1.20}$$

Similarly, the outage probability under Nakagami fading is given by

$$P_{\text{out}} = 1 - \frac{\Gamma\left(m, \frac{m}{\Omega}P_0\right)}{\Gamma(m)}, \tag{1.21}$$

where $\Gamma(\cdot, \cdot)$ is the incomplete gamma function.

1.1.4 Channel Correlation

In the design and analysis of different wireless transmission technologies, we are also interested in the correlation of the wireless channel gains over space or time. To study spatial correlation, we focus on the autocorrelation of the bandpass channel gain given by

$$\text{Re}\{z(t)e^{j2\pi f_c t}\} = z_I(t)\cos(2\pi f_c t) - z_Q(t)\sin(2\pi f_c t). \tag{1.22}$$

With the assumption of uniform scattering, i.e., the angle formed by the incident wave with the moving direction is uniformly distributed over $[0, 2\pi]$, we can show that the autocorrelation and cross-correlation function of the in-phase and quadrature components of the complex channel gain satisfy

$$A_{z_Q}(\tau) = A_{z_I}(\tau) = \frac{P_r}{2} J_0(2\pi f_D \tau);$$
$$A_{z_I, z_Q}(\tau) = 0, \tag{1.23}$$

where $J_0(x)$ is the zero-order Bessel function. It follows that the autocorrelation function of $\text{Re}\{z(t)e^{j2\pi f_c t}\}$ is

$$A_z(\tau) = A_{z_I}(\tau) \cos(2\pi f_c \tau),$$

which is approximately zero if $f_D \tau \geq 0.4$, i.e., $v\tau \geq 0.4\lambda$. Based on this result, we arrive at a rule of thumb that channel gain becomes uncorrelated over a distance of a half-wavelength.

The time-domain variation of wireless channel gain is mainly due to the relative motion of transmitters and receivers. We use the so-called channel coherence time T_c to characterize the rate of such variation. The channel coherence time is defined as the time duration that the channel response remains highly correlated. In terms of the time-domain correlation of the channel frequency response, defined as $A_C(\Delta t) = \mathbf{E}[C^*(f,t)C(f,t+\Delta t)]$, T_c should satisfy $A_C(\Delta t) \approx 1$ for all $\Delta t < T_c$. It can be shown, also as one would intuitively expect, that T_c is inversely proportional to the maximum Doppler shift $f_D = v/\lambda$ and approximately given by $T_c \approx 0.4/f_D$. If T_c is much greater than the symbol period T_s, then we claim that the transmitted signal experiences slow fading. Otherwise, the fading is considered to be fast. To meet the increasing demand for high-data-rate applications, most emerging wireless systems will be operating in a slow fading environment. In this context, a block fading channel model is often adopted in the design and analysis of various transmission technologies. In particular, the channel response is assumed to remain constant for the duration in the order of a channel coherence time and it becomes independent afterwards. While serving as an inaccurate approximation of the reality, the block fading channel model greatly facilitates the description and understanding of wireless transmission technologies, especially those based on channel estimation and feedback. We will adopt the same model in the following chapters unless otherwise noted.

1.2 Digital Communications over Fading Channels

Most modern communication systems employ digital modulation schemes. Digital modulation schemes offer the following major advantages: (1) facilitate source/channel coding for efficient transmission and error protection; (2) provide better immunity to additive noise and interference; and (3) achieve higher spectrum efficiency with guaranteed error performance through adaptive transmission. The modulation process can be viewed as a mapping from bit/bit sequences (obtained after digitization if necessary and source/channel coding) to different sinusoidal waveforms, i.e.,

$$\{d_j\}_{j=1}^n \Longrightarrow A(i)\cos(2\pi f(i)t + \theta(i)), i = 1, 2, \ldots, 2^n. \tag{1.24}$$

Different modulation schemes lead to different trade-offs among spectrum efficiency, power efficiency, error performance, as well as implementation complexity. The desired properties of modulation schemes for wireless systems include: (1) high spectral efficiency to better explore the limited spectrum resource; (2) high power efficiency to preserve the valuable energy of wireless terminals; (3) robustness to the impairments introduced by multipath fading; and (4) low implementation complexity to reduce the overall system cost. Usually, these are conflicting requirements. Therefore, the best choice would be the one resulting in the most desirable trade-off.

1.2.1 Linear Bandpass Modulation

In this context, we focus on the class of linear bandpass modulation schemes, which are widely used in modern wireless systems. With these modulation schemes, the information is carried using either the amplitude and/or phase of the sinusoidal. In particular, the modulated symbols over the ith symbol period can be written as

$$s(t) = A(i)\cos(2\pi f_c t + \theta(i)), \ (i-1)T_s \leq t \leq iT_s, \tag{1.25}$$

where f_c is the carrier frequency, $A(i)$ and $\theta(i)$ are information-carrying amplitude and phase. Both amplitude shift keying (ASK), with constant $\theta(i)$, and phase shift keying (PSK), with constant $A(i)$, are special cases of this general modulation type. Applying the trigonometric relationship, we can also write the modulation symbols of the linear modulation scheme into the in-phase/quadrature representation as

$$s(t) = s_I(i)\cos 2\pi f_c t - s_Q(i)\sin 2\pi f_c t, \ (i-1)T_s \leq t \leq iT_s, \tag{1.26}$$

where $s_I(i) = A(i)\cos\theta(i)$ is the in-phase component and $s_Q(i) = A(i)\sin\theta(i)$ is the quadrature component. The in-phase/quadrature representation is convenient for the understanding of the modulator structure, which is shown in Fig. 1.5. Note that different modulation schemes will differ only in the mapping from bits to in-phase/quadrature components. Such properties will greatly facilitate the implementation of adaptive modulation schemes. Finally, the modulation symbol can be written into the complex envelope format as

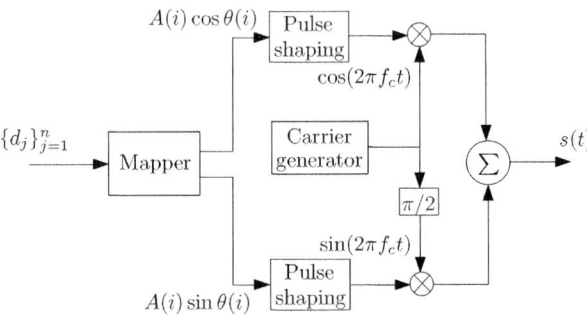

Figure 1.5 Demodulator structure of linear modulation schemes.

$$s(t) = \text{Re}\{(s_I(i) + js_Q(i))e^{j2\pi f_c t}\}, \ (i-1)T_s \le t \le iT_s, \tag{1.27}$$

where $s_I(i) + js_Q(i)$ is the complex baseband symbol and varies from one symbol period to another.

The signal space concept is useful to understand the demodulation process of digital modulation schemes. The modulated symbols of all linear bandpass modulation schemes share the same basis, which are given by

$$\phi_1(t) = \sqrt{\frac{2}{T_s}}\cos(2\pi f_c t), \text{ and } \phi_2(t) = \sqrt{\frac{2}{T_s}}\sin(2\pi f_c t). \tag{1.28}$$

The modulated symbols become points in the two-dimensional plane defined by these two-orthonormal bases. The collection of all possible symbol points forms a constellation. Different modulation schemes differ by their constellation structure. As an illustration, the constellations of square quadrature amplitude modulation (QAM) with $M = 4$ and $M = 16$ are plotted in Fig. 1.6. The coordinates of the constellation points are given by

$$s_{iI} = \sqrt{\frac{T_s}{2}}A(i)\cos\theta(i), \ s_{iQ} = \sqrt{\frac{T_s}{2}}A(i)\sin\theta(i), \tag{1.29}$$
$$i = 1, 2, \cdots M.$$

Noting that the energy of the ith symbol can be calculated as

$$E_{s_i} = \int_0^{T_s} s^2(t)dt = \frac{A(i)^2 T_s}{2},$$

the coordinates simplify to

$$s_{iI} = \sqrt{E_{s_i}}\cos\theta(i), \ s_{iQ} = \sqrt{E_{s_i}}\sin\theta(i). \tag{1.30}$$

The demodulator will first calculate the projection of the received signal over one symbol period, $r(t)$, $(i-1)T_s \le t \le iT_s$, onto the signal space defined by the orthonormal

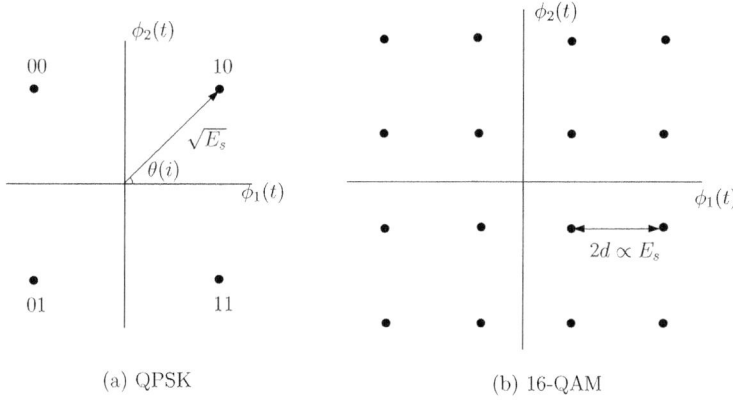

Figure 1.6 Sample symbol constellation for linear bandpass modulation.

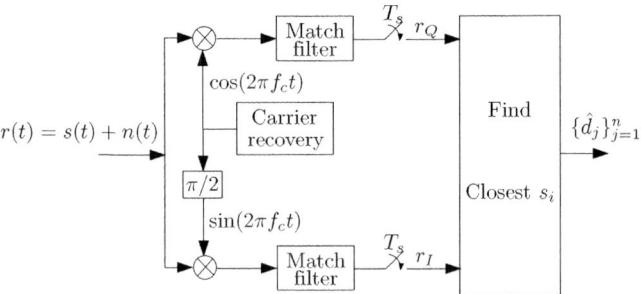

Figure 1.7 Demodulator structure of linear modulation schemes.

basis and obtain the sufficient statistics for the detection of the transmit symbol, which are given by

$$r_I = \int_0^{T_s} r(t)\phi_1(t)dt, \quad r_Q = \int_0^{T_s} r(t)\phi_2(t)dt. \tag{1.31}$$

The maximum likelihood detection rule is to detect the symbol s_i if the point (r_I, r_Q) is the closest to (s_{iI}, s_{iQ}). The structure of the demodulator is shown in Fig. 1.7.

Detection error occurs when the additive noise causes the received symbol to be closer to a symbol different from the transmitted one. Therefore, the error performance of the digital modulation scheme depends heavily on the ratio of received signal power over noise power over signal bandwidth B. For additive white Gaussian noise (AWGN) channels, where the received signal $r(t)$ is related to the transmitted signal $s(t)$ by

$$r(t) = s(t) + n(t), \tag{1.32}$$

where $n(t)$ is a Gaussian random process with mean zero and power spectral density (PSD) $N_0/2$, the received signal power is $P_r = \frac{E_s}{T_s}$ and the noise power $N = N_0/2 \cdot 2B = N_0 B$. It follows that the signal-to-noise ratio (SNR) of the received signal can be determined as

$$\text{SNR} = \frac{P_r}{N} = \frac{E_s}{N_0 B T_s}. \tag{1.33}$$

The product of BT_s varies only with the pulse-shaping function, and as such can be treated as a constant. This means the main figure of merit is the ratio of $\frac{E_s}{N_0}$, which is commonly referred to as received SNR per symbol and denoted by γ_s. For example, it can be shown that the error rate of the binary PSK (BPSK) modulation scheme over the AWGN channel is given by

$$P_b(E) = Q\left(\sqrt{2\gamma_s}\right), \tag{1.34}$$

where $Q(\cdot)$ is the Gaussian Q-function. In addition, the symbol error probability of quadrature phase shift keying (QPSK), or equivalently 4-QAM, over AWGN channel can be shown to be given by

$$P_s(E) = 1 - \left[1 - Q\left(\sqrt{2\gamma_b}\right)\right]^2 \approx 2Q\left(\sqrt{2\gamma_b}\right), \tag{1.35}$$

where $\gamma_b = \frac{E_b}{N_0}$ is the received SNR per bit and related to γ_s as $\gamma_b = \gamma_s/\log M$. Finally, the symbol error probability of square M-QAM can be calculated as

$$P_s(E) = 1 - \left[1 - \frac{2(M-1)}{M} Q\left(\sqrt{\frac{3\gamma_s}{M^2-1}}\right)\right]^2.$$

1.2.2 Performance Analysis over Fading Channels

The complex baseband channel model for flat fading channel is given by

$$r(t) = z(t)s(t) + n(t), \quad (1.36)$$

where $z(t)$ is the time-varying complex channel gain, the statistical characterization of which was discussed in the previous sections. The instantaneous received signal power can be shown to be given by $P_r = |z(t)|^2 \frac{E_s}{T_s}$ and will also vary linearly with $|z(t)|^2$. It follows that the instantaneous SNR given by

$$\gamma_s = |z(t)|^2 \frac{E_s}{N_0} \quad (1.37)$$

becomes a random variable. In this context, the instantaneous system performance will not reflect the overall system performance. We need to apply average performance metrics, including the outage probability and the average error rate, to characterize the transmission performance over fading channels.

Outage occurs when the instantaneous received signal power is too low for reliable information transmission. Due to the fading effect, the system may experience outage even when the average received SNR, after considering path loss and shadowing effects, is very large. An outage event can also be defined in terms of the instantaneous SNR as $\gamma_s < \gamma_{th}$, where γ_{th} is the SNR threshold. Mathematically, the outage probability, denoted by P_{out}, is given by

$$P_{out} = \Pr[\gamma_s < \gamma_{th}]. \quad (1.38)$$

For the flat fading scenario, the outage probability can be calculated using the CDF of the instantaneous SNR. For example, the outage probability for the Rician fading channel is given by

$$P_{out} = 1 - Q_1\left(\sqrt{2K}, \sqrt{2(1+K)\gamma_{th}}\right), \quad (1.39)$$

where $Q_1(\cdot,\cdot)$ is the Marcum Q-function, and for Nakagami fading by

$$P_{out} = 1 - \frac{\Gamma(m, m\gamma_{th})}{\Gamma(m)}, \quad (1.40)$$

where $\Gamma(\cdot,\cdot)$ is the incomplete gamma function.

The average error rate performance can be evaluated by averaging the instantaneous error rate over the distribution of the received SNR. Note that at any time instant, the fading channel can be viewed as an AWGN channel with SNR given in (1.37). Therefore, the average error rate of a modulation scheme over the flat fading channel can be calculated by averaging the instantaneous error rate, which is the error rate of this modulation scheme over the AWGN channel with SNR γ_s, over the distribution function of γ_s. Mathematically, the average error rate, denoted by $\overline{P}_E$, is given by

$$\overline{P}_E = \int_0^\infty P_E(\gamma) p_{\gamma_s}(\gamma) d\gamma,$$

where $P_E(\gamma)$ is the error rate of the chosen modulation scheme over the AWGN channel with SNR γ and $p_{\gamma_s}(\gamma)$ is the PDF of γ_s.

As an example, let us consider the average error rate performance of the BPSK modulation scheme over Rayleigh fading channels. The instantaneous error rate of BPSK over AWGN channel with SNR γ is equal to $Q\left(\sqrt{2\gamma}\right)$. For the Rayleigh fading environment, the PDF of the received SNR γ_s can be shown to be given by

$$p_{\gamma_s}(\gamma) = \frac{1}{\overline{\gamma}} \exp\left(-\frac{\gamma}{\overline{\gamma}}\right), \gamma \geq 0,$$

where $\overline{\gamma} = 2\sigma^2 E_s/N_0$ is the average received SNR. Therefore, the average error rate of BPSK over Rayleigh fading can be calculated as

$$\overline{P}_b = \int_0^\infty Q\left(\sqrt{2\gamma}\right) \frac{1}{\overline{\gamma}} \exp\left(-\frac{\gamma}{\overline{\gamma}}\right) d\gamma$$
$$= \frac{1}{2}\left(1 - \sqrt{\frac{\overline{\gamma}}{1+\overline{\gamma}}}\right).$$

For other fading channel models and/or modulation schemes, we need to plug in different distribution functions and/or instantaneous error rate expressions and then perform the integration. It is worth noting that in most cases, a closed-form result may not be feasible, partly because the instantaneous error rate expression usually involves the Gaussian Q-function and its square. We typically have to evaluate the integration through a numerical method after proper truncation of the integration upper limit. In this context, a new analytical framework based on the MGF of random variables was proposed and extensively used for the accurate evaluation of the average error rate over general fading channels [3].

The MGF of a nonnegative random variable γ is defined as

$$\mathcal{M}_\gamma(s) = \int_0^\infty p_\gamma(\gamma) e^{s\gamma} d\gamma, \gamma \geq 0,$$

where s is a complex dummy variable and $p_\gamma(\cdot)$ is the PDF of γ. The MGF is related to the Laplace transform of the PDF as $\mathcal{M}_\gamma(-s) = \mathcal{L}\{p_\gamma(\gamma)\}$. The MGF is so named because we can easily calculate the moments of random variable γ from its MGF as

$$\mathbf{E}[\gamma^n] = \frac{d^n}{ds^n}\mathcal{M}_\gamma(s)|_{s=0}. \qquad (1.41)$$

Fortunately, the MGFs of the received SNRs for most fading channel models are readily available in a compact closed form. For example, the MGF of Rayleigh faded SNR is

$$\mathcal{M}_{\gamma_s}(s) = (1 - s\overline{\gamma}_s)^{-1}, \qquad (1.42)$$

1.2 Digital Communications over Fading Channels

whereas those for Rician and Nakagami fading are given by

$$\mathcal{M}_{\gamma_s}(s) = \frac{1+K}{1+K-s\bar{\gamma}_s} e^{\frac{s\bar{\gamma}_s K}{1+K-s\bar{\gamma}_s}}, \quad (1.43)$$

and

$$\mathcal{M}_{\gamma_s}(s) = \left(1 - \frac{s\bar{\gamma}_s}{m}\right)^{-m}, \quad (1.44)$$

respectively.

The key starting point of the MGF-based average error rate analysis is the alternative expressions of the Gaussian Q-function and its square, which were obtained by Craig in the early 1990s [5], given by

$$Q(x) = \frac{1}{\pi} \int_0^{\pi/2} \exp\left[\frac{-x^2}{2\sin^2 \phi}\right] d\phi, x > 0. \quad (1.45)$$

$$Q^2(x) = \frac{1}{\pi} \int_0^{\pi/4} \exp\left[\frac{-x^2}{2\sin^2 \phi}\right] d\phi, x > 0. \quad (1.46)$$

Applying these alternative expressions, we can calculate the average error rate of most modulation schemes of interest over general fading channel models through the finite integration of basic functions. As an example, let us consider a generic class of modulation schemes whose instantaneous error rate expression takes the form $P_E(\gamma) = aQ\left(\sqrt{b\gamma}\right)$, where a and b are modulation-specific constants. With the conventional approach, the average error rate of this modulation scheme should be calculated as

$$\bar{P}_s = \int_0^\infty aQ\left(\sqrt{b\gamma_s}\right) p_{\gamma_s}(\gamma) d\gamma. \quad (1.47)$$

Substituting the alternative expression of the Q-function and changing the order of integration, we can rewrite the average error rate as

$$\bar{P}_s = \frac{a}{\pi} \int_0^{\pi/2} \mathcal{M}_{\gamma_s}\left(\frac{-b}{2\sin^2 \phi}\right) d\phi, \quad (1.48)$$

where $\mathcal{M}_{\gamma_s}(\cdot)$ is the MGF of the received SNR. Note that the resulting expression only involves the integration with respect to ϕ over the interval of $[0, \pi/2]$.

Fig. 1.8 plots the average error rate of coherent BPSK and noncoherent BFSK over fading channels as the function of the average received SNR. For reference, the error rate of these modulation schemes over the AWGN channel with received SNR equal to the average received SNR over fading channels is also plotted on the same figure. As we can see, the error performance degrades dramatically due to fading. In particular, the error rate decreases at a log rate over the AWGN channel as the SNR increases but at a linear rate over fading channels. As such, to achieve the same average error rate over fading channels, we need to maintain a much higher average SNR. An intuitive explanation of this phenomenon is that fading causes the frequent occurrence of very low received signal power, also known as deep fade, due to destructive addition of different multipath signals. It is exactly from this perspective that diversity techniques try to improve the performance of wireless systems. Diversity has become one of the most essential

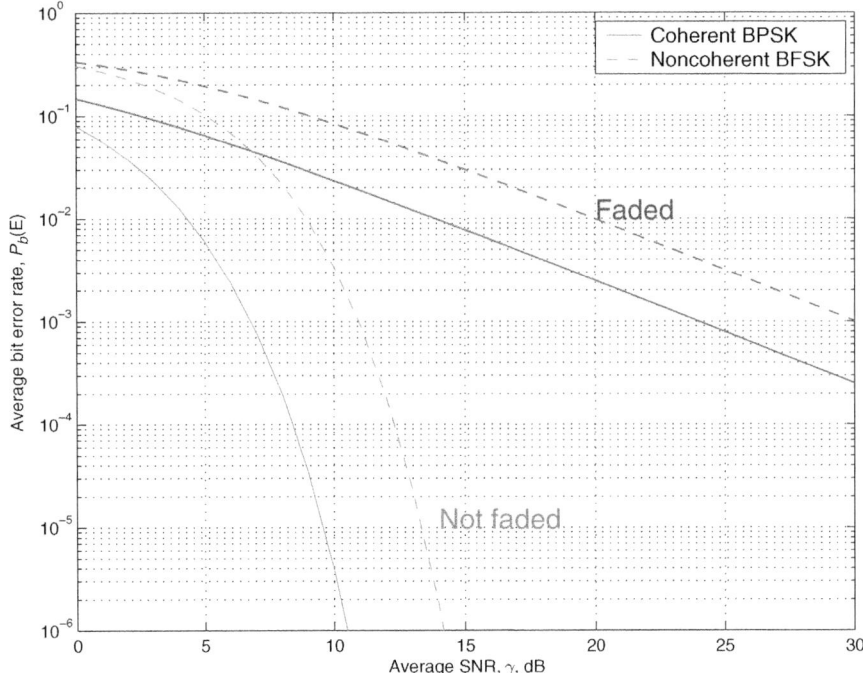

Figure 1.8 Error rate comparison between fading and nonfading environments.

concepts in wireless system design and implementation. We will elaborate more on it in the following chapters.

1.3 Effect of Frequency-Selective Fading

The flat fading scenario applies when the channel delay spread is negligible compared to the transmit symbol period, i.e., $\sigma_T \ll T_s$. To satisfy the growing demand for high-data-rate wireless services, wireless communication systems use an increasingly smaller symbol period or a larger channel bandwidth. For example, the typical channel bandwidth of first-generation cellular systems is 30 kHz, whereas the signal bandwidth of most popular second-generation cellular systems, GSM, is 200 kHz. The signal bandwidth increased to 1–2 MHz in the third-generation cellular systems, and as high as 20 MHz in the fourth-generation systems. Meanwhile, the channel delay spread σ_T and the corresponding channel coherence bandwidth T_c depend on the radio propagation environment and remain more or less unchanged. As such, the channel delay spread becomes significant compared to the symbol period, i.e., $\sigma_T \gtrsim T_s$, or equivalently, the channel coherence bandwidth B_c is smaller than the transmitted signal bandwidth B_s. The transmitted signals of these wireless systems will experience frequency-selective fading.

The received complex baseband signal over frequency-selective fading channels is given by

$$v(t) = \sum_{n=1}^{N} \alpha_n e^{j\phi_n} u(t - \tau_n) + \tilde{n}(t), \qquad (1.49)$$

1.3 Effect of Frequency-Selective Fading

where $\alpha_n e^{j\phi_n}$ is the complex gain and τ_n is the delay, both for the nth path; $u(t)$ is the transmitted complex baseband signal; and $\tilde{n}(t)$ is the complex baseband additive noise. Essentially, the channel acts as a discrete filter with impulse response given by

$$h(t) = \sum_{n=1}^{N} \alpha_n e^{j\phi_n} \delta(t - \tau_n), \quad (1.50)$$

where $\delta(\cdot)$ is the unit impulse signal. As such, the channel will modify the spectrum of transmitted signals. In particular, the spectrum of the received complex baseband signal $v(t)$ is determined as

$$V(f) = \mathcal{F}\{v(t)\} = \mathcal{F}\{u(t) * h(t) + \tilde{n}(t)\} = H(f) \times U(f) + N(f), \quad (1.51)$$

where $\mathcal{F}\{\cdot\}$ denotes the Fourier transform operation, $U(f)$ is the spectrum of $u(t)$, $H(f)$ is the channel frequency response, and $N(f)$ is the noise spectrum. The spectrum of received RF signal $r(t)$ is obtained by shifting that of $V(f)$ to the carrier frequency f_c, based on the relationship $r(t) = \text{Re}\{v(t)e^{j2\pi f_c t}\}$.

The filtering effect of the wireless channel is illustrated in Fig. 1.9. Here, the spectrum of the transmitted signal is assumed to be constant over the signal bandwidth B_s. The wireless channel introduces frequency-selective fading when the channel coherence bandwidth B_c is less than B_s. As such, the frequency response of the channel $H(f)$ varies considerably over the signal bandwidth B_s, as shown in Fig. 1.9. The spectrum of received signal is equal to the spectrum of the transmitted signal multiplied by the channel frequency response. Note that the spectrum of the received signal becomes much different from that of the transmitted signal. As such, the direct detection of the transmitted signal based on the received signal will lead to poor performance, even when the noise is negligible.

We now examine the effects of frequency-selective fading in the time domain. Let us assume that the wireless system transmits a data symbol s_i over each symbol period after proper pulse-shaping. The transmitted baseband signal is given by

$$u(t) = \sum_{i=-\infty}^{+\infty} s_i g(t - iT_s), \quad (1.52)$$

where s_i is the transmitted symbol over the ith symbol period and $g(t)$ is the pulse shape. Applying the tapped delay line model for the selective fading channel from

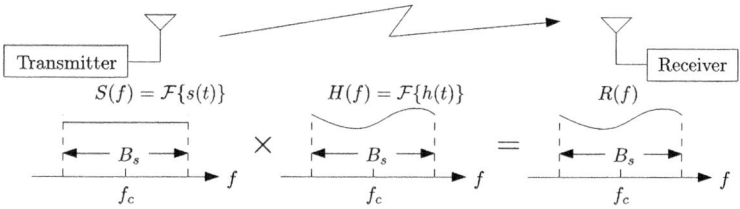

Figure 1.9 The filtering effect of selective fading channel. Reproduced with permission from [6]. ©2017 IET.

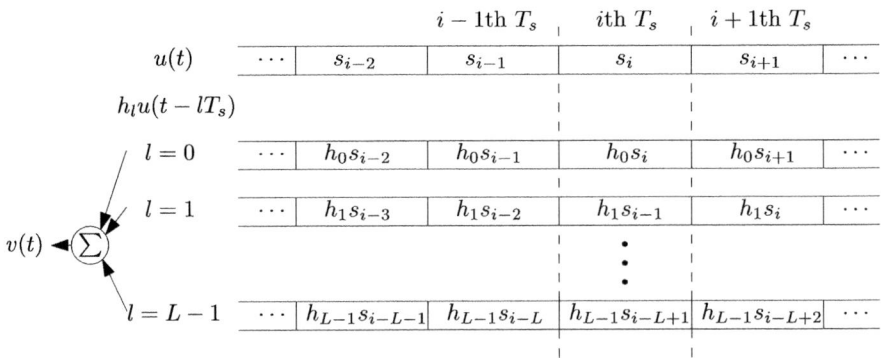

Figure 1.10 Intersymbol interference due to selective fading. Reproduced with permission from [6]. ©2017 IET.

the previous section and setting $\Delta \tau = T_s$, the received complex baseband signal is given by

$$v(t) = \sum_{l=0}^{L-1} z_l \left(\sum_{i=-\infty}^{+\infty} s_i g(t - lT_s - iT_s) \right) + \tilde{n}(t), \tag{1.53}$$

where h_l denotes the gain of the lth tap. The receiver applies matched filter detection with filter response $g^*(T_s - t)$. We assume that $g(t)$ is properly chosen such that the overall response of pulse shaping and matched filter, $g(t) * g^*(T_s - t)$, satisfies the Nyquist criterion. It can be shown, while referring to the illustration in Fig. 1.10, that the received baseband symbol over the ith symbol period is given by

$$r_i = \sum_{l=0}^{L-1} z_l s_{i-l} + n_i. \tag{1.54}$$

In particular, the received symbol consists of the linear convolution of the transmitted symbols with the channel tap gains. The effect of the composite baseband channel, consisting of transmit pulse shaping/RF front end, selective fading channel, and receiver RF front end/matched filtering, can be characterized by the discrete channel response z_l, $l = 0, 1, 2, \ldots, L - 1$.

The received symbol can be rewritten as

$$r_i = z_0 s_i + \sum_{l=1}^{L-1} z_l s_{i-l} + n_i. \tag{1.55}$$

If one tried to detect the transmitted symbol s_i, then the term $\sum_{l=1}^{L-1} h_l s_{i-l}$ would act as interference. Such interference from previously transmitted symbols, usually referred to as intersymbol interference (ISI), is inherent to digital transmission over frequency-selective fading channels. If not properly mitigated, ISI will severely degrade the performance of digital wireless transmission. The signal to interference plus noise ratio (SINR) for the detection of s_i based on r_i is given by

$$\gamma = \frac{|z_0|^2 E_s}{\sum_{i=1}^{L-1} |z_i|^2 E_s + N_0}, \qquad (1.56)$$

where E_s is the symbol energy and N_0 is the noise power spectrum density. Note that increasing the transmission power can slightly improve the SINR but cannot effectively reduce the effect of ISI. We will discuss transmission technologies over selective fading channels in chapter 3.

1.4 Summary

In this chapter, we reviewed the basics of digital wireless communications, including channel modeling, digital bandpass modulation, and performance analysis over fading channels. Specifically, we introduced the key performance metrics and the general approaches for their accurate evaluation, which also apply to the analysis of advanced wireless technologies in later chapters.

1.5 Further Reading

For more detailed discussion on channel modeling, the reader can refer to the popular textbook of A. Goldsmith [1]. Simon and Alouini's book [3] provides thorough and in-depth coverage of digital transmission schemes for fading channels and their performance evaluation, especially the MGF-based approach.

References

[1] A. J. Goldsmith, *Wireless Communications*. Cambridge University Press, 2005.
[2] G. L. Stüber, *Principles of Mobile Communications*, 2nd ed. Kluwer Academic Publishers, 2000.
[3] M. K. Simon and M.-S. Alouini, *Digital Communications Over Generalized Fading Channels*, 2nd ed. Wiley, 2004.
[4] W. C. Jakes, *Microwave Mobile Communication*, 2nd ed. IEEE Press, 1994.
[5] J. W. Craig, "A new simple and exact result for calculating the probability of error for two dimensional signal constellations," in *IEEE MILCOM '91 Conf. Rec.*, Boston, MA, pp. 24.5.1–25.5.5.
[6] H.-C. Yang, *Introduction to Digital Wireless Communications*, IET, 2017.

2 Transmission Technologies for Flat Fading Channels

Wireless channels introduce frequency flat fading to the transmitted signal when the channel delay spread is negligible compared to the symbol period. The effect of flat-fading channel is characterized by a complex channel gain. The flat fading channel model also applies to the transmission over each subcarrier of a multicarrier system. As such, improving the transmission performance over flat fading channels is of fundamental importance.

In this chapter, we introduce basic transmission technologies for flat fading channels. We first present diversity combining techniques, one of the most important fading mitigation solutions. Both antenna reception diversity and transmit diversity scenarios are considered. After that, we discuss fundamental adaptive transmission technologies, including rate and power adaptation. The chapter concludes with a discussion on the fundamental results of MIMO transmission.

2.1 Diversity Combining Techniques

Diversity combining techniques can effectively improve the performance of wireless communication systems over fading channels. The basic idea is to transmit/receive the same information-bearing signal over multiple independent channels. Since the probability that these independent channels simultaneously experience deep fade is low, by properly combining these signal copies together, we can improve the quality of the received signal and achieve better performance. There are several ways to create independent channels, including using different frequencies, using different time slots, using different antennas, or using different codewords. In general, the antenna diversity approach is more attractive as they provide diversity benefits without requiring extra spectral–temporal resources. We will base the following discussion mainly on the antenna reception diversity.

2.1.1 Antenna Reception Diversity

The generic structure of a diversity combiner is shown in Fig. 2.1. Specifically, the diversity combiner will generate its output signal by properly combining the signal replicas received from L diversity branches. Note that the detection will be carried out using combiner output signal. As such, the statistics of the combiner output signal-to-noise ration

2.1 Diversity Combining Techniques

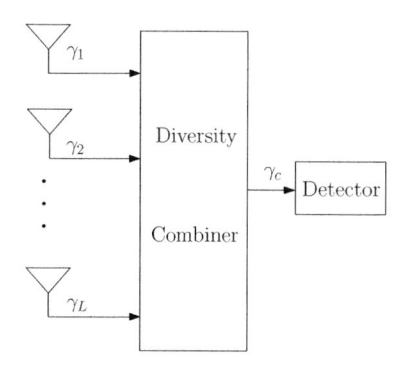

Figure 2.1 Antenna reception diversity system.

(SNR), denoted by γ_c, dictate the overall system performance. The design objective of the diversity combiner is to maximize the quality of the combined signal under a given complexity constraint. There is a trade-off of performance versus complexity among different combining schemes. To quantify these trade-offs, we need to accurately evaluate the performance of each combining scheme. In the following, we will briefly explain the basic design of each scheme and analyze the performance of the resulting system using the statistics of γ_c.

For the sake of clarity, we assume in the following that the diversity branches experience independent and identically distributed (i.i.d.) flat fading. The complex channel gains of the ith diversity branch, denoted by $z_i = a_i e^{j\theta_i}$, vary independently over time. The corresponding received SNR, given by $\gamma_i = a_i^2 \frac{E_b}{N_0}$, can serve as the quality indicator of each branch. The performance of the diversity system depends on the statistics of the combiner output SNR γ_c. In particular, the outage probability becomes the probability that γ_c is smaller than a threshold γ_{th}, given by

$$P_{\text{out}} = \Pr[\gamma_c < \gamma_{\text{th}}] = F_{\gamma_c}(\gamma_{\text{th}}), \tag{2.1}$$

where $F_{\gamma_c}(\cdot)$ denotes the cumulative distribution function (CDF) of the combined SNR γ_c. The average error rate performance of diversity systems can be evaluated by averaging the instantaneous bit error rate (BER) of the chosen modulation scheme $P_E(\gamma)$ over the distribution of γ_c as

$$\overline{P}_E = \int_0^\infty P_E(\gamma) p_{\gamma_c}(\gamma) d\gamma, \tag{2.2}$$

where $p_{\gamma_c}(\cdot)$ denotes the probability density function (PDF) of the combined SNR γ_c. As such, we need to derive the statistics of the combiner output SNR based on the combiner mode of operation.

Selection combining (SC) is the most popular low-complexity combining scheme. With SC, the diversity branch with the highest received SNR is used for data detection. The mode of operation of SC is illustrated in Fig. 2.2. The combiner needs to estimate the received SNRs of all available diversity branches and select the best one. The combiner output SNR with SC is then mathematically given by $\gamma_c = \max\{\gamma_1, \gamma_2, \ldots, \gamma_L\}$.

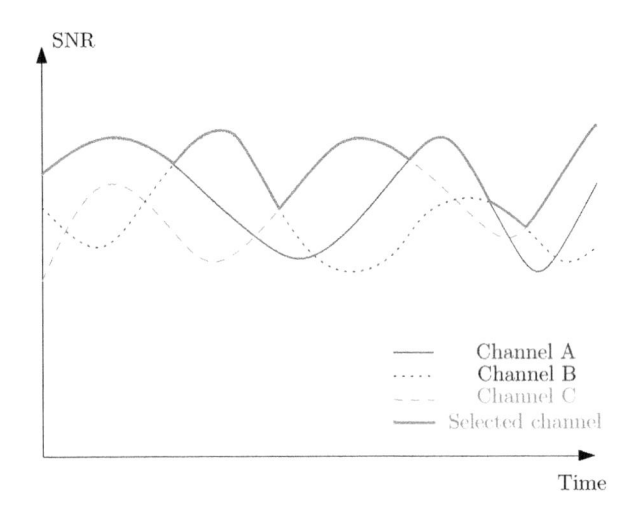

Figure 2.2 Selection combining over three diversity branches.

Therefore, the combiner output SNR is the largest one of L independent random variables. The statistics of γ_c can be easily obtained from the basic order statistic results. In particular, the PDF of γ_c can be obtained, with i.i.d. fading assumptions on different diversity branches, as

$$p_{\gamma_c}(\gamma) = L[F_\gamma(\gamma)]^{L-1} p_\gamma(\gamma), \tag{2.3}$$

where $F_\gamma(\cdot)$ and $p_\gamma(\cdot)$ denote the common CDF and PDF of the branch SNRs. The outage probability can be calculated as

$$P_{\text{out}} = [F_\gamma(\gamma_{\text{th}})]^L. \tag{2.4}$$

While the receiver with SC only needs to process the signal from one diversity branch, it always requires the estimation of all available diversity branches.

Maximum ratio combining (MRC) is the optimal linear combining scheme in a noise-limited environment. The basic idea of MRC is to generate the combiner output signal as a linear combination of the received signal from different diversity branches such that the SNR of the combined signal is maximized. Mathematically speaking, the combined signal is given by

$$r_c(t) = \sum_{i=1}^{L} w_i r_i(t) = \sum_{i=1}^{L} w_i a_i e^{j\theta_i} s(t) + \sum_{i=1}^{L} w_i n_i(t), \tag{2.5}$$

where $r_i(t) = a_i e^{j\theta_i} s(t) + n_i(t), i = 1, 2, \ldots, L$ is the received signal from the ith branch, and w_i is the weight for the ith branch. It can be shown that the optimal weights w_i should be proportional to the complex conjugate of the complexity channel gain of the ith branch, i.e., $a_i e^{-j\theta_i}$, in order to maximize the combiner output SNR. The maximum output SNR with the optimal w_i is given by

2.1 Diversity Combining Techniques

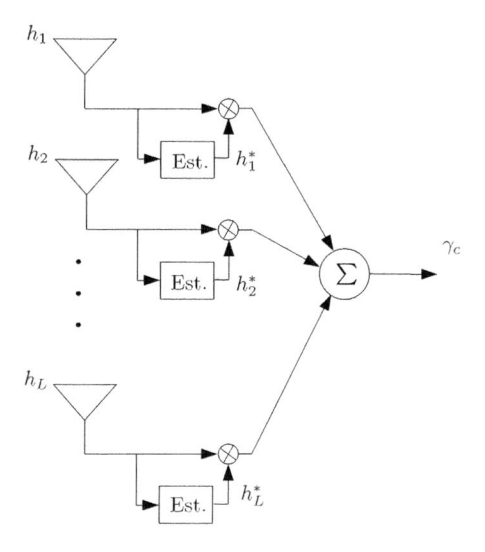

Figure 2.3 Structure of an MRC-based diversity combiner.

$$\gamma_c = \sum_{i=1}^{L} \gamma_i. \tag{2.6}$$

The structure of an MRC combiner is shown in Fig. 2.3.

For the performance evaluation of an MRC scheme, we need to determine the statistics of the sum of L random variables. When the branch SNRs are independent random variables, the moment generating function (MGF) based approach can readily apply. Note that the MGF of the sum of independent random variables is the product of the MGFs of individual random variables. As such, the MGF of the combined SNR with MRC over independent fading branches can be written as

$$\mathcal{M}_{\gamma_c}(s) = \prod_{i=1}^{L} \mathcal{M}_{\gamma_i}(s),$$

where $\mathcal{M}_{\gamma_i}(s)$ is the MGF of the ith branch SNR. For convenience, we summarize in Table 2.1 the distribution functions of the individual branch SNRs under three popular fading channel models. In Table 2.1, $\overline{\gamma}$ is the common average SNR per branch, $\Gamma(\cdot)$ is the gamma function [1, Sec. 8.31], $I_0(\cdot)$ is the modified Bessel function of the first kind with zero order [1, Sec. 8.43], $\Gamma(\cdot, \cdot)$ is the incomplete gamma function [1, Sec. 8.35], and $Q_1(\cdot, \cdot)$ is the first-order Marcum Q-function [2]. As an example, for the i.i.d. Rayleigh fading scenario, the MGF of combined SNR with MRC is given by

$$\mathcal{M}_{\gamma_c}(s) = \prod_{i=1}^{L} (1 - s\overline{\gamma})^{-1}, \tag{2.7}$$

Table 2.1 Statistics of the fading signal SNR γ for the three fading models under consideration.

Model	Rayleigh	Rice	Nakagami-m
Parameter	·	$K \geq 0$	$m \geq \frac{1}{2}$
PDF, $p_\gamma(x)$	$\frac{1}{\bar{\gamma}} e^{-\frac{x}{\bar{\gamma}}}$	$\frac{(1+K)}{\bar{\gamma}} e^{-K - \frac{1+K}{\bar{\gamma}} x} I_0 \left(2 \sqrt{\frac{1+K}{\bar{\gamma}} K x}\right)$	$\left(\frac{m}{\bar{\gamma}}\right)^m \frac{x^{m-1}}{\Gamma(m)} e^{-\frac{mx}{\bar{\gamma}}}$
CDF, $F_\gamma(x)$	$1 - e^{-\frac{x}{\bar{\gamma}}}$	$1 - Q_1 \left(\sqrt{2K}, \sqrt{\frac{2(1+K)}{\bar{\gamma}} x}\right)$	$1 - \frac{\Gamma\left(m, \frac{m}{\bar{\gamma}} x\right)}{\Gamma(m)}$
MGF, $\mathcal{M}_\gamma(s)$	$(1 - s\bar{\gamma})^{-1}$	$\frac{1+K}{1+K-s\bar{\gamma}} e^{\frac{s\bar{\gamma} K}{1+K-s\bar{\gamma}}}$	$\left(1 - \frac{s\bar{\gamma}}{m}\right)^{-m}$

which leads to the PDF of the combined SNR, after proper inverse Laplace transform, as

$$p_{\gamma_c}(\gamma) = \frac{\gamma^{L-1} e^{-\gamma/\bar{\gamma}}}{\bar{\gamma}^L (L-1)!}. \tag{2.8}$$

Both results can apply to the average error rate analysis over Rayleigh fading channels.

The equal gain combining (EGC) scheme has slightly lower complexity than MRC. With EGC, the combined signal is still the linear combination of the signals received from different branches. But the weight for the ith branch with EGC becomes $e^{-j\theta_i}$, which is simpler than that for MRC. As such, the receiver with EGC needs only to estimate the channel phase of each diversity branch. In general, it is more challenging to estimate the channel phase than the channel amplitude. Therefore, EGC entails higher implementation complexity than SC. Furthermore, SC needs only to process the currently selected branch whereas the MRC and EGC receiver needs to process all L available branches.

2.1.2 Threshold Combining and Its Variants

Threshold combining is another low-complexity combining scheme. With threshold combining, the receiver needs only to monitor the quality of the currently used branch and as such entails even lower complexity than SC. Dual-branch switch and stay combining (SSC) is the most well-known threshold combining scheme. With SSC, the receiver estimates the SNR of the currently used branch and compares it with a fixed threshold, denoted by γ_T. If the estimated SNR is greater or equal to γ_T, then the receiver continues to use the current branch. Otherwise, the receiver will switch to the other branch and use it for data reception, regardless of its quality. The mode of operation of SSC is illustrated in Fig. 2.4 and can be mathematically summarized, assuming γ_1 is the SNR of the currently used branch, as

$$\gamma_c = \begin{cases} \gamma_1, & \gamma_1 \geq \gamma_T; \\ \gamma_2, & \gamma_1 < \gamma_T. \end{cases} \tag{2.9}$$

2.1 Diversity Combining Techniques

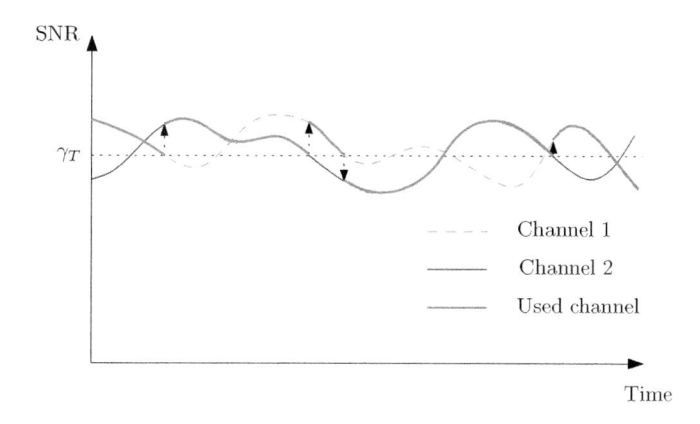

Figure 2.4 Mode of operation of dual-branch switch and stay combining.

To derive the statistics of the combined SNR with SSC, we can start by applying the total probability theorem and writing the CDF of the combined SNR as

$$F_{\gamma_c}(\gamma) = \Pr[\gamma_1 < \gamma, \gamma_1 \geq \gamma_T] + \Pr[\gamma_2 < \gamma, \gamma_1 < \gamma_T]. \tag{2.10}$$

With the i.i.d. fading branch assumption, the CDF can be rewritten, in terms of the common CDF of individual branch SNR, as

$$F_{\gamma_c}(\gamma) = \begin{cases} F_\gamma(\gamma_T)F_\gamma(\gamma) + F_\gamma(\gamma) - F_\gamma(\gamma_T), & \gamma \geq \gamma_T; \\ F_\gamma(\gamma_T)F_\gamma(\gamma), & \gamma < \gamma_T. \end{cases} \tag{2.11}$$

It follows that the PDF of the combined SNR with SSC is given by

$$p_{\gamma_c}(\gamma) = \begin{cases} F_\gamma(\gamma_T)p_\gamma(\gamma) + p_\gamma(\gamma), & \gamma \geq \gamma_T; \\ F_\gamma(\gamma_T)p_\gamma(\gamma), & \gamma < \gamma_T. \end{cases} \tag{2.12}$$

The generic expression of the average error rate for a certain modulation scheme can be calculated as

$$\overline{P}_E = F_\gamma(\gamma_T)\int_0^\infty P_E(\gamma)p_\gamma(\gamma)d\gamma + \int_{\gamma_T}^\infty P_E(\gamma)p_\gamma(\gamma)d\gamma. \tag{2.13}$$

The statistics of the combined SNR for a more general scenario with correlated and/or unbalanced branches can also be obtained by applying a Markov chain-based approach [3].

In the multiple branch scenario, SSC was generalized to switch and examine combining (SEC) [4], which can exploit the benefit of extra diversity branches. The receiver with SEC will examine the quality of the switched-to branch and switch again if it finds the quality of that branch is unacceptable. This process will continue until either an acceptable branch is found or all branches have been examined. The mode of operation of L-branch SEC can be summarized with the following mathematical relationship, assuming the first branch is the currently used branch:

$$\gamma_c = \begin{cases} \gamma_1, & \gamma_1 \geq \gamma_T; \\ \gamma_2, & \gamma_1 < \gamma_T, \gamma_2 \geq \gamma_T; \\ \gamma_3, & \gamma_1 < \gamma_T, \gamma_2 < \gamma_T, \gamma_3 \geq \gamma_T; \\ \vdots & \vdots \\ \gamma_L, & \gamma_i < \gamma_T, i = 1, 2, \ldots, L-1. \end{cases} \quad (2.14)$$

Following the i.i.d. fading assumption, the CDF of SEC output SNR is given by

$$F_{\gamma_c}(x) = \begin{cases} [F_\gamma(\gamma_T)]^{L-1} F_\gamma(x), \ x < \gamma_T; \\ \sum_{j=0}^{L-1} [F_\gamma(x) - F_\gamma(\gamma_T)][F_\gamma(\gamma_T)]^j \\ + [F_\gamma(\gamma_T)]^L, \ x \geq \gamma_T. \end{cases} \quad (2.15)$$

The PDF and MGF of the combined SNR with L-branch SEC can then be routinely obtained and applied to its performance analysis over fading channels.

In general, there exists an optimal value for the switching threshold γ_T for switched combining schemes in terms of minimizing the average error rate. In particular, if γ_T is set too small, then the receiver will not switch branches most of the time. The receiver essentially performs nondiversity combining. On the other hand, when γ_T is set too large, the switched diversity receiver will switch branches all the time, as the current branch is never acceptable. The receiver becomes equivalent to a non-diversity receiver. With the statistics of the combined SNR, we can derive the average error rate expression for the chosen fading channel model, which allows us determine the optimal value of γ_T that minimizes the average error rate.

Fig. 2.5 compares the average error rate performance of SC, SSC/SEC with optimal threshold, and MRC. As we can see, MRC achieves the best performance and the performance advantage increases as the number of diversity branches increases. The performance gain of MRC comes with higher hardware complexity and power consumption. Note that the receiver needs to know the complete complex channel gain, including the amplitude and phase, for each diversity branch in order to determine the optimal branch weights for MRC. With SC, the receiver only requires the amplitude or power gain of each branch. We can also see that the complexity saving of threshold combining over the SC scheme comes at the cost of a high average error rate, even with the optimal switching threshold value.

Another variant of threshold combining is the so-called switch and examine combining with post-examining selection (SECps) [5]. Similar to SSC/SEC schemes, the receiver with SECps tries to use an acceptable diversity branch by examining as many branches as necessary. But when no acceptable branch is found after examining all available ones, the receiver with SECps will use the best unacceptable one. The operation of SECps for the dual-branch case is illustrated in Fig. 2.6. Note that compared with the dual SSC scheme, the SECps scheme will lead to better output signal when both branches are unacceptable. The mode of operation of L-branch SECps can be summarized as

2.1 Diversity Combining Techniques 27

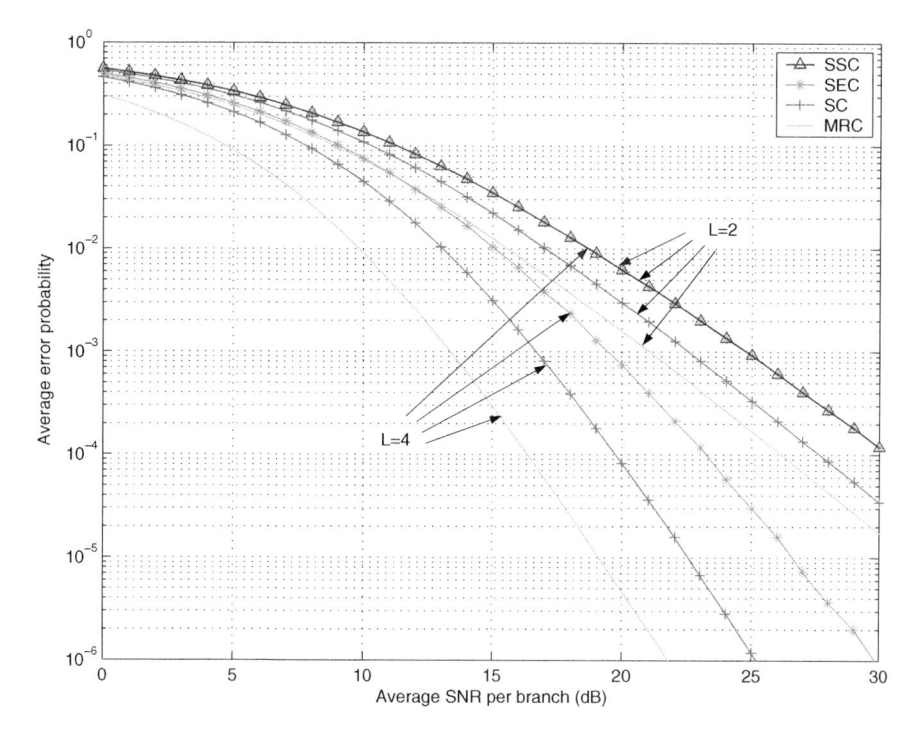

Figure 2.5 Average error rate comparison of conventional diversity combining schemes. Reproduced with permission from [4]. ©2003 IEEE.

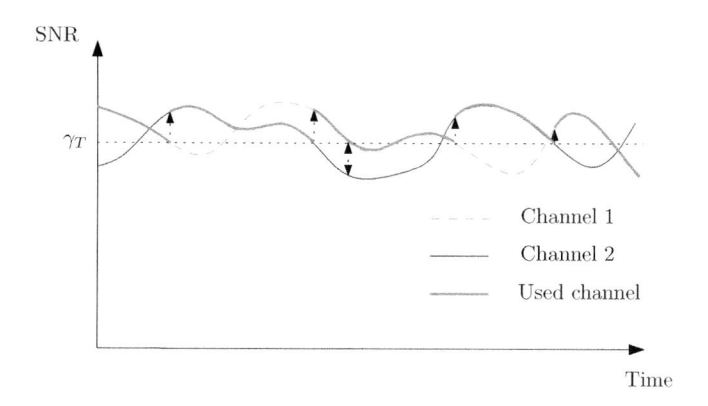

Figure 2.6 Mode of operation of SECps scheme over dual diversity branches.

$$\gamma_c = \begin{cases} \gamma_1, & \gamma_1 \geq \gamma_T; \\ \gamma_2, & \gamma_1 < \gamma_T, \gamma_2 \geq \gamma_T; \\ \gamma_3, & \gamma_1 < \gamma_T, \gamma_2 < \gamma_T, \gamma_3 \geq \gamma_T; \\ \vdots & \vdots \\ \max\{\gamma_1, \gamma_2, \ldots, \gamma_L\}, & \gamma_i < \gamma_T, i = 1, 2, \ldots, L-1. \end{cases} \quad (2.16)$$

Consequently, the CDF of the combined SNR with SECps over i.i.d. fading branches can be obtained as

$$F_{\gamma_c}(x) = \begin{cases} 1 - \sum_{i=0}^{L-1}[F_\gamma(\gamma_T)]^i[1 - F_\gamma(x)], & x \geq \gamma_T; \\ [F_\gamma(x)]^L, & x < \gamma_T, \end{cases} \quad (2.17)$$

which can be applied to the outage evaluation. Fig. 2.7 shows the outage performance of SECps in a multibranch scenario. As we can see, as the number of diversity branches increases from 2 to 4, the outage performance of SECps improves considerably. While comparing the outage performance of four-branch SECps with four-branch SEC and SC, we can see that SECps has the same outage probability as SC when $\gamma_{th} \leq \gamma_T$ and as SEC when $\gamma_{th} > \gamma_T$. Intuitively, when $\gamma_{th} \leq \gamma_T$, the receiver with SECps experiences outage only when the SNR of all diversity branches falls below γ_T and their maximum is smaller than γ_{th}, which is equivalent to the case of SC. On the other hand, when $\gamma_{th} > \gamma_T$, the receiver with SECps experiences outage when the SNR of the first acceptable branch is smaller than γ_{th} or when the SNR of all diversity branches falls below γ_T, which lead to the same outage performance as conventional SEC.

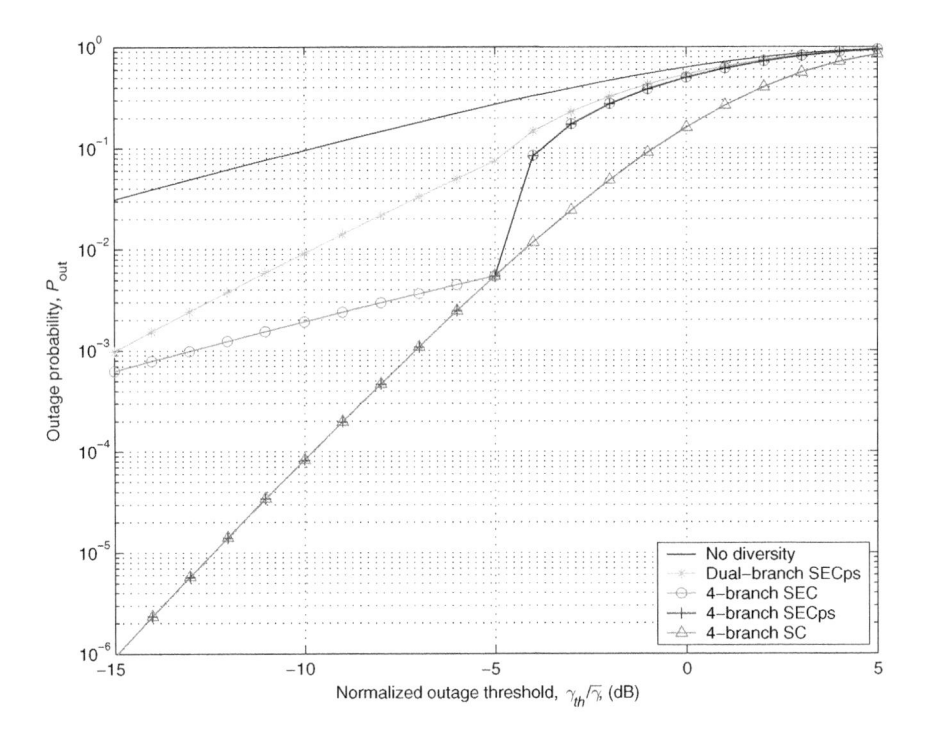

Figure 2.7 Outage probability of L-branch SECps as a function of normalized outage threshold $\gamma_{th}/\overline{\gamma}$ in comparison with SEC and SC ($\gamma_T/\overline{\gamma} = -5$ dB). Reproduced with permission from [5]. ©2006 IEEE.

2.1 Diversity Combining Techniques

Table 2.2 Complexity comparison of dual-branch SECps, SSC, and SC in terms of operations needed for combining decision over Rayleigh fading branches

Schemes	Average number of channel estimations	Probability of branch switching
Dual-branch SC	2	0.5
Dual-branch SSC	1	$1 - e^{-\frac{\gamma_T}{\bar{\gamma}}} \in [0, 1)$
Dual-branch SECps	$2 - e^{-\frac{\gamma_T}{\bar{\gamma}}} \in [1, 2)$	$\frac{1}{2} - \frac{1}{2}e^{-2\frac{\gamma_T}{\bar{\gamma}}}$

To summarize this section, we quantitatively compare the complexity of SC, SEC, and SECps in terms of the average number of branch estimations and the probability of branch switching. Based on its mode of operation, the average number of branch estimations needed by SECps is given by

$$N_E^{\text{SECps}} = 1 + \sum_{i=1}^{L-1}[F_\gamma(\gamma_T)]^i = \frac{1 - [F_\gamma(\gamma_T)]^L}{1 - F_\gamma(\gamma_T)}. \tag{2.18}$$

Note that the SC scheme always needs L-branch estimations, i.e., $N_E^{\text{SC}} = L$. The receiver with conventional SEC needs to estimate at most $L - 1$ branches because when the first $L - 1$ branches are found unacceptable, the receiver will use the last branch for reception. Therefore, SEC requires fewer branch estimations than SECps, i.e., $N_E^{\text{SEC}} = 1 + \sum_{i=1}^{L-2}[P_\gamma(\gamma_T)]^i$. As a result, we have $N_E^{\text{SEC}} \leq N_E^{\text{SECps}} \leq N_E^{\text{SC}}$. On the other hand, the probability of branch switching with SECps, denoted by P_{SW}^{SECps}, can be shown to be given by

$$P_{SW}^{\text{SECps}} = F_\gamma(\gamma_T) - \int_0^{\gamma_T} p_\gamma(x)[F_\gamma(x)]^{L-1} dx.$$

It is easy to see that the receiver with SC switches branches with a probability of $(L - 1)/L$ over identically faded diversity branches, i.e., $P_{SW}^{\text{SC}} = (L - 1)/L$. Meanwhile, the receiver with L-branch SEC switches branches whenever the current branch becomes unacceptable (i.e., $\gamma_1 \leq \gamma_T$). As such, the switching probability with SEC is $P_{SW}^{\text{SEC}} = P_\gamma(\gamma_T)$. Consequently, we have $P_{SW}^{\text{SECps}} < P_{SW}^{\text{SEC}}$ and $P_{SW}^{\text{SECps}} < P_{SW}^{\text{SEC}}$. Table 2.2 summarizes these results for the $L = 2$ case.

2.1.3 Transmit Diversity

When there are multiple transmit antennas and a single receive antenna, as is often the case in cellular downlink transmission, we can apply transmit diversity techniques to obtain diversity benefit. If the appropriate channel state information corresponding to different transmit antennas is available on the transmitter side, then we readily apply the conventional combining schemes discussed in the previous subsection. For example, if the complete complex channel gain from the ith transmit antenna to the receive antenna $h_i = a_i e^{j\theta_i}$ is available at the transmitter, we can implement the so-called transmit MRC

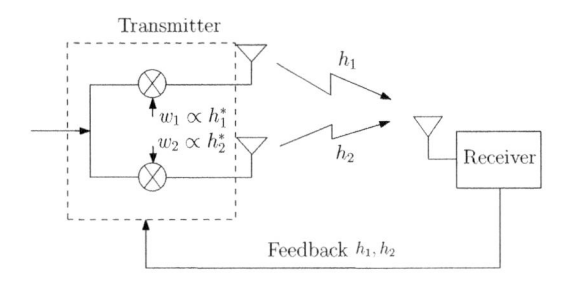

Figure 2.8 Closed-loop transmit diversity based on MRC.

by multiplying the transmit signal by a weight w_i, which is proportional to $h_i^* a_i e^{-j\theta_i}$, before transmitting it on the ith antenna. To satisfy the total transmit constraint, we can choose the amplitude of w_i as

$$|w_i| = \frac{a_i}{\sqrt{\sum_{j=1}^L a_j^2}}, \qquad (2.19)$$

which leads to $\sum_{i=1}^L |w_i|^2 = 1$. After propagating through the fading channels, the signal transmitted from different antennas will add up coherently at the receiver and each is weighted proportional to the amplitude of the corresponding channel gain. As such, the SNR of the received signal is the same as the combiner output SNR with receive MRC, i.e., $\gamma_c = \sum_{i=1}^L \gamma_i$. Fig. 2.8 illustrates the structure of the MRC-based transmit diversity system with perfect CSI. Note that the provision of complete channel state information on the transmitter side can be challenging in practice systems, considering the fact that they usually entail the channel estimation and feedback from the receiver. From this perspective, the low-complexity combining schemes including SC and threshold combining are more attractive as much less channel information is required for their implementation. For example, the receiver with transmit SC just needs to inform the transmitter which antenna leads to the highest receive SNR and therefore should be used, which results in $\lceil \log_2 L \rceil$ bits of feedback load. With transmitter SSC, the receiver just needs to feed back one bit of information to indicate whether the transmitter should switch antennas or not.

When no channel state information can be made available on the transmitter side, we can still explore the diversity benefit inherent in multiple transmit antennas through a class of linear coding schemes, termed as space-time block codes [6]. While the code design for a scenario of more than two transmit antennas is available, the particular design for the dual transmit antenna case, widely recognized as Alamouti's scheme [7], is of great practical importance. Unlike other designs, Alamouti's scheme achieves full diversity gain over fading channels without incurring any rate loss. The scheme transmits two complex data symbols (s_1, s_2) over two symbol periods using two antennas as follows: s_1 from antenna 1 and s_2 from antenna 2 over the first symbol period and $-s_2^*$ from antenna 1 and s_1^* from antenna 2 over the second symbol period. Assuming that channel gains $h_i = a_i e^{j\theta_i}$, $i = 1, 2$ remain constant over two consecutive symbol periods, the received symbols over two symbol periods are

$$y_1 = h_1 s_1 + h_2 s_2 + n_1$$
$$y_2 = h_1(-s_2^*) + h_2 s_1^* + n_2, \qquad (2.20)$$

where n_1 and n_2 are additive Gaussian noise collected by the receiver. After taking the complex conjugate of y_2, we can rewrite the two equations in a matrix form as

$$\begin{bmatrix} y_1 \\ y_2^* \end{bmatrix} = \begin{bmatrix} h_1 & h_2 \\ h_2^* & -h_1^* \end{bmatrix} \begin{bmatrix} s_1 \\ s_2 \end{bmatrix} + \begin{bmatrix} n_1 \\ n_2^* \end{bmatrix}. \qquad (2.21)$$

With the estimated channel gains, the receiver can perform detection on

$$\begin{bmatrix} z_1 \\ z_2 \end{bmatrix} = \begin{bmatrix} h_1 & h_2 \\ h_2^* & -h_1^* \end{bmatrix}^H \begin{bmatrix} y_1 \\ y_2^* \end{bmatrix}, \qquad (2.22)$$

where $[\cdot]^H$ denotes the Hermitian transpose. Due to the special structure of the matrix, we have

$$\begin{bmatrix} z_1 \\ z_2 \end{bmatrix} = (a_1^2 + a_2^2) \begin{bmatrix} s_1 \\ s_2 \end{bmatrix} + \begin{bmatrix} \tilde{n}_1 \\ \tilde{n}_2 \end{bmatrix}, \qquad (2.23)$$

where $\tilde{n}_i$ can be shown to be zero-mean Gaussian with variance $(a_1^2 + a_2^2)N_0$. As such, the effective SNR for the detection of s_i based on $z_i, i = 1, 2$, is

$$\gamma_i = (a_1^2 + a_2^2)\frac{E_s}{2N_0} = \frac{1}{2}(\gamma_1 + \gamma_2). \qquad (2.24)$$

Therefore, Alamouti's scheme achieves the same diversity gain as receive MRC except for a 3-dB power loss due to the fact that each symbol is transmitted twice. The key advantage of Alamouti's scheme is that absolutely no channel state information is required at the transmitter side. As a result, this scheme has been included in several wireless standards.

2.2 Channel Adaptive Transmission

Adaptive transmission can achieve high spectral and power efficiency over wireless fading channels with guaranteed error rate performance [9, 10]. The basic idea of adaptive transmission is to vary the transmission schemes/parameters, such as modulation mode, coding rate, or transmit power, with the prevailing fading channel conditions as illustrated in Fig. 2.9, The system will exploit favorable channel condition with higher data rate transmission at lower transmit power levels and response to channel degradation with reduced data rate or increased power level. As a result, the overall system throughput is maximized with controlled power consumption while maintaining a certain desired error rate performance. As such, adaptive transmission techniques have received great interest from both academics and industry. Several adaptive transmission schemes have been incorporated in cellular systems and wireless LAN systems.

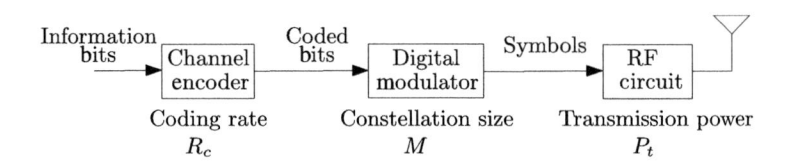

Figure 2.9 Generic structure of a digital wireless transmitter. Reproduced with permission from [8]. ©2017 IET.

The availability of a certain channel state information (CSI) at the transmitter is a fundamental requirement of adaptive transmission techniques. It has been shown that with perfect CSI at the transmitter, Continuous rate adaptation can achieved the ergodic capacity of fading channels and optimal power adaptation can further enhance the capacity [8]. However, the provision of perfect CSI at the transmitter is a very challenging task even for the most advanced wireless systems. In addition, implementing continuous rate adaptation entails prohibitively high complexity. As a result, while acknowledging a certain performance gap compared to the ideal scheme, most current wireless standards adopt discrete rate/power adaptation, which requires only limited CSI at the transmitter, achieved either through feedback signaling or by exploring the channel reciprocity property.

2.2.1 Power Adaptation

With power adaptive transmission schemes, the transmit power P_t is adjusted with the instantaneous channel condition, while the information data rate R_b remains constant. A typical objective of power adaptation is to mitigate the effect of channel power gain variation due to fading and to maintain constant receive SNR for a certain BER target. While definitely suboptimal, such an objective leads to a low-complexity transceiver structure. Note that the received SNR is related to the transmission power as

$$\gamma = \frac{P_t \cdot g}{N}, \tag{2.25}$$

where g denotes the channel power gain and N is the noise power at the receiver. As such, the power adaptation policy is typically specified by the function $P_t(g)$.

Full channel inversion is one of the most fundamental power adaptation strategies. The object of the full channel inversion power adaptation strategy is to maintain a constant received SNR, denoted by γ_t over all possible channel realizations. The transmit power $P_t(g)$ is set such that

$$\frac{P_t(g) \cdot g}{N} = \gamma_t. \tag{2.26}$$

As such, the transmit power is proportional to the inverse of the channel power gain, i.e.,

$$P_t(g) = \frac{\gamma_t \cdot N}{g}. \tag{2.27}$$

Meanwhile, the transmit power should satisfy the average power constraint that entails

$$\int_0^\infty P_t(g) p_g(g) dg \leq \overline{P_t}, \tag{2.28}$$

where $\overline{P_t}$ denotes the maximum average transmit power. After proper substitution and some manipulation, we can shown that γ_t is upper-bounded as

$$\gamma_t \leq \frac{\overline{P_t}}{N \int_0^\infty (1/g) p_g(g) dg} = \frac{\overline{P_t}}{N \cdot \mathbf{E}[1/g]}, \tag{2.29}$$

where $\mathbf{E}[\cdot]$ denotes the statistical averaging operation.

Truncated channel inversion is a variant of full channel inversion that avoids the high transmission power when the channel gain is small. With the truncated channel inversion power adaptation strategy, the transmission system maintains a constant received SNR γ_t when the channel gain g is large enough. When the channel gain is too small, the transmitter will hold the transmission until the channel condition improves. The transmit power $P_t(g)$ with truncated channel inversion is set as

$$P_t(g) = \begin{cases} \frac{\gamma_t \cdot N}{g}, & g \geq g_T; \\ 0, & g < g_T, \end{cases} \tag{2.30}$$

where g_T is the cutoff value of the channel power gain. Apparently, when the channel gain is larger than or equal to g_T, the received SNR will always be equal to γ_t. Meanwhile, the system will enter outage when g is less than g_T. The cutoff value g_T can be determined from the desired outage requirement, i.e.,

$$P_{\text{out}} = \Pr[g < g_T] \leq \epsilon. \tag{2.31}$$

2.2.2 Rate Adaptation

With rate adaptive transmission, the transmitter adapts the information data rate R_b by varying the coding rate R_c and/or the constellation size M, with the instantaneous channel condition, while using a constant transmit power P_t. Such rate adaptation schemes are often referred to as adaptive modulation and coding (AMC) in various wireless standards. The basic premise of such adaptation schemes is that the channel encoder and digital modulator should be reconfigurable on the fly. Since most channel encoders and decoders are implemented using digital circuits, coding rate adaptation is readily implementable. Fortunately, most linear modulation schemes, including PSK (phase shift keying) and QAM (quadrature amplitude modulation) schemes, share similar modulator and demodulator structures, which facilitates constellation adjustment.

Typically, a finite set of coding schemes and/or modulation types is used in real-world systems. As such, the transmission rate is adapted in a discrete manner with a finite number of possible values. The modulation and coding scheme (MCS) is selected based on the instantaneous quality of the wireless channel, characterized by the received SNR γ. Typically, the MCS that achieves the highest spectrum efficiency while maintaining acceptable instantaneous BER performance is chosen, as illustrated in Fig. 2.10.

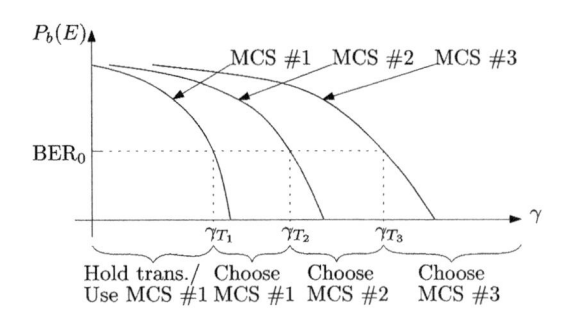

Figure 2.10 Modulation and coding scheme selection. Reproduced with permission from [8]. ©2017 IET.

It may happen, in the worst case scenario, that none of the available MCS can satisfy the target BER requirement. In this case, the system may either choose to hold the transmission until the channel condition improves or to transmit in violation of the target BER requirement.

The SNR thresholds for different MCSs play a critical role in the operation of rate-adaptive transmission systems. These thresholds are determined based on the instantaneous BER requirement. Specifically, the nth threshold is selected to be the smallest SNR value required by the nth MCS to achieve an instantaneous BER value no greater than the target BER value, denoted by BER_0. To better explain, we consider the constant-power variable-rate adaptive M-QAM scheme, which is of both theoretical and practical interest.

Constant-power variable-rate adaptive M-QAM
With this scheme, the system adaptively selects one of N different M-QAM modulation schemes based on the fading channel condition while using a fixed power level for transmission. Different M-QAM schemes use distinct constellation sizes. For squared M-QAM schemes, the constellation sizes are $M = 2^n$, $n = 1, 2, \ldots, N$, with size 2^n corresponding to a spectral efficiency of n bps/Hz according to the Nyquist criterion. The modulation schemes are chosen to achieve the highest spectral efficiency while maintaining the instantaneous error rate below a the target value. Specifically, the value range of the received SNR, is divided into $N + 1$ regions, with threshold values denoted by $0 < \gamma_{T_1} < \gamma_{T_2} < \ldots < \gamma_{T_N} < \infty$. When the received SNR γ falls into the nth region, i.e., $\gamma_{T_n} \leq \gamma < \gamma_{T_{n+1}}$, the QAM scheme with constellation size 2^n will be used for transmission. For practical implementation, the modulation mode selection is carried out at the receiver after estimating the received SNR. The receiver will then feed back the mode selection result to the transmitter over the control channel, which entails a feedback load of only $\lceil \log_2 N \rceil$ bits per channel coherence time.

The threshold values are set such that the instantaneous error rate of the chosen modulation mode is below BER_0. For example, the instantaneous BER of 2^n-QAM with two-dimensional Grey coding over the AWGN channel with SNR γ can be approximately calculated as [10, Eq. (28)]

2.2 Channel Adaptive Transmission

Table 2.3 Threshold values in dB for 2^n-ary QAM to satisfy 1%, 0.1%, and 0.01% BER.

n	$\text{BER}_0 = 10^{-2}$	$\text{BER}_0 = 10^{-3}$	$\text{BER}_0 = 10^{-4}$
2	4.32	6.79	8.34
3	7.07	9.65	11.30
4	7.88	10.52	12.21
5	10.84	13.58	15.30
6	11.95	14.77	16.52
7	15.40	18.01	19.79
8	16.40	19.38	21.20

$$\text{BER}_n(\gamma) = \frac{1}{5} \exp\left(-\frac{3\gamma}{2(2^n - 1)}\right), \quad n = 1, 2, \ldots, N. \tag{2.32}$$

As such, the threshold values can be calculated, for a target BER value of BER_0, as

$$\gamma_{T_n} = -\frac{2}{3} \ln(5 \, \text{BER}_0)(2^n - 1); \quad n = 0, 1, 2, \ldots, N. \tag{2.33}$$

Alternatively, we can solve for the threshold values by inverting the exact BER expression for square M-QAM [11] as

$$\gamma_{T_n} = \text{BER}_n^{-1}(\text{BER}_0), \tag{2.34}$$

where $\text{BER}_n^{-1}(\cdot)$ is the inverse BER expression. The threshold values for the target BERs of 10^{-2}, 10^{-3}, and 10^{-4} cases are summarized in Table 2.3.

Performance Analysis Over Fading Channels

The performance of an adaptive M-QAM system can be evaluated in terms of the average spectral efficiency and the average error rate. In particular, the average spectral efficiency of the adaptive M-QAM scheme under consideration can be calculated as

$$\eta = \sum_{n=1}^{N} n \, P_n, \tag{2.35}$$

where P_n is the probability of using the nth constellation size. With the application of the appropriate fading channel model, P_n can be calculated using the distribution function of the received SNR as

$$P_n = \int_{\gamma_{T_n}}^{\gamma_{T_{n+1}}} p_\gamma(x) dx, \tag{2.36}$$

where $p_\gamma(\cdot)$ denotes the PDF of the received SNR. The average BER of the adaptive modulation system can be calculated as

$$<BER> = \frac{1}{\eta} \sum_{n=1}^{N} n \, \overline{BER}_n, \tag{2.37}$$

where $\overline{BER}_n$ is the average error rate when using 2^n-ary QAM, given by

$$\overline{BER}_n = \int_{\gamma_{T_n}}^{\gamma_{T_{n+1}}} BER_n(x) p_\gamma(x) dx, \tag{2.38}$$

where $BER_n(\gamma)$ is the instantaneous BER of 2^n-QAM scheme with SNR γ, the approximate expression of which was given in (2.32).

2.3 MIMO Transmission

Multiple-antenna transmission and reception (i.e., MIMO) techniques can considerably improve the performance and/or efficiency of wireless communication systems. Since first introduced in the mid-1990s [12, 13], MIMO transmission techniques have been an area of active research and have found applications in various wireless systems. There is already a rich literature on MIMO wireless communications. Several books have been published on this general subject (see, for example, [6, 15]). In this section, we provide a general overview of early results on MIMO transmission, focusing on point-to-point links where both transmitter and receiver have multiple antennas. The main objective is to demonstrate the huge potential of MIMO techniques in terms of providing spatial diversity gain and spatial multiplexing gain.

2.3.1 MIMO Channel Model

Let us consider a point-to-point link where the transmitter has M_t antennas and the receiver has M_r antennas, as shown in Fig. 2.11. We assume that the channel for the jth transmit antenna to the ith receiver antenna experiences frequency flat fading, with complex channel gain h_{ij}, modeled as i.i.d. zero mean unit variance complex Gaussian random variables. As such, the received signal on the ith receive antenna is given by

$$y_i = h_{i1}x_1 + h_{i2}x_2 + \ldots + h_{iM_t}x_{M_t} + n_i, \qquad (2.39)$$

where x_j is the transmitted symbol from the jth transmit antenna and n_i is the independent Gaussian noise with zero mean and variance $\sigma^2 = N_0/2$. The discrete time MIMO channel model can be rewritten in matrix form as

$$\begin{bmatrix} y_1 \\ y_2 \\ \vdots \\ y_{M_r} \end{bmatrix} = \begin{bmatrix} h_{11} & h_{12} & \cdots & h_{1M_t} \\ h_{21} & h_{22} & \cdots & h_{2M_t} \\ \vdots & \vdots & \ddots & \vdots \\ h_{M_r 1} & h_{M_r 2} & \cdots & h_{M_r M_t} \end{bmatrix} \begin{bmatrix} x_1 \\ x_2 \\ \vdots \\ x_{M_t} \end{bmatrix} + \begin{bmatrix} n_1 \\ n_2 \\ \vdots \\ n_{M_r} \end{bmatrix},$$

or in matrix notation

$$\mathbf{y} = \mathbf{H}\mathbf{x} + \mathbf{n}. \qquad (2.40)$$

The inherent spatial degree of freedom of the above MIMO system can be exploited for both spatial multiplexing gain and diversity gain.

2.3.2 Diversity Over MIMO Channels

Multiple antennas at the transmitter and the receiver can be explored for diversity benefit. If certain information about the channel matrix $\mathbf{H}$ can be made available at the

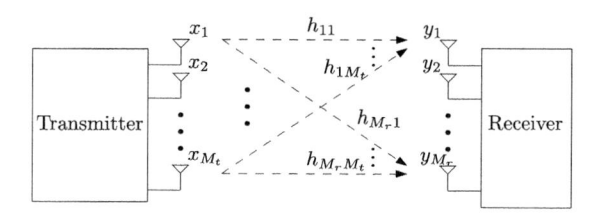

Figure 2.11 Generic MIMO channel model. Reproduced with permission from [8]. ©2017 IET.

transmitter, i.e., channel state information at the transmitter (CSIT) scenario, we can apply conventional diversity transmission/reception schemes. The most straightforward strategy to explore diversity benefit is to apply selection combining over all transmit and receive antenna pairs. Since there are $M_t M_r$ antenna pairs to choose from, such a full selection strategy essentially implements an $M_t M_r$-branch selection combining scheme. With the conventional transmit or receive antenna diversity solution, we will need to deploy a total of $M_t M_r$ antennas to achieve the same diversity benefit. With MIMO implementation, we only need to deploy $M_t + M_r$ antennas. An alternative strategy is to perform antenna selection at the transmitter and maximum ratio combining (MRC) on the receiver side. This so-called transmit antenna selection with receiver MRC (TAS/MRC) strategy entails the same feedback load as the full selection combining scheme above, but achieves better performance as intuitively expected.

The performance of MIMO transmission can be further improved when all M_t transmit antennas are used for transmission. Assuming a linear structure, data symbols are transmitted from all M_t antennas after being weighted by vector **v**, $\|\mathbf{v}\| \leq 1$. The receiver will linearly combine the received symbol over all M_r antennas with normalized weighting vector **w**. The maximum diversity benefit can be achieved by optimally designing vectors **v** and **w** to maximize the SNR of the combined signal. It can be shown that when **v** and **w** are proportional to the singular vectors corresponding to the largest singular value of **H**, the combined SNR is maximized and given by

$$\gamma = \frac{\lambda_{\max} E_s}{\sigma^2}, \qquad (2.41)$$

where $\lambda_{\max}$ denotes the largest singular value of **H**. When the transmitter has absolutely no knowledge about the channel (CSI at the receiver only scenario), then space-time coded transmission combined with reception diversity can be employed to extract diversity benefit.

The performance improvement through diversity combining, in terms of the reduction in average error rate and outage probability, is also characterized by the diversity order of the system. Diversity order is defined as the rate of average error decrease as the average received SNR increases. It has been shown that MRC achieves full diversity order in diversity reception systems. Space-time coded transmission also achieves full diversity order over the transmit diversity system without requiring CSIT. Finally, space-time coded transmission combined with MRC reception achieves a full diversity order of $M_t M_r$ over MIMO channels.

2.3.3 Multiplexing Over MIMO Channels

Multiple antennas at the transmitter and the receiver can also be exploited for spatial multiplexing gain. The most convenient way to demonstrate the capacity benefit of a MIMO system is parallel decomposition. Specifically, the MIMO channel can be decomposed into multiple parallel channels as follows. After applying a singular value decomposition to $\mathbf{H}$, we have

$$\mathbf{H}_{M_r \times M_t} = \mathbf{U}_{M_r \times M_r} \mathbf{\Sigma}_{M_r \times M_t} \mathbf{V}_{M_t \times M_t}^H,$$

where $\mathbf{U}$ and $\mathbf{V}$ are unitary matrices and $\mathbf{\Sigma}$ is a diagonal matrix of the singular values of $\mathbf{H}$, denoted by σ_i. In general, $\mathbf{H}$ will have $R_\mathbf{H}$ nonzero singular values, where $R_\mathbf{H} \in [1, \min\{M_t, M_r\}]$ (also known as the rank of $\mathbf{H}$). Therefore, the MIMO channel can be transformed into $R_\mathbf{H}$ parallel independent channels if the channel input is precoded by $\mathbf{V}$ and the channel output is preprocessed by $\mathbf{U}$. The ith channel will have a channel gain σ_i with additive noise. Effectively, the system data rate can increase $R_\mathbf{H}$ times. The process of MIMO channel decomposition is illustrated in Fig. 2.12.

The spatial multiplexing gain of MIMO wireless channels can be readily achieved when the CSI is available at both the transmitter and the receiver. Specifically, with the knowledge of the right singular matrix $\mathbf{V}$ at the transmitter and the left singular matrix $\mathbf{U}$ at the receiver, the system can decompose the MIMO channel into $R_\mathbf{H}$ parallel channels and transmit independent data streams over each channel. Different decomposed channels may use different coding and modulation schemes to transmit at different information rates with a common symbol rate. Depending on whether the transmitter has the additional knowledge of the singular values σ_i or not, the transmitter may apply either equal power allocation or optimal power allocation, which further enhances the overall transmission rate over these parallel channels.

To demonstrate the effect of transmit power allocation, we first note that the capacity of the MIMO wireless channel can be determined as the sum capacity of the individual decomposed channels. Let P_i denote the power allocation to the ith decomposed channel, which satisfies a total power constraint of $\sum_i P_i \leq P_T$. The MIMO capacity with perfect CSIT can be calculated by solving the following optimization problem:

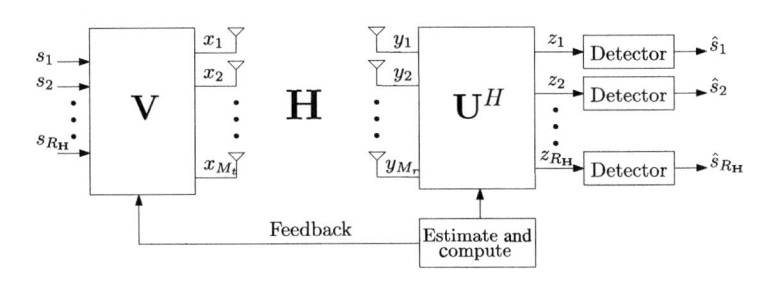

Figure 2.12 Parallel decomposition of MIMO channels. Reproduced with permission from [8]. ©2017 IET.

$$\max_{\{P_i\}_{i=1}^{R_\mathbf{H}}} \sum_{i=1}^{R_\mathbf{H}} B \log_2\left(1 + \frac{\sigma_i^2 P_i}{\sigma^2}\right),$$
$$\text{s.t.} \sum_i P_i \leq P_T.$$

The maximum is achieved with the well-known water-filling solution, given by

$$P_i = \begin{cases} \frac{1}{\gamma_0} - \frac{1}{\gamma_i}, & \gamma_i \geq \gamma_0, \\ 0, & \gamma_i < \gamma_0, \end{cases} \quad (2.42)$$

where $\gamma_i = \sigma_i^2/\sigma^2$ and γ_0 is the cutoff value based on satisfying the total power constraint $\sum_{i:\gamma_i \geq \gamma_0} P_i = P_T$. The resulting instantaneous capacity of MIMO channel is equal to

$$C = \sum_{i:\gamma_i \geq \gamma_0} B \log_2\left(\frac{\gamma_i}{\gamma_0}\right). \quad (2.43)$$

As the result of multipath fading, h_{ij} varies over time. The ergodic capacity of the MIMO channel over fading channels can be obtained by averaging the instantaneous capacity over the fading distribution of channel matrix **H**:

$$C_f = \mathbf{E}_\mathbf{H}[C(\mathbf{H})], \quad (2.44)$$

or equivalently, over the distribution of the singular values of **H**, σ_i^2 as

$$C_f = \mathbf{E}_{\sigma_i^2}\left[\max_{P_i:\sum_i P_i \leq P} \sum_i B \log_2(1 + P_i \gamma_i)\right]. \quad (2.45)$$

The expectation can be solved in some cases while noting that σ_i^2 is the ith largest eigenvalue of $\mathbf{HH}^H$, which is of the Wishart type.

2.4 Summary

In this chapter we reviewed several basic transmission technologies for flat fading channels, including diversity combining, transmit diversity, adaptive transmission, and MIMO transmission. While most of the materials in this chapter were introduced as background materials for later chapters, we have presented a thorough trade-off analysis on different threshold combining schemes.

2.5 Further Reading

For more detailed discussion on most of the material in this chapter, readers can refer to [8, 14]. The reader can refer to [6] for more general space-time code designs. A thorough treatment of MIMO transmission is available in [13]. The Markov chain-based analysis of the different implementation options of the SSC scheme is available in [3].

References

[1] I. S. Gradshteyn and I. M. Ryzhik, *Table of Integrals, Series, and Products*, 5th ed., Academic Press, 1994.

[2] A. H. Nuttall, "Some integrals involving the Q_M function," *IEEE Trans. Information Theory*, 1, pp. 95–96, 1975.

[3] H.-C. Yang and M.-S. Alouini, "Markov chain and performance comparison of switched diversity systems," *IEEE Trans. Commu.*, 52, no. 7, pp. 1113–1125, 2004.

[4] H.-C. Yang and M.-S. Alouini, "Performance analysis of multibranch switched diversity systems," *IEEE Trans. Commun.*, vol. 51, no. 5, pp. 782–794, 2003.

[5] H.-C. Yang and M.-S. Alouini, "Improving the performance of switched diversity with post-examining selection," *IEEE Trans. Wireless Commun.*, 5, no. 1, pp. 67–71, 2006.

[6] A. Paulraj, R. Nabar, and D. Gore, *Introduction to Space-Time Wireless Communications*, Cambridge University Press, 2003.

[7] S. M. Alamouti, "A simple transmitter diversity scheme for wireless communications," *IEEE J. Select. Areas Commun.*, 16, pp. 1451–458, 1998.

[8] H.-C. Yang, *Introduction to Digital Wireless Communications*, The Institution of Engineering and Technology, 2017.

[9] A. J. Goldsmith and S.-G. Chua, "Adaptive coded modulation for fading channels," *IEEE Trans. Commun.*, 46, no. 5, pp. 595–602, 1998.

[10] M.-S. Alouini and A. J. Goldsmith, "Adaptive modulation over Nakagami fading channels," *Kluwer J. Wireless Commun.*, 13, no. 1–2, pp. 119–143, 2000.

[11] K. Cho and D. Yoon, "On the general BER expression of one- and two-dimensional amplitude modulation," *IEEE Trans. Commun.*, 50, no. 7, pp. 1074–1080, 2002.

[12] G. J. Foschini and M. Gans, "On limits of wireless communications in a fading environment using multiple antennas," *Wireless Pers. Commun.*, 6, pp. 311–335, 1998.

[13] E. Telatar, "Capacity of multi-antenna Gaussian channels," *Euro. Trans. Telecommun.*, 10, pp. 585–596, 1999.

[14] A. J. Goldsmith, *Wireless Communications*. Cambridge University Press, 2005.

[15] E. Biglieri, R. Calderbank, A. Constantinides, et al., *MIMO Wireless Communications*, Cambridge University Press, 2007.

3 Transmission Technologies for Selective Fading Channels

Wireless channel introduces frequency selective fading when channel coherence bandwidth is smaller than signal bandwidth. The filtering effect in the frequency domain manifests intersymbol interference (ISI) in the time domain, which will seriously deteriorate the performance of digital wireless transmission if not properly mitigated. In this chapter we present popular transmission technologies for frequency selective fading channels, including equalization, multicarrier transmission, and spread spectrum transmission. Equalization tries to remove the effects of selective fading through advanced signal processing. Multicarrier transmission proactively avoids the channel filtering effect by converting selective fading channel into parallel flat fading channels, also known as subcarriers. Spread spectrum transmission explores multipath diversity by intentionally operating in a selective fading environment.

We discuss the design ideas, salient features, and implementation challenges of these technologies to establish the foundation for advanced wireless technologies in the following chapters. Specifically, with the wide adoption of orthogonal frequency division multiplexing (OFDM), an efficient discrete implementation of multicarrier transmission, in modern wireless communication systems, those transmission technologies for flat fading channels can readily apply to individual subcarriers. RAKE receivers in spread spectrum systems can employ advanced diversity combining schemes introduced in later chapters.

3.1 Equalization

Equalization generally refers to various signal processing techniques applied at the receiver or the transmitter to mitigate the effect of frequency selective fading. Due to the channel filtering effect, direct detection based on received signal over a selective fading channel will lead to poor performance. The basic idea of equalization is first to preprocess the received signal or even the transmitted signal, with the knowledge of the channel response, to remove the channel filtering effect and then perform detection.

3.1.1 Equalizing Receiver

Equalizing receiver preprocesses the received signal in order to remove the effect of the selective fading channel and then applies symbol-by-symbol detection. Fig. 3.1 presents a receiver structure with equalization. In particular, the receiver implements

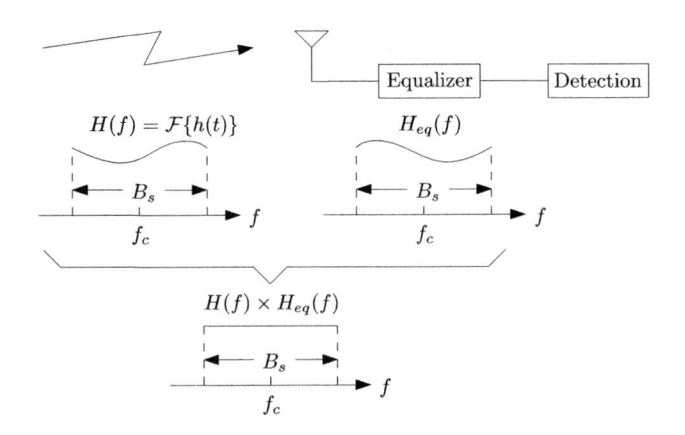

Figure 3.1 The filtering effect of the selective fading channel. Reproduced with permission from [1]. ©2017 IET.

a preprocessing filter, usually referred to as an *equalizer*. The frequency response of the equalizer is designed to be the inverse of the channel frequency response over the signal bandwidth, i.e.,

$$H_{eq}(f) = \frac{c}{H(f)}, \qquad (3.1)$$

where c is a normalizing constant, depending on the output power constraint of the equalizer. As a result, the overall frequency response of the channel and the preprocessing filter becomes flat over the signal bandwidth, hence the name "equalization." The spectrum of the signal component at the equalizer output will have the same shape as that of the transmitted signal, and as such, the channel filtering effect is successfully removed. It is important to note that the equalizer design requires accurate knowledge of the channel response, which can be obtained at the receiver through channel estimation.

While capable of eliminating the effect of selective fading on the received signal, the equalizer will necessarily introduce a serious side-effect known as *noise coloring/enhancement*. The receiver front end will collect additive white Gaussian noise in addition to the received signal, both of which will be processed by the equalizer. Specifically, the spectrum of the received signal is given by

$$V(f) = H(f)U(f) + N(f)$$

where $U(f)$ is the spectrum of transmitted signal and $N(f)$ is that of the noise. The spectrum of the complex signal at the equalizer output is given by

$$Y(f) = H_{eq}(f) \times V(f) = c \times U(f) + H_{eq}(f) \times N(f). \qquad (3.2)$$

The spectrum of the additive noise is modified by the filter, resulting in a colored Gaussian noise with power spectral density proportional to $|H_{eq}(f)|^2$. Specifically, if the noise spectral density of white noise $N(f)$ is $N_0/2$, then that of the colored noise $N'(f) = H_{eq}(f) \times N(f)$ is given by $c^2 N_0/(2|H(f)|^2)$. If $|H(f)|$ is small at certain frequencies, then the equalizing filter will greatly enhance the noise power at those frequencies.

3.1 Equalization

Such noise coloring/enhancement may seriously affect the detection performance of the receiver.

Modern wireless transmission systems implement the equalizing operation digitally. Equalizers are designed using on the complex baseband input/output relationship for selective fading channels developed in the previous chapter. In particular, the received symbol over the ith symbol period is given by

$$r_i = \sum_{l=0}^{L-1} h_l s_{i-l} + n_i = h_0 s_i + \sum_{l=1}^{L-1} h_l s_{i-l} + n_i, \qquad (3.3)$$

where s_i is the transmitted symbols and h_l, $l = 0, 1, 2, \ldots, L-1$, characterize the channel response. Note that the received symbol over the current symbol period involves those transmitted over the current and previous $L-1$ symbol periods. If not properly mitigated, the ISI term $\sum_{l=1}^{L-1} h_l s_{i-l}$ will negatively affect the detection performance and result in an irreducible floor of the average bit error rate (BER), even when the noise power becomes negligible. Many equalizer structures have been proposed in the literature. Zero-forcing (ZF) and minimum mean-square error (MMSE) are two of the most popular linear equalization schemes [2]. A ZF equalizer is designed to force ISI equal to zero, but suffers the same noise coloring/enhancement problem as analog equalizers. MMSE equalizer achieves better performance than ZF equalizer by properly balancing ISI mitigation and noise enhancement. Nonlinear equalization solutions include decision-feedback equalization (DFE) and maximum likelihood sequence estimation (MLSE). Different equalizer structures achieve different trade-offs of performance versus complexity.

3.1.2 Adaptive Implementation

The equalizer design that we presented in the previous subsection assumes that the channel response is known to the receiver and remains unchanged for the duration of transmission. In practice, the channel knowledge is not readily available and has to be obtained through channel estimation. Furthermore, the channel response will change over time for wireless transmission and some wireline transmission, e.g., power-line communication systems. Therefore, the equalizer implementation in such scenarios will involve periodic equalizer training/update, also known as adaptive equalization.

Equalizer training/update should be performed at least every channel coherence time T_c. Typically, the transmitter will send some training sequence, i.e., data symbols known to the receiver, at the beginning of each T_c. The receiver will use the corresponding received symbols to train or update the equalizer. Various signal processing algorithms can apply, leading to different trade-offs between performance and converging speed. In particular, weight updating using the least mean square (LMS) algorithm results in lower complexity than weight calculation using the Weiner filter, but with worse performance. It is worth noting that certain equalizer designs do not require channel knowledge; typically referred to as *blind equalization*. Such an approach uses decoded data symbols for training/updating purposes. While certainly more efficient, blind equalizers tend to suffer error propagation.

The time duration required for training symbol transmission, if needed, and equalizer training/updating should be much smaller than the channel coherence time T_c. Otherwise, the system will have poor transmission efficiency due to the excess overhead for adaptive equalization. The number of training symbols to be transmitted over each T_c depends on the number of equalizer weights to be updated as well as the updating algorithms adopted. Meanwhile, different training/updating algorithms require different amounts of time to converge. If the channel coherence time of a selective fading channel is relatively small such that the time required for training symbol transmission is comparable with T_c, then equalization will not be a suitable countermeasure to ISI. We can apply other transmission technologies to mitigate ISI, such as multicarrier transmission.

3.2 Multicarrier Transmission/OFDM

Multicarrier transmission is another digital wireless transmission technology for frequency selective fading channels. The basic idea is to divide the wideband selective channel into many parallel narrowband subchannels and to transmit a low-rate substream over each subchannel. If the bandwidth of a subchannel is smaller or comparable to the channel coherence bandwidth, then each substream will experience frequency flat fading. Compared to the equalization approach, multicarrier transmission is a more proactive transmission solution for frequency selective fading channels. Note that the equalizer is trying to mitigate the negative effect of selective fading by processing the received signal, and therefore can be deemed more of a reactive approach. By jointly designing the transmitter and receiver, multicarrier transmission ensures that the transmitted signal will not experience the negative effect of selective fading at all.

Consider a digital wireless transmission system with symbol rate R_s. We assume that the transmitter uses raised cosine pulse shape with roll-off factor β, $0 \leq \beta \leq 1$ for each modulated symbol. The bandwidth of the modulated signal is then equal to

$$B_s = (1 + \beta)R_s. \tag{3.4}$$

If the coherence bandwidth of the channel B_c is less than B_s (or equivalently, the symbol period T_s is comparable or smaller than the channel RMS delay spread σ_T), then the transmitted signal will experience frequency selective fading. Now we divide the symbol stream to be transmitted into N parallel substreams. The symbol rate of each substream becomes R_s/N and symbol period NT_s. If N is large enough such that the bandwidth of modulated substream $B_N = (1 + \beta)R_s/N$ is less than B_c (or equivalently, the symbol period NT_s is larger than σ_T), then each substream will experience frequency flat fading.

While capable of completely eliminating ISI, multicarrier transmission usually suffers a certain bandwidth penalty and therefore becomes less spectral efficient than single-carrier transmission. With the conventional multicarrier receiver structure presented above, additional guard bandwidths between subchannels are required to facilitate the practical bandpass filter and to accommodate time-limiting pulse shapes.

3.2 Multicarrier Transmission/OFDM

The bandwidth penalty associated with conventional multicarrier transmission originates from the requirement for nonoverlapping subchannels in the frequency domain. For digital wireless transmission with matched filter detection, the interference between substreams can be avoided as long as subcarrier frequencies are orthogonal over the subchannel symbol period $T_N = N/R_s$. While achieving much higher spectrum efficiency than conventional non-overlapping implementation, such matched filter implementation is very sensitive to timing and frequency offsets between the transmitter and the receiver, which compromise the orthogonality. Nevertheless, both multicarrier implementations require N parallel carrier modulators and demodulators at the transmitter and the receiver, respectively, which may lead to prohibitively high system complexity and energy consumption when the required number of substreams is large. This observation motivates the discrete implementation of multicarrier transmission, which only requires a single carrier-modulator/demodulator and as such has much lower transceiver complexity.

3.2.1 Discrete Implementation/OFDM

The discrete implementation of multicarrier transmission eliminates the need for multiple carrier-modulator/demodulator pairs with the application of fundamental DSP operations: the discrete Fourier transform (DFT) and inverse DFT (IDFT). Since both DFT and IDFT can be efficiently calculated with fast Fourier transform (FFT) algorithms, the resulting design, typically termed as OFDM, is widely used in wireless transmission systems.

Consider the transmission of N data symbols, $s_0, s_1, \ldots, s_{N-1}$, over a frequency selective fading channel with impulse response, h_n, $n = 0, 1, \ldots, L-1$. The direct transmission of these data symbols will lead to the linear convolution of data symbols and channel response. Each symbol will experience ISI from the previously transmitted symbols. Multicarrier transmission presented in the previous subsection eliminates ISI by dividing the selective fading channel into parallel flat fading channels in the frequency domain. Effectively, multicarrier transmission achieves the input/output relation given by

$$y_i = H(i) \cdot s_i + n_i, \quad i = 0, 1, \ldots, N-1, \tag{3.5}$$

where y_i is the decision statistics corresponding to data symbol s_i, n_i is the additive Gaussian noise, and $H(i)$ is the channel frequency response at frequency $f_i = f_0 + i/T_N$. Each transmitted data symbol experiences frequency flat fading with effective channel gain $H(i)$. Mathematically, $H(i)$ can be determined as the ith sample of the N-point DFT of the channel impulse response h_n, as

$$H(i) = \text{DFT}\{h_n\} = \sum_{n=0}^{L-1} h_n e^{-j2\pi ni/N}, \quad i = 0, 1, \ldots, N-1. \tag{3.6}$$

Let us focus on the signal component of y_i and neglect the additive noise. Applying N-point IDFT to y_i with the application of the convolution property of DFT, we have

$$r_n \triangleq \text{IDFT}\{y_i\} = \text{IDFT}\{H(i) \cdot s_i\} = h_n \circledast x_n, \tag{3.7}$$

where x_n denotes the N-point IDFT of s_i, defined as

$$x_n = \text{IDFT}\{s_i\} = \sum_{i=0}^{N-1} s_i e^{j2\pi ni/N}, \ n = 0, 1, \ldots, N-1, \tag{3.8}$$

and $\circledast$ represents the circular convolution of two sequences, defined as

$$r_n = \sum_{k=0}^{L-1} h_k x_{(n-k)_N}, \tag{3.9}$$

where $(n-k)_N$ denotes $n-k$ modulo N. Essentially, r_n becomes the circular convolution of the IDFT of data symbols with the channel response h_n. From the above analysis, we arrive at the basic idea of the discrete implementation of multicarrier transmission summarized as:

The desired input/output relation for multicarrier transmission can be achieved if we (1) transmit the IDFT of data symbols, x_n, instead of data symbols s_i themselves; (2) manage to create the circular convolution of x_n and channel response h_n, i.e. r_n given in (3.9), at the receiver; and (3) perform DFT on r_n at the receiver.

The main challenge when implementing discrete multicarrier transmission is how to create the circular convolution of the transmitted channel symbols x_n and channel response h_n at the receiver. Note that the channel output is typically the linear convolution of transmitted symbols and channel vector. In particular, when N symbols are transmitted over a channel with length L, the channel output is of length $N + L - 1$ due to linear convolution. Furthermore, if we transmit another block of N channel symbols immediately, the earlier transmitted symbol block will affect the channel output corresponding to the later transmitted symbol block, leading to so-called interblock interference (IBI).

Introducing a cyclic prefix can effectively address such a challenge. In particular, adding a cyclic prefix can create the circular convolution of two finite-length sequences when they are linearly convoluted together. A cyclic prefix can also help eliminate IBI. In particular, if the length of the cyclic prefix added to the beginning of a channel symbol block is long enough, then IBI will only affect the reception of the cyclic prefix, not the results of circular convolution between data symbols and channel response. Finally, we arrive at the discrete implementation of multicarrier transmission, often referred to as an OFDM transceiver, as shown in Fig. 3.2.

3.2.2 Challenges of OFDM Transmission

OFDM technology achieves a practical multicarrier transmission solution with the application of IDFT and DFT operations. OFDM transmission systems face several critical challenges, including the overhead of cyclic prefix, high peak-to-average power ratio (PAPR), sensitivity of frequency and timing offset, and severe subcarrier fading. Some of these challenges are unique to discrete implementations while others are common to all multicarrier systems.

3.2 Multicarrier Transmission/OFDM

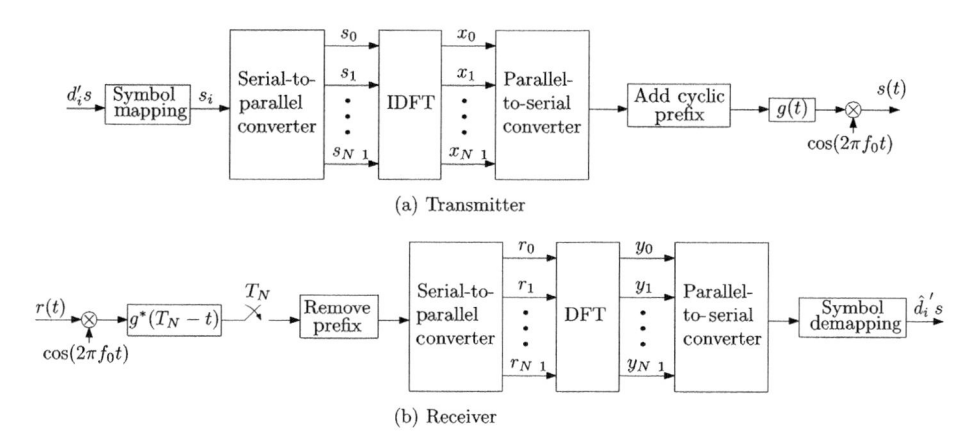

Figure 3.2 Transceiver structure for discrete implementation of multicarrier transmission. Reproduced with permission from [1]. ©2017 IET.

Discrete multicarrier implementation introduces a cyclic prefix to remove IBI. For the transmission of every N data symbols, the system actually transmits $N + M$ channel symbols, where M is the number of cyclic prefix symbols. The transmission of these redundant prefix symbols, although necessary, will consume extra bandwidth resources. Specifically, if the data symbol rate is R_s, then the channel symbol rate of OFDM systems will be $(N + M)R_s/N$. As such, the bandwidth requirement of OFDM systems increases by a factor of $(N+M)/N$ compared to conventional single-carrier transmission systems.

The PAPR, characterizing the dynamic range of the transmitted signal power, has a significant effect on the power efficiency of communication systems. The PAPR is defined as the ratio of the peak power over the average power of the signal. For digitally modulated signals, the PAPR can be calculated as

$$\text{PAPR} \triangleq \frac{\max_n\{|x_n|^2\}}{\mathbf{E}[|x_n|^2]}, \tag{3.10}$$

where $|x_n|^2$ denotes the power of transmitted symbols and $\mathbf{E}[\cdot]$ denotes the statistical averaging operation. Here we neglect the effect of the pulse shaping function. Based on the above definition, the PAPR of a linearly modulated signal with phase shift keying (PSK) modulation schemes will have a PAPR close to 1. Meanwhile, the transmitted signal of an OFDM system typically has a high PAPR due to the inverse DFT operation at the transmitter. Note that the transmitted channel symbols of OFDM systems are determined as

$$x_n = \sum_{i=0}^{N-1} s_i e^{j2\pi ni/N}, \ n = 0, 1, \ldots, N-1. \tag{3.11}$$

As such, the channel symbols x_n may have a high PAPR, even when modulated data symbols s_i have identical power. Numerous research efforts have been carried out to develop effective solutions. Candidate PAPR reduction solutions include signal clipping, peak cancelation with complementary signal, and dynamic null placement.

OFDM systems achieve orthogonal multicarrier transmission. When the block size N is used, the subcarrier spacing is $B_s/N = 1/T_N$. As such, neighboring subchannels overlap, which requires perfect orthogonality to eliminate inter-subchannel interference. In practical systems, the orthogonality may be imperfect due to, for example, mismatched oscillators, Doppler frequency shift, and/or timing synchronization errors. In general, the larger the block size N, and as such the smaller the subcarrier spacing, the more significant the inter-subchannel interference due to frequency and timing offset. The inter-subchannel interference due to frequency offset is typically much more severe than that due to timing offset.

3.3 Spread Spectrum Transmission

Spread spectrum transmission refers to the class of transmission technologies in which transmitted signal bandwidth is much greater than that required by the modulated signal. The transmitted signal, generated by spreading the spectrum of the modulated signal, will necessarily experience frequency selective fading, while the direct transmission of the modulated signal may experience flat fading. Utilizing more bandwidth than necessary creates several important advantages for spread spectrum transmission systems. First of all, spread spectrum transmission is resistant to intentional narrowband jamming. Second, the transmitted signal can hide below the noise floor and make eavesdropping very difficult. These two features of spread spectrum transmission make it especially suitable for military applications. Combined with a RAKE receiver, spread spectrum transmission can explore path diversity by combining different multipath components coherently and improve transmission reliability. Spread spectrum transmission also facilitates the sharing of the radio spectrum among multiple users in the code domain, leading to the code division multiple access (CDMA) scheme. These two features are exploited in various commercial wireless communication systems. Finally, the wide spectrum of spread spectrum signal facilitates certain localization applications.

3.3.1 Direct-Sequence Spread Spectrum

Spread spectrum transmission uses data-independent spreading code to spread the spectrum of a modulated signal. With direct-sequence implementation, the spectrum is spread by directly multiplying the modulated signal with a spreading signal generated using a certain spreading code. The structure of the direct-sequence spread spectrum transmitter is shown in Fig. 3.3. While mostly linear modulation schemes can apply, we assume binary phase shift keying (BPSK) for the sake of presentation clarity. As such, the modulated symbols over each symbol period, denoted by s_l, will take on values of $+1$ and -1. The baseband modulated signal over the lth symbol period after pulse shaping is given by

$$x(t) = s_l \cdot g(t), \ 0 \le t \le T_s,$$

3.3 Spread Spectrum Transmission

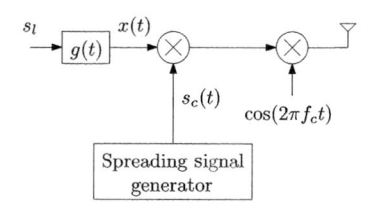

Figure 3.3 Direct-sequence spread spectrum transmitter. Reproduced with permission from [1]. ©2017 IET.

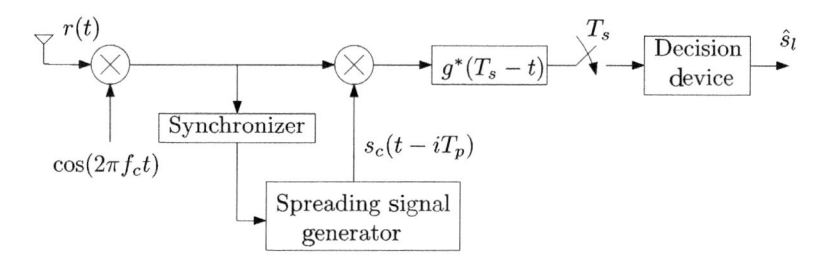

Figure 3.4 Direct-sequence spread spectrum receiver. Reproduced with permission from [1]. ©2017 IET.

where T_s is the symbol period and $g(t)$ is the rectangular shaping pulse. The modulated signal is then multiplied by the spreading signal $s_c(t)$, which consists of a sequence of short chips, given by

$$s_c(t) = \sum_{n=0}^{G-1} d_n \, g_c(t - nT_p), \ 0 \le t \le T_s, \quad (3.12)$$

where T_p is the chip period, $d_n = \pm 1$ are chip values, and $g_c(t)$ is the chip shaping pulse. Here, the constant G, equal to T_s/T_p, is usually referred to as the *processing gain* of a spread spectrum system. Typically, G is chosen to be much greater than 1. Note that the bandwidth of $x(t)$ is proportional to $1/T_s$ and that of $s_c(t)$ to $1/T_p$, which is G times larger than the bandwidth of $x(t)$. As such, the bandwidth of the transmitted bandpass signal, given modulation, by

$$x_c(t) = x(t) s_c(t) \cos(2\pi f_c t),$$

will be proportional to $1/T_p$, and much larger than that of $x(t)$.

The structure of a direct-sequence spread spectrum receiver is shown in Fig. 3.4. The received signal is first down-converted to baseband with a locally generated carrier. Then, the signal is multiplied with a locally generated spreading signal, which is synchronized to a particular propagation path, usually the first one with a large enough path power gain. After that, the receiver applies a matched filter to perform symbol-by-symbol detection. To elaborate further, let us assume that the receiver is synchronized to the path with delay iT_p and examine the matched filter output corresponding to the signal received over the synchronized path. The baseband received signal over the lth symbol period corresponding to the signal copy received over the synchronized path is given by

$$z_i x(t - iT_p) s_c(t - iT_p) = z_i s_l \, g(t - iT_p) s_c(t - iT_p). \tag{3.13}$$

The corresponding matched filter output, after multiplying the locally generated spreading signal $s_c(t - iT_p)$, is given by

$$\frac{1}{T_s} \int_{iT_p}^{T_s + iT_p} z_i \cdot s_l \, s_c^2(t - iT_p) dt = z_i \cdot s_l. \tag{3.14}$$

The signal on the synchronized path is perfectly despread for detection. Meanwhile, the baseband received signal corresponding to those unsynchronized paths is given by

$$\sum_{j=0, j \neq i}^{L-1} z_j x(t - jT_p) s_c(t - jT_p). \tag{3.15}$$

After multiplying the locally generated spreading signal $s_c(t - iT_p)$ and applying the matched filter, we can show that the corresponding sample at the matched filter output is given by

$$\sum_{j=0, j \neq i}^{L-1} \frac{1}{T_s} \int_{iT_p}^{T_s + iT_p} z_j s_l s_c(t - jT_p) s_c(t - iT_p) dt = \sum_{j=0, j \neq i}^{L-1} z_j s_l \rho((i-j)T_p), \tag{3.16}$$

where $\rho(\tau)$ is an autocorrelation function of the spreading signal $s_c(t)$, defined by

$$\rho(\tau) = \frac{1}{T_s} \int_0^{T_s} s_c(t) s_c(t - \tau) dt. \tag{3.17}$$

From these results, we can see that multipath interference will be greatly reduced if we can design the spreading signal such that $\rho((i-j)T_p) \approx 0, j \neq i$.

A direct-sequence spread spectrum receiver may experience certain narrowband interference. Such interference may originate from narrowband systems using the same frequency band or from an intentionally hostile jammer. The received signal on the synchronized path with narrowband interference is given by

$$r(t) = z_i x_c(t - iT_p) + n(t) + I(t), \tag{3.18}$$

where $I(t)$ denotes the narrowband interference collected by the receiver. We assumed $I(t)$ to be of the form

$$I(t) = I_b(t) \cos(2\pi f_c t), \tag{3.19}$$

where $I_b(t)$ is a baseband signal of bandwidth B_I. The direct-sequence spread spectrum receiver will first down-convert it to baseband and multiply with a locally generated spreading signal. The resulting signal, given by $I_b(t) s_c(t - iT_p)$, will have a bandwidth of $B_I + B_p$, where B_p is the bandwidth of the spreading signal. As such, the spectrum of the interference signal is spread by the despreading process at the receiver. The receiver will then apply matched filter detection, which generates matched filter output over the lth symbol period as

$$I_l = \frac{1}{T_s} \int_{iT_p}^{T_s+iT_p} I_b(t)s_c(t - iT_p)dt. \quad (3.20)$$

The matched filter essentially acts as a low-pass filter with a bandwidth of $B_s \propto 1/T_s$. As such, the percentage of interference signal power that will affect the signal detection is $B_s/(B_I+B_p)$. When $B_p \gg B_I$, $B_s/(B_I+B_p)$ is approximately equal to $1/G$, the inverse of the processing gain of the spread spectrum system. The interference power is reduced by a factor of $1/G$ with the spread spectrum receiver.

3.3.2 RAKE Receivers

A basic direct-sequence spread spectrum receiver synchronizes to a single path and performs detection on the signal received on that path while treating signals from other paths as interference. The synchronized path signal is typically the first one found during the synchronization process. Meanwhile, multipath signals travel through different propagation mechanisms and carry the same information signal to the receiver. Such potential diversity benefit, usually referred to as path diversity, can be fully explored with more complex receiver structure. One intuitive approach is to modify the synchronization process such that the receiver is synchronized to the strongest multipath signal among available ones. Such an approach will essentially achieve selection combining over multipath signals but incur the complexity of estimating and comparing power gains of all resolvable paths.

A more complex receiver will implement multiple branches, with each branch synchronized to a different resolvable path. The resulting receiver structure is shown in Fig. 3.5 and widely known as a RAKE receiver due to its resemblance to a garden rake. In particular, the jth branch, $j = 1, 2, \ldots, J$, is synchronized with the multipath signal with delay τ_j, and as such multiplies the received signal with $s_c(t - \tau_j)$. Typically, different branches are synchronized to signal paths with delay difference greater than the chip period T_p to extract maximum diversity benefit. The matched filter output of the jth branch over the lth symbol period is given by

$$r_l^{(j)} = z_j s_l + I_m^{(j)} + n_l^{(j)}, \; j = 1, 2, \ldots, J, \quad (3.21)$$

where $I_m^{(j)}$ denotes the residual multipath interference and $n_l^{(j)}$ is the sample of the additive white Gaussian noise on the jth branch. These outputs are then combined together before being passed to the decision device.

Various combining techniques can apply while maximum ratio combining (MRC) is the most common in practice. Specifically, with properly designed spreading code, the residual multipath interference $I_m^{(j)}$ can be treated as additional noise. With MRC,

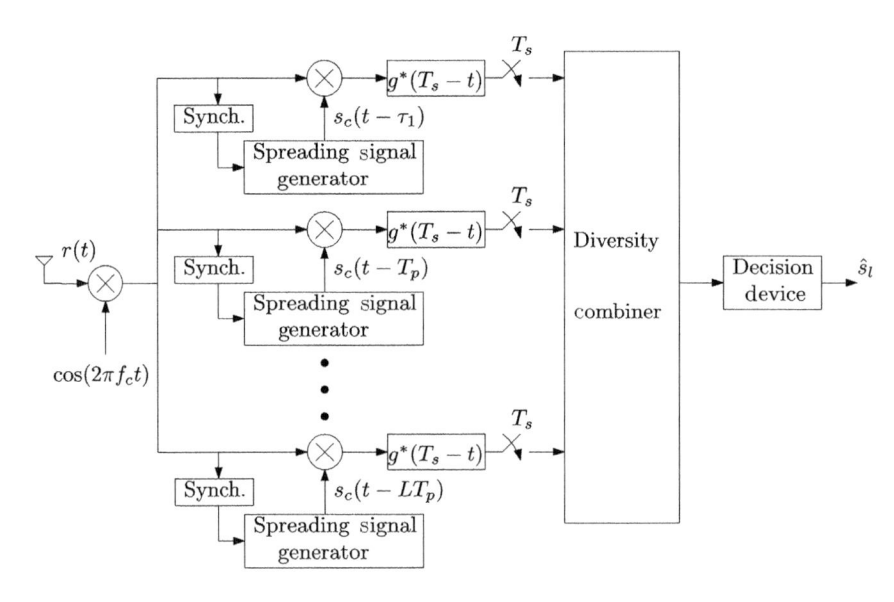

Figure 3.5 Structure of a RAKE receiver. Reproduced with permission from [1]. ©2017 IET.

the matched filter output $r_l^{(i)}$ will be weighted by a weight proportional to z_i^* and then summed together, leading to the decision statistics given by

$$\hat{s}_l = \sum_{j=1}^{J} |z_j|^2 s_l + \sum_{j=1}^{J} z_j^* \tilde{n}_l^{(j)}, \qquad (3.22)$$

where $\tilde{n}_l^{(j)} = I_m^{(j)} + n_l^{(j)}$. The receiver can achieve similar performance as an L-branch MRC diversity receiver when $\tilde{n}_l^{(j)}$ are independent and identically distributed across different branches. If the multipath interference cannot be approximated by white noise or if there exists narrowband interference, MRC is no longer optimal. Alternatively, we can apply the optimal combining scheme.

3.3.3 Frequency-Hopping Spread Spectrum

The frequency-hopping spread spectrum implementation spreads the spectrum of the modulated signal by changing its carrier frequency over a wide bandwidth according to the spreading signal $s_c(t)$. If there are N carrier frequencies available for hopping and the modulated signal bandwidth is B_s, then the bandwidth of the frequency hopping transmission system is NB_s. When T_p is larger than the symbol period T_s, i.e., $T_p = kT_s$, where k is an integer, the hopping is deemed relatively slow, leading to the so-called slow frequency-hopping (SFH) system. On the other hand, when $T_p = T_s/k$, i.e., the carrier frequency changes multiple times per symbol, we have a fast frequency-hopping (FFH) system. Fast frequency-hopping systems can exploit frequency diversity within each symbol period, whereas SFH systems need to rely on coding/interleaving schemes to extract frequency diversity.

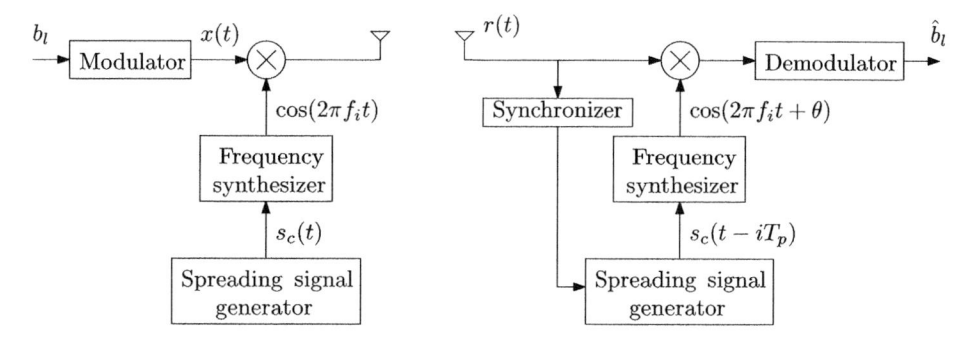

Figure 3.6 Structure of a frequency-hopping transmitter and receiver. Reproduced with permission from [1]. ©2017 IET.

The structure of a frequency-hopping transmitter and receiver is shown in Fig. 3.6. The spreading signal $s_c(t)$ is fed into the frequency synthesizer to control the carrier frequency. Unlike direct-sequence spread spectrum implementation, the values of $s_c(t)$ over each chip period T_p will be the index of carrier frequencies. Due to the difficulty of maintaining phase coherence, noncoherent or differentially coherent modulation schemes are typically used in frequency-hopping systems. The receiver will first synchronize the local spreading signal generator. The synchronized spreading signal is then used to control the frequency synthesizer, which generates local carrier used for down-conversion.

3.4 Summary

In this chapter we reviewed the essential transmission technologies for the selective fading environment. We first present the basic principles and implementation challenges of equalization technology. We then introduce multicarrier transmission technology and its discrete implementation – orthogonal frequency division multiplexing (OFDM), which has been widely adopted in modern wireless communication systems. We conclude the chapter with a succinct introduction to spread spectrum transmission technology. These technologies will serve as the fundamentals for advanced wireless technologies presented in later chapters.

3.5 Further Reading

For more detailed discussion on equalization, the reader can refer to the popular textbooks of G. Stuber [2] and A. Goldsmith [3]. [4] gives a complete overview of the OFDM technology and its applications in wireless communication systems. A unique presentation of OFDM is available in [1]. Dedicated books on spread spectrum transmission include [5, 6].

References

[1] H.-C. Yang, *Introduction to Digital Wireless Communications*, The Institution of Engineering and Technology, 2017.
[2] G. L. Stüber, *Principles of Mobile Communications*, 2nd ed., Kluwer Academic Publishers, 2000.
[3] A. J. Goldsmith, *Wireless Communications*, Cambridge University Press, 2005.
[4] R. Prasad, *OFDM for Wireless Communication Systems*, Artech House Publishers, 2004.
[5] A. J. Viterbi, *CDMA: Principles of Spread Spectrum Communication*, Addison-Wesley, 1995.
[6] A. J. Viterbi, *Principles of Spread-Spectrum Communication Systems*, 3rd ed., Springer, 2015.

4 Advanced Diversity Techniques

Diversity combining technique can greatly improve the performance of wireless transmission systems over fading channels. In general, the potential diversity gain increases as the number of diversity branches increases, although with a diminishing gain. As such, there is considerable interest in wireless transmission systems, where a large number of diversity paths exist. Examples include wideband code division multiple access (WCDMA) systems [1], ultra wideband (UWB) systems [2], millimeter (MM)-wave systems [3], and massive MIMO systems [4]. Applying the optimal maximum ratio combining (MRC) scheme in the resulting diversity-rich environment will entail very high system complexity and power consumption. It is critical to exploit the potential diversity benefit of these systems in a highly efficient manner.

In this chapter, we present several advanced diversity combining schemes for diversity-rich environments, including generalized selection combining (GSC), GSC with threshold test per branch (T-GSC), generalized switch and examine combining (GSEC), and GSEC with post-examining selection (GSECps). The general principle of these schemes is to select a subset of good diversity paths among available ones and combine them in the optimal MRC fashion. Specifically, these schemes will determine the path subset using either best-selection or thresholding or a combination of them. As intuitively expected, different schemes will lead to different performance-versus-complexity trade-offs. We obtain the exact statistics of the combiner output signal-to-noise ratio (SNR) with these schemes, which are then utilized to quantify their performance and complexity.

4.1 Generalized Selection Combining (GSC)

GSC is one of the most widely studied low-complexity combining schemes for diversity-rich environments. It is also known as a hybrid selection and maximum ratio combining (HS/MRC) scheme. The basic idea of GSC is to select a subset of the best paths and then combine them in the MRC fashion. The rationale is that applying MRC to bad paths will bring little additional performance benefit, while entailing extra hardware complexity of additional RF chains. In addition, the channel estimation of weak paths can be unreliable, which further limits the benefit of applying MRC to them, as imperfect channel estimation will considerably degrade the performance of MRC [4, 5]. The structure of a GSC combiner is shown in Fig. 4.1, which consists of the concatenation

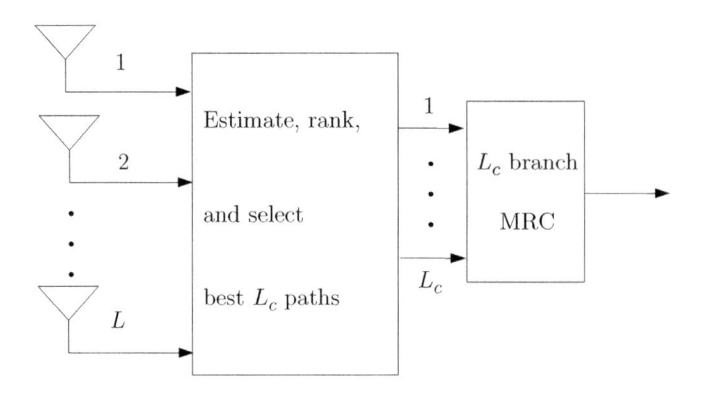

Figure 4.1 Structure of a GSC-based diversity combiner.

of a multibranch selection combiner and a conventional MRC combiner. The immediate benefit of the preselection stage is that only an L_c-branch MRC combiner needs to be implemented.

The mode of operation of a GSC combiner can be summarized as follows: (1) The receiver estimates the SNRs of all L available diversity paths, which may correspond to the resolved paths for UWB and WCDMA systems or the antenna branches in MM-wave systems. (2) The receiver ranks the SNRs and selects the L_c strongest paths, i.e., those with the highest SNR. (3) The receiver determines the MRC weights for those selected paths, which entails the full channel estimation (amplitude and phase) of the selected paths, and starts actual data reception. Note that unlike the conventional MRC combiner, where the receiver needs to estimate the complex channel gains for all diversity paths, the receiver with GSC only needs to estimate the complete channel gains of those selected paths. The path selection process only requires the received signal power on each diversity path.

4.1.1 Statistics of Output SNR

To evaluate the performance of GSC, we need the statistics of the combined SNR γ_c, which is given by

$$\gamma_c = \sum_{i=1}^{L_c} \gamma_{i:L}, \tag{4.1}$$

where $\gamma_{i:L}$ denotes the ith largest one among the total L SNRs. Applying the order statistics results in Section A.2.1 of the Appendix, we can obtain the closed-form expression of the moment generating function (MGF) of the combined SNR with GSC for the i.i.d. Rayleigh fading scenario as

$$\mathcal{M}_{\gamma_c}(s) = (1 - s\overline{\gamma})^{-L_c} \prod_{l=L_c+1}^{L} \left(1 - \frac{s\overline{\gamma}L_c}{l}\right)^{-1}. \tag{4.2}$$

4.1 Generalized Selection Combining (GSC)

With the MGF of γ_c, we can readily evaluate the average error rate performance of GSC over i.i.d Rayleigh fading with the MGF-based approach. In particular, the average error rate of M-ary phase-shift-keying (M-PSK) modulation scheme is given by [6]

$$\overline{P}_s = \frac{1}{\pi} \int_0^{\frac{(M-1)\pi}{M}} \mathcal{M}_{\gamma_c}\left(-\frac{g_{\text{PSK}}}{\sin^2 \phi}\right) d\phi, \tag{4.3}$$

where $g_{\text{PSK}} = \sin^2\left(\frac{\pi}{M}\right)$. The probability density function (PDF) and cumulative distribution function (CDF) of γ_c with GSC can be routinely derived after taking proper inverse Laplace transform. In particular, the CDF of γ_c with GSC is given in the following closed-form expression [7]:

$$F_{\gamma_c}(x) = \frac{L!}{(L-L_c)!L_c!} \left\{ 1 - e^{-\frac{x}{\overline{\gamma}}} \sum_{k=0}^{L_c-1} \frac{1}{k!}\left(\frac{x}{\overline{\gamma}}\right)^k \right.$$

$$+ \sum_{l=1}^{L-L_c} (-1)^{L_c+l-1} \frac{(L-L_c)!}{(L-L_c-l)!l!} \left(\frac{L_c}{l}\right)^{L_c-1}$$

$$\times \left[\left(1+\frac{l}{L_c}\right)^{-1}\left[1 - e^{-\left(1+\frac{l}{L_c}\right)\frac{x}{\overline{\gamma}}}\right]\right.$$

$$\left.\left. - \sum_{m=0}^{L_c-2} \left(-\frac{l}{L_c}\right)^m \left(1 - e^{-\frac{x}{\overline{\gamma}}} \sum_{k=0}^{m} \frac{1}{k!}\left(\frac{x}{\overline{\gamma}}\right)^k\right)\right]\right\}, \tag{4.4}$$

which can be directly applied to the outage performance analysis of GSC.

For general fading channel models, the distribution function of γ_c can still be obtained, although a closed-form expression may not be feasible. Specifically, let $\gamma_{1:L} \geq \gamma_{2:L} \geq \ldots \geq \gamma_{L:L}$ denote the ordered path SNRs in descending order. The combiner output SNR with GSC can be viewed as the sum of two correlated random variables: the sum of the first $L_c - 1$ largest SNRs $\sum_{i=1}^{L_c-1} \gamma_{i:L}$, denoted by Γ_{L_c-1}, and the L_cth largest path SNR $\gamma_{L_c:L}$. It follows that the PDF of the combiner output SNR γ_c can be written in terms of the joint PDF of Γ_{L_c-1} and $\gamma_{L_c:L}$ as

$$p_{\gamma_c}(x) = \int_0^\infty p_{\Gamma_{L_c-1},\gamma_{L_c:L}}(x-y, y) dy. \tag{4.5}$$

The generic expression of the joint PDF $p_{\Gamma_{L_c-1},\gamma_{L_c:L}}(\cdot,\cdot)$ is given by

$$p_{\Gamma_{L_c-1},\gamma_{L_c:L}}(z,y) = p_{\sum_{j=1}^{l-1}\gamma_j^+}(z) \frac{L!}{(L-l)!(l-1)!}[F_\gamma(y)]^{L-l}[1-F_\gamma(y)]^{l-1}p_\gamma(y), \tag{4.6}$$

where $p_{\sum_{j=1}^{l-1}\gamma_j^+}(z)$ denotes the PDF of the sum of the $l-1$ truncated random variable from left at y. See Section A.2.2 of the Appendix for further discussion.

The MGF of the combined SNR with GSC for independent and nonidentical distributed (i.n.d.) fading can be obtained as

Advanced Diversity Techniques

$$\mathcal{M}_{\gamma_c}(s) = \sum_{\substack{n_1,\cdots,n_{L_c-1} \\ n_1<n_2<\cdots<n_{L_c-1}}} \sum_{n_{L_c}} \int_0^\infty e^{-sx} p_{n_{L_c}}(x)$$

$$\times \left[\prod_{l=1}^{L_c-1} \mathcal{M}_{n_l}(s,x) \right] \left[\prod_{l'=L_c+1}^{L} F_{n'_l}(x) \right] dx, \qquad (4.7)$$

where $n_i \in \{1, 2, \ldots, L\}$, $i = 1, \ldots, L_c$, are the index of the selected diversity paths, $p_{n_{L_c}}(x)$ is the PDF of the L_cth selected path SNR, $F_{n'_l}(x)$ is the SNR CDF for the remaining branches, and $\mathcal{M}_{n_l}(s,x)$ is the truncated MGF of the lth selected branch SNR. Note that summation $\sum_{\substack{n_1,\cdots,n_{L_c-1} \\ n_1<n_2<\cdots<n_{L_c-1}}}$ is carrying over all possible index sets of the largest $L_c - 1$ branch SNRs out of the total L branches and $\sum_{n_{L_c}}$ over the possible indexes of the L_cth selected branch.

4.1.2 Numerical Results

Fig. 4.2 shows the outage probability of GSC as a function of the normalized average SNR per path $\overline{\gamma}/\gamma_{\text{th}}$ for fixed $L = 5$ and varying values of combined paths L_c. We

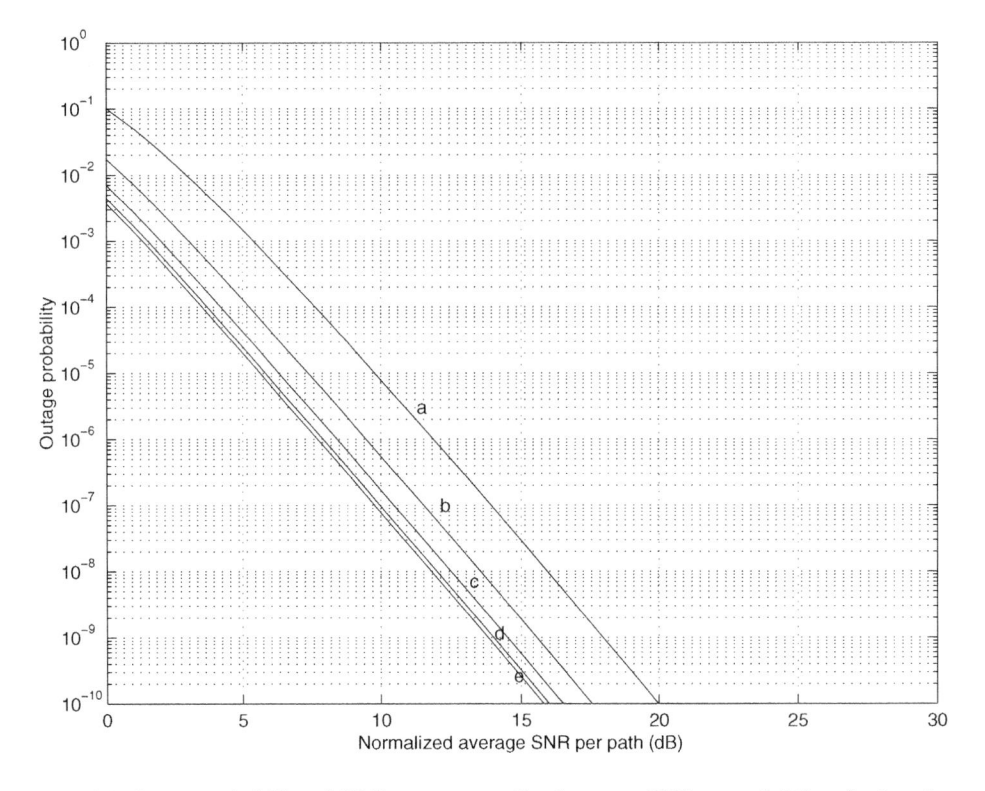

Figure 4.2 Outage probability of GSC versus normalized average SNR per path $\overline{\gamma}/\gamma_{\text{th}}$ for $L = 5$: (a) $L_c = 1$ (SC); (b) $L_c = 2$; (c) $L_c = 3$; (d) $L_c = 4$; and (e) $L_c = 5$. Reprint with permission from [7]. ©2000 IEEE.

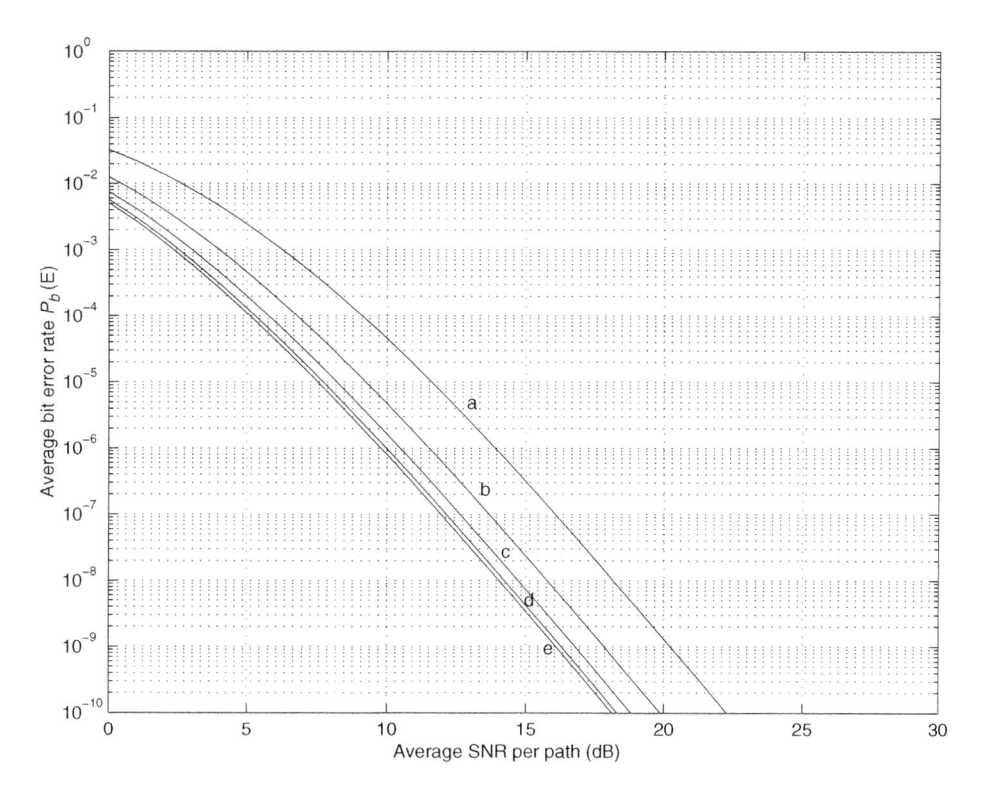

Figure 4.3 Average BER of BPSK with GSC versus average SNR per path $\bar{\gamma}$ for $L = 5$: (a) $L_c = 1$ (SC); (b) $L_c = 2$; (c) $L_c = 3$; (d) $L_c = 4$; and (e) $L_c = 5$. Reprint with permission from [7]. ©2000 IEEE.

can see that for a fixed number of available diversity paths a diminishing diversity gain is obtained as the number of combined paths increases. The same conclusion can be reached from Fig. 4.3, where we plot the average bit error rate (BER) of BPSK with GSC versus the average SNR per path for varying values of L_c.

Fig. 4.4 illustrates the effect of L and L_c on the average BER of BPSK with GSC reception over i.n.d. Rayleigh fading paths with an exponentially decaying average received SNR, i.e., $\bar{\gamma}_l = \bar{\gamma}_1 \exp(-\delta(l - 1))$, $l = 1, 2, \ldots, L$, where δ is the power decaying factor. These curves confirm again that diminishing returns are obtained as the number of ongest combined paths increases. We also observe a significant performance improvement can be gained by increasing the number of available diversity paths.

4.2 GSC with Threshold Test per Branch (T-GSC)

The GSC scheme discussed in the previous section can be viewed as a natural combination of SC and MRC. While maintaining a fixed low hardware complexity, GSC

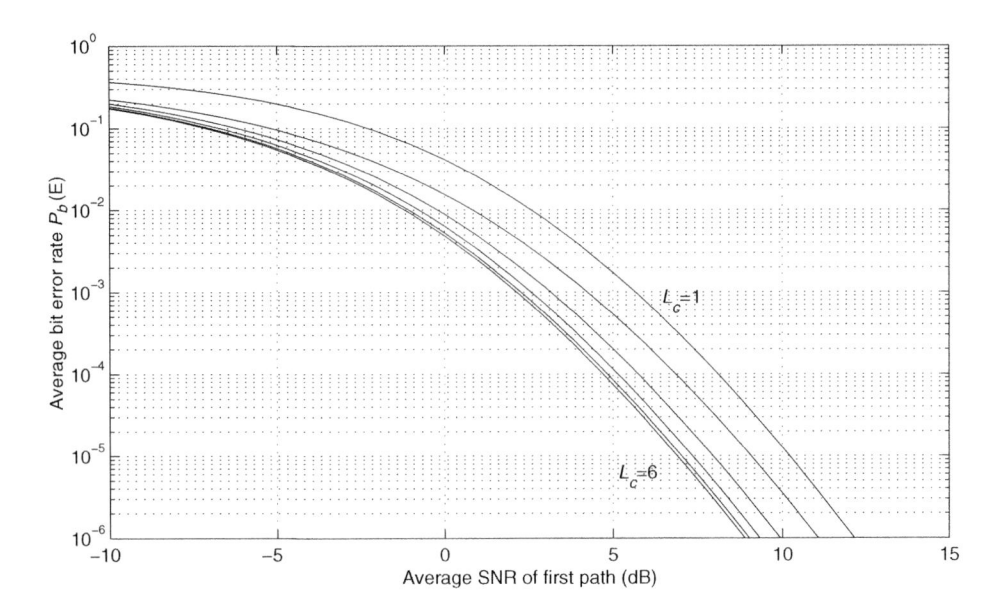

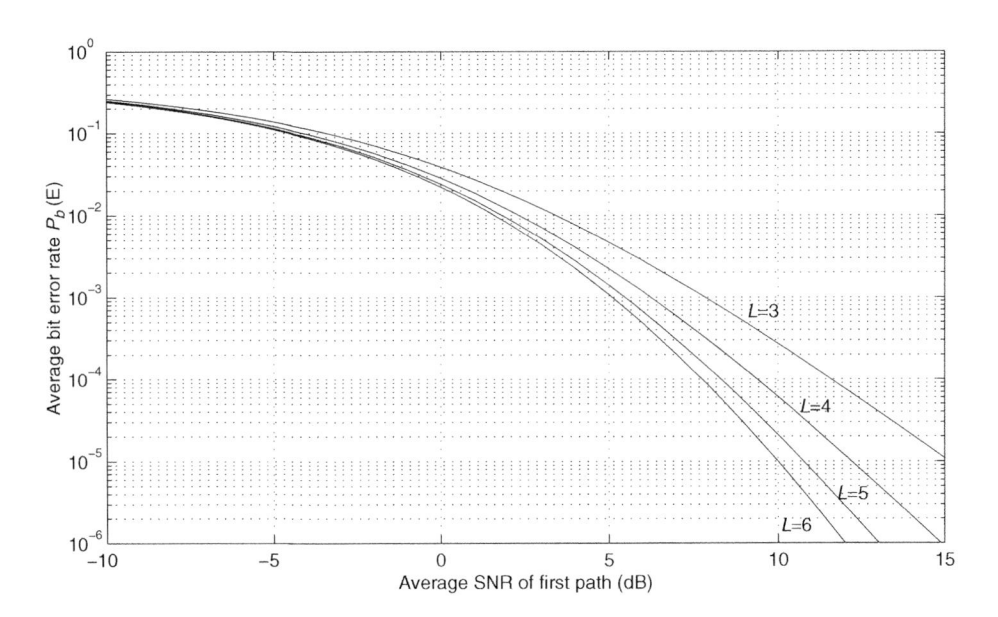

Figure 4.4 Average BER of BPSK versus the average SNR of the first path $\overline{\gamma}_1$ over an exponentially decaying power delay profile. (a) $\delta = 0.1$ and $L = 6$ and (b) $\delta = 0.3$ and $L_c = 3$. Reprint with permission from [7]. ©2000 IEEE.

may discard diversity paths with good quality or include some weak paths. The GSC with threshold test per branch (T-GSC) scheme alleviates the above-mentioned shortcomings of the conventional GSC scheme [8, 9]. With T-GSC, the combining decision

4.2 GSC with Threshold Test per Branch (T-GSC)

on a particular path is based on the comparison result of its SNR against a preselected threshold. Specifically, if the path SNR is above the threshold, the path will be combined in the MRC fashion. Otherwise, it will be discarded. As such, T-GSC will combine a variable number of diversity paths over time. Note that as the T-GSC combiner may need to combine all L diversity paths in the MRC fashion, T-GSC has the same hardware complexity as the conventional MRC scheme. But T-GSC can save the receiver processing power by only combining those paths with good-enough quality.

Depending on how the SNR threshold is chosen, there are two T-GSC schemes, i.e., absolute threshold GSC (AT-GSC) and normalized threshold GSC (NT-GSC). With AT-GSC, the ith diversity path is combined if $\gamma_i \geq \gamma_T$, where γ_T is a fixed SNR threshold. With NT-GSC, however, the threshold γ_T is determined using the best path SNR as

$$\gamma_T = \eta \max_l \{\gamma_l\},$$

where $0 < \eta < 1$ is the normalized threshold [10, 11]. Note that with AT-GSC, it may happen in the worst-case scenario that no path is combined, whereas with NT-GSC such a problem is avoided as at least the best path will be selected. On the other hand, the NT-GSC scheme requires the additional complexity of determining best path SNR, which involves the comparison of different estimated path SNRs. In the following, we obtain the statistics of the combiner output SNR with both T-GSC schemes.

4.2.1 Statistics of Output SNR with AT-GSC

Based on the mode of operation of the AT-GSC scheme, the combiner can be viewed as an L-branch MRC combiner with input SNRs given by

$$\gamma_l' = \begin{cases} \gamma_l, & \gamma_l \geq \gamma_{\text{th}}; \\ 0, & 0 \leq \gamma_l < \gamma_{\text{th}}, \end{cases} \quad (4.8)$$

where γ_l is the SNR of the lth diversity path. The combiner output SNR is equal to the sum of these L input SNRs, i.e., $\gamma_c = \sum_{l=1}^{L} \gamma_l'$. With the i.i.d. assumption, we can easily calculate the MGF of γ_c as the product of the MGFs of γ_l'. It can be shown that the PDF of γ_l' is given by

$$p_{\gamma_l'}(\gamma) = \begin{cases} F_\gamma(\gamma_T)\delta(\gamma), & \gamma = 0; \\ p_\gamma(\gamma), & \gamma \geq \gamma_T, \end{cases} \quad (4.9)$$

where $F_\gamma(\cdot)$ and $p_\gamma(\cdot)$ are the common CDF and PDF of the path SNRs, respectively. It follows that the MGF of γ_l' can be obtained as

$$\mathcal{M}_{\gamma_l'}(s) = F_\gamma(\gamma_T) + (1 - F_\gamma(\gamma_T))\mathcal{M}_{\gamma^+}(s), \quad (4.10)$$

Table 4.1 MGFs of truncated random variable γ^+ for the three fading models under consideration.

$\mathcal{M}_{\gamma^+}(s)$ where γ^+ has PDF $p_{\gamma^+}(x) = \frac{p_\gamma(x)}{1-P_\gamma(\gamma_T)}$, $x \geq \gamma_T$.	
Rayleigh	$\frac{1}{1-s\bar{\gamma}} e^{s\gamma_T}$
Rice	$\frac{1+K}{1+K-s\bar{\gamma}} e^{\frac{s\bar{\gamma}K}{1+K-s\bar{\gamma}}} \frac{Q_1\left(\sqrt{\frac{2K(1+K)}{1+K-s\bar{\gamma}}}, \sqrt{2(1+K-s\bar{\gamma})\frac{\gamma_T}{\bar{\gamma}}}\right)}{Q_1\left(\sqrt{2K}, \sqrt{2(1+K)\frac{\gamma_T}{\bar{\gamma}}}\right)}$
Nakagami-m	$\left(1-\frac{s\bar{\gamma}}{m}\right)^{-m} \frac{\Gamma\left(m, \frac{m\gamma_T}{\bar{\gamma}} - s\gamma_T\right)}{\Gamma\left(m, \frac{m\gamma_T}{\bar{\gamma}}\right)}$

where $\mathcal{M}_{\gamma_l^+}(s)$ denotes the MGF of the truncated random variable introduced in Theorem A.2 of the Appendix, defined as

$$\mathcal{M}_{\gamma^+}(s) = \frac{1}{1-F_\gamma(\gamma_T)} \int_{\gamma_T}^{\infty} p_\gamma(\gamma) e^{s\gamma} d\gamma. \quad (4.11)$$

For most popular fading channel models, the closed-form expression of $\mathcal{M}_{\gamma^+}(\cdot)$ is available. Table 4.1 summarizes these results for the three popular fading channel models under consideration. Finally, the MGF of the combined SNR with AT-GSC is given by

$$\mathcal{M}_{\gamma_c}(s) = [F_{\gamma_l}(\gamma_T) + (1 - F_{\gamma_l}(\gamma_T))\mathcal{M}_{\gamma_l^+}(s)]^L. \quad (4.12)$$

The PDF and CDF of the combined SNR can be routinely obtained after taking proper inverse Laplace transform and carrying out integration. With the statistics of the combined SNR, we can readily evaluate the performance of AT-GSC over fading channels.

4.2.2 Statistics of Output SNR with NT-GSC

To derive the statistics of the output SNR with NT-GSC, we consider L mutually exclusive events depending on how many diversity paths are combined [10]. Let $\gamma_{1:L}$ denote the largest one among all L path SNRs, i.e., $\gamma_{1:L} = \max_l\{\gamma_l\}$. It follows that L_c paths are combined with NT-GSC if and only if $\gamma_{1:L} \geq \gamma_{2:L} \geq \cdots \geq \gamma_{L_c:L} \geq \eta\gamma_{1:L} \geq \gamma_{L_c+1:L} \geq \cdots \geq \gamma_{L:L}$, which leads to the combiner output SNR being

$$\gamma_{c,L_c} = \sum_{i=1}^{L_c} \gamma_{i:L}, \quad L_c = 1, 2, \ldots, L. \quad (4.13)$$

Applying the total probability theorem, the overall distribution function of the combined SNR with NT-GSC can be calculated as the sum of the distribution functions for each individual event. In particular, the MGF of γ_c can be written as

$$\mathcal{M}_{\gamma_c}(s) = \sum_{L_c=1}^{L} \mathcal{M}_{\gamma_{c,L_c}}(s), \quad (4.14)$$

4.2 GSC with Threshold Test per Branch (T-GSC)

where $\mathcal{M}_{\gamma_c, L_c}(s)$ is the MGF of the output SNR for the event that L_c paths are combined and is given by

$$\mathcal{M}_{\gamma_c, L_c}(s) = \int_0^\infty d\gamma_{1:L} \int_{\eta\gamma_{1:L}}^{\gamma_{1:L}} d\gamma_{2:L} \cdots \int_{\eta\gamma_{1:L}}^{\gamma_{L_c-1:L}} d\gamma_{L_c:L} \int_0^{\eta\gamma_{1:L}} d\gamma_{L_c+1:L}$$
$$\int_0^{\gamma_{L_c+1:L}} d\gamma_{L_c+2:L} \cdots \int_0^{\eta\gamma_{L-1:L}} e^{s \sum_{i=1}^{L_c} \gamma_{i:L}} \quad (4.15)$$
$$p_{\gamma_{1:L}, \gamma_{2:L}, \ldots, \gamma_{L:L}}(\gamma_{1:L}, \gamma_{2:L}, \ldots, \gamma_{L:L}) d\gamma_{L:L},$$

where $p_{\gamma_{1:L}, \gamma_{2:L}, \cdots, \gamma_{L:L}}(\cdot)$ is the joint PDF of the ordered path SNRs, given in the Appendix. After proper substitution and carrying out the integration with the help of the definition of partial MGF,

$$\mathcal{M}_\gamma(s, x) = \int_x^\infty p_\gamma(\gamma) e^{s\gamma} d\gamma, \quad (4.16)$$

we can obtain the MGF for the event that L_c paths are combined in the following compact form, involving a single integral, as

$$\mathcal{M}_{\gamma_c, L_c}(s) = L_c \binom{L}{L_c} \int_0^\infty e^{s\gamma} p_\gamma(\gamma) [F_\gamma(\eta\gamma)]^{L-L_c} [\mathcal{M}_\gamma(s, \gamma) - \mathcal{M}_\gamma(s, \eta\gamma)]^{L_c - 1} d\gamma. \quad (4.17)$$

Note that the partial MGF $\mathcal{M}_\gamma(s, x)$ is related to the MGF defined in (4.11) as $\mathcal{M}_\gamma(s, \gamma_T) = \mathcal{M}_{\gamma_+}(s)(1 - F_\gamma(\gamma_T))$. As such, we can easily obtain the closed-form expression of the partial MGF for three popular fading channel models from Table 4.1. The PDF and CDF of the combined SNR with NT-GSC can be routinely obtained.

4.2.3 Complexity Analysis

Both GSC and T-GSC will reduce the combiner complexity in terms of the number of active MRC branches. With GSC, the number of active MRC branches is fixed to $L_c < L$, whereas with T-GSC, the number of active branches is randomly varying. In particular, the probability that i diversity branches will be active with AT-GSC is equal to

$$\Pr[N_c = i] = \binom{L}{i} [1 - F_\gamma(\gamma_T)]^i [F_\gamma(\gamma_T)]^{L-i}. \quad (4.18)$$

As such, the average number of combined paths with AT-GSC is given by

$$\overline{N}_c = \sum_{i=0}^L i \Pr[N_c = i] = L[1 - F_\gamma(\gamma_T)]. \quad (4.19)$$

According to the mode of operation of NT-GSC, the probability that i diversity branches are active is equal to $\Pr[\gamma_{i:L} \ge \eta\gamma_{1:L} \ge \gamma_{i+1:L}]$, which can be calculated using the joint PDF of $\gamma_{1:L}$, $\gamma_{i:L}$, and $\gamma_{i+1:L}$ as

$$\Pr[N_c = i] = \int_0^\infty \int_{\eta x}^x \int_0^{\eta x} p_{\gamma_{1:L}, \gamma_{i:L}, \gamma_{i+1:L}}(x, y, z) dx dy dz. \quad (4.20)$$

After slightly generalizing (A.4), the joint PDF $p_{\gamma_{1:L},\gamma_{i:L},\gamma_{i+1:L}}(\cdot,\cdot,\cdot)$ can be obtained as

$$p_{\gamma_{1:L},\gamma_{i:L},\gamma_{i+1:L}}(x,y,z) = \frac{L!}{(i-2)!(L-i-1)!}p_\gamma(x) \\ \times \left(F_\gamma(x) - F_\gamma(y)\right)^{i-2} p_\gamma(y)p_\gamma(z)\left(F_\gamma(z)\right)^{L-i-1}. \quad (4.21)$$

Therefore, the average number of combined paths with NT-GSC can be calculated.

4.2.4 Numerical Results

Fig. 4.5 plots the average BER of BPSK with AT-GSC versus average SNR for $L = 6$ and $\overline{N}_c = L_c$ value ranging from 1 to 6. In particular, γ_T value is set to reach target $\overline{N}_c$. The curve for $\overline{N}_c = 6$ corresponds to MRC. We observe that for sufficiently high SNR, the AT-GSC scheme, even with only $\overline{N}_c = 1$, outperforms MRC. This result follows from the fact that the asymptotic BER behavior of AT-GSC varies inversely with $L^{\overline{\gamma}}$. Equivalently, on a logarithmic scale, MRC has an inverse linear asymptotic performance (with slope proportional to L), whereas AT-GSC with $\overline{N}_c < L$ has an inverse exponential asymptotic performance.

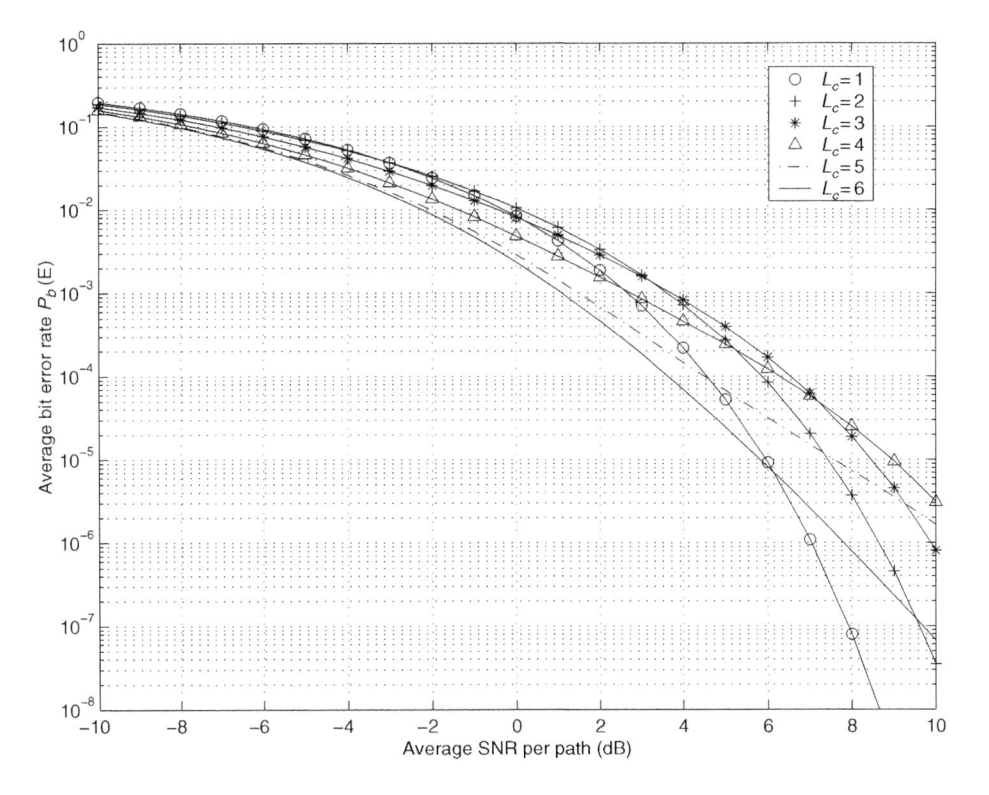

Figure 4.5 Average BER of BPSK with AT-GSC versus average SNR with average number of paths combined $\overline{N}_c = L_c$ as a parameter ($L = 6$). Reprint with permission from [9]. ©2002 IEEE.

4.2 GSC with Threshold Test per Branch (T-GSC)

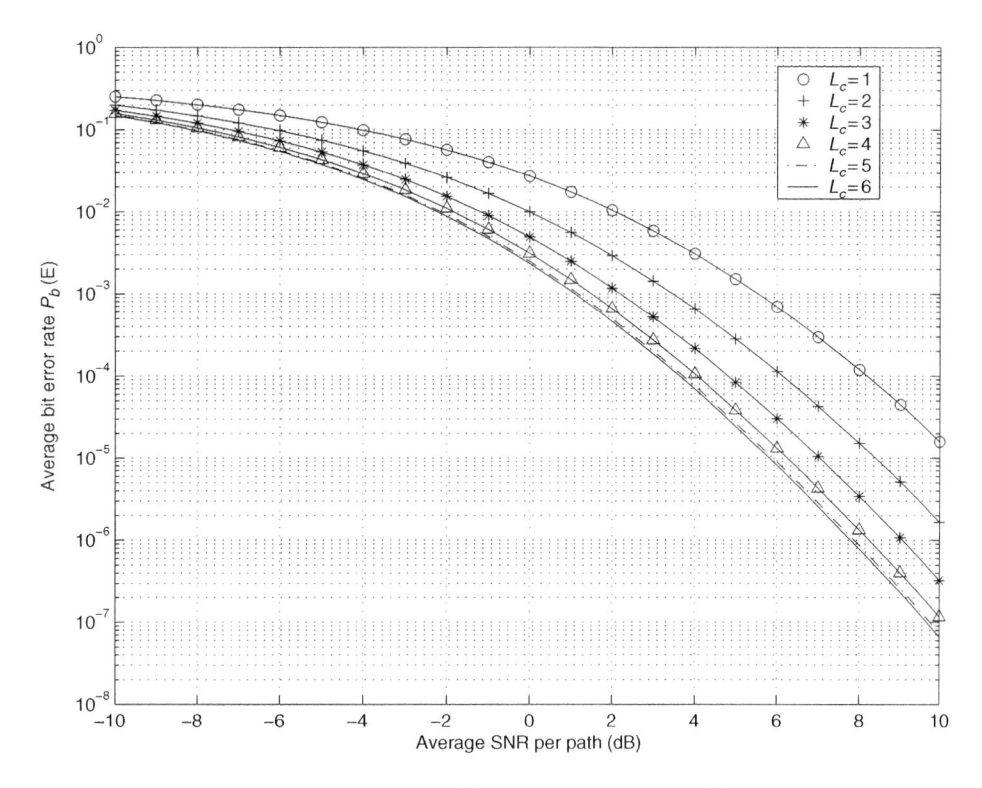

Figure 4.6 Average BER of BPSK with NT-GSC versus average SNR with average number of paths combined $\overline{N}_c = L_c$ as a parameter ($L = 6$). Reprint with permission from [9]. ©2002 IEEE.

Fig. 4.6 plots the average BER of BPSK with NT-GSC versus average SNR for $L = 6$ and $\overline{N}_c = L_c$ value ranging from one to six. Unlike AT-GSC, where the relation between the threshold γ_T and $\overline{N}_c$ is invertible, for NT-GSC the solution for η as a function of $\overline{N}_c$ must be determined numerically. We observe from Fig. 4.6 that the average BER of BPSK with NT-GSC has a much more regular behavior, resembling that of conventional GSC. Specifically, for any value of $\overline{\gamma}$, the average BER of NT-GSC improves monotonically with increasing $\overline{N}_c$ spanning between that of SC at $\overline{N}_c = 1$ and that of MRC at $\overline{N}_c = 6$.

Fig. 4.7 compares the average BER performance of GSC, AT-GSC, and NT-GSC for $L = 6$ and an average number of combined paths $\overline{N}_c$ equal to five. We see that traditional GSC slightly outperforms NT-GSC from an average BER standpoint. This, of course, comes at the expense of a slightly higher complexity since GSC requires the ranking of all diversity branch strengths, whereas NT-GSC just needs the knowledge of the branch relative strengths and, therefore, does not require full ranking. On another front, we see from Fig. 4.7 that AT-SC clearly outperforms GSC, NT-GSC, and actually even MRC over the high SNR region. This might be surprising at first glance, but recall that the average BER of AT-GSC is computed only for the fraction of time that one or more

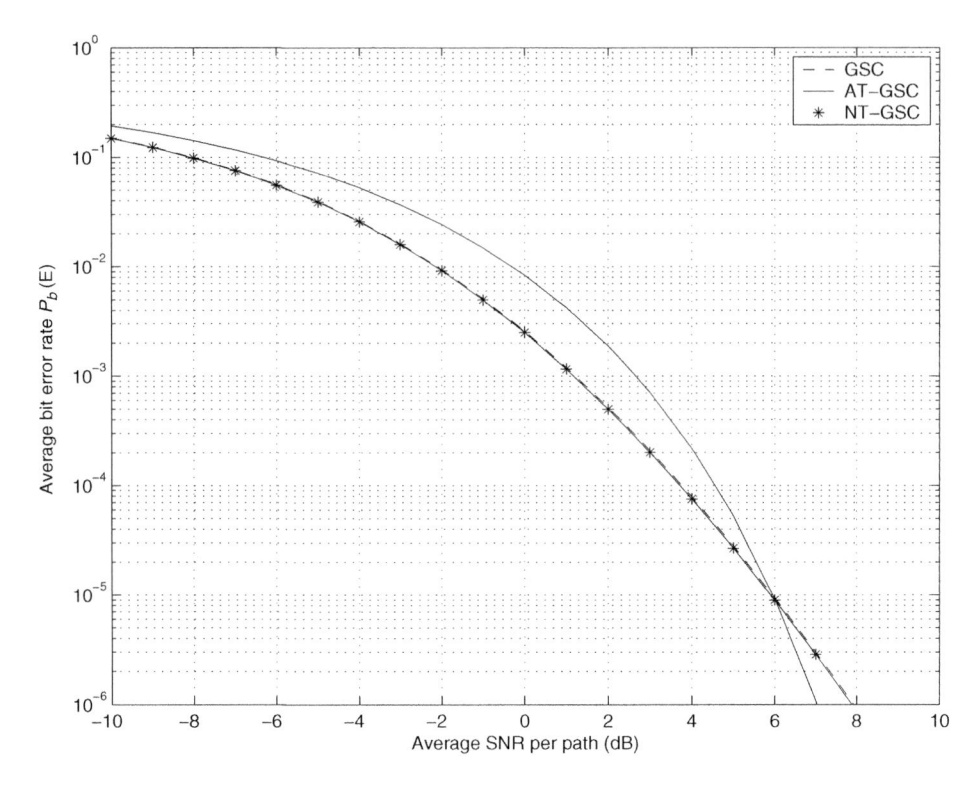

Figure 4.7 Comparison of average BER of BPSK with GSC, AT-GSC, and NT-GSC for $\overline{N}_c = L_c = 5$ and $L = 6$. Reprint with permission from [9]. ©2002 IEEE.

diversity paths are combined. The superior BER performance comes at the cost of its poor outage performance.

4.3 Generalized Switch and Examine Combining

Both GSC variants discussed in the previous section require the estimation of all available diversity paths, regardless of whether they are eventually combined or not. The path estimation complexity can be further reduced with a switching-based mechanism during the path selection stage. GSEC is another low-complexity combining scheme that was proposed for diversity-rich environments in this context [12]. The basic idea of GSEC is to extend the notion of conventional switch and examine combining to the diversity-rich scenario and use it for multiple path selection before cascading with a traditional MRC combiner. Note that the main complexity saving of threshold combining schemes over selection combining is fewer path estimations. As such, the GSEC scheme can also be viewed as a hybrid switch and examine/maximum ratio combining scheme.

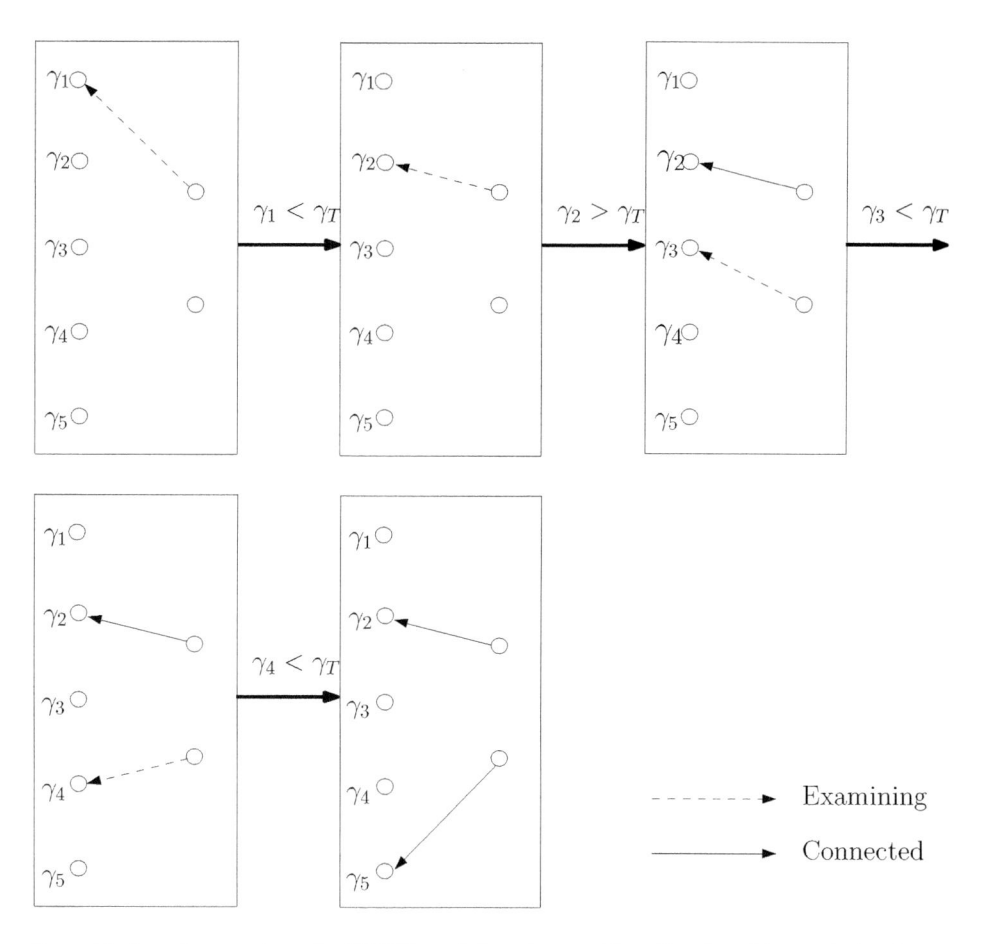

Figure 4.8 Sample operation of a GSEC-based diversity combiner. Reprint with permission from [12]. ©2002 IEEE.

Specifically, the receiver with GSEC tries to select L_c acceptable paths, i.e., whose instantaneous SNRs are above a preselected fixed threshold, out of the total L available ones for subsequent MRC processing. Fig. 4.8 illustrates a sample path selection process with GSEC. Specifically, the receiver with GSEC will sequentially estimate and compare path SNRs against γ_T until it finds L_c acceptable ones. After that, the receiver will stop path estimation. As such, the receiver with GSEC does not always need to estimate all L diversity paths, which is the major complexity saving of GSEC over GSC and T-GSC. If there are not enough acceptable paths even after examining all paths, the receiver will randomly select some unacceptable paths for MRC processing. Therefore, similar to GSC, GSEC always combines a fixed number of diversity branches. On the other hand, the path selection process with GSEC is much simpler than that of GSC as the receiver does not need to rank all available diversity paths. Note that with GSC, the receiver needs to compare two estimated SNRs, whereas with GSEC, only comparison of an estimated SNR with a fixed threshold is required.

4.3.1 Statistics of Output SNR

We need the statistical characterization of the combined SNR with GSEC to quantify its performance. In the following, we first derive the generic expression of the MGF. Other distribution functions of the combined SNR can be routinely obtained. According to the mode of operation of GSEC, the number of acceptable diversity paths that are eventually combined with GSEC takes values from 0 to L_c. Since they are exclusive and disjoint events, we can apply the total probability theorem and write the MGF of combiner output γ_c in the following weighted sum form:

$$\mathcal{M}_{\gamma_c}(s) = \sum_{i=0}^{L_c} \pi_i \, \mathcal{M}_{\gamma_c}^{(i)}(s), \tag{4.22}$$

where $\mathcal{M}_{\gamma_c}^{(i)}(\cdot)$ is the conditional MGF of γ_c given that exactly i out of L_c combined paths are acceptable and π_i is the probability that there are exactly i acceptable paths. With the i.i.d. fading assumption, we can show that π_i are mathematically given by

$$\pi_i = \begin{cases} \binom{L}{i} [F_\gamma(\gamma_T)]^{L-i}[1 - F_\gamma(\gamma_T)]^i, & i = 0, \ldots, L_c - 1; \\ \sum_{j=L_c}^{L} \binom{L}{j} [F_\gamma(\gamma_T)]^{L-j}[1 - F_\gamma(\gamma_T)]^j, & i = L_c, \end{cases} \tag{4.23}$$

where $P_\gamma(\cdot)$ is the common CDF of path SNRs and is given in Table 2.1 for the three fading models under consideration.

The conditional MGF $\mathcal{M}_{\gamma_c}^{(i)}(\cdot)$ can be written as

$$\mathcal{M}_{\gamma_c}^{(i)}(s) = [\mathcal{M}_{\gamma^+}(s)]^i [\mathcal{M}_{\gamma^-}(s)]^{L_c - i}, \tag{4.24}$$

where $\mathcal{M}_{\gamma^+}(\cdot)$ denotes the MGF of path SNR given that it is greater or equal to γ_T and $\mathcal{M}_{\gamma^-}(\cdot)$ denotes the MGF of the single path SNR given that it is less than γ_T, both of which can be calculated in closed forms. The closed-form expressions of $\mathcal{M}_{\gamma^+}(\cdot)$ for the three fading models under consideration are given in Table 4.1 and those for $\mathcal{M}_{\gamma^-}(\cdot)$ are summarized in Table 4.2.

Finally, after substituting (4.23) and (4.24) into (4.22), the generic expression for the MGF of combined SNR γ_c is given by

$$\mathcal{M}_{\gamma_c}(s) = \sum_{i=0}^{L_c-1} \binom{L}{i} [P_\gamma(\gamma_T)]^{L-i}[1 - P_\gamma(\gamma_T)]^i [\mathcal{M}_{\gamma^+}(s)]^i [\mathcal{M}_{\gamma^-}(s)]^{L_c-i}$$
$$+ \sum_{j=L_c}^{L} \binom{L}{j} [P_\gamma(\gamma_T)]^{L-j}[1 - P_\gamma(\gamma_T)]^j [\mathcal{M}_{\gamma^+}(s)]^{L_c}. \tag{4.25}$$

When $L_c = L$, it can be shown, with the help of the relationship

$$[1 - F_\gamma(\gamma_T)]\mathcal{M}_{\gamma^+}(s) + F_\gamma(\gamma_T)\mathcal{M}_{\gamma^-}(s) = \mathcal{M}_\gamma(s), \tag{4.26}$$

4.3 Generalized Switch and Examine Combining

Table 4.2 MGFs of truncated random variable γ^- for the three fading models under consideration.

$M_{\gamma^-}(s)$ where γ^- has PDF $p_{\gamma^-}(x) = \frac{p_\gamma(x)}{P_\gamma(\gamma_T)}$, $0 < x < \gamma_T$.
Rayleigh $\quad \frac{1}{1-s\overline{\gamma}} \frac{1-e^{s\gamma_T - \frac{\gamma_T}{\overline{\gamma}}}}{1-e^{-\frac{\gamma_T}{\overline{\gamma}}}}$
Rice $\quad \frac{1+K}{1+K-s\overline{\gamma}} e^{\frac{s\overline{\gamma}K}{1+K-s\overline{\gamma}}} \frac{1-Q_1\left(\sqrt{\frac{2K(1+K)}{1+K-s\overline{\gamma}}}, \sqrt{2(1+K-s\overline{\gamma})\frac{\gamma_T}{\overline{\gamma}}}\right)}{1-Q_1\left(\sqrt{2K}, \sqrt{2(1+K)\frac{\gamma_T}{\overline{\gamma}}}\right)}$
Nakagami-m $\quad \left(1-\frac{s\overline{\gamma}}{m}\right)^{-m} \frac{1-\Gamma\left(m, \frac{m\gamma_T}{\overline{\gamma}} - s\gamma_T\right)/\Gamma(m)}{1-\Gamma\left(m, \frac{m\gamma_T}{\overline{\gamma}}\right)/\Gamma(m)}$

where $\mathcal{M}_\gamma(\cdot)$ is the common MGF of the received SNR, that (4.25) reduces to $[\mathcal{M}_\gamma(s)]^L$, i.e., the GSEC combiner is equivalent to an L-branch MRC combiner, as expected. It can also be shown that when $L_c = 1$, (4.25) simplifies to the MGF of the output SNR of a traditional L-branch SEC combiner [13, Eq. 35].

With the analytical expression of the MGF, we can readily derive the PDF and CDF of the combined SNR with GSEC through inverse Laplace transform and integration. As an example, the CDF γ_c over i.i.d. Rayleigh fading channel can be obtained as

$$F_{\gamma_c}(x) = \sum_{i=0}^{L_c-1} \binom{L}{i} \left[1 - \exp\left(-\frac{\gamma_T}{\overline{\gamma}}\right)\right]^{L-L_c} \sum_{j=0}^{\min[L_c, \lfloor \frac{x}{\gamma_T} \rfloor]-i}$$
$$\times \binom{L_c - i}{j}(-1)^j \exp\left(-\frac{(i+j)\gamma_T}{\overline{\gamma}}\right)$$
$$\times \left[1 - \exp\left(-\frac{x-(i+j)\gamma_T}{\overline{\gamma}}\right) \sum_{k=0}^{L_c-1} \frac{1}{k!}\left(\frac{x-(i+j)\gamma_T}{\overline{\gamma}}\right)^k\right]$$
$$+ I_{L_c\gamma_T}(x) \sum_{j=L_c}^{L} \binom{L}{j}\left[1 - \exp\left(-\frac{\gamma_T}{\overline{\gamma}}\right)\right]^{L-j} \exp\left(-\frac{j\gamma_T}{\overline{\gamma}}\right)$$
$$\times \left[1 - \exp\left(-\frac{x - L_c\gamma_T}{\overline{\gamma}}\right) \sum_{k=0}^{L_c-1} \frac{1}{k!}\left(\frac{x - L_c\gamma_T}{\overline{\gamma}}\right)^k\right], \quad (4.27)$$

where $I_{L_c\gamma_T}(x)$ is an indicator function, which is equal to 1 if $x \geq L_c\gamma_T$ and 0 otherwise.

4.3.2 Complexity Analysis

We now quantify the complexity of GSEC in terms of the average number of path estimations. Because of the subsequent MRC combining, the receiver with GSEC needs always to estimate at least L_c diversity paths. In the case that there are fewer than L_c acceptable paths among the total L ones, based on the mode of operation of GSEC, whenever the receiver encounters $L - L_c$ unacceptable paths it will apply MRC to the remaining paths, which necessitates their full estimations. As a result, the receiver needs to estimate all L available diversity paths. It can be shown that the probability that there are fewer than L_c acceptable paths is given by

$$P_B = \sum_{L-L_c+1}^{L} \binom{L}{k} [F_\gamma(\gamma_T)]^{L-k} [1 - F_\gamma(\gamma_T)]^k. \quad (4.28)$$

On the other hand, if there are at least L_c acceptable diversity paths, GSEC combiner will stop examining paths whenever it has found L_c acceptable ones. In this case, the number of channel estimates during a guard period takes values from L_c to L. Note that the probability that k channel estimates are performed in a guard period, denoted by $P_A^{(k)}$, is equal to the probability that the L_cth acceptable path is the kth path examined, or equivalently, exactly $k - L_c$ ones of the first $k - 1$ examined paths are unacceptable whereas the kth examined path is acceptable. It can be shown, with the assumption of i.i.d. diversity paths, that $P_A^{(k)}$ is mathematically given by

$$P_A^{(k)} = \binom{k-1}{k-L_c} [1 - F_\gamma(\gamma_T)]^{L_c} [F_\gamma(\gamma_T)]^{k-L_c}. \quad (4.29)$$

Finally, by combining these two mutually exclusive cases, we obtain the expression for the overall average number of channel estimates needed by GSEC during a guard period as

$$\begin{aligned} N_E &= \sum_{L_c}^{L} k \, P_A^{(k)} + L \, P_B \\ &= \sum_{k=L_c}^{L} k \, [1 - F_\gamma(\gamma_T)]^{L_c} \binom{k-1}{k-L_c} [F_\gamma(\gamma_T)]^{k-L_c} \\ &\quad + L \sum_{L-L_c+1}^{L} \binom{L}{k} [F_\gamma(\gamma_T)]^{L-k} [1 - F_\gamma(\gamma_T)]^k. \end{aligned} \quad (4.30)$$

In the case of $\gamma_T = 0$, since $F_\gamma(\gamma_T) = 0$, it can be shown that $N = L_c$, as expected. On the other hand, when $\gamma_T \to \infty$ and thus $F_\gamma(\gamma_T) = 1$, it can be similarly shown that $N = L$, also as expected.

4.3.3 Numerical Results

Fig. 4.9 plots the average bit error probability of BPSK with GSEC as a function of the switching threshold for $L = 4$ and different values of L_c. It can be observed that there exists an optimal choice of the switching threshold in the minimum average error probability sense except for the $L_c = 4$ case, which corresponds to a traditional four-branch MRC. The performance advantage of the optimal threshold diminishes and the value of the optimal thresholds decreases when L_c increases.

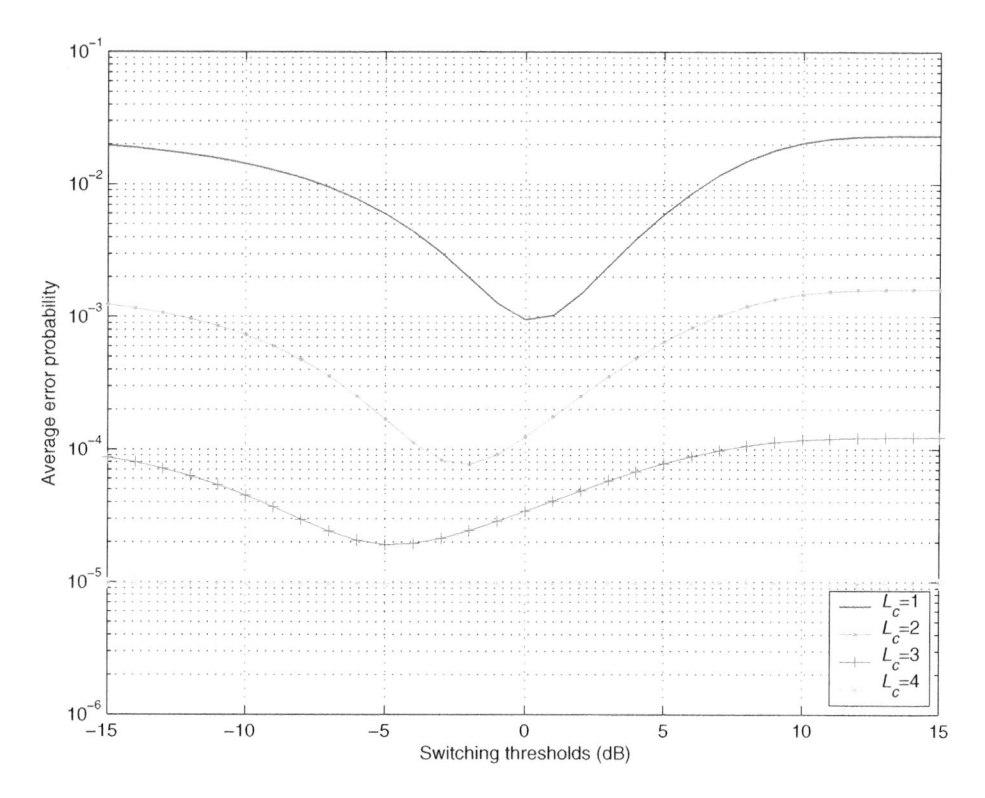

Figure 4.9 Average BER of BPSK with GSEC over $L = 4$ i.i.d. Rayleigh fading paths as a function of the common switching threshold γ_T ($\overline{\gamma} = 10$ dB). Reprint with permission from [12]. ©2002 IEEE.

Fig. 4.10 compares the error performance of GSEC and GSC in a diversity-rich environment with $L = 12$ available diversity paths. GSEC is not as good as GSC performance-wise. We also note that as L_c increases, the performance gap between GSEC and GSC reduces significantly. This implies that GSEC can take advantage of additional diversity paths with a relative lower complexity compared to GSC. For instance, for a BER of 10^{-6}, there is only a 0.6 dB difference when $L_c = 6$. Notice that, in this scenario, the GSC combiner needs to perform approximately $L_c \times L = 12 \times 6 = 72$ comparisons in each guard period to select the six strongest paths, whereas GSEC requires at most $L = 12$ comparisons.

Fig. 4.11 plots the average number of path estimates of GSEC with $L = 4$ as a function of the switching threshold for the Rayleigh fading case. As we can see, as the switching threshold γ_T increases, the number of path estimates N increases from L_c to L. Intuitively, that is because if the threshold is large, it becomes more difficult for the receiver to find acceptable paths and as such more paths need to be estimated. We also marked the optimal operating points for GSEC in terms of minimizing the average error rate, which were read from Fig. 4.9. Note that even for the best performance, GSEC

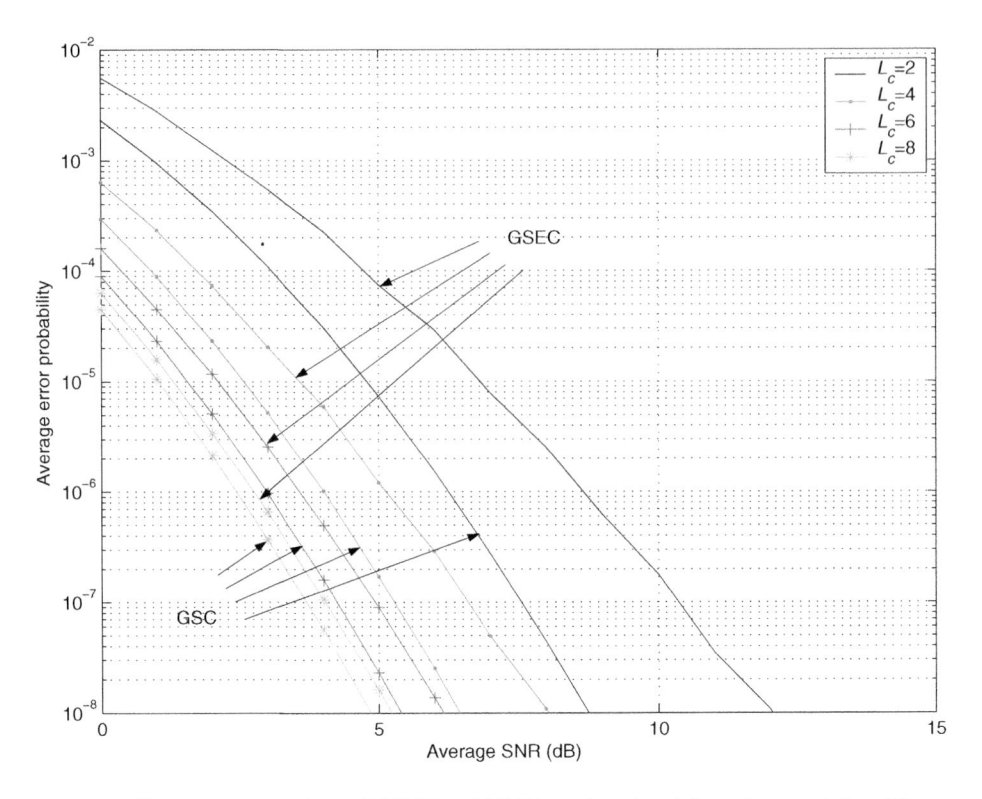

Figure 4.10 Error rate comparison of GSEC and GSC in a diversity-rich environment, $L = 12$. Reprint with permission from [12]. ©2002 IEEE.

does not need to estimate all the diversity paths, whereas for GSC all L diversity paths need to be estimated in each guard period.

4.4 GSEC with Post-Examining Selection (GSECps)

The GSEC scheme works well when the channel condition is favorable and the receiver can find enough acceptable paths. When there are not enough acceptable paths, however, GSEC will combine all the acceptable paths and some "randomly" selected unacceptable paths. In this case, the selected unacceptable paths may have low instantaneous SNR, which degrades the performance of GSEC, especially in the low SNR region. Noting that in this case all diversity paths have anyway been estimated, a preferred alternative is to combine those unacceptable paths with the highest instantaneous SNRs.

With this observation in mind, an improved version of GSEC, called generalized switch and examine combining with post-examine selection (GSECps), was proposed [14]. The new scheme operates in exactly the same way as GSEC when there are enough acceptable paths. When the number of acceptable paths is smaller than the number of paths required, instead of combining some "randomly" selected unacceptable

4.4 GSEC with Post-Examining Selection (GSECps)

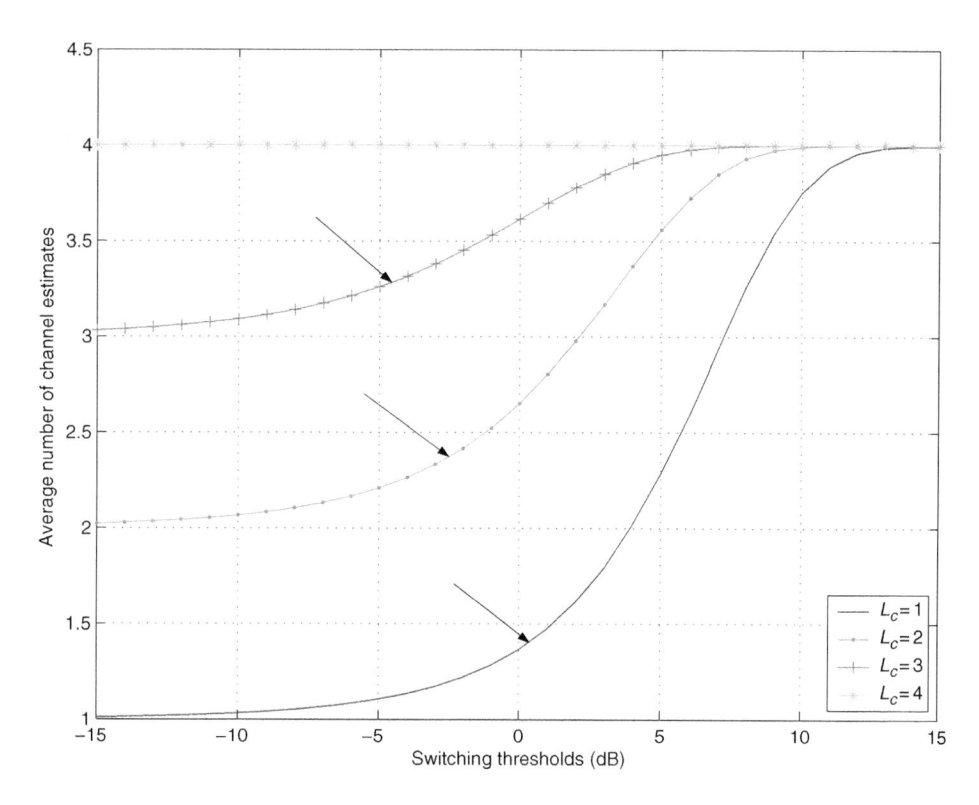

Figure 4.11 The average number of channel estimates of GSEC with $L = 4$ as a function of the common switching threshold γ_T ($\bar{\gamma} = 10$ dB). Reprint with permission from [12]. ©2002 IEEE.

paths, as with GSEC, GSECps combines the best unacceptable paths together with all acceptable paths. Intuitively, we expect that GSECps can offer better performance than GSEC through the occasional selection of the best unacceptable paths while maintaining roughly the same complexity. Note that GSECps can also be viewed as the generalization of switch and examine combining with post-examine selection (SECps) scheme studied in [15], since GSECps reduces to SECps when the number of paths to be combined is one.

Specifically, the mode of operation of GSEC can be summarized as follows. The receiver will sequentially estimate the instantaneous SNR of available diversity paths and compare them with the fixed threshold γ_T to determine their acceptance. Let L_a stand for the number of acceptable paths. If at least L_c paths are found acceptable, i.e., $L_a \geq L_a$, the receiver will combine the first L_c of them in the MRC fashion. Otherwise, the receiver will combine all the L_a acceptable paths, where $L_a < L_c$, and the strongest $L_c - L_a$ unacceptable paths. As we can see, when there are enough acceptable paths, i.e., $L_a \geq L_c$, GSECps will be equivalent to GSEC. On the other hand, when $L_a < L_c$, the receiver is essentially combining the L_c best paths of the L available paths, as is the case with GSC. Fig. 4.12 presents a sample operation of the GSECps scheme.

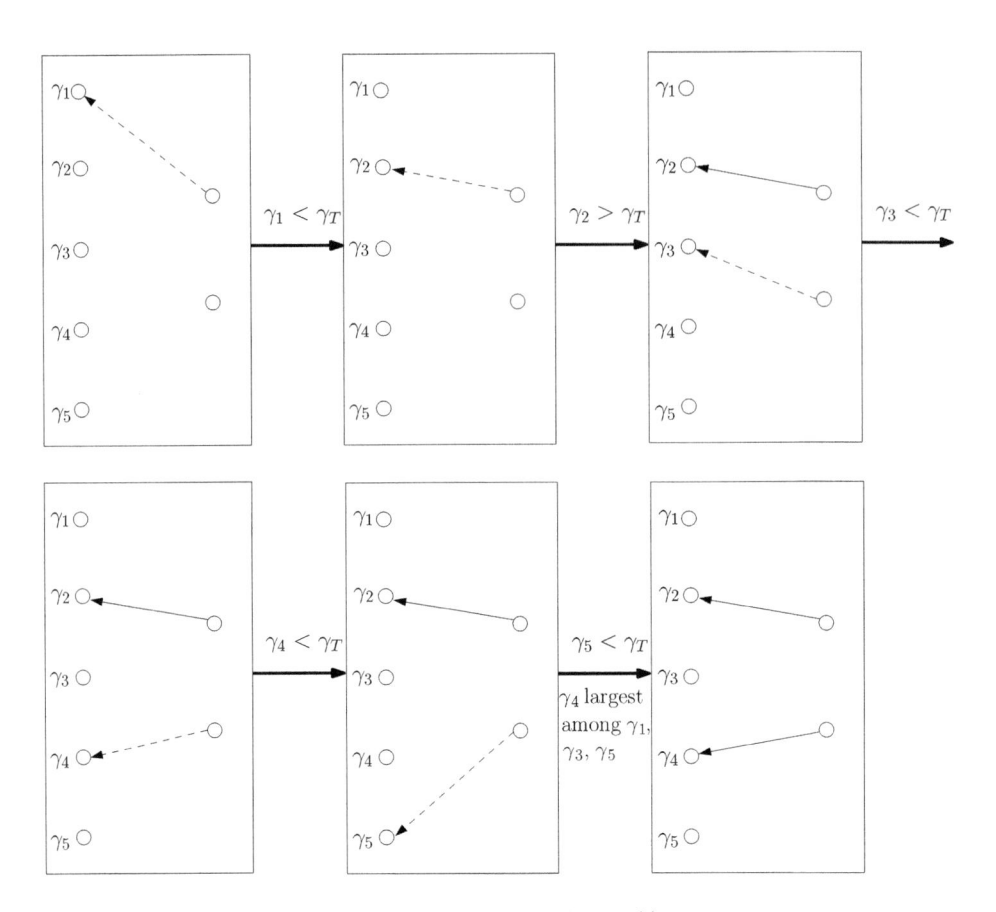

Figure 4.12 Sample operation of a GSECps-based diversity combiner.

4.4.1 Statistics of Output SNR

We focus on the statistics of the combined SNR with GSECps over the i.i.d. Rayleigh fading environment. Conditioning on whether there are L_c acceptable paths or not, we can apply the total probability theorem and write the CDF of γ_c in the following form:

$$F_{\gamma_c}(x) = \Pr[\gamma_c < x] \quad (4.31)$$
$$= \pi_{L_c} \cdot \Pr[\gamma_c < x | L_a \geq L_c] + \Pr[\gamma_c < x, L_a < L_c],$$

where π_{L_c} is the probability that there are at least L_c acceptable paths, which was given in (4.23), $\Pr[\gamma_c < x | L_a \geq L_c]$ is the conditional probability of the event $\gamma_c < x$ given that there are at least L_c acceptable paths, and $\Pr[\gamma_c < x, L_a < L_c]$ is the joint probability of the events $\gamma_c < x$ and $L_a < L_c$. Note that when $L_a \geq L_c$, GSECps combines L_c acceptable paths. As such, γ_c is the sum of the L_c path SNRs that are greater than γ_T. Therefore, $\Pr[\gamma_c < x | L_a \geq L_c]$ can be shown to be given by [12, Eq. 38]

4.4 GSEC with Post-Examining Selection (GSECps)

$$\Pr[\gamma_c < x | L_a \geq L_c] = 1 - e^{-\frac{x-L_c\gamma_T}{\bar{\gamma}}} \sum_{k=0}^{L_c-1} \frac{1}{k!}\left(\frac{x-L_c\gamma_T}{\bar{\gamma}}\right)^k, \quad x \geq L_c\gamma_T. \tag{4.32}$$

Note that since $L_a \geq L_c$, the combined SNR γ_c is always greater than $L_c\gamma_T$.

To calculate $\Pr[\gamma_c < x, L_a < L_c]$, we first note that since there are fewer than L_c acceptable paths, the receiver will in effect combine L_c strongest paths. Therefore, we have

$$\gamma_c = \sum_{i=1}^{L_c} \gamma_{i:L}, \tag{4.33}$$

where $\gamma_{i:L}$ is the ith largest SNR of all L path SNRs. Let Γ_{L_c-1} denote the sum of the first $L_c - 1$ largest SNR, i.e., $\Gamma_{L_c-1} = \sum_{j=1}^{L_c-1} \gamma_{j:L}$. We can rewrite $\Pr[\gamma_c < x, L_a < L_c]$, while noting that $L_a < L_c$ holds if and only if the L_cth largest SNR, $\gamma_{L_c:L}$, is less than γ_T, as

$$\Pr[\gamma_c < x, L_a < L_c] = \Pr[\Gamma_{L_c-1} + \gamma_{L_c:L} < x, \gamma_{L_c:L} < \gamma_T], \tag{4.34}$$

which can be computed using the joint PDF of $\gamma_{L_c:L}$ and Γ_{L_c-1}, denoted by $p_{\gamma_{L_c:L},\Gamma_{L_c-1}}(y,z)$, as

$$\Pr[\Gamma_{L_c-1} + \gamma_{L_c:L} < x, \gamma_{L_c:L} < \gamma_T] = \int_0^{\min\{\gamma_T, \frac{x}{L_c}\}} \int_{(L_c-1)y}^{x-y} p_{\gamma_{L_c:L},\Gamma_{L_c-1}}(y,z)\,dz\,dy. \tag{4.35}$$

Based on the order statistics results in the Appendix, $p_{\gamma_{L_c:L},\Gamma_{L_c-1}}(y,z)$ for the i.i.d. Rayleigh fading environment can be obtained as [16, Eq. 11]

$$p_{\gamma_{L_c:L},\Gamma_{L_c-1}}(y,z) = \sum_{j=0}^{L-L_c} \frac{(-1)^j L! [z-(L_c-1)y]^{L_c-2}}{(L-L_c-j)!(L_c-1)!(L_c-2)!j!\bar{\gamma}^{L_c}} e^{-\frac{z+(j+1)y}{\bar{\gamma}}}, \tag{4.36}$$

$$y \geq 0,\ z \geq (L_c-1)y.$$

After substituting (4.36) into (4.35), carrying out the integration and appropriate simplifications, we have

$$\Pr[\gamma_c < x, L_a < L_c] = \begin{cases} F_{\gamma_c}^{GSC}(L_c\gamma_T) + \sum_{j=1}^{L-L_c}\sum_{k=0}^{L_c-2} A_{(j,k)} \\ \quad \times \left[e^{-\frac{j\gamma_T + L_c\gamma_T}{\bar{\gamma}}}\left(1 - \Gamma\left(L_c-1-k, \frac{x-L_c\gamma_T}{\bar{\gamma}}\right)\right) \right. \\ \quad \left. - \left(\Gamma\left(L_c-1-k, \frac{L_c\gamma_T}{\bar{\gamma}}\right) - \Gamma\left(L_c-1-k, \frac{x}{\bar{\gamma}}\right)\right)\right] \\ \quad + \binom{L}{L_c}\left[\left(\Gamma\left(L_c, \frac{L_c\gamma_T}{\bar{\gamma}}\right) - \Gamma\left(L_c, \frac{x}{\bar{\gamma}}\right)\right)\right. \\ \quad \left. - e^{-\frac{L_c\gamma_T}{\bar{\gamma}}}\left(1 - \Gamma\left(L_c, \frac{x-L_c\gamma_T}{\bar{\gamma}}\right)\right)\right],\ x \geq L_c\gamma_T; \\ F_{\gamma_c}^{GSC}(x),\quad x < L_c\gamma_T, \end{cases} \tag{4.37}$$

where

$$F_{\gamma_c}^{GSC}(x) = \binom{L}{L_c}\left\{1 - \Gamma(L_c, x/\overline{\gamma}) + \sum_{l=1}^{L-L_c}(-1)^{L_c+l-1}\binom{L-L_c}{l}\left(\frac{L_c}{l}\right)^{L_c-1}\right.$$
$$\left.\times\left[\frac{1-e^{-(1+l/L_c)(x/\overline{\gamma})}}{1+l/L_c} - \sum_{m=0}^{L_c-2}\left(\frac{-l}{L_c}\right)^m\left(1-\Gamma(m+1,x/\overline{\gamma})\right)\right]\right\}, \quad (4.38)$$

$$A_{(j,k)} = \frac{(-1)^j L! L_c^k}{(L-L_c-j)!(L_c-1)! j!(-j)^{k+1}}, \quad (4.39)$$

and $\Gamma(n,x)$ is the incomplete gamma function defined by

$$\Gamma(n,x) = \frac{1}{(n-1)!}\int_x^\infty t^{n-1}e^{-t}dt. \quad (4.40)$$

Combining (4.23), (4.32), and (4.37), we obtain the closed-form expression of the CDF of the output SNR γ_c and, equivalently, the outage probability of GSECps.

Starting from the CDF of the combined SNR with GSECps, we can routinely derive the PDF and MGF. Specifically, after differentiation and some manipulations, the PDF of γ_c over i.i.d. Rayleigh fading channels is given by

$$p_{\gamma_c}(x) = \quad (4.41)$$
$$\begin{cases} \binom{L}{L_c}\left[\frac{x^{L_c-1}e^{-x/\overline{\gamma}}}{\overline{\gamma}^{L_c}(L_c-1)!} + \frac{1}{\overline{\gamma}}\sum_{l=1}^{L-L_c}(-1)^{L_c+l-1}\binom{L-L_c}{l}\left(\frac{L_c}{l}\right)^{L_c-1}\right. \\ \left.\times e^{-x/\overline{\gamma}}\left(e^{-lx/L_c\overline{\gamma}} - \sum_{m=0}^{L_c-2}\frac{1}{m!}\left(\frac{-lx}{L_c\overline{\gamma}}\right)^m\right)\right], \quad x < L_c\gamma_T \\ \pi_{L_c}\left[\frac{1}{(L_c-1)!\overline{\gamma}^{L_c}}(x-L_c\gamma_T)^{L_c-1}e^{-\frac{x-L_c\gamma_T}{\overline{\gamma}}}\right] \\ + \sum_{j=1}^{L-L_c}\sum_{k=0}^{L_c-2}\frac{A_{(j,k)}\overline{\gamma}^{k+1-L_c}}{(L_c-2-k)!}e^{-x/\overline{\gamma}}\left(e^{\frac{-j\gamma_T}{\overline{\gamma}}}(x-L_c\gamma_T)^{L_c-2-k} - x^{L_c-2-k}\right) \\ + \binom{L}{L_c}\frac{1}{(L_c-1)!\overline{\gamma}_c^L}e^{-\frac{x}{\overline{\gamma}}}\left(x^{L_c-1} - (x-L_c\gamma_T)^{L_c-1}\right), \quad x \geq L_c\gamma_T, \end{cases}$$

where π_{L_c} and $A_{(j,k)}$ were defined in (4.23) and (4.39), respectively. The MGF of γ_c, denoted by $M_{\gamma_c}(t)$, can be shown to be given, for the i.i.d. Rayleigh fading case, by

$$\mathcal{M}_{\gamma_c}(t) = \binom{L}{L_c}\left\{\frac{1-\Gamma(L_c,(1-t\overline{\gamma})L_c\gamma_T/\overline{\gamma})}{(1-t\overline{\gamma})^{L_c}} + \sum_{l=1}^{L-L_c}(-1)^{L_c+l-1}\binom{L-L_c}{l}\right.$$
$$\left.\times\left(\frac{L_c}{l}\right)^{L_c-1}\left[I(l,t) - \sum_{m=0}^{L_c-2}\left(\frac{-l}{L_c}\right)^m\frac{(1-\Gamma(m+1,(1-t\overline{\gamma})L_c\gamma_T/\overline{\gamma}))}{(1-t\overline{\gamma})^{m+1}}\right]\right\}$$

4.4 GSEC with Post-Examining Selection (GSECps)

$$+ \sum_{j=1}^{L-L_c}\sum_{k=0}^{L_c-2} A_{(j,k)}\left[e^{-\frac{L_c\gamma_T+j\gamma_T}{\bar{\gamma}}}\frac{e^{L_c\gamma_T t}}{(1-t\bar{\gamma})^{L_c-1-k}} - \frac{\Gamma(L_c-1-k,(1-t\bar{\gamma})L_c\gamma_T/\bar{\gamma})}{(1-t\bar{\gamma})^{L_c-1-k}}\right]$$

$$+ \binom{L}{L_c}\left[\frac{\Gamma(L_c,(1-t\bar{\gamma})L_c\gamma_T/\bar{\gamma})}{(1-t\bar{\gamma})^{L_c}} - \frac{e^{L_c\gamma_T(t-\frac{1}{\bar{\gamma}})}}{(1-t\bar{\gamma})^{L_c}}\right] + \pi_{L_c}e^{L_c\gamma_T t}\left(\frac{1}{1-t\bar{\gamma}}\right)^{L_c}, \tag{4.42}$$

where

$$I(l,t) = \frac{1}{\bar{\gamma}t - \frac{L_c+l}{L_c}}\left[\exp\left(\frac{L_c\gamma_T(tL_c\bar{\gamma}-L_c-l)}{L_c\bar{\gamma}}\right) - 1\right], \tag{4.43}$$

and π_{L_c} and $A_{(j,k)}$ were defined in (4.23) and (4.39), respectively. With these closed-form results, we can readily evaluate the error performance of GSECps over Rayleigh fading environment.

4.4.2 Complexity Analysis

In this subsection, we calculate the average number of path estimations and comparisons needed by GSECps during the path selection process over the i.i.d Rayleigh fading channel. This study allows a thorough complexity comparison among GSC, GSEC, and GSECps.

Note that compared with GSEC, GSECps adds a ranking process when $L_a < L_c$. Since the ranking process only involves comparisons, the average number of path estimations needed by GSECps, denoted by N, will be the same as GSEC, which was given in (4.30) in the previous section [12, Eq. 27]. For Rayleigh fading channels, (4.30) specializes to

$$N_E = \sum_{k=L_c}^{L} k\binom{k-1}{k-L_c}\left[1-e^{-\frac{\gamma_T}{\bar{\gamma}}}\right]^{k-L_c}e^{-\frac{\gamma_T L_c}{\bar{\gamma}}}$$

$$+ L\sum_{k=L-L_c+1}^{L}\binom{L}{k}\left[1-e^{-\frac{\gamma_T}{\bar{\gamma}}}\right]^{k}e^{-\frac{\gamma_T(L-k)}{\bar{\gamma}}}. \tag{4.44}$$

To calculate the average number of comparisons of GSECps, we first note that every path examination involves one path estimation and one comparison. If $L_a \geq L_c$, according to the operation of GSECps, the receiver stops examining paths whenever it has found L_c acceptable paths. Therefore, in this case the number of comparisons will be equal to the number of path examinations. Hence, the average number of comparisons for the $L_a \geq L_c$ case, denoted by N_A, can be shown to be given by

$$N_A = \sum_{k=L_c}^{L} k\binom{k-1}{k-L_c}\left[F_\gamma(\gamma_T)\right]^{k-L_c}\left[1-F_\gamma(\gamma_T)\right]^{L_c}. \tag{4.45}$$

On the other hand, when $L_a < L_c$, GSECps will first examine all L available paths. Therefore, the number of comparisons associated with path examinations is L. Then GSECps ranks the $L - L_a$ unacceptable paths to select $L_c - L_a$ paths with the highest instantaneous SNRs. The number of comparisons associated with the ranking process is $\sum_{k=1}^{L_c-L_a}(L-L_a-k)$. Therefore, for any L_a, $0 \le L_a < L_c$, the number of comparisons is $L + \sum_{k=1}^{L_c-L_a}(L-L_a-k)$. Hence, the average number of comparisons for the $L_a < L_c$ case, denoted by N_B, is given by

$$N_B = \sum_{j=0}^{L_c-1} \Pr[L_a = j]\left[L + \sum_{k=1}^{L_c-j}(L-j-k)\right], \qquad (4.46)$$

where $\Pr[L_a = j]$ is the probability that there are exactly j acceptable paths. It can be shown, with the assumption of i.i.d. fading paths, that $\Pr[L_a = j]$ is mathematically given by

$$\Pr[L_a = j] = \binom{L}{j}[F_\gamma(\gamma_T)]^{L-j}[1-F_\gamma(\gamma_T)]^j. \qquad (4.47)$$

By combining these two mutually exclusive cases, we obtain an expression for the overall average number of comparisons needed by GSECps over i.i.d. fading channels, denoted by N_C, as

$$\begin{aligned}N_C &= N_A + N_B \\ &= \sum_{k=L_c}^{L} k \binom{k-1}{k-L_c}[F_\gamma(\gamma_T)]^{k-L_c}[1-F_\gamma(\gamma_T)]^{L_c} \\ &+ \sum_{L_a=0}^{L_c-1}\left[L + \sum_{k=1}^{L_c-L_a}(L-L_a-k)\right]\binom{L}{L_a}[F_\gamma(\gamma_T)]^{L-L_a}[1-F_\gamma(\gamma_T)]^{L_a}.\end{aligned} \qquad (4.48)$$

For Rayleigh fading channels, N_C specializes to

$$\begin{aligned}N_C &= \sum_{k=L_c}^{L} k \binom{k-1}{k-L_c}[1-e^{-\frac{\gamma_T}{\bar\gamma}}]^{k-L_c} e^{-\frac{\gamma_T L_c}{\bar\gamma}} \\ &+ \sum_{L_a=0}^{L_c-1}\left[L + \sum_{k=1}^{L_c-L_a}(L-L_a-k)\right]\binom{L}{L_a}[1-e^{-\frac{\gamma_T}{\bar\gamma}}]^{L-L_a} e^{-\frac{\gamma_T L_a}{\bar\gamma}}.\end{aligned} \qquad (4.49)$$

For GSEC, there is no ranking process and, as such, the average number of comparisons needed is equal to the number of path examinations, which has been given in (4.44). For GSC, the number of comparisons is always $\sum_{k=1}^{L_c}(L-k)$ to select the strongest L_c paths.

4.4.3 Numerical Results

In Fig. 4.13 we study the operation complexity of GSECps in comparison with GSEC and GSC. We set $L = 4$ and $L_c = 2$. In Fig. 4.13(a), we plot the average number of path estimations needed by the three schemes as a function of normalized switching threshold, $\gamma_T/\bar\gamma$. As expected, GSECps needs the same number of path estimations as GSEC and both schemes require fewer path estimations than GSC when the threshold is

4.4 GSEC with Post-Examining Selection (GSECps)

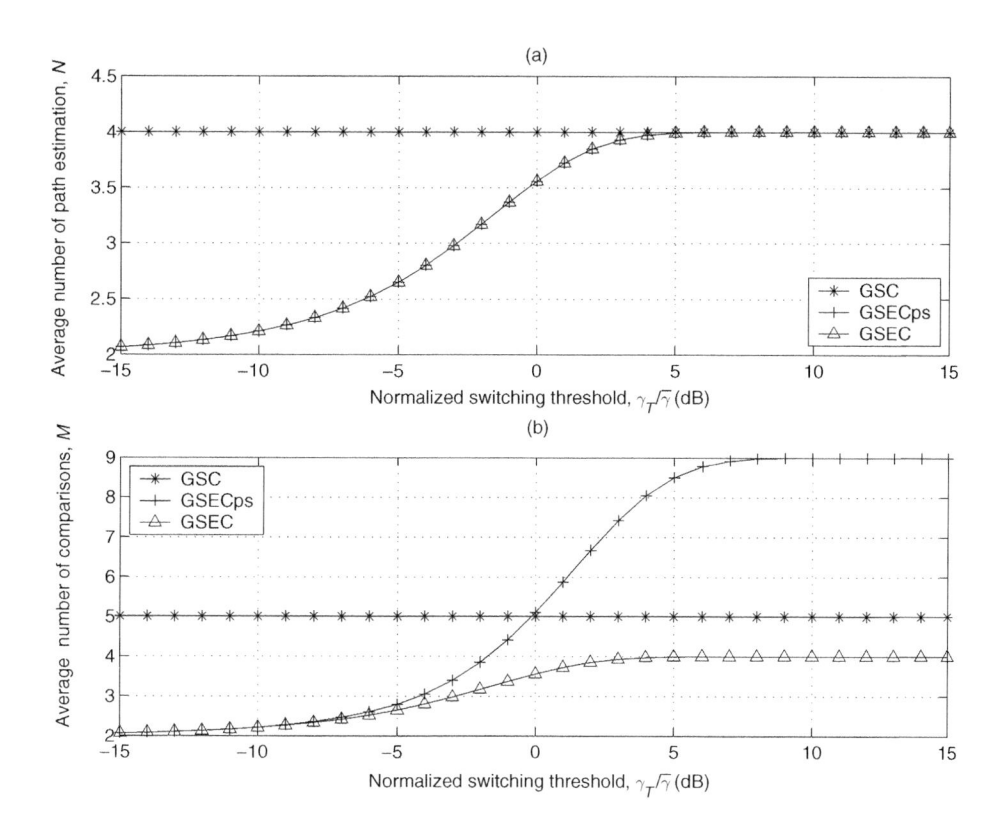

Figure 4.13 Operation complexity of GSECps in comparison with GSC and GSEC ($L = 4$ and $L_c = 2$): (a) average number of path estimations, (b) average number of comparisons. Reprint with permission from [14]. ©2002 IEEE.

not too large. Only when γ_T is larger than $\overline{\gamma}$ by more than 5 dB will GSECps and GSEC need to estimate all four paths.

In Fig. 4.13(b), the average number of comparisons needed by these schemes during the path selection process is plotted as a function of the normalized switching threshold, $\gamma_T/\overline{\gamma}$. Note that GSC always performs roughly five comparisons to find the two best paths among the four available ones while the receiver with GSEC performs at most four comparisons between the path SNR and the switching threshold. The number of comparisons needed by GSECps, however, increases from two to nine in the worst case, which includes four comparisons between the instantaneous path SNR and the switching threshold and five more comparisons to find the two best unacceptable paths. While GSECps may need four more comparisons than GSC in the worst case, these extra comparisons involve a fixed threshold and therefore are easy to implement. Moreover, as will be shown in the next subsection, GSECps can provide nearly the same performance as GSC when γ_T is slightly less than $\overline{\gamma}$, where the number of comparisons needed by GSECps is much less than GSC.

Fig. 4.14 plots the average error rate of BPSK with GSEC, GSECps, and GSC as a function of average SNR per path with $\gamma_T = 3$ dB. Again, we set $L = 4$ and $L_c = 2$. It is

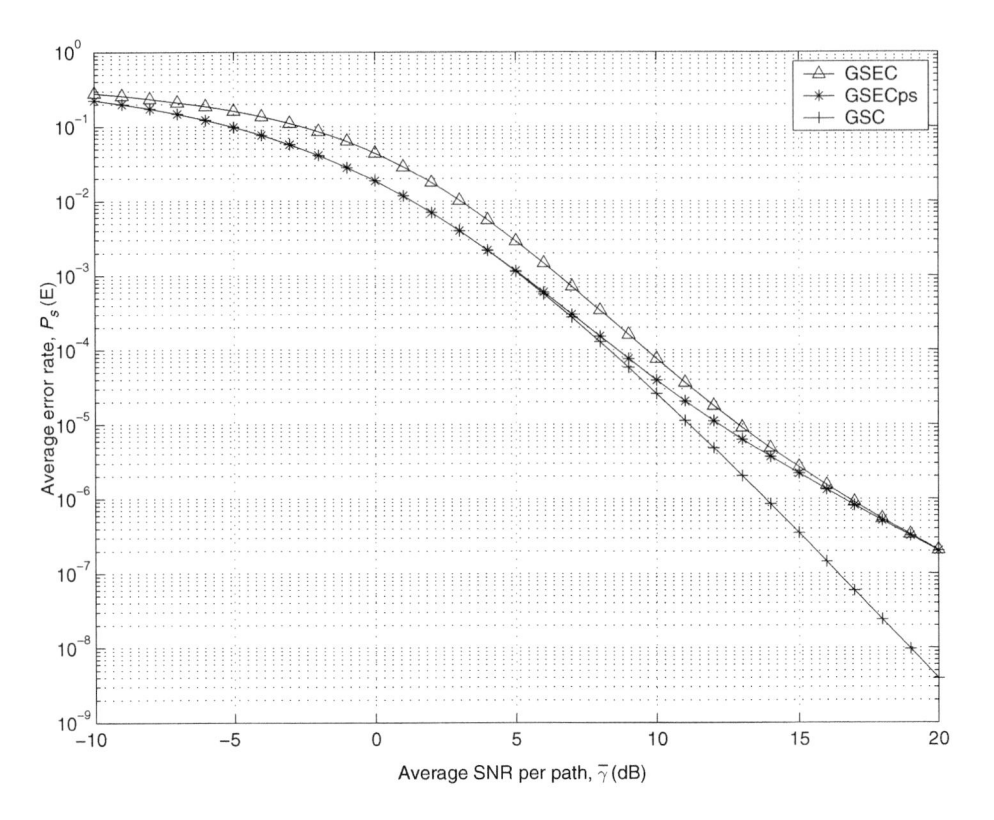

Figure 4.14 Average error rate of BPSK with GSC, GSEC, and GSECps over $L = 4$ i.i.d. Rayleigh fading channels as a function of the average SNR per path ($\gamma_T = 3$ dB and $L_c = 2$). Reprint with permission from [14]. ©2002 IEEE.

easy to see that when the channel condition is poor, i.e., $\overline{\gamma}$ is small, GSECps has the same error performance as GSC. That is because it is more likely that there are not enough acceptable paths and GSECps operates in the same way as GSC, by selecting the best unacceptable paths. As $\overline{\gamma}$ becomes larger, that is, the channel condition is improving, the performance advantage of GSECps over GSEC decreases. Note that in this case, the probability of $L_a \geq L_c$ increases and GSECps behaves more like GSEC.

In Fig. 4.15 we plot the average error rate of BPSK with the three schemes under consideration as a function of the normalized switching threshold, $\gamma_T/\overline{\gamma}$. It indicates that when γ_T is small, GSECps has the same error performance as GSEC because there are usually enough acceptable paths and GSECps operates more like GSEC. As γ_T increases, the performance of GSEC degrades while the performance of GSECps becomes the same as that of GSC.

Considering Fig. 4.15 together with Fig. 4.13, we can study the trade-off involved among GSECps, GSEC, and GSC. Comparing GSECps with GSEC, we observe that both schemes require the same number of path estimations while GSECps offers better error performance at the cost of slightly more comparison operations. In particular, if $\gamma_T/\overline{\gamma}$ is set to be -7 dB, where GSEC reaches its best error performance, GSECps

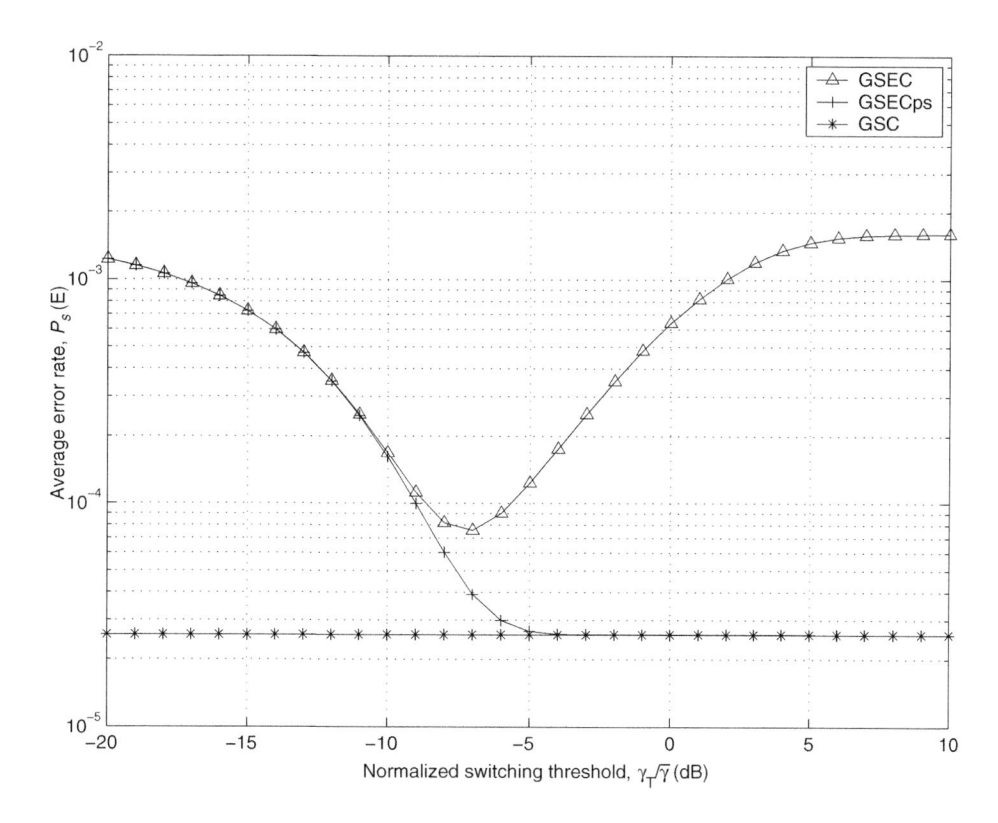

Figure 4.15 Average error rate of BPSK with GSC, GSEC, and GSECps over $L = 4$ i.i.d. Rayleigh fading channels as a function of the normalized switching threshold $\gamma_T/\bar{\gamma}$ ($L_c = 2$). Reprint with permission from [14]. ©2002 IEEE.

still offers a 60% decrease in average error rate while requiring 0.05 more comparisons on average. Comparing GSECps with GSC, we can see that if we set $\gamma_T/\bar{\gamma} = -3$ dB, GSECps has nearly the same error performance as GSC but needs fewer path estimations and comparisons. In conclusion, GSECps achieves a better performance–complexity trade-off than both GSEC and GSC.

4.5 Summary

In this chapter, we studied four advanced diversity combining schemes for a diversity-rich environment. The goal of these schemes is to reduce the combiner hardware complexity and/or processing power by only combining a proper-selected path subset in the optimal MRC fashion. We derived the exact statistics of their combiner output SNR, which in turn are applied to their performance analysis over fading channels. Whenever feasible, we have also quantified the associated complexity of each scheme in terms of the average number of active MRC branches and the average number of path estimations/comparisons.

4.6 Further Reading

[17] presents a virtual branch-based method to analyze the performance of the GSC scheme over fading channels. [18, 19] address the analysis of GSC over general correlated fading. [20] presents a modified GSC scheme using the log-likelihood ratio of each diversity branch. [21] applies the idea of a hybrid combining scheme to an EGC scheme. Xiao and Dong propose a modified version of the NT-GSC scheme and present a new analytical method for the performance analysis of both GSC and NT-GSC [11]. Femenias [22] proposes a generalized sort, switch, and examine combining (GSSEC) scheme and compares it with GSC and GSEC schemes.

References

[1] N. Kong, T. Eng, and L. B. Milstein, "A selection combining scheme for RAKE receivers," in *Proc. IEEE Int. Conf. Univ. Personal Comm. ICUPC'95, Tokyo, Japan*, November 1995, pp. 426–429.

[2] M. Z. Win and Z. A. Kostić, "Virtual path analysis of selective Rake receiver in dense multipath channels," *IEEE Commun. Letters*, 3, no. 11, pp. 308–310, 1999.

[3] Y. Roy, J.-Y. Chouinard, and S. A. Mahmoud, "Selection diversity combining with multiple antennas for MM-wave indoor wireless channels," *IEEE J. Select. Areas Commun.*, 14, no. 4, pp. 674–682, 1998.

[4] M. J. Gans, "The effect of Gaussian error in maximal ratio combiners," *IEEE Trans. Commun. Technol.*, 19, no. 4, pp. 492–500, 1971.

[5] B. R. Tomiuk, N. C. Beaulieu, and A. A. Abu-Dayya, "General forms for maximal ratio diversity with weighting errors," *IEEE Trans. Commun.*, 47, no. 4, pp. 488–492, 1999.

[6] M. K. Simon and M.-S. Alouini, *Digital Communications over Generalized Fading Channels: A Unified Approach to Performance Analysis*, Wiley, 2000.

[7] M.-S. Alouini and M. K. Simon, "An MGF-based performance analysis of generalized selective combining over Rayleigh fading channels," *IEEE Trans. Commun.*, 48, no. 3, pp. 401–415, 2000.

[8] A. I. Sulyman and M. Kousa, "Bit error rate performance of a generalized diversity selection combining scheme in Nakagami fading channels," in *Proc. IEEE Wireless Commun. and Networking Conf. (WCNC'00), Chicago, Illinois*, September 2000, pp. 1080–1085.

[9] M. K. Simon and M.-S. Alouini, "Performance analysis of generalized selection combining with threshold test per branch (T-GSC)," *IEEE Trans. Veh. Technol.*, 51, no. 5, pp. 1018–1029, 2002.

[10] X. Zhang and N. C. Beaulieu, "SER and outage of threshold based hybrid selection/maximal-ratio combining over generalized fading channels," *IEEE Trans. Commun.*, 52, no. 12, pp. 2143–2153, 2004.

[11] L. Xiao and X. Dong, "Unified analysis of generalized selection combining with normalized threshold test per branch," *IEEE Trans. Wireless Commun.*, 5, no. 8, pp. 2153–2163, 2006.

[12] H.-C. Yang and M.-S. Alouini, "Generalized switch and examine combining (GSEC): a low-complexity combining scheme for diversity rich environments," *IEEE Trans. Commun.*, 52, no. 10, pp. 1711–1721, 2004.

References

[13] H.-C. Yang and M.-S. Alouini, "Performance analysis of multibranch switched diversity systems," *IEEE Trans. Commun.*, 51, no. 5, pp. 782–794, 2003.

[14] H.-C. Yang and L. Yang, "Tradeoff analysis of performance and complexity on GSECps diversity combining scheme," *IEEE Trans. Wireless Commun.*, 7, no. 1, pp. 32–36, 2008.

[15] H.-C. Yang and M.-S. Alouini, "Improving the performance of switched diversity with post-examining selection," *IEEE Trans. Wireless Commun.*, 5, no. 1, pp. 67–71, 2006.

[16] H.-C. Yang, "New results on ordered statistics and analysis of minimum-selection generalized selection combining (GSC)," *IEEE Trans. Wireless Commun.*, 5, no. 7, 2006.

[17] M. Z. Win and J. H. Winters, "Virtual branch analysis of symbol error probability for hybrid selection/maximal-ratio combining in Rayleigh fading," *IEEE Trans. Commun.*, 49, no. 11, pp. 1926–1934, 2001.

[18] Y. Ma and S. Pasupathy, "Efficient performance evaluation for generalized selection combining on generalized fading channels," *IEEE Trans. Wireless. Commun.*, 3, no. 1, pp. 29–34, 2004.

[19] R. K. Mallik and M. Z. Win, "Analysis of hybrid selection/maximal-ratio combining in correlated Nakagami fading," 50, no. 8, pp. 1372–1383, 2002.

[20] S. W. Kim, Y. G. Kim, and M. K. Simon, "Generalized selection combining based on the log-likelihood ratio," *IEEE Trans. Commun.*, 52, no. 4, pp. 521–524, 2004.

[21] Y. Ma and J. Jin, "Unified performance analysis of hybrid-selection/equal-gain combining," *IEEE Trans. Veh. Technol.*, 56, no. 4, pp. 1866–1873, 2007.

[22] G. Femenias, "Performance analysis of generalized sort, switch, and examine combining," *IEEE Trans. Commun.*, 54, no. 12, pp. 2137–2143, 2006.

5 Adaptive Diversity Combining

Most conventional diversity combining schemes apply a fixed amount of operations to the received signal from different diversity paths to generate an output signal with better quality. This basic design principle manifests itself in advanced combining schemes in previous chapter as well as classical combining schemes. Meanwhile, such an approach may lead to a waste of receiver processing resource when the channel is favorable and fewer favorable and fewer combining operations can satisfy the performance requirement. Adaptive diversity combining tries to save receiver processing power by performing just enough combining operations such that the quality of the combined signal becomes acceptable. The generic structure of the adaptive diversity receiver is shown in Fig. 5.1. Adaptive combining schemes differ from those threshold-based combining schemes discussed in the previous chapter, e.g., T-GSC and GSEC, in that the threshold checking is performed on the combiner output, rather than on each individual branch. Conventional switch and stay combining (SSC) can be viewed as the very first adaptive combining scheme.

In this chapter, we study various adaptive combining schemes and their applications in different transmission scenarios. Specifically, we first present three representative adaptive combining schemes, namely output threshold maximum ratio combining (OT-MRC), minimum selection GSC (MS-GSC), and output threshold GSC (OT-GSC), and carry out a thorough performance versus complexity trade-off analysis on them. After that, we apply the adaptive combining idea to transmit diversity systems and RAKE finger management systems. Finally, we investigate the joint design of adaptive diversity combining with rate-adaptive transmission, noting that both rate-adaptive transmission and adaptive combining utilize some predetermined thresholds in their operations.

5.1 Output-Threshold Maximum Ratio Combining

The OT-MRC scheme can be viewed as an adaptive implementation of the conventional MRC scheme [1]. The diversity combiner in Fig. 5.1 is similar to a classical L-branch MRC combiner, but some branches may not be active. With OT-MRC, the combiner tries to raise its output signal-to-noise ratio (SNR) γ_c above an output threshold denoted by γ_T by gradually activating and combining additional diversity branches. For the sake of presentation convenience and clarity, we adopt a discrete-time implementation in

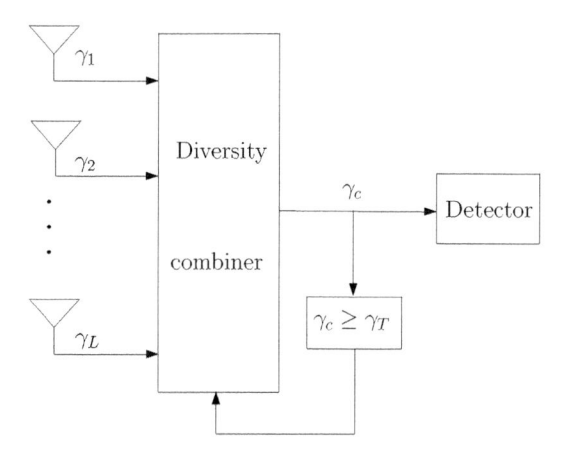

Figure 5.1 Structure of an adaptive diversity receiver.

the following. More specifically, pilot symbols are periodically (usually at the rate of channel coherence time) inserted into the transmitted signal. Using these pilot symbols, the receiver performs the necessary channel estimations and determines the appropriate combining operation for the subsequent data reception.

5.1.1 Mode of Operation

Starting from a single branch, the OT-MRC combiner successively estimates additional diversity paths, activates MRC branches, and applies MRC to the estimated paths in order to raise the combined SNR above the threshold γ_T. The flow chart in Fig. 5.2 illustrates the mode of operation of OT-MRC. In particular, the combiner checks the SNR of the first available diversity path, denoted by γ_1. If it is above γ_T, then the combiner simply uses this path for data reception, i.e., $\gamma_c = \gamma_1$. This is exactly the same as the no-diversity case. If $\gamma_1 < \gamma_T$, the combiner estimates a second diversity path, whose SNR is denoted by γ_2, activates another MRC branch, and applies MRC to these two paths. If the resulting combined SNR $\gamma_c = \gamma_1 + \gamma_2$, is above γ_T, then the combiner just acts as a dual-branch MRC. Otherwise (i.e., $\gamma_1 + \gamma_2 < \gamma_T$), a third path is estimated and another MRC branch is activated. This process stops whenever the combined SNR γ_c exceeds γ_T. Note that an L-branch MRC receiver will be used only in the worst case, when the combined SNR γ_c is still below γ_T after the activation of $L - 1$ MRC branches.

According to the mode of operation of OT-MRC, the receiver does not always combine all L available diversity paths. If the MRC-combined SNR after the activation of the first l MRC branches is above the threshold, the receiver does not need to estimate the remaining $L - l$ diversity paths. Correspondingly, only l MRC branches need to be active during data reception. These constitute the main power-saving features of OT-MRC over the traditional MRC combiner, where all L diversity paths are always estimated and all L MRC branches are always active during data reception. Such power

Adaptive Diversity Combining

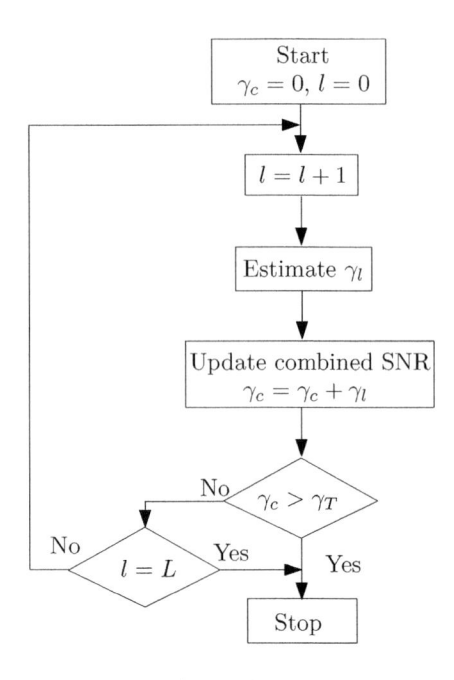

Figure 5.2 Flow chart for OT-MRC mode of operation.

saving comes, of course, at the cost of certain performance loss, which will be accurately quantified next.

5.1.2 Statistics of Output SNR

We note from the mode of operation of OT-MRC that the events that l diversity paths (or equivalently, an l-branch MRC) are used for data reception, $l = 1, 2, \ldots, L$, are mutually exclusive. Applying the total probability theorem, we can write the cumulative distribution function (CDF) of the combined SNR, $F_{\gamma_c}(\cdot)$, in a summation form as

$$F_{\gamma_c}(x) = \Pr[\gamma_c < x]$$
$$= \sum_{l=1}^{L} \Pr\left[\gamma_c = \sum_{j=1}^{l} \gamma_j \ \& \ \gamma_c < x\right], \qquad (5.1)$$

where $\gamma_c = \sum_{j=1}^{l} \gamma_j$ denotes the event that an l-branch MRC scheme is used for data reception. We observe that with OT-MRC (1) a single-branch receiver is used when the SNR of the first diversity path is above the threshold (i.e., $\gamma_1 \geq \gamma_T$); (2) an l-branch MRC (with $2 \leq l \leq L - 1$) is used when the combined SNR of an $l - 1$-branch MRC is below the threshold, whereas the combined SNR of an l-branch MRC is above the threshold (i.e., $\sum_{j=1}^{l-1} \gamma_j < \gamma_T$ and $\sum_{j=1}^{l} \gamma_j \geq \gamma_T$); and (3) an L-branch MRC is used when the combined SNR of an $L - 1$-branch MRC is below the threshold (i.e., $\sum_{j=1}^{L-1} \gamma_j < \gamma_T$).

5.1 Output-Threshold Maximum Ratio Combining

Therefore, we can rewrite the CDF of the combined SNR with OT-MRC after some manipulation as

$$F_{\gamma_c}(x) = \Pr[\gamma_T \leq \gamma_1 \ \& \ \gamma_1 < x]$$
$$+ \sum_{l=2}^{L-1} \Pr\left[\sum_{j=1}^{l-1} \gamma_j < \gamma_T \leq \sum_{j=1}^{l} \gamma_j \ \& \ \sum_{j=1}^{l} \gamma_j < x\right]$$
$$+ \Pr\left[\sum_{j=1}^{L-1} \gamma_j < \gamma_T \ \& \ \sum_{j=1}^{L} \gamma_j < x\right] \quad (5.2)$$

$$= \begin{cases} \Pr[\gamma_T \leq \gamma_1 < x] \\ \quad + \sum_{l=2}^{L-1} \Pr\left[\sum_{j=1}^{l-1} \gamma_j < \gamma_T \leq \sum_{j=1}^{l} \gamma_j < x\right] \\ \quad + \Pr\left[\sum_{j=1}^{L-1} \gamma_j < \gamma_T \ \& \ \sum_{j=1}^{L} \gamma_j < x\right], & \gamma_T < x; \\ \Pr\left[\sum_{j=1}^{L} \gamma_j < x\right], & \gamma_T \geq x. \end{cases}$$

Under the assumption of i.i.d. fading paths, we can rewrite (5.2) in terms of the CDF of single-branch SNR, $F_\gamma(\cdot)$, and the probability density function (PDF) of an l-branch MRC output SNR (i.e., $\sum_{j=1}^{l} \gamma_j$), $p_{\gamma_c}^{l-MRC}(\cdot)$, as

$$F_{\gamma_c}(x) = \begin{cases} F_\gamma(x) - F_\gamma(\gamma_T) \\ \quad + \sum_{l=2}^{L-1} \int_0^{\gamma_T} p_{\gamma_c}^{(l-1)-MRC}(y)(F_\gamma(x-y) - F_\gamma(\gamma_T - y)) dy \\ \quad + \int_0^{\gamma_T} p_{\gamma_c}^{(L-1)-MRC}(y) F_\gamma(x-y) dy, & \gamma_T < x; \\ \int_0^x p_{\gamma_c}^{L-MRC}(y) dy, & \gamma_T \geq x. \end{cases} \quad (5.3)$$

Differentiating $F_{\gamma_c}(x)$ in (5.3) with respect to x, we obtain a generic formula for the PDF of the combined SNR as

$$p_{\gamma_c}(x) = \begin{cases} p_\gamma(x) + \sum_{l=2}^{L} \int_0^{\gamma_T} p_{\gamma_c}^{(l-1)-MRC}(y) F_\gamma(x-y) \, dy, & \gamma_T < x; \\ p_{\gamma_c}^{L-MRC}(x), & \gamma_T \geq x, \end{cases} \quad (5.4)$$

where $p_\gamma(\cdot)$ is the PDF of single-branch SNR. Applying the CDF and PDF of an l-branch MRC output SNR summarized in Table 5.1, we can obtain the CDF and PDF of the combined SNR with OT-MRC for various fading models.

As an example, for the Rayleigh fading case, the CDF of the combined SNR with OT-MRC is given, after integration and some simplifications, by the following closed-form expression

$$F_{\gamma_c}(x) = \begin{cases} 1 - A e^{-\frac{x}{\bar{\gamma}}}, & \gamma_T < x; \\ 1 - \sum_{l=1}^{L} \frac{1}{(l-1)!} \left(\frac{x}{\bar{\gamma}}\right)^{l-1} e^{-\frac{x}{\bar{\gamma}}}, & \gamma_T \geq x, \end{cases} \quad (5.5)$$

Table 5.1 Closed-form expressions of $p_{\gamma_c}^{l-MRC}(x)$ and $F_{\gamma_c}^{l-MRC}(x)$ for the three popular fading models of interest

Model	$p_{\gamma_c}^{l-MRC}(x)$	$F_{\gamma_c}^{l-MRC}(x)$
Rayleigh	$\frac{1}{(l-1)!} \frac{x^{l-1}}{\bar{\gamma}^l} e^{-\frac{x}{\bar{\gamma}}}$	$1 - e^{-\frac{x}{\bar{\gamma}}} \sum_{i=0}^{l-1} \frac{1}{i!} \left(\frac{x}{\bar{\gamma}}\right)^i$
Rician	$\frac{K+1}{\bar{\gamma}} e^{-lK - \frac{(K+1)}{\bar{\gamma}} x}$ $\left(\frac{K+1}{lK\bar{\gamma}} x\right)^{\frac{l-1}{2}} I_{l-1}\left(2\sqrt{\frac{lK(K+1)}{\bar{\gamma}} x}\right)$	$1 - Q_l\left(\sqrt{2lK}, \sqrt{\frac{2(K+1)}{\bar{\gamma}} x}\right)$
Nakagami	$\left(\frac{m}{\bar{\gamma}}\right)^{lm} \frac{x^{lm-1}}{\Gamma(lm)} e^{-\frac{m}{\bar{\gamma}} x}$	$1 - \frac{\Gamma(lm, \frac{m}{\bar{\gamma}} x)}{\Gamma(lm)}$

where

$$A = \sum_{l=1}^{L} \frac{1}{(l-1)!} \left(\frac{\gamma_T}{\bar{\gamma}}\right)^{l-1}. \tag{5.6}$$

Similarly, the PDF of the combined SNR with OT-MRC is obtained as

$$p_{\gamma_c}(x) = \begin{cases} \frac{A}{\bar{\gamma}} e^{-\frac{x}{\bar{\gamma}}}, & \gamma_T < x; \\ \frac{1}{(L-1)!} \frac{x^{L-1}}{\bar{\gamma}^L} e^{-\frac{x}{\bar{\gamma}}}, & \gamma_T \geq x. \end{cases} \tag{5.7}$$

With the statistics of the combined SNR of OT-MRC, we can accurately evaluate the performance of OT-MRC. Specifically, the average error rate of OT-MRC can be calculated by averaging the conditional error rate of modulation scheme of interest, $P_E(\gamma_c)$, over the PDF of the combined SNR γ_c, $f_{\gamma_c}(x)$, given in (5.4). For the important special case of binary phase shift keying (BPSK) over Rayleigh fading, we can show, after proper substitution and manipulations, that the average BER with OT-MRC is given by the following closed-form expression:

$$\overline{P}_b = \int_0^\infty Q(\sqrt{2x}) p_{\gamma_c}(x) dx \tag{5.8}$$

$$= \frac{1}{2} - \sqrt{\frac{\bar{\gamma}}{1+\bar{\gamma}}} \sum_{l=0}^{L-1} \frac{1}{l!} \left\{ \left(\frac{\gamma_T}{\bar{\gamma}}\right)^l Q\left(\sqrt{2 \frac{1+\bar{\gamma}}{\bar{\gamma}} \gamma_T}\right) \right.$$
$$\left. + \frac{1}{2\sqrt{\pi}} \left(\frac{1}{1+\bar{\gamma}}\right)^l \left[\Gamma\left(l+\frac{1}{2}\right) - \Gamma\left(l+\frac{1}{2}, \frac{1+\bar{\gamma}}{\bar{\gamma}} \gamma_T\right)\right] \right\},$$

where $Q(\cdot)$ is the Gaussian Q-function and $\Gamma(\cdot,\cdot)$ is the incomplete Gamma function.

5.1.3 Complexity Analysis

We quantify the power savings of the OT-MRC scheme by calculating the average number of active MRC branches during data reception.

With OT-MRC, a varying number of MRC branches may be active during data reception. The average number of active MRC branches, denoted by N_A, can be calculated in summation form as

$$N_A = \sum_{l=1}^{L} l \, \Pr\left[\gamma_c = \sum_{j=1}^{l} \gamma_j\right], \quad (5.9)$$

where $\Pr[\gamma_c = \sum_{j=1}^{l} \gamma_j]$ is the probability that an l-branch MRC is used for the subsequent data reception and can be calculated as

$$\Pr\left[\gamma_c = \sum_{j=1}^{l} \gamma_j\right] = \begin{cases} \Pr[\gamma_1 \geq \gamma_T], & l = 1; \\ \Pr[\sum_{j=1}^{l-1} \gamma_j < \gamma_T \ \& \ \sum_{j=1}^{l} \gamma_j \geq \gamma_T], & 2 \leq l \leq L-1; \\ \Pr[\sum_{j=1}^{L-1} \gamma_j < \gamma_T], & l = L. \end{cases} \quad (5.10)$$

With the help of the assumption of i.i.d. fading paths, (5.10) can be rewritten in terms of the common CDF of SNR per branch $F_\gamma(\cdot)$ and the CDF of combined SNR with an l-branch MRC $F_{\gamma_c}^{l-MRC}(\cdot)$ as

$$\Pr\left[\gamma_c = \sum_{j=1}^{l} \gamma_j\right] = \begin{cases} 1 - F_\gamma(\gamma_T), & l = 1; \\ F_{\gamma_c}^{(l-1)-MRC}(\gamma_T) - F_{\gamma_c}^{l-MRC}(\gamma_T), & 2 \leq l \leq L-1; \\ F_{\gamma_c}^{(L-1)-MRC}(\gamma_T), & l = L. \end{cases} \quad (5.11)$$

Substituting (5.11) into (5.9), the average number of active MRC branches with OT-MRC can be finally obtained after some simplifications as

$$N_A = 1 + \sum_{l=1}^{L-1} F_{\gamma_c}^{l-MRC}(\gamma_T). \quad (5.12)$$

Note that (5.12) also gives the average number of path estimations needed by OT-MRC during each coherence time period.

For the Rayleigh fading environment, using appropriate expression for $F^{(l)}(\cdot)$ from Table 5.1, we obtain the average number of active MRC branches needed with OT-MRC as

$$N_A = 1 + \sum_{l=1}^{L-1} \left[1 - e^{-\frac{\gamma_T}{\bar{\gamma}}} \sum_{i=0}^{l-1} \frac{1}{i!}\left(\frac{\gamma_T}{\bar{\gamma}}\right)^i\right]. \quad (5.13)$$

5.1.4 Numerical Results

Fig. 5.3(a) shows the average bit error rate (BER) of BPSK with OT-MRC (given in (5.8)) as a function of the average SNR per path with the threshold γ_T set to 6 dB. As we can see, when the average SNR per path is small relative to γ_T, OT-MRC has nearly the same error performance as conventional MRC. However, when the average SNR increases, the error rate of OT-MRC decreases at a smaller rate than conventional MRC. OT-MRC still offers considerable performance gain over the no-diversity case. Fig. 5.3(b) shows the outage probability of OT-MRC as a function of

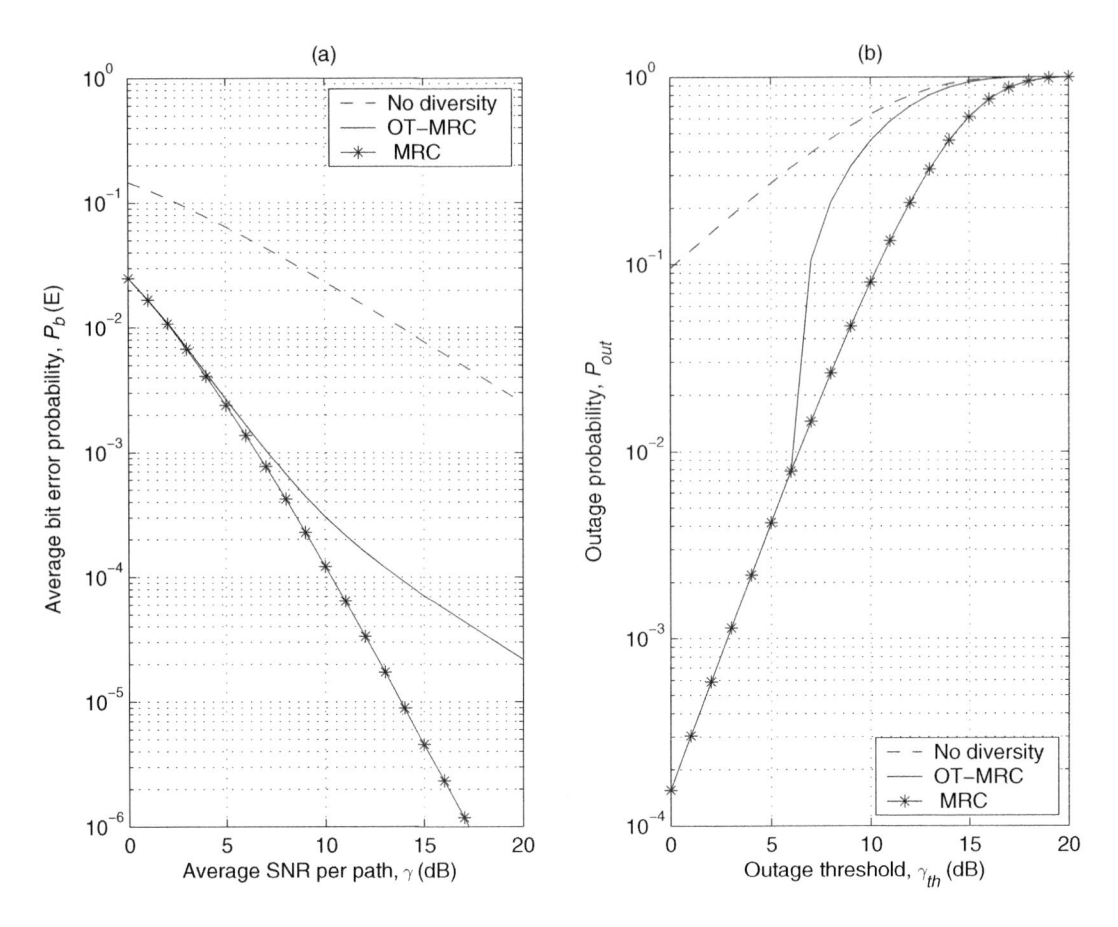

Figure 5.3 Performance of OT-MRC in the Rayleigh fading environment: (a) Average BER of BPSK as a function of the average SNR per path; (b) outage probability as a function of outage threshold with $\overline{\gamma} = 10$ dB ($L = 3$ and $\gamma_T = 6$ dB). Reprint with permission from [1]. ©2005 IEEE.

the outage threshold γ_{th}, where γ_T is set equal to 6 dB. For comparison purposes the outage probabilities of no-diversity and conventional MRC cases are also plotted. We can see that when $\gamma_{th} \leq \gamma_T$, OT-MRC has the same outage probability as conventional MRC. But its outage performance degrades when γ_{th} becomes larger than γ_T, as expected by intuition. Indeed, when $\gamma_{th} \leq \gamma_T$, the combined SNR with OT-MRC γ_c might be smaller than γ_{th} during data reception only if the combined SNR of an L-branch MRC is smaller than γ_T. On the other hand, when $\gamma_T < \gamma_{th}$, if the combined SNR of an l-branch MRC $\gamma_c = \sum_{j=1}^{l} \gamma_j$ falls in the interval (γ_T, γ_{th}), an l-branch MRC will be used for data reception while the system is actually experiencing outage. Therefore, a natural choice of γ_T from the outage probability perspective is γ_{th}. This choice gives an optimal outage performance without introducing unnecessary power consumption.

Fig. 5.4(a) shows the average number of active MRC branches of OT-MRC in Rayleigh fading environment with $L = 5$. We can see that as the SNR threshold γ_T increases, the average number of active MRC branches increases from one to five, just as expected. When the threshold γ_T is small, single-branch SNR will usually be larger than γ_T. As the threshold γ_T increases, it becomes more difficult to raise the combined SNR above γ_T. Note that for conventional MRC, all five branches are always active. So with a moderate choice of γ_T, OT-MRC can provide considerable power saving over conventional MRC by activating less branches on average. In Fig. 5.4(b), we plot the average BER of BPSK with OT-MRC in the Rayleigh fading environment as a function of the SNR threshold γ_T. For reference purposes, the average error rate of the no-diversity and conventional MRC cases are also plotted. As we can see, when the threshold γ_T is small, OT-MRC gives roughly the same performance as the no-diversity case. Conversely, when the threshold γ_T increases, the error rate of OT-MRC decreases and becomes eventually the same as that of conventional MRC. This is because as γ_T increases, more paths need to be combined to raise the output SNR above γ_T, which leads to better performance.

The choice of the SNR threshold γ_T is of critical importance in balancing this new trade-off of receiver power consumption versus error rate performance with OT-MRC. By looking at Fig. 5.4(a) and 5.4(b) simultaneously, we can see that if γ_T is set to 10 dB in this scenario, OT-MRC achieves the same error performance as the conventional MRC but requires less than four active MRC branches on average. Therefore, we can conclude that OT-MRC can achieve optimal MRC performance while offering a notable amount of power saving.

5.2 Minimum Selection GSC

Minimum selection GSC (MS-GSC) was first proposed in [2] as a power-saving implementation of the GSC scheme. With MS-GSC, the receiver combines the least number of best diversity paths such that the combiner output SNR is above a certain threshold. Both the simulation study in [2] and the preliminary theoretical analysis in [3, 4] show that MS-GSC can save a considerable amount of processing power by keeping fewer MRC branches active on average. In what follows, we present the mode of operation of MS-GSC and derive the statistics of its combined SNR using order statistics results [5]. These statistics are then applied to the exact performance analysis of MS-GSC over fading channels.

5.2.1 Mode of Operation

With MS-GSC, the receiver combines the smallest number of best paths such that the SNR of the combined signal, denoted by γ_c, is greater than a preselected output threshold, γ_T. The flow chart in Fig. 5.5 explains the mode of operation of MS-GSC, where $\gamma_{l:L}$ denotes the SNR of the lth strongest path. In particular, the receiver first estimates

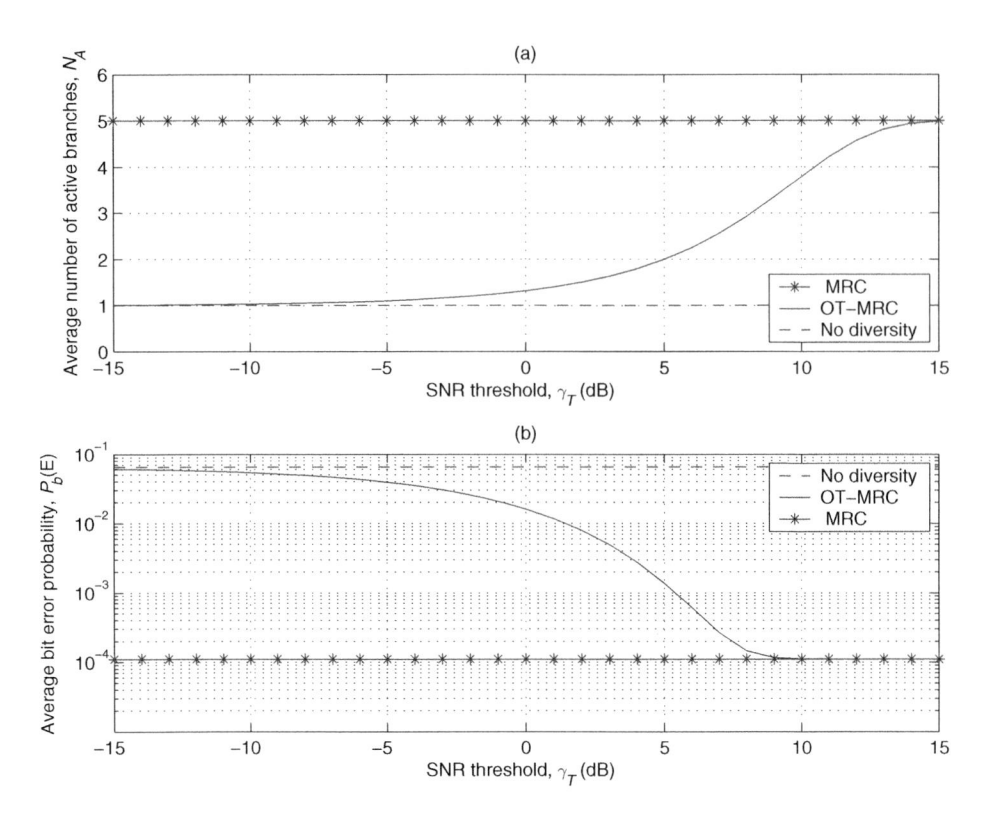

Figure 5.4 Illustration of the average error rate vs. average power consumption trade-off of OT-MRC: (a) Average number of active MRC branches as function of the SNR threshold; (b) average bit error rate of BPSK as function of the SNR threshold ($L = 5$ and $\bar{\gamma} = 5$ dB). Reprint with permission from [1]. ©2005 IEEE.

and ranks the SNR of all available diversity paths. Then, starting from the best path, i.e., the path with SNR $\gamma_{1:L}$, the receiver tries to increase the combined SNR γ_c above the threshold γ_T by combining an increasing number of diversity paths. More specifically, if $\gamma_{1:L} \geq \gamma_T$, then only the strongest diversity path will be used, i.e., $\gamma_c = \gamma_{1:L}$. In this case, the receiver acts as an L-branch selection combiner. If $\gamma_{1:L} < \gamma_T$, then the receiver will activate another MRC branch, find the second strongest path, and apply MRC to two strongest paths. If the output SNR, now equal to $\gamma_{1:L} + \gamma_{2:L}$, is greater than γ_T, only two-branch MRC will be used for data reception, i.e., $\gamma_c = \gamma_{1:L} + \gamma_{2:L}$. Similarly, if the output SNR of $i - 1$-branch MRC $\sum_{j=1}^{i-1} \gamma_{j:L}$ is less than γ_T and the output SNR of i-branch MRC $\sum_{j=1}^{i} \gamma_{j:L}$ is greater than γ_T, then i strongest paths will be combined for data reception. This process is continued until either the combined SNR is above γ_T or all L_c MRC branches are activated. In the later case, the receiver will act as a traditional L/L_c GSC combiner, i.e., L_c strongest paths are combined in an MRC fashion.

5.2 Minimum Selection GSC

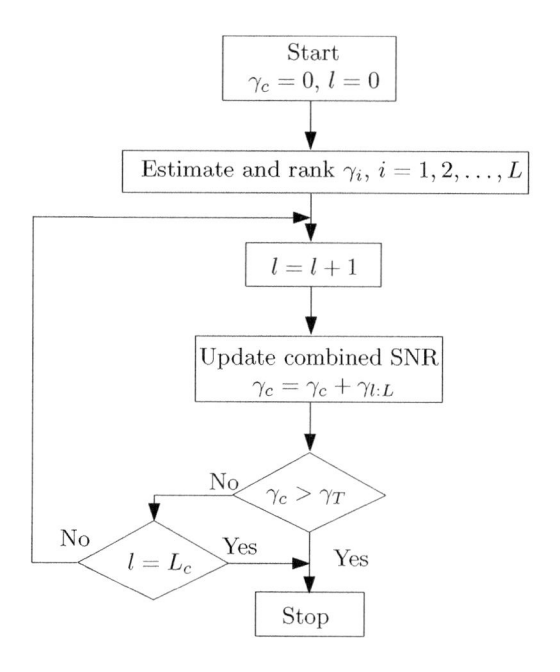

Figure 5.5 Flow chart for MS-GSC mode of operation. Reprint with permission from [5]. ©2006 IEEE.

The mode of operation of MS-GSC can be mathematically summarized as

$$\gamma_c = \gamma_{1:L} \text{ iff } \gamma_{1:L} \geq \gamma_T;$$
$$\gamma_c = \gamma_{1:L} + \gamma_{2:L} \text{ iff } \gamma_{1:L} < \gamma_T \ \& \ \gamma_{1:L} + \gamma_{2:L} \geq \gamma_T;$$
$$\vdots$$
$$\gamma_c = \sum_{j=1}^{i} \gamma_{j:L} \text{ iff } \sum_{j=1}^{i-1} \gamma_{j:L} < \gamma_T \ \& \ \sum_{j=1}^{i} \gamma_{j:L} \geq \gamma_T;$$
$$\vdots$$
$$\gamma_c = \sum_{j=1}^{L_c} \gamma_{j:L} \text{ iff } \sum_{j=1}^{L_c-1} \gamma_{j:L} < \gamma_T. \qquad (5.14)$$

5.2.2 Statistics of Output SNR

We now derive the statistics of the output SNR with MS-GSC. From the mode of operation of MS-GSC, we can see that the events $\gamma_c = \sum_{j=1}^{i} \gamma_{j:L}$ are mutually exclusive. Applying the total probability theorem, we can write the CDF of γ_c as

$$F_{\gamma_c}(x) = \Pr[\gamma_c < x]$$
$$= \sum_{i=1}^{L_c} \Pr[\gamma_c = \sum_{j=1}^{i} \gamma_{j:L} \ \& \ \gamma_c < x]. \quad (5.15)$$

Applying the mode of operation of MS-GSC as summarized in (5.14), we have

$$F_{\gamma_c}(x) = \Pr[\gamma_{1:L} \geq \gamma_T \ \& \ \gamma_c = \gamma_{1:L} < x]$$
$$+ \sum_{i=2}^{L_c-1} \Pr[\sum_{j=1}^{i-1} \gamma_{j:L} < \gamma_T \ \& \ \gamma_c = \sum_{j=1}^{i} \gamma_{j:L} \geq \gamma_T \ \& \ \gamma_c < x]$$
$$+ \Pr[\sum_{j=1}^{L_c-1} \gamma_{j:L} < \gamma_T \ \& \ \gamma_c = \sum_{j=1}^{L_c} \gamma_{j:L} < x]. \quad (5.16)$$

Note that only when $\gamma_c = \sum_{j=1}^{L_c} \gamma_{j:L}$ can the combined SNR be smaller than γ_T. We can rewrite the CDF of γ_c as

$$F_{\gamma_c}(x) = \begin{cases} F_{\Gamma_1}(x) - F_{\Gamma_1}(\gamma_T) \\ \quad + \sum_{i=2}^{L_c} F_{\gamma_c}^{(i)}(x) + F_{\Gamma_{L_c}}(\gamma_T), & x \geq \gamma_T; \\ F_{\Gamma_{L_c}}(x), & 0 \leq x < \gamma_T, \end{cases} \quad (5.17)$$

where $F_{\Gamma_i}(\cdot)$ is the CDF of the sum of the first i largest path SNRs and $F_{\gamma_c}^{(i)}(x)$ denotes the probability $\Pr[\sum_{j=1}^{i-1} \gamma_{(j)} < \gamma_T \ \& \ \gamma_T \leq \sum_{j=1}^{i} \gamma_{(j)} < x]$. The statistics of the sum of the first i largest path SNRs were extensively studied in the previous chapter during the discussion of the conventional GSC scheme. Specifically, the closed-form expression of $F_{\Gamma_i}(\cdot)$ for i.i.d. Rayleigh fading case is given by [6, Eq. 9.332]

$$F_{\Gamma_i}(x) = \frac{L!}{(L-i)!i!} \left\{ 1 - e^{-\frac{x}{\bar{\gamma}}} \sum_{k=0}^{i-1} \frac{1}{k!} \left(\frac{x}{\bar{\gamma}}\right)^k \right.$$
$$+ \sum_{l=1}^{L-i} (-1)^{i+l-1} \frac{(L-i)!}{(L-i-l)!l!} \left(\frac{i}{l}\right)^{i-1}$$
$$\times \left[\left(1 + \frac{l}{i}\right)^{-1} \left[1 - e^{-\left(1+\frac{l}{i}\right)\frac{x}{\bar{\gamma}}}\right] \right.$$
$$\left. \left. - \sum_{m=0}^{i-2} \left(-\frac{l}{i}\right)^m \left(1 - e^{-\frac{x}{\bar{\gamma}}} \sum_{k=0}^{m} \frac{1}{k!} \left(\frac{x}{\bar{\gamma}}\right)^k \right) \right] \right\}. \quad (5.18)$$

Let Γ_{i-1} denote the sum of the first $i-1$ largest SNRs, i.e., $\Gamma_{i-1} = \sum_{j=1}^{i-1} \gamma_{j:L}$. Then, the probability $F_{\gamma_c}^{(i)}(x)$ can be rewritten as

$$F_{\gamma_c}^{(i)}(x) = \Pr[\Gamma_{i-1} < \gamma_T \ \& \ \gamma_T \leq \gamma_{i:L} + \Gamma_{i-1} < x], \quad (5.19)$$

5.2 Minimum Selection GSC

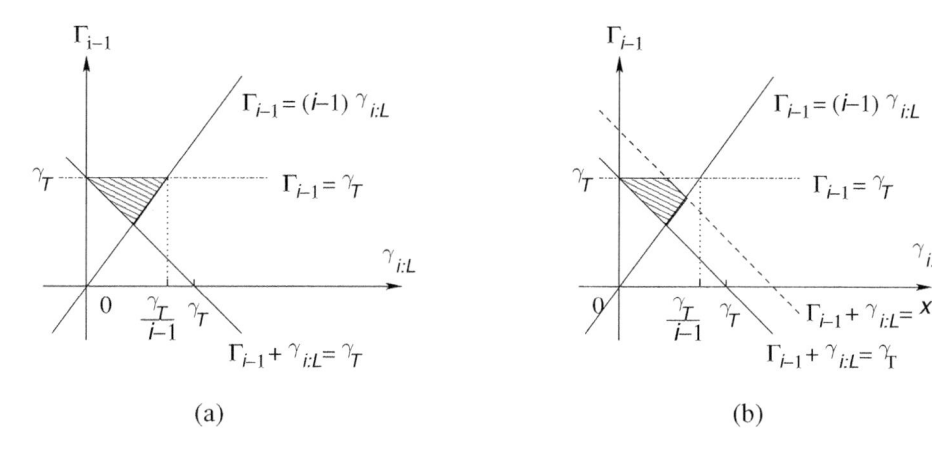

Figure 5.6 The integration region of the joint PDF of Γ_{i-1} and $\gamma_{i:L}$ for calculating (a) $F_{\gamma_c}^{(i)}(x)$, $x \geq i\gamma_T/(i-1)$ and $\Pr[N=i]$; (b) $F_{\gamma_c}^{(i)}(x)$, $\gamma_T \leq x < i\gamma_T/(i-1)$. Reprint with permission from [5]. ©2006 IEEE.

which can be calculated using the joint PDF of $\gamma_{i:L}$ and Γ_{i-1}. $p_{\gamma_{i:L},\Gamma_{i-1}}(\cdot,\cdot)$. Specifically, with the help of Fig. 5.6, $F_{\gamma_c}^{(i)}(x)$ can be calculated as

$$F_{\gamma_c}^{(i)}(x) = \begin{cases} \int_{\frac{i-1}{i}\gamma_T}^{\frac{i-1}{i}x} \int_{\gamma_T-y}^{\frac{y}{i-1}} p_{\gamma_{i:L},\Gamma_{i-1}}(z,y)dzdy \\ \quad + \int_{\frac{i-1}{i}x}^{\gamma_T} \int_{\gamma_T-y}^{x-y} p_{\gamma_{i:L},\Gamma_{i-1}}(z,y)dzdy, & \gamma_T \leq x < \frac{i}{i-1}\gamma_T; \\ F_{\Gamma_{i-1}}(\gamma_T) - F_{\Gamma_i}(\gamma_T), & \frac{i}{i-1}\gamma_T \leq x. \end{cases} \quad (5.20)$$

After substituting (5.20) into (5.17), we obtain the generic expression of the CDF of the combined SNR γ_c with MS-GSC as

$$F_{\gamma_c}(x) = \begin{cases} F_{\Gamma_{L_c}}(x), & 0 \leq x < \gamma_T; \\ F_{\Gamma_1}(x) - F_{\Gamma_1}(\gamma_T) + F_{\Gamma_{L_c}}(\gamma_T) \\ \quad + \sum_{i=2}^{L_c} \left(\int_{\frac{i-1}{i}\gamma_T}^{\frac{i-1}{i}x} \int_{\gamma_T-y}^{\frac{y}{i-1}} p_{\gamma_{i:L},\Gamma_{i-1}}(z,y)dzdy \right. \\ \quad \left. + \int_{\frac{i-1}{i}x}^{\gamma_T} \int_{\gamma_T-y}^{x-y} p_{\gamma_{i:L},\Gamma_{i-1}}(z,y)dzdy \right), & \gamma_T \leq x < \frac{L_c}{L_c-1}\gamma_T; \\ \vdots \\ F_{\Gamma_1}(x) - F_{\Gamma_1}(\gamma_T) + F_{\Gamma_l}(\gamma_T) \\ \quad + \sum_{i=2}^{l} \left(\int_{\frac{i-1}{i}\gamma_T}^{\frac{i-1}{i}x} \int_{\gamma_T-y}^{\frac{y}{i-1}} p_{\gamma_{i:L},\Gamma_{i-1}}(z,y)dzdy \right. \\ \quad \left. + \int_{\frac{i-1}{i}x}^{\gamma_T} \int_{\gamma_T-y}^{x-y} p_{\gamma_{i:L},\Gamma_{i-1}}(z,y)dzdy \right), & \frac{l+1}{l}\gamma_T \leq x < \frac{l}{l-1}\gamma_T; \\ \vdots \\ F_{\Gamma_1}(x), & 2\gamma_T < x. \end{cases} \quad (5.21)$$

The generic expression for the PDF of the combined SNR with MS-GSC can be obtained by differentiating (5.21) with respect to x and noting that

$$\frac{d}{dx}F_{\gamma_c}^{(i)}(x) = \int_{\frac{i-1}{i}x}^{\gamma_T} p_{\gamma_{i:L},\Gamma_{i-1}}(x-y,y)dy, \quad \gamma_T \le x < \frac{i}{i-1}\gamma_T, \quad (5.22)$$

as

$$p_{\gamma_c}(x) = \begin{cases} p_{\Gamma_1}(x) + \sum_{i=2}^{L_c} \int_{\frac{i-1}{i}x}^{\gamma_T} p_{\gamma_{i:L},\Gamma_{i-1}}(x-y,y)dy \\ \quad \times (\mathcal{U}(x-\gamma_T) - \mathcal{U}(x-\frac{i}{i-1}\gamma_T)), & x \ge \gamma_T; \\ p_{\Gamma_{L_c}}(x), & 0 \le x < \gamma_T, \end{cases} \quad (5.23)$$

where $\mathcal{U}(\cdot)$ is the unit step function. It follows that the generic expression of the MGF of the combined SNR with MS-GSC is given by

$$\mathcal{M}_{\gamma_c}(s) = \int_0^{\gamma_T} p_{\Gamma_{L_c}}(x)e^{sx}dx + \int_{\gamma_T}^{\infty} p_{\Gamma_1}(x)e^{sx}dx$$

$$+ \sum_{i=1}^{L_c} \int_{\gamma_T}^{\frac{i}{i-1}\gamma_T} \left[\int_{\frac{i-1}{i}x}^{\gamma_T} p_{\gamma_{i:L},\Gamma_{i-1}}(x-y,y)dy \right] e^{sx}dx. \quad (5.24)$$

The joint PDF of $p_{\gamma_{i:L},\Gamma_{i-1}}(\cdot,\cdot)$ is studied in the Appendix as one of the new order statistics results. Specifically, for i.i.d. Rayleigh fading environments, its closed-form expression is given by

$$p_{\gamma_{l:L},y_l}(x,y) = \sum_{j=0}^{L-l} \frac{(-1)^j L!}{(L-l-j)!(l-1)!(l-2)!j!\bar{\gamma}^l}[y-(l-1)x]^{(l-2)}e^{-\frac{y+(j+1)x}{\bar{\gamma}}},$$

$$x \ge 0, \ y \ge (l-1)x. \quad (5.25)$$

After proper substitution and integration, we can obtain the statistics of the combined SNR with MS-GSC over i.i.d. Rayleigh fading in closed forms. For example, we can show that the summand in (5.23) is given, after successive integration by parts, by

$$\int_{\frac{i-1}{i}x}^{\gamma_T} p_{\gamma_{i:L},\Gamma_{i-1}}(x-y,y)dy = \frac{L!}{(L-i)!(i-1)!i!\bar{\gamma}}e^{-\frac{x}{\bar{\gamma}}}\left(\frac{i\gamma_T - (i-1)x}{\bar{\gamma}}\right)^{i-1} \quad (5.26)$$

$$+ \sum_{j=1}^{L-i} \frac{L!(-1)^{j-i+1}}{(L-i-j)!i!j!\bar{\gamma}}\left(\frac{i}{j}\right)^{i-1} e^{-\frac{(i+j)x}{i\bar{\gamma}}}$$

$$\times \left[1 - e^{-\frac{j((i-1)x-i\gamma_T)}{i\bar{\gamma}}} \sum_{k=0}^{i-2} \frac{1}{k!}\left(\frac{j((i-1)x-i\gamma_T)}{i\bar{\gamma}}\right)^k \right],$$

$$\gamma_T \le x < \frac{i}{i-1}\gamma_T.$$

Note that the PDF of the sum of the first i largest path SNRs $p_{\Gamma_i}(\cdot)$ is given by [6, Eq. 9.325]

5.2 Minimum Selection GSC

$$p_{\Gamma_i}(x) = \frac{L!}{(L-i)!i!}e^{-\frac{x}{\bar{\gamma}}}\left[\frac{x^{i-1}}{\bar{\gamma}^i(i-1)!}\right.$$
$$+\frac{1}{\bar{\gamma}}\sum_{l=1}^{L-i}(-1)^{i+l-1}\frac{(L-i)!}{(L-i-l)!l!}\left(\frac{i}{l}\right)^{i-1}$$
$$\left.\times\left(e^{-\frac{lx}{i\bar{\gamma}}}-\sum_{m=0}^{i-2}\frac{1}{m!}\left(-\frac{lx}{i\bar{\gamma}}\right)^m\right)\right]. \quad (5.27)$$

Therefore, the PDF of the combined SNR γ_c with MS-GSC over i.i.d. Rayleigh fading paths can be shown, by specializing (5.23) with (5.26) and (5.27), to be given in the following closed-form expression

$$p_{\gamma_c}(x) = \begin{cases} \frac{L}{\bar{\gamma}}\left(1-e^{-\frac{x}{\bar{\gamma}}}\right)^{L-1}e^{-\frac{x}{\bar{\gamma}}} \\ +\sum_{i=2}^{L_c}\left\{\frac{L!}{(L-i)!(i-1)!i!\bar{\gamma}}e^{-\frac{x}{\bar{\gamma}}}\left(\frac{i\gamma_T-(i-1)x}{\bar{\gamma}}\right)^{i-1}\right. \\ +\sum_{j=1}^{L-i}\frac{L!(-1)^{j-i+1}}{(L-i-j)!i!j!\bar{\gamma}}\left(\frac{i}{j}\right)^{i-1}\exp\left(-\frac{(i+j)x}{i\bar{\gamma}}\right) \\ \left.\times\left[1-e^{-\frac{j((i-1)x-i\gamma_T)}{i\bar{\gamma}}}\sum_{k=0}^{i-2}\frac{1}{k!}\left(\frac{j((i-1)x-i\gamma_T)}{i\bar{\gamma}}\right)^k\right]\right\} \\ \times(\mathcal{U}(x-\gamma_T)-\mathcal{U}(x-\frac{i}{i-1}\gamma_T)), \quad x \geq \gamma_T; \\ \frac{L!}{(L-L_c)!L_c!}e^{-\frac{x}{\bar{\gamma}}}\left[\frac{x^{L_c-1}}{\bar{\gamma}^{L_c}(L_c-1)!}\right. \\ +\frac{1}{\bar{\gamma}}\sum_{l=1}^{L-L_c}(-1)^{L_c+l-1}\frac{(L-L_c)!}{(L-L_c-l)!l!}\left(\frac{L_c}{l}\right)^{L_c-1} \\ \left.\times\left(e^{-\frac{lx}{L_c\bar{\gamma}}}-\sum_{m=0}^{L_c-2}\frac{1}{m!}\left(-\frac{lx}{L_c\bar{\gamma}}\right)^m\right)\right], \quad 0 \leq x < \gamma_T. \end{cases} \quad (5.28)$$

Furthermore, after substituting the closed-form expression for the Rayleigh PDF of γ_c given in (5.28) into (5.24) and carrying out the integration with the help of the definition of the incomplete gamma function [7, Sec. 8.35], we obtain the closed-form expression for the MGF of γ_c as

$$\mathcal{M}_{\gamma_c}(s) = L\sum_{j=0}^{L-1}\binom{L-1}{j}(-1)^j\frac{e^{-\left(\frac{1+j}{\bar{\gamma}}-s\right)\gamma_T}}{1+j-\bar{\gamma}s}$$
$$+\sum_{i=2}^{l}\binom{L}{i}\left[\sum_{m=0}^{i-1}\frac{(1-i)^m}{(i-1-m)!}\left(\frac{i\gamma_T}{\bar{\gamma}}\right)^{i-1-m}\mathcal{G}_{i,0}(s)\right.$$
$$+\sum_{j=1}^{L-i}\binom{L-i}{j}(-1)^{j-i+1}\left(\frac{i}{j}\right)^{i-1}\left(\frac{e^{-\left(\frac{1+j/i}{\bar{\gamma}}-s\right)\gamma_T}-e^{-\left(\frac{1+j/i}{\bar{\gamma}}-s\right)\frac{i}{i-1}\gamma_T}}{1+j/i-\bar{\gamma}s}\right.$$

$$-\sum_{k=0}^{i-2}\sum_{m=0}^{k}\frac{\left(\frac{i-1}{i}j\right)^{m}\left(-j\frac{\gamma_T}{\overline{\gamma}}\right)^{k-m}\mathcal{G}_{i,j}(s)}{(k-m)!}\right)\right]+\binom{L}{l}\left[\mathcal{F}_l(s)+\sum_{i=1}^{L-l}\binom{L-l}{i}\right.$$

$$\left.\times(-1)^{l+i-1}\left(\frac{l}{i}\right)^{l-1}\left(\frac{1-e^{-\left(\frac{1+i/l}{\overline{\gamma}}-s\right)\gamma_T}}{1+i/l-\overline{\gamma}s}-\sum_{m=0}^{l-2}\left(-\frac{i}{l}\right)^{m}\mathcal{F}_{m+1}(s)\right)\right], \quad (5.29)$$

where

$$\mathcal{G}_{i,j}(s)=e^{j\frac{\gamma_T}{\overline{\gamma}}}\frac{\Gamma\left(m+1,\left(\frac{1+j}{\overline{\gamma}}-s\right)\gamma_T\right)-\Gamma\left(m+1,\left(\frac{1+j}{\overline{\gamma}}-s\right)\left(\frac{i}{i-1}\right)\gamma_T\right)}{m!(1+j-\overline{\gamma}s)^{m+1}},$$

$$\mathcal{F}_l(s)=\frac{\Gamma(l)-\Gamma(l,(1/\overline{\gamma}-s)\gamma_T)}{(l-1)!(1-\overline{\gamma}s)^l},$$

and $\Gamma(\cdot)$ and $\Gamma(\cdot,\cdot)$ are the complete and the incomplete Gamma functions, respectively. It can be shown that when $\gamma_T=\infty$, (5.29) reduces to the MGF of the original GSC over i.i.d. Rayleigh fading paths, as given in [6, Eq. 9.430]. These closed-form results can be readily applied to the error rate analysis of MS-GSC over the Rayleigh fading environment.

5.2.3 Complexity Analysis

The receiver with MS-GSC needs only to find enough best paths such that the combined SNR is above the output threshold γ_T. Meanwhile, the receiver with conventional GSC always searches for the L_c best paths. Although the MS-GSC receiver introduces an extra threshold check at the combiner output, the threshold checks are much easier to perform than path SNR ordering. As such, MS-GSC requires less computing power than the original GSC for path selection. In addition, the receiver with MS-GSC needs to keep fewer MRC branches active on average during data reception stage. As such, valuable processing power of the receiver is saved with MS-GSC. These features of MS-GSC constitute its main complexity savings over conventional GSC. As a quantification of its power savings, we derive the average number of active MRC branches with MS-GSC in this section.

The number of active MRC branches with MS-GSC, denoted by n, takes a value from 1 to L_c. From the mode of operation of MS-GSC, we can see that $n=1$ if and only if $\gamma_{1:L}>\gamma_T$, $n=i$, $2\leq i\leq L_c-1$ if and only if $\sum_{j=1}^{i-1}\gamma_{j:L}<\gamma_T$ and $\sum_{j=1}^{i}\gamma_{j:L}\geq\gamma_T$, and $n=L_c$ if $\sum_{j=1}^{L_c-1}\gamma_{j:L}<\gamma_T$. Therefore, the probability mass function (PMF) of n can be written as

$$\Pr[n=i]=\begin{cases}\Pr[\gamma_{1:L}>\gamma_T], & i=1;\\ \Pr[\sum_{j=1}^{i-1}\gamma_{j:L}<\gamma_T \ \& \ \sum_{j=1}^{i}\gamma_{j:L}\geq\gamma_T], & 2\leq i\leq L_c-1;\\ \Pr[\sum_{j=1}^{L_c-1}\gamma_{j:L}<\gamma_T], & i=L_c.\end{cases} \quad (5.30)$$

Note that the probability $\Pr[\sum_{j=1}^{i-1} \gamma_{j:L} < \gamma_T \ \& \ \sum_{j=1}^{i} \gamma_{j:L} \geq \gamma_T]$ is equivalent to $\Pr[\Gamma_{i-1} < \gamma_T \ \& \ \Gamma_{i-1} + \gamma_{i:L} \geq \gamma_T]$. Therefore, this probability can be calculated by integrating the joint PDF of Γ_{i-1} and $\gamma_{i:L}$ over the shaded region shown in Fig. 5.6(a). Since the region of integration is exactly the difference of the regions $\Gamma_{i-1} < \gamma_T$ and $\Gamma_{i-1} + \gamma_{i:L} < \gamma_T$, we can instead calculate this probability by using the CDFs of Γ_{i-1} and $\Gamma_{i-1} + \gamma_{i:L}$ as the following

$$\Pr[\Gamma_{i-1} < \gamma_T \ \& \ \Gamma_{i-1} + \gamma_{i:L} \geq \gamma_T] = F_{\Gamma_{i-1}}(\gamma_T) - F_{\Gamma_{i-1}+\gamma_{i:L}}(\gamma_T)$$
$$= F_{\Gamma_{i-1}}(\gamma_T) - F_{\Gamma_i}(\gamma_T), \quad (5.31)$$

where $F_{\Gamma_i}(\cdot)$ is the CDF of the sum of the first i largest path SNRs out of L available ones. Consequently, the PMF of n is given by

$$\Pr[N = i] = \begin{cases} 1 - F_{\Gamma_1}(\gamma_T), & i = 1; \\ F_{\Gamma_{i-1}}(\gamma_T) - F_{\Gamma_i}(\gamma_T), & 2 \leq i \leq L_c - 1; \\ F_{\Gamma_{L_c-1}}(\gamma_T), & i = L_c. \end{cases} \quad (5.32)$$

The average number of active MRC branches with MS-GSC can then be calculated as

$$N_A = \sum_{i=1}^{L_c} i \Pr[N = i]$$
$$= 1 + \sum_{i=1}^{L_c-1} F_{\Gamma_i}(\gamma_T). \quad (5.33)$$

Note that the closed-form expression of $F_{\Gamma_i}(\cdot)$ for i.i.d. Rayleigh fading was given in (5.18). As such, both the PMF and the average of the number of active MRC branches can be calculated in closed form.

5.2.4 Numerical Results

Fig. 5.7 plots the outage probability of MS-GSC over eight i.i.d. Rayleigh fading paths for different numbers of MRC branches. The outage probability of conventional 8/4-GSC is also plotted for comparison. As we can see, the outage performance improves as the number of MRC branches increases, especially when γ_{th} is smaller than γ_T. In fact, comparing the outage performance of MS-GSC of the $L_c = 4$ case with that of conventional GSC, we can see that they have the same outage performance if $\gamma_{\text{th}} \leq \gamma_T$. In this case, the combined SNR with MS-GSC can only be smaller than γ_T if the combined SNR of the corresponding GSC scheme is smaller than γ_T. When $\gamma_{\text{th}} > \gamma_T$, the outage performance of MS-GSC degrades rapidly in comparison with the original GSC scheme because the MS-GSC combiner only tries to increase the combined SNR above γ_T. We can conclude from this figure that the SNR threshold γ_T for MS-GSC should be chosen to be larger than or equal to the outage threshold γ_{th} in practice. As

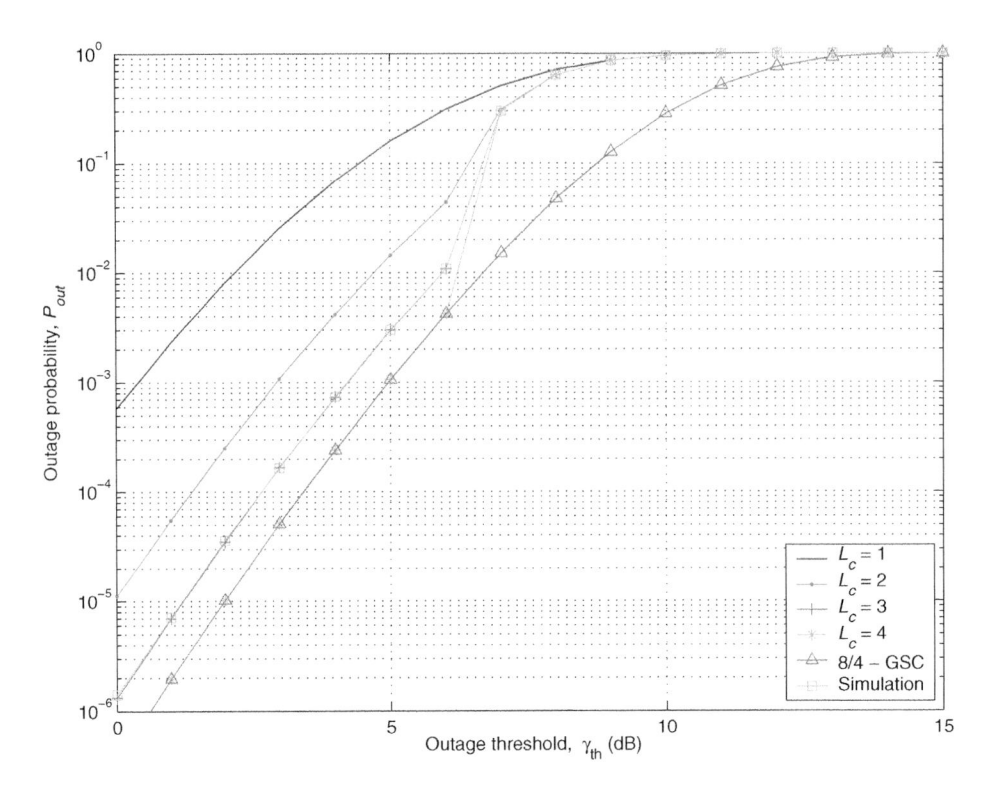

Figure 5.7 Outage probability, or equivalently CDF of the combined SNR, with MS-GSC over i.i.d. Rayleigh fading paths ($L = 8$, $\overline{\gamma} = 3$ dB, and $\gamma_T = 6$ dB). Reprint with permission from [5]. ©2006 IEEE.

a validation of the analytical results, simulation results for the outage probability of MS-GSC for $L_c = 3$ case are also presented. Note that these simulation results match perfectly with analytical results.

Fig. 5.8 shows the average error rate of BPSK with MS-GSC over six i.i.d. Rayleigh fading paths for different numbers of MRC branches. The error rate of the conventional 6/4-GSC is also plotted for reference. As we can see, the error performance improves as L_c increases, especially in the low average SNR region. Comparing the error rate of MS-GSC for the $L_c = 4$ case and that of conventional GSC, we observe that they have the same error performance in the low average SNR region, but the performance of MS-GSC is not as good in the medium to large SNR region. This BER behavior of MS-GSC can be explained by the following. When the average SNR is small compared to the output threshold, the receiver will always combine the L_c best paths because combining fewer paths will not result in a combined SNR above the required output threshold. As a result, the combiner is actually working as a conventional GSC combiner. On the other hand, when the average SNR is large compared to the output threshold, the receiver will most likely combine only the best paths because it can satisfy the output

5.2 Minimum Selection GSC

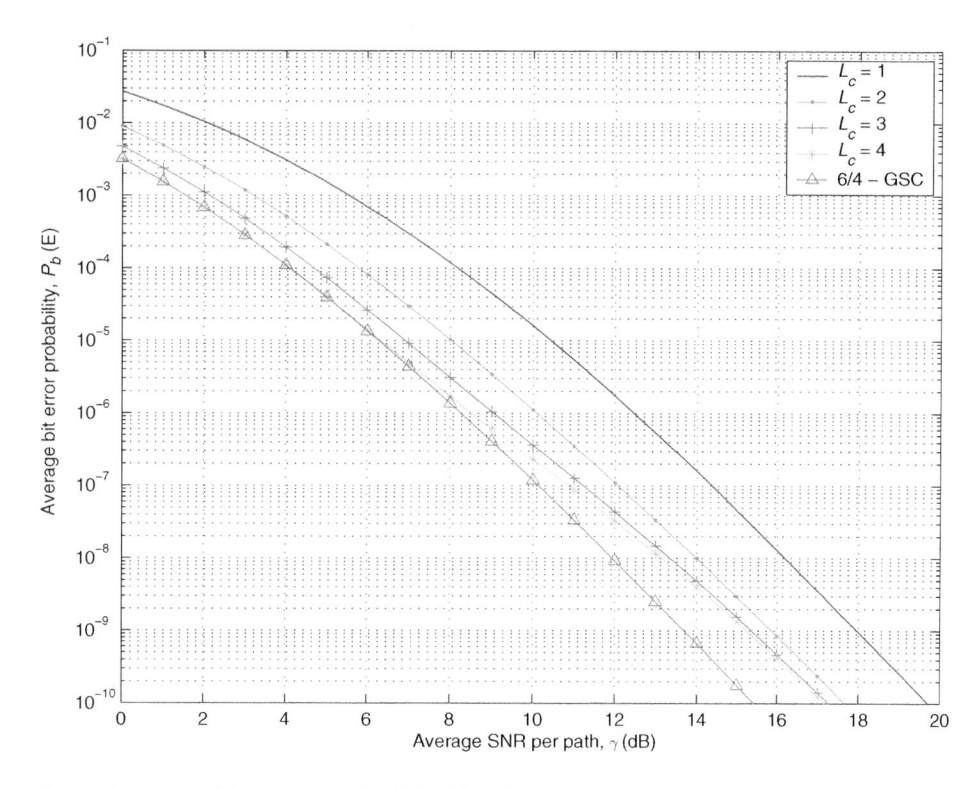

Figure 5.8 Average bit error rate of BPSK with MS-GSC over i.i.d. Rayleigh fading paths ($L = 6$ and $\gamma_T = 10$ dB). Reprint with permission from [5]. ©2006 IEEE.

threshold requirement. As a result, the combiner is actually working as a selection combiner.

In Figs. 5.9 and 5.10, we study the performance versus power consumption trade-off involved in MS-GSC. Fig. 5.9 shows the average error rate of MS-GSC as a function of the output threshold with four MRC branches and different numbers of diversity paths. As expected, the performance of MS-GSC improves as the number of available diversity paths L increases. In addition, the error rate of MS-GSC decreases from that of L-branch SC to that of $L/4$-GSC as γ_T increases. Fig. 5.10 shows the corresponding average number of active MRC branches with MS-GSC as the function of the output threshold. As we can see, the average number of active MRC branches decreases as the number of available diversity paths increases, but increases as the output threshold increases since it becomes more difficult to raise the combined SNR above the output threshold. Considering both Figs. 5.9 and 5.10, we can see that if the output threshold γ_T is set to 10 dB, MS-GSC has the same error rate performance as the conventional GSC while requiring less active branches on average. In particular, for the $L = 6$ and $L_c = 4$ case, MS-GSC needs on average 3.3 active MRC branches, which translates to a 17.5% power saving compared to conventional GSC.

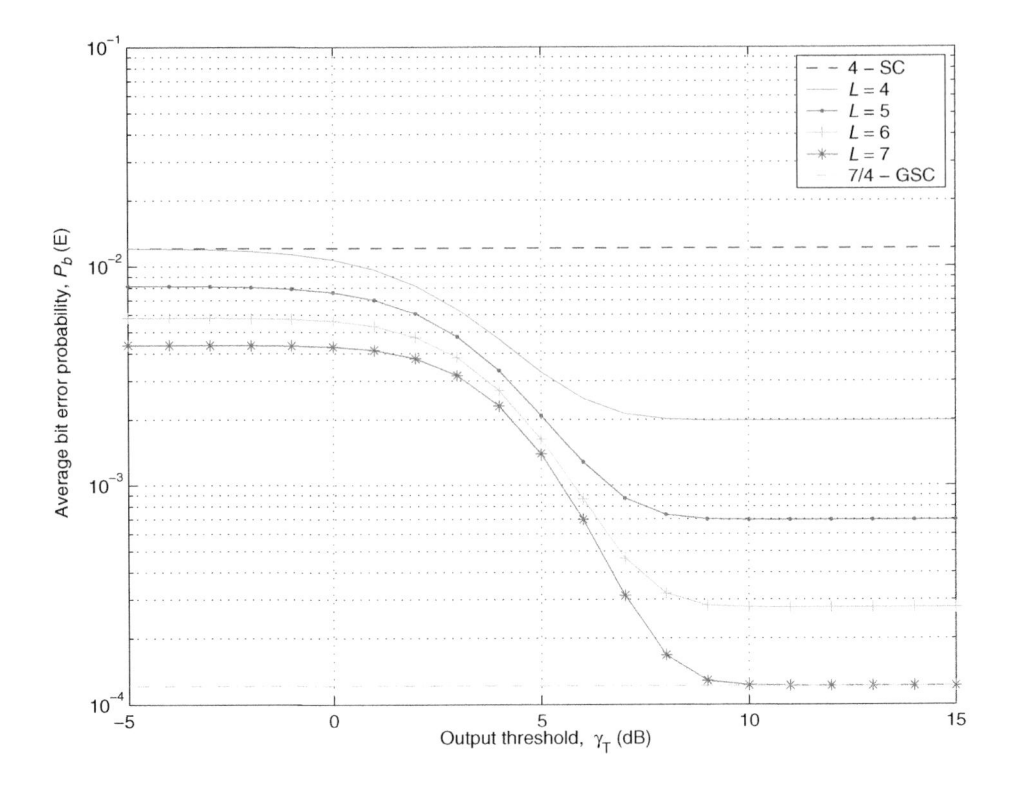

Figure 5.9 Average bit error rate of BPSK with MS-GSC over i.i.d. Rayleigh fading paths as a function of output threshold γ_T ($L_c = 4$ and $\overline{\gamma} = 3$ dB). Reprint with permission from [5]. ©2006 IEEE.

5.3 Output-Threshold GSC

The OT-MRC scheme saves the receiver processing power by estimating and combining a minimum number of paths [1]. It still requires the same hardware complexity as conventional MRC. On the other hand, the MS-GSC scheme can reduce the receiver hardware complexity with the GSC principle and the number of combining paths through minimum best selection. But the MS-GSC receiver always needs to estimate all L available diversity paths. In this section, we present another power-saving variant of GSC [1], which essentially combines the desired properties of both OT-MRC and MS-GSC schemes. The resulting output-threshold GSC (OT-GSC) scheme provides considerable processing power saving over conventional GSC while maintaining the same diversity benefit.

5.3.1 Mode of Operation

The OT-GSC scheme estimates and combines additional diversity paths only if necessary. When the number of estimated paths is larger than the number of available MRC

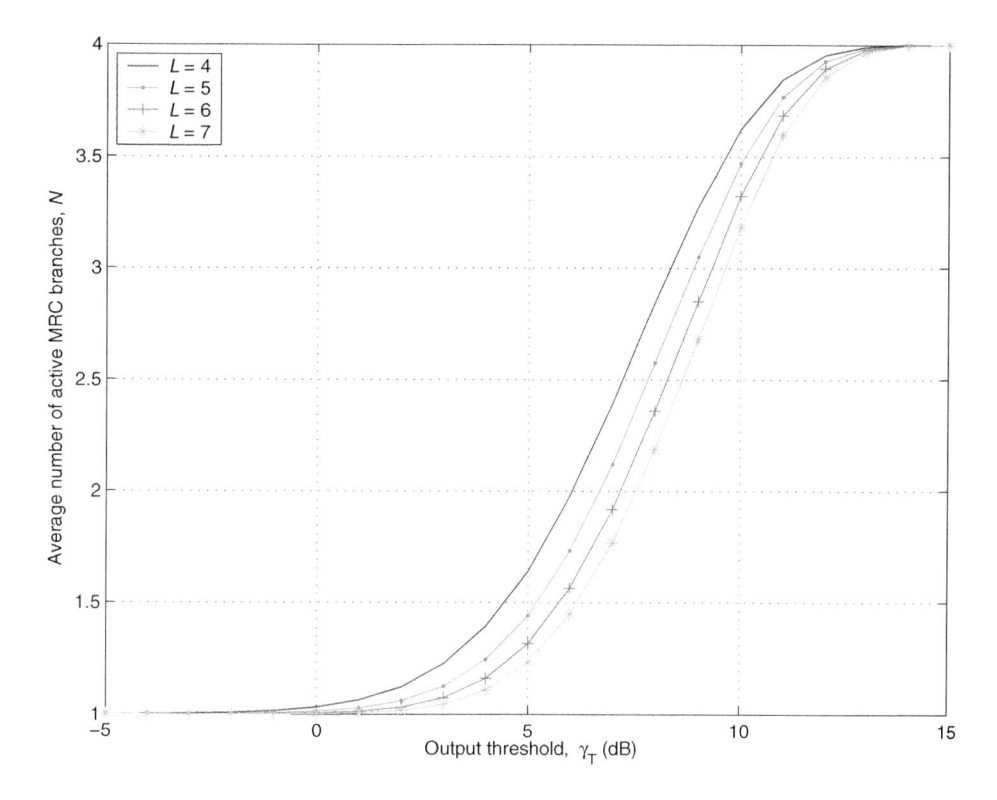

Figure 5.10 Average number of active MRC branches with MS-GSC as a function of output threshold γ_T ($L_c = 4$ and $\overline{\gamma} = 3$ dB). Reprint with permission from [5]. ©2006 IEEE.

branches, the receiver applies best selection to reduce the hardware complexity. The flow chart in Fig. 5.11 illustrates the mode of operation of OT-GSC. In particular, the operation of the OT-GSC combiner can be divided into two stages: an MRC stage followed by a GSC stage.

MRC stage: During the MRC stage, the combiner acts as an L_c-branch OT-MRC combiner. In particular, the combiner tries to improve the output SNR γ_c above the threshold γ_T by gradually changing its configuration from a single-branch MRC to an L_c-branch MRC. If all L_c MRC branches have been activated but the output SNR is still below γ_T, then the combiner operation enters the GSC stage.

GSC stage: During the GSC stage, the combiner tries to improve the combined SNR γ_c above the SNR threshold γ_T by gradually changing its configuration from an L_c+1/L_c GSC to an L/L_c GSC, where l/L_c GSC denotes a combiner that combines in an MRC fashion the L_c strongest paths out of the l available paths. In particular, knowing that the combined SNR after MRC combining of the first L_c diversity paths is still below γ_T, the combiner checks the output SNR with $L_c + 1/L_c$ GSC. As such, the combiner first estimates another diversity path and then checks if its SNR is greater than the minimum SNR of the first L_c combined paths. If this is the case, the newly estimated path will replace the path that has minimum SNR. If the resulting combined SNR is above γ_T, the

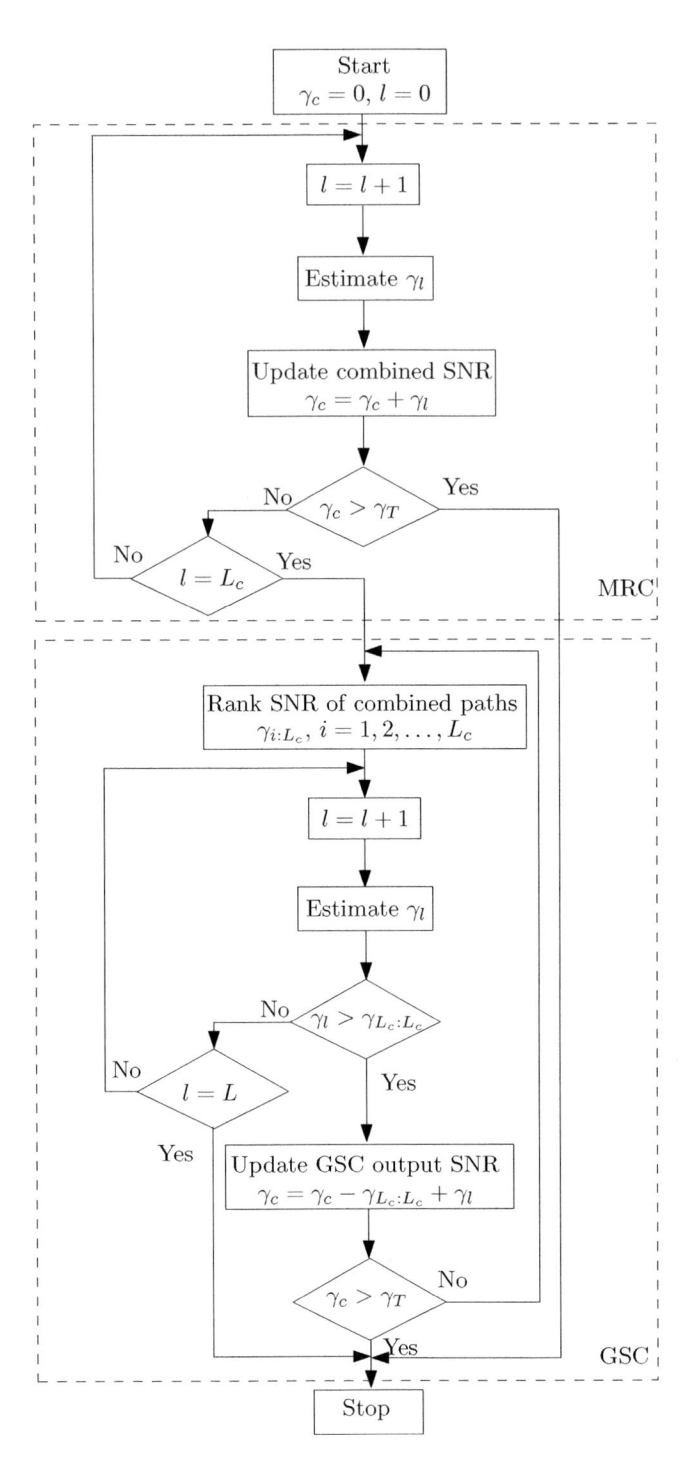

Figure 5.11 Flow chart for the OT-GSC mode of operation. Reprint with permission from [1]. ©2005 IEEE.

combiner stops the path update. On the other hand, if either the SNR of the newly estimated path is less than the minimum SNR of the currently used paths or if the resulting combined SNR is still below the threshold after path replacement, then the combiner looks into the combined SNR of $L_c + 2/L_c$ GSC by estimating another diversity path. This process is continued until either the combined SNR γ_c is above γ_T or all L available diversity paths have been estimated. Only in the worst case will does the receiver employs L/L_c GSC for data reception.

From the operation of OT-GSC, we can see that unlike conventional GSC, where all L diversity paths need to be estimated and ranked in every guard period and all L_c MRC branches need to be active all the time, the receiver with OT-GSC uses the available resources adaptively. In particular, both path estimation and MRC branch activation are performed only if necessary. These features of OT-GSC constitute its main power saving over the conventional GSC scheme.

5.3.2 Complexity Analysis

In this section, we accurately quantify the power saving of OT-GSC by deriving closed-form expressions for both the average number of path estimations and the average number of active MRC branches.

Average Number of Path Estimations

According to the mode of operation of OT-GSC, the number of path estimations performed during a guard period varies from 1 to L. Therefore, we can write the average number of path estimations, denoted by N_E, in the following summation form:

$$N_E = \sum_{l=1}^{L} \pi_l\, l, \qquad (5.34)$$

where π_l is the probability that exactly l diversity paths are estimated during a guard period and it is equal to the probability that either an l-branch MRC for $l \leq L_c$ or an l/L_c-GSC for $l > L_c$ is used for data reception. Since a single-branch receiver is used if the SNR of the first diversity path is above γ_T while an l-branch MRC receiver is used if the combined SNR of an $l - 1$-branch MRC is below γ_T whereas the combined SNR of an l-branch MRC is above γ_T, we have under the i.i.d. fading paths assumption:

$$\pi_l = \begin{cases} 1 - F_\gamma(\gamma_T), & l = 1; \\ F_{\gamma_c}^{(l-1)-MRC}(\gamma_T) - F_{\gamma_c}^{l-MRC}(\gamma_T), & 1 < l \leq L_c, \end{cases} \qquad (5.35)$$

where $F_{\gamma_c}^{l-MRC}(\cdot)$ is the CDF of the combined SNR with l-branch MRC combiner, the closed-form expressions of which for three popular fading channel models of interest are given in Table 5.1. On the other hand, for $L_c < l \leq L - 1$, an l/L_c-branch GSC is used if the combined SNR of an $(l - 1)/L_c$-branch GSC is below γ_T, whereas the combined SNR of an l/L_c-branch GSC is above γ_T. In addition, if the combined SNR of an $(L-1)/L_c$-branch GSC is below γ_T, then an L/L_c-branch GSC is used. Based on these observations and under the i.i.d. fading path assumption, it can be shown that

$$\pi_l = \begin{cases} F_{\gamma_c}^{(l-1)/L_c-GSC}(\gamma_T) - F_{\gamma_c}^{l/L_c-GSC}(\gamma_T), & L_c < l < L; \\ F_{\gamma_c}^{(L-1)/L_c-GSC}(\gamma_T), & l = L, \end{cases} \quad (5.36)$$

where $F_{\gamma_c}^{l/L_c-GSC}(\cdot)$ is the CDF of the combined SNR with l/L_c branch GSC, i.e., the CDF of the sum of the L_c largest path SNR out of l estimated ones. Substituting (5.35) and (5.36) into (5.34), it can be shown after some manipulation and simplifications that the average number of path estimations is given by

$$N_E = 1 + \sum_{l=1}^{L_c} F_{\gamma_c}^{l-MRC}(\gamma_T) + \sum_{l=L_c+1}^{L-1} F_{\gamma_c}^{l/L_c-GSC}(\gamma_T). \quad (5.37)$$

For the i.i.d. Rayleigh fading case, substituting $F_{\gamma_c}^{l-MRC}(\cdot)$ from Table 5.1 (for the Rayleigh case) and $F_{\gamma_c}^{l/L_c-GSC}(\cdot)$ as given in (5.18) into (5.37) leads to the following closed-form expression for N_E:

$$N_E = 1 + \sum_{l=1}^{L_c} \left[1 - e^{-\frac{\gamma_T}{\bar{\gamma}}} \sum_{i=0}^{l-1} \frac{1}{i!} \left(\frac{\gamma_T}{\bar{\gamma}}\right)^i \right]$$

$$+ \sum_{l=L_c+1}^{L-1} \binom{l}{L_c} \left\{ 1 - e^{-\frac{\gamma_T}{\bar{\gamma}}} \sum_{i=0}^{L_c-1} \frac{1}{i!} \left(\frac{\gamma_T}{\bar{\gamma}}\right)^i \right.$$

$$+ \sum_{i=1}^{l-L_c} (-1)^{L_c+i-1} \binom{l-L_c}{i} \left(\frac{L_c}{i}\right)^{L_c-1}$$

$$\times \left[\left(1 + \frac{i}{L_c}\right)^{-1} \left[1 - e^{\left(1+\frac{i}{L_c}\right)\frac{\gamma_T}{\bar{\gamma}}}\right]\right.$$

$$\left. - \sum_{m=0}^{L_c-2} \left(-\frac{i}{L_c}\right)^m \left(1 - e^{-\frac{\gamma_T}{\bar{\gamma}}} \sum_{k=0}^{m} \frac{1}{k!} \left(\frac{\gamma_T}{\bar{\gamma}}\right)^k\right) \right] \right\}. \quad (5.38)$$

In Fig. 5.12, we show the average number of path estimations with OT-GSC as a function of the normalized SNR threshold, $\gamma_T/\bar{\gamma}$. Note that the average number of path estimations N_E increases from 1 to L as the threshold increases, since it becomes more difficult to raise the combined SNR above the threshold. Note also that N_E decreases as the number of available MRC branches L_c increases, but the difference becomes negligible after L_c becomes greater than 2.

Average Number of Active MRC Branches
With OT-GSC, if the combined SNR exceeds the SNR threshold during the MRC stage, less than L_c MRC branches will be active during data reception. As a result, a considerable amount of receiver processing power will be saved. We can calculate

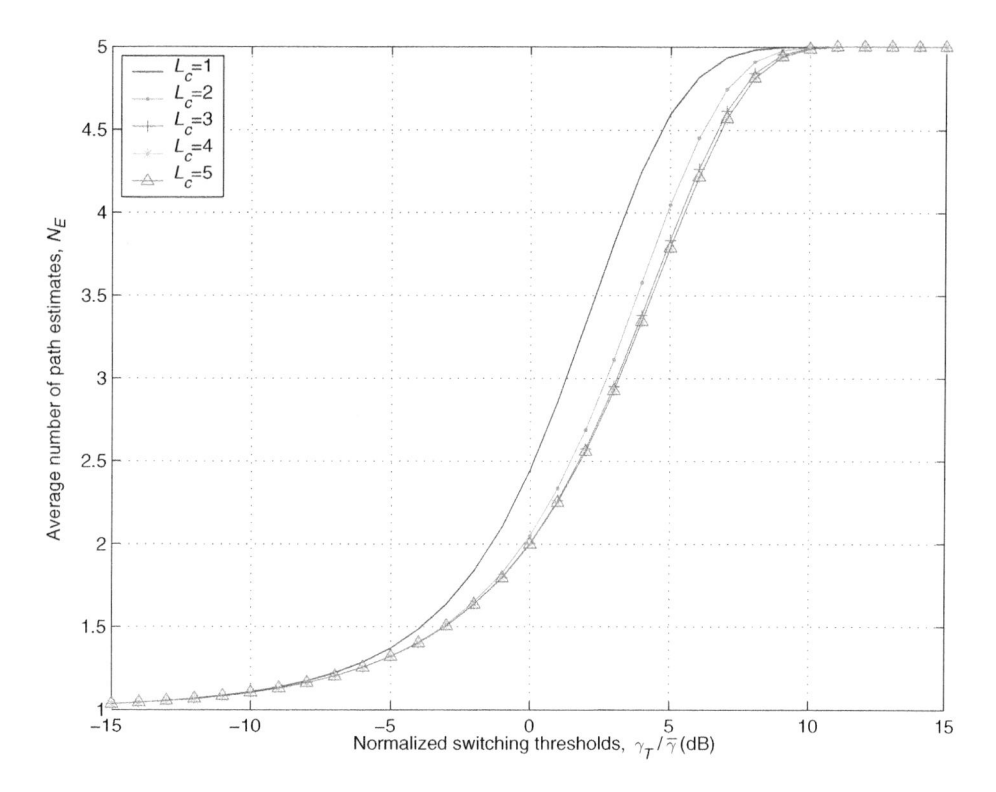

Figure 5.12 Average number of path estimations with OT-GSC as a function of the normalized SNR threshold, $\gamma_T/\overline{\gamma}$ ($L = 5$). Reprint with permission from [1]. ©2005 IEEE.

the average number of active MRC branches, denoted N_A, in the following summation form:

$$N_A = \sum_{l=1}^{L_c} \pi_l \, l, \qquad (5.39)$$

where π_l is the probability that exactly l MRC branches are active during data reception. For $l < L_c$, this is equal to the probability that an l-branch MRC is used during data reception. Based on the results of the previous subsection and noting that once combining operation enters the GSC stage, all L_c MRC branches need to be active, it can be shown that N_A is given by

$$N_A = 1 + \sum_{l=1}^{L_c-1} F_{\gamma_c}^{l-MRC}(\gamma_T). \qquad (5.40)$$

For the i.i.d. Rayleigh fading case, after substituting the Rayleigh case of $F_{\gamma_c}^{l-MRC}(\cdot)$ from Table 5.1 into (5.40), we arrive at the closed-form expression for N_A given by

$$N_A = 1 + \sum_{l=1}^{L_c-1} \left[1 - e^{-\frac{\gamma_T}{\overline{\gamma}}} \sum_{i=0}^{l-1} \frac{1}{i!} \left(\frac{\gamma_T}{\overline{\gamma}} \right)^i \right]. \qquad (5.41)$$

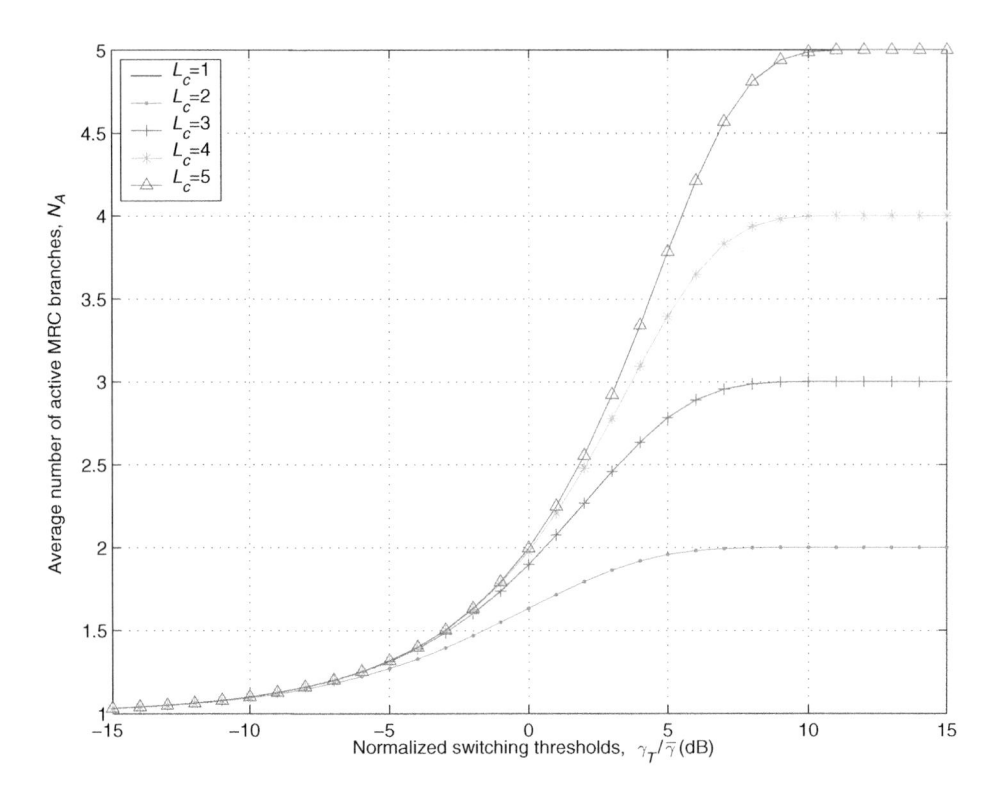

Figure 5.13 Average number of active MRC branches with OT-GSC as a function of the normalized SNR threshold, $\gamma_T/\overline{\gamma}$ ($L = 5$). Reprint with permission from [1]. ©2005 IEEE.

Fig. 5.13 show the average number of active MRC branches during data reception as a function of the normalized SNR threshold, $\gamma_T/\overline{\gamma}$. Note that the average number of active MRC branches N_A increases from 1 to L_c as the normalized SNR threshold increases, as expected by intuition.

5.3.3 Statistics of Output SNR

We now derive the statistics of the combined SNR with OT-GSC, which will facilitate its performance analysis over fading channels. We again focus on the i.i.d. Rayleigh fading environment. Note that the events that L_e, $L_e = 1, 2, \ldots, L$, diversity paths are estimated are mutually exclusive. Applying the total probability theorem, we can write the CDF of the combined SNR γ_c in a summation form as

$$F_{\gamma_c}(x) = \Pr[\gamma_c < x] = \sum_{k=1}^{L} \Pr[\gamma_c < x, L_e = k], \qquad (5.42)$$

where $\Pr[\gamma_c < x, L_e = k]$ is the joint probability of the events that $\gamma_c < x$ and the receiver has estimated exactly k paths. According to the mode of operation of OT-GSC, we note that when $L_e \leq L_c$, L_e paths are estimated, if and only if the MRC-combined

5.3 Output-Threshold GSC

SNR of the first $L_e - 1$ branches is less than the output threshold γ_T but the combined SNR of the first L_e branches is above γ_T. Therefore, $\Pr[\gamma_c < x, L_e = k]$, $k \le L_c$, can be shown to be given by

$$\Pr[\gamma_c < x, L_e = k] = \begin{cases} \Pr[\gamma_T \le \gamma_c = \gamma_1 < x], & k = 1; \\ \Pr[\gamma_T \le \gamma_c = \sum_{j=1}^{k} \gamma_j < x, \sum_{j=1}^{k-1} \gamma_j < \gamma_T], & 2 \le k \le L_c. \end{cases} \quad (5.43)$$

When $L_e > L_c$, an L_e/L_c-GSC scheme is used when the combined SNR of L_e/L_c-GSC is above the output threshold γ_T, whereas the output SNR of $(L_e - 1)/L_c$-GSC is less than γ_T. Finally, $L_e = L$ if and only if the combined SNR of $(L - 1)/L_c$-GSC is below the output threshold. Mathematically speaking, we have

$$\Pr[\gamma_c < x, L_e = k] =$$
$$\begin{cases} \Pr[\gamma_T \le \gamma_c = \sum_{j=1}^{L_c} \gamma_{j:k} < x, \sum_{j=1}^{L_c} \gamma_{j:k-1} < \gamma_T], & k \le L - 1; \\ \Pr[\gamma_c = \sum_{j=1}^{L_c} \gamma_{j:L} < x, \sum_{j=1}^{L_c} \gamma_{j:L-1} < \gamma_T], & k = L, \end{cases} \quad (5.44)$$

where $\gamma_{j:k}$ denotes the jth largest SNR among k estimated path SNRs. Substituting (5.43) and (5.44) into (5.42), we can rewrite the CDF of the combined SNR with OT-GSC as

$$F_{\gamma_c}(x) = \begin{cases} \underbrace{\Pr[\sum_{j=1}^{L_c} \gamma_{j:L} < \gamma_T] + \Pr[\gamma_T \le \gamma_1 < x] + \sum_{k=2}^{L_c} \Pr[\gamma_T \le \sum_{j=1}^{k} \gamma_j < x, \sum_{j=1}^{k-1} \gamma_j < \gamma_T]}_{F_1(x)} \\ + \underbrace{\sum_{k=L_c+1}^{L} \Pr[\gamma_T \le \sum_{j=1}^{L_c} \gamma_{j:k} < x, \sum_{j=1}^{L_c} \gamma_{j:k-1} < \gamma_T]}_{F_2(x)}, & x \ge \gamma_T \\ \Pr[\sum_{j=1}^{L_c} \gamma_{j:L} < x], & x < \gamma_T. \end{cases} \quad (5.45)$$

Note that the probability $\Pr[\sum_{j=1}^{L_c} \gamma_{j:L} < x]$ for i.i.d Rayleigh fading channels is given in (5.18). The term $F_1(x)$ can be rewritten in terms of the CDF of single-branch SNR $F_\gamma(\cdot)$ and the PDF of an l-branch MRC output SNR $f_{\gamma_c}^{l-MRC}(\cdot)$ as

$$F_1(x) = F_\gamma(x) - F_\gamma(\gamma_T) + \sum_{k=2}^{L_c} \int_0^{\gamma_T} f_{\gamma_c}^{k-MRC}(y)\bigl(F_\gamma(x-y) - F_\gamma(\gamma_T - y)\bigr) dy. \quad (5.46)$$

For Rayleigh fading, after proper substitution from Table 5.1, carrying out the integration and some manipulation, $F_1(x)$ can be shown to be given by

$$F_1(x) = \sum_{k=1}^{L_c} \frac{\gamma_T^{k-1}}{(k-1)! \overline{\gamma}^{k-1}} \bigl[e^{-\frac{\gamma_T}{\overline{\gamma}}} - e^{-\frac{x}{\overline{\gamma}}}\bigr]. \quad (5.47)$$

To derive $F_2(x)$, we first note that $\sum_{j=1}^{L_c} \gamma_{j:k-1}$ can be written as $\sum_{j=1}^{L_c} \gamma_{j:k-1} = \Gamma_{L_c-1/k-1} + \gamma_{L_c:k-1}$, where $\gamma_{L_c:k-1}$ is the L_cth largest SNR among $k-1$ estimated path SNRs and $\Gamma_{L_c-1/k-1} = \sum_{j=1}^{L_c-1} \gamma_{j:k-1}$, i.e., $\Gamma_{L_c-1/k-1}$ is the sum of the largest $L_c - 1$ branch SNRs. Based on the mode of operation of OT-GSC, the weakest path is replaced at the GSC stage if and only if the instantaneous SNR of the newly examined branch γ_k is larger than $\gamma_{L_c:k-1}$. Also note that in this case $\Gamma_{L_c-1/k-1}$ remains unchanged. We can write the summand of $F_2(x)$ as

$$\Pr[\gamma_T \leq \sum_{j=1}^{L_c} \gamma_{j:k} < x, \sum_{j=1}^{L_c} \gamma_{j:k-1} < \gamma_T] \tag{5.48}$$
$$= \Pr[\gamma_T \leq \Gamma_{L_c-1/k-1} + \gamma_k < x, \Gamma_{L_c-1/k-1} + \gamma_{L_c:k-1} < \gamma_T].$$

Because all branch SNRs are i.i.d. random variables, γ_k is independent of both $\Gamma_{L_c-1/k-1}$ and $\gamma_{L_c:k-1}$. As such, we can compute the probability in (5.48) with the joint PDF of $\Gamma_{L_c-1/k-1}$ and $\gamma_{L_c:k-1}$, denoted by $p_{\gamma_{L_c:k-1},\Gamma_{L_c-1/k-1}}(y,z)$, and the CDF of γ_k, $F_\gamma(\cdot)$. In particular, $F_2(x)$ can be written as

$$F_2(x) = \sum_{k=L_c+1}^{L} \int_0^{\frac{\gamma_T}{L_c}} \int_{(L_c-1)y}^{\gamma_T-y} p_{\gamma_{L_c:k-1},\Gamma_{L_c-1:k-1}}(y,z) \tag{5.49}$$
$$\times (F_\gamma(x-z) - F_\gamma(\gamma_T - z)) dz dy.$$

For i.i.d. Rayleigh fading channels, it can be shown that the joint PDF $p_{\gamma_{L_c:L},\Gamma_{L_c-1/L}}(y,z)$ was given in (5.25). After substituting (5.25) into (5.49), carrying out the integration and appropriate simplifications, we obtain the closed-form expression of $F_2(x)$ as

$$F_2(x) = [e^{-\frac{\gamma_T}{\bar{\gamma}}} - e^{-\frac{x}{\bar{\gamma}}}] \sum_{k=L_c+1}^{L} \sum_{j=0}^{k-1-L_c} A(k-1,j) \tag{5.50}$$
$$\times \left[B(L_c,j) + (-\frac{\bar{\gamma} L_c}{j+1})^{L_c} e^{-\frac{(j+1)\gamma_T}{\bar{\gamma} L_c}} \right],$$

where

$$A(k,j) = \frac{(-1)^j k!}{L_c(k-L_c-j)!(L_c-1)! j! \bar{\gamma}^{L_c}}, \tag{5.51}$$

and

$$B(L_c,j) = \sum_{i=0}^{L_c-1} \frac{(-1)^i \gamma_T^{L_c-1-i}}{(L_c-1-i)!(\frac{j+1}{\bar{\gamma} L_c})^{i+1}}. \tag{5.52}$$

Finally, after substituting (5.47), (5.50), and (5.18) into (5.45), we arrive at the closed-form expression of the CDF of OT-GSC output SNR, or equivalently, the outage probability of OT-GSC, over i.i.d. Rayleigh fading channels.

Starting from the CDF of γ_c with OT-GSC, we can routinely obtain its PDF and MGF. In particular, after differentiation and some manipulations, the PDF of γ_c with OT-GSC for i.i.d. Rayleigh fading channels is given by

$$p_{\gamma_c}(x) = \begin{cases} \dfrac{1}{\overline{\gamma}}e^{-\frac{x}{\overline{\gamma}}}\left\{\sum_{k=1}^{L_c}\dfrac{\gamma_T^{k-1}}{(k-1)!\overline{\gamma}^{k-1}} + \sum_{k=L_c+1}^{L}\sum_{j=0}^{k-1-L_c} A(k-1,j)\right. \\ \qquad \left.\times\left[B(L_c,j) + (-\dfrac{\overline{\gamma}L_c}{j+1})^{L_c}e^{-\frac{(j+1)\gamma_T}{\overline{\gamma}L_c}}\right]\right\}, & x \geq \gamma_T \\[1em] \binom{L}{L_c}\left[\dfrac{x^{L_c-1}e^{-x/\overline{\gamma}}}{\overline{\gamma}^{L_c}(L_c-1)!} + \dfrac{1}{\overline{\gamma}}\sum_{l=1}^{L-L_c}(-1)^{L_c+l-1}\binom{L-L_c}{l}\right. \\ \qquad \left.\times\left(\dfrac{L_c}{l}\right)^{L_c-1}e^{-x/\overline{\gamma}}\left(e^{-lx/L_c\overline{\gamma}} - \sum_{m=0}^{L_c-2}\dfrac{1}{m!}\left(\dfrac{-lx}{L_c\overline{\gamma}}\right)^m\right)\right], & x < \gamma_T \end{cases}$$
(5.53)

where $A(k,j)$ and $B(L_c,j)$ are defined in (5.51) and (5.52), respectively. It follows that the MGF of γ_c can be obtained as

$$M_{\gamma_c}(t) = \binom{L}{L_c}\left\{\dfrac{1 - \Gamma(L_c, (1-t\overline{\gamma})\gamma_T/\overline{\gamma})}{(1-t\overline{\gamma})^{L_c}} + \sum_{l=1}^{L-L_c}(-1)^{L_c+l-1}\binom{L-L_c}{l}\right.$$
$$\left.\times\left(\dfrac{L_c}{l}\right)^{L_c-1}\left[I(l,t) - \sum_{m=0}^{L_c-2}\left(\dfrac{-l}{L_c}\right)^m \times \dfrac{(1-\Gamma(m+1,(1-t\overline{\gamma})\gamma_T/\overline{\gamma}))}{(1-t\overline{\gamma})^{m+1}}\right]\right\}$$
$$+ \dfrac{e^{-\frac{(1-\overline{\gamma}t)\gamma_T}{\overline{\gamma}}}}{1-\overline{\gamma}t}\left\{\sum_{k=1}^{L_c}\dfrac{\gamma_T^{k-1}}{(k-1)!\overline{\gamma}^{k-1}} + \sum_{k=L_c+1}^{L}\sum_{j=0}^{k-1-L_c} A(k-1,j)\right.$$
$$\left.\times\left[B(L_c,j) + (-\dfrac{\overline{\gamma}L_c}{j+1})^{L_c}e^{-\frac{(j+1)\gamma_T}{\overline{\gamma}L_c}}\right]\right\},$$
(5.54)

where

$$I(l,t) = \dfrac{1}{\overline{\gamma}t - \frac{L_c+l}{L_c}}\left[\exp\left(\dfrac{\gamma_T(tL_c\overline{\gamma} - L_c - l)}{L_c\overline{\gamma}}\right) - 1\right].$$

These closed-form results can be readily applied to the performance analysis of OT-GSC over i.i.d. Rayleigh fading channels.

5.3.4 Numerical Results

Fig. 5.14 compares the average error rate of BPSK with OT-GSC, MS-GSC, and GSC as a function of average SNR per branch with the output threshold γ_T for both OT-GSC and MS-GSC set to be 8 dB. As a verification of the analytic results, simulation results for the average error rate of OT-GSC are also presented. Note that these simulation results match the analytic results perfectly. We can see that, when the channel condition is poor, i.e., $\overline{\gamma}$ is small compared to γ_T, these three schemes have the same error performance. This behavior can be explained as follows. When the average branch SNR is small compared to the output threshold, the receiver with OT-GSC and MS-GSC will usually estimate all available branches and combine the best L_c ones because applying MRC/GSC to a subset of available branches will not be able to raise the combined SNR

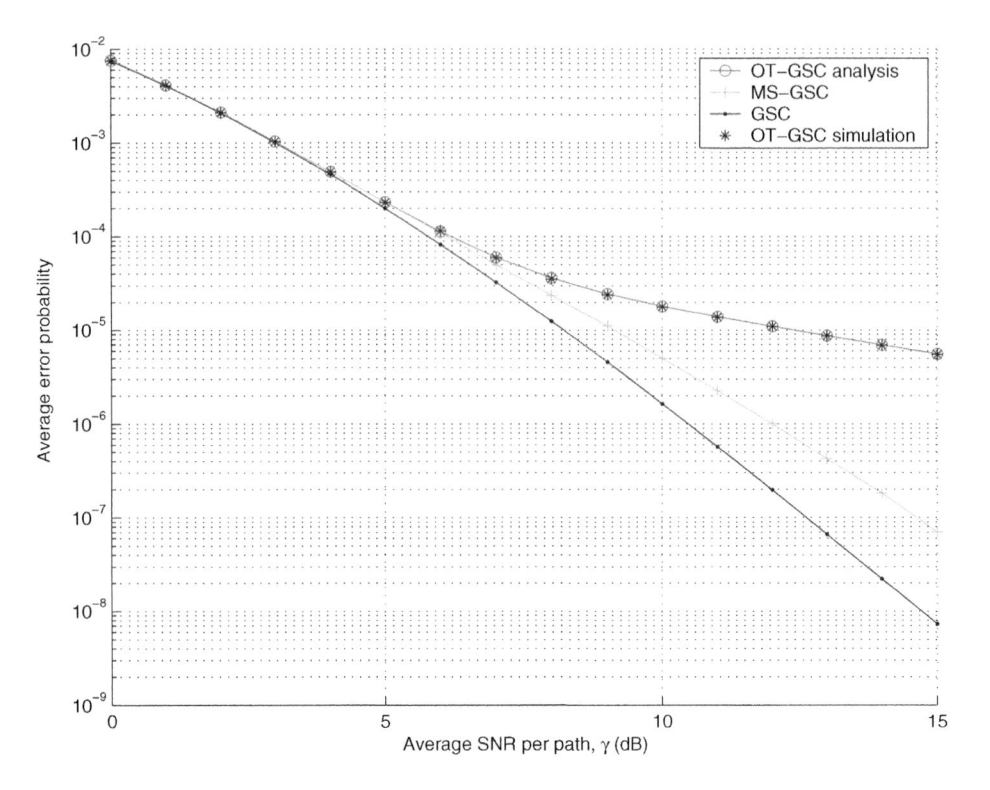

Figure 5.14 Average error rate of BPSK with OT-GSC, MS-GSC, and GSC over $L = 5$ i.i.d. Rayleigh fading channels as a function of the average SNR per branch ($L_c = 3$ and $\gamma_T = 8$ dB). Reprint with permission from [8]. ©2007 IEEE.

above γ_T. Therefore, both MS-GSC and OT-GSC are actually working in the same way as a conventional GSC. As $\overline{\gamma}$ becomes larger, that is, the channel condition is improving, OT-GSC has higher error probability than both GSC and MS-GSC. This can be explained by noting that in this case, it is more likely that the instantaneous SNR of the first branch is greater than γ_T and, as such, OT-GSC reduces to a single-branch receiver while MS-GSC tends to work as a selection combiner.

In Fig. 5.15, we plot the average error rate of BPSK with OT-GSC, MS-GSC, and GSC schemes as a function of the normalized output threshold $\gamma_T/\overline{\gamma}$. As we can see, the error rate of OT-GSC decreases from that of a single-branch MRC to that of a conventional 6/3-GSC when the normalized output threshold increases from -12 dB to 12 dB, whereas that of MS-GSC degrades from a six-branch SC. To better illustrate the new trade-off of complexity versus performance offered by OT-GSC, we present in Fig. 5.16 the average number of path estimations and combined paths with OT-GSC, MS-GSC, and GSC schemes based on the analytical results given in previous sections. It can be seen that while both GSC and MS-GSC always require L path estimations, OT-GSC requires much less path estimations, especially when the normalized SNR is small. Considering Figs. 5.15 and 5.16 together, we can observe that the OT-GSC scheme can offer the same performance as GSC with fewer combined paths and path estimations.

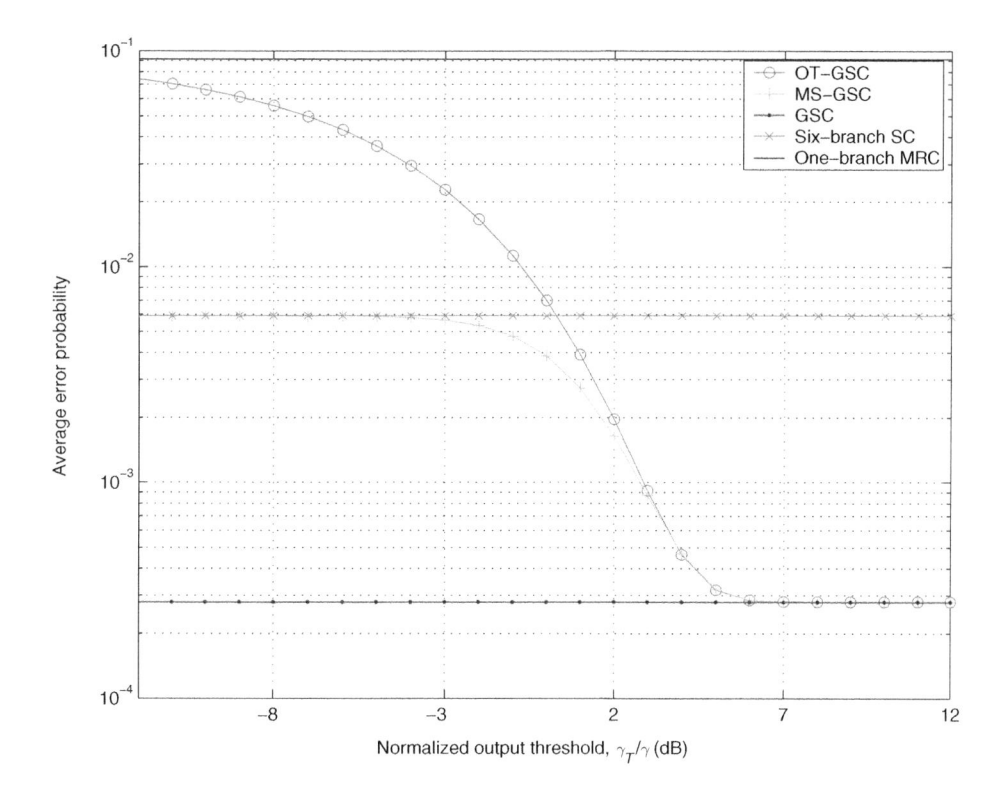

Figure 5.15 Average error rate of BPSK with OT-GSC, MS-GSC, and GSC over $L = 6$ i.i.d. Rayleigh fading channels as function of the normalized output-threshold $\gamma_T/\overline{\gamma}(L_c = 3$ and $\overline{\gamma} = 3$dB). Reprint with permission from [8]. ©2007 IEEE.

In comparison with MS-GSC, OT-GSC still saves a certain number of path estimations while combining more paths to achieve the same error performance.

5.4 Adaptive Transmit Diversity

The idea of adaptive diversity combining can also apply to transmit diversity systems. The goal of transmit diversity is to explore the potential diversity gain of multiple transmit antennas. Depending on whether the channel state information is required at the transmitter or not, transmit diversity schemes can be classified into closed-loop and open-loop schemes. Most classical diversity combining schemes can apply to the closed-loop setting when the required channel knowledge is available at transmitter. For example, transmit antenna selection uses the antenna leading to the highest received SNR for transmission. Meanwhile, the transmitter needs to know which antenna leads to the highest received SNR. The most popular open-loop schemes are those based on orthogonal space-time block coding (OSTBC)[9–11]. This class of transmit diversity schemes can achieve full diversity gain with no channel information at transmitter and simple linear processing at receiver, and as such, is of great practical interest.

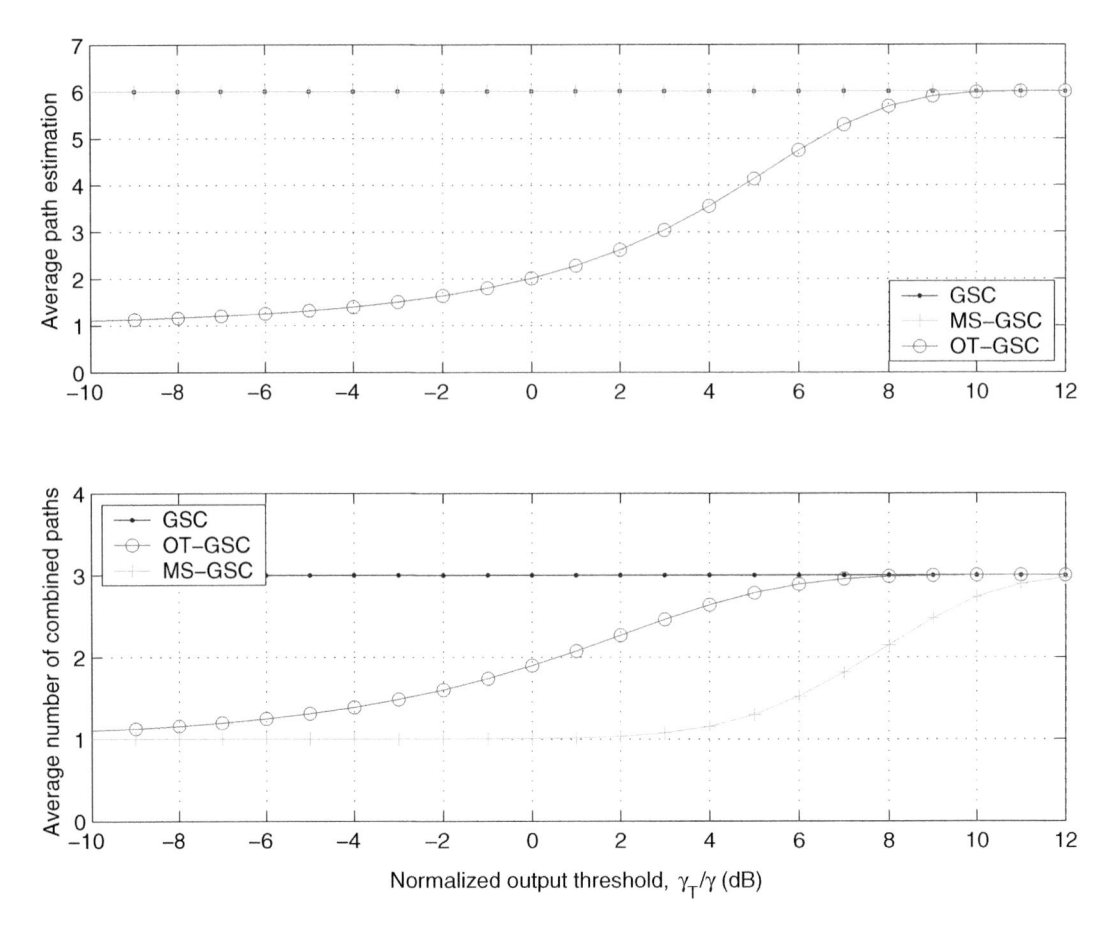

Figure 5.16 Average number of path estimations and combined paths with OT-GSC, MS-GSC, and GSC over $L = 6$ i.i.d. Rayleigh fading channels as a function of normalized output threshold $\gamma_T/\overline{\gamma}$ ($L_c = 3$ and $\overline{\gamma} = 3$ dB). Reprint with permission from [8]. ©2007 IEEE.

On the other hand, OSTBC suffers rate loss when the number of transmit antennas is greater than two. Moreover, the power spreading over multiple transmit antennas causes a certain SNR loss. A viable solution to these issues of OSTBC transmission is antenna subset selection at the transmitter side [12–15]. The most popular antenna subset selection scheme follows the same principle of GSC and selects multiple transmit antennas that lead to the highest receive SNR. While requiring a certain amount of channel state information, transmit antenna subset selection schemes incur much less rate/power loss than conventional OSTBC-based transmit diversity systems. Nevertheless, GSC-type antenna subset selection requires the estimation of wireless channels corresponding to every transmit antenna and the receiver needs to feed back the index of all selected antennas.

In this section, we apply the idea of adaptive combining to transmit diversity systems and develop a new class of closed-loop transmit diversity system [16–18]. Specifically, a subset of the available transmit antennas are used for data transmission with OSTBC. The transmitter will replace the antenna, leading to the lowest path SNRs with unused ones if the output SNR of the space-time decoder is below a certain threshold. The resulting OSTBC transmission with the transmit antenna replacement scheme offers performance comparable to the GSC-based transmit diversity systems much reduced feedback load.

5.4.1 Mode of Operation

The system model of transmit diversity with antenna replacement is illustrated in Fig. 5.17. The transmitter is equipped with a total of $N + L$ antennas, where N antennas are used for orthogonal space-time transmission and L antennas are used for replacement. The receiver compares the output SNR of the space-time decoder with a fixed SNR threshold, denoted by γ_T. If the output SNR $\sum_{i=1}^{N} \gamma_i$ is larger than or equal to γ_T, no antenna replacement is necessary and the receiver informs the transmitter to keep using the current antennas. On the other hand, if $\sum_{i=1}^{N} \gamma_i$ is smaller than the threshold γ_T, the receiver asks the transmitter to replace the L antennas leading to the weakest received signal with the L remaining antennas. Therefore, the receiver needs to feed back either one bit of information to indicate no need for antenna replacement or $1 + \log_2(\frac{N!}{(N-L)!L!})$ bits of information to inform the transmitter of the index of the transmitting antennas to be replaced. We can show that the average feedback load of this scheme is

$$N_F = 1 + \log_2\left(\frac{N!}{(N-L)!L!}\right) \cdot \Pr\left[\sum_{i=1}^{N} \gamma_i < \gamma_T\right]. \quad (5.55)$$

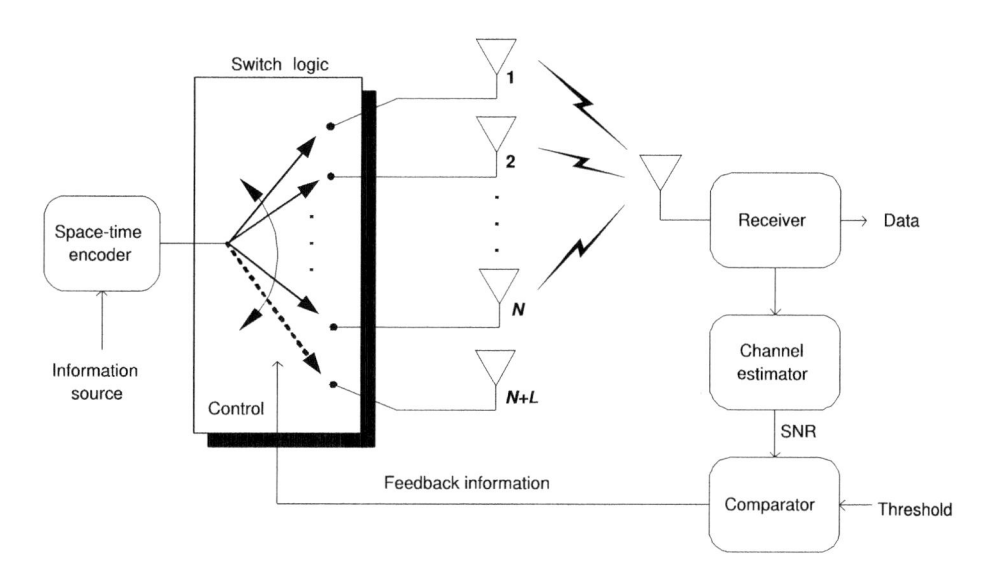

Figure 5.17 System model of transmit diversity systems. Reprint with permission from [17]. ©2008 IEEE.

Adaptive Diversity Combining

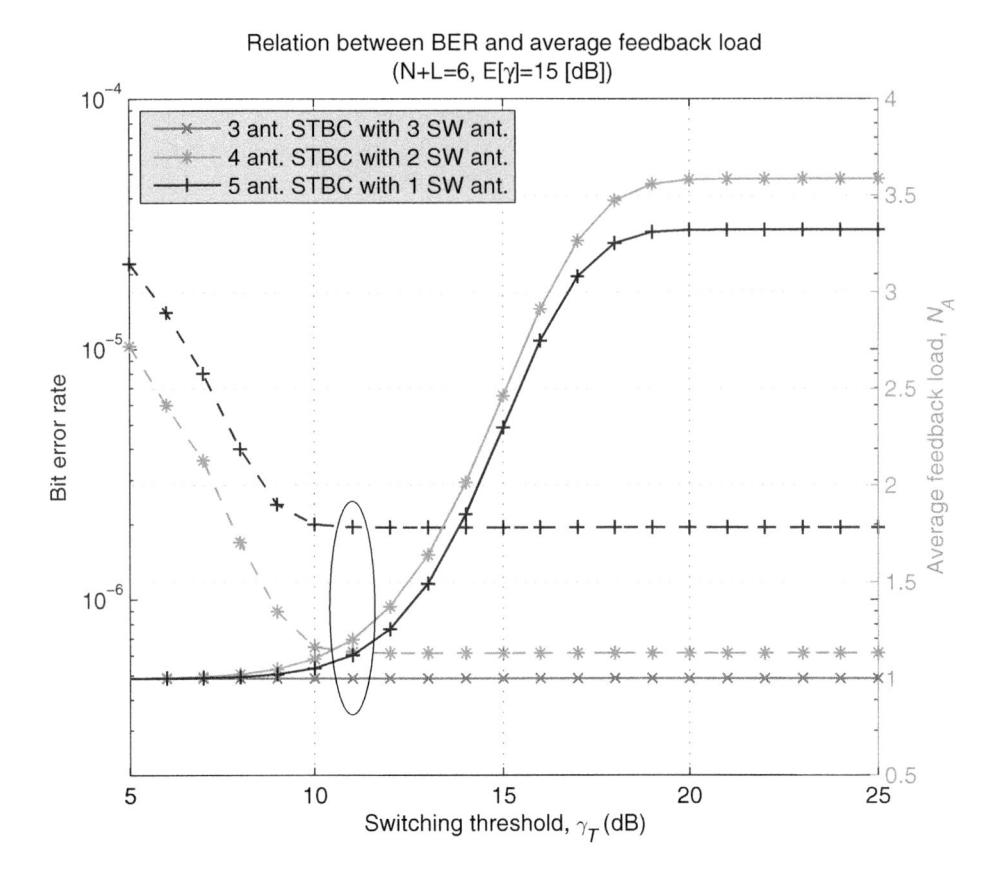

Relation between BER and average feedback load
(N+L=6, E[γ]=15 [dB])

Figure 5.18 Comparison of average feedback load of the proposed systems with $N + L = 6$ transmit antennas over i.i.d. Rayleigh fading channels. Reprint with permission from [18]. ©2008 IEICE.

For the i.i.d. Rayleigh fading scenario, it can be shown that the average feedback load specializes to

$$N_F = 1 + \log_2\left(\frac{N!}{(N-L)!L!}\right)\left(1 - e^{-\frac{\gamma_T}{\bar{\gamma}}}\sum_{i=0}^{N-1}\frac{1}{i!}\left(\frac{\gamma_T}{\bar{\gamma}}\right)^i\right). \qquad (5.56)$$

We can see that the antenna replacement scheme has the maximum feedback load when the number of switching antennas $L \approx \frac{N}{2}$. In Fig. 5.18, we plot the average feedback load of the proposed system as a function of γ_T for different values of L and fixed average SNR, $\bar{\gamma} = 15$ dB, while keeping the total number of transmit antennas $(N + L)$ fixed to 6. As we can see, for $L < N$ cases, the feedback load increases from 1 bit to $\log_2(\frac{N!}{(N-L)!L!}) + 1$ bits when the threshold γ_T increases, as expected. When $N = L = 3$, the systems need only one bit of feedback information to indicate whether all transmit antennas shall be replaced or not. We note that different threshold values will lead to different trade-offs of error performance and feedback load. To illustrate this, we also plot the average error rate of the transmit diversity system on the same figure.

As we can see, the error performance improves and eventually saturates to a constant value as the threshold γ_T increases. Based on this figure, we can see if we select the threshold properly, the antenna replacement scheme can achieve its best possible error performance with very low feedback load.

5.4.2 Statistics of Received SNR

We now derive the statistics of the received SNR of the OSTBC transmission with transmit antenna replacement scheme. We first consider the case of $L = N$. In this case, all transmit antennas will be replaced when the decoder output SNR $\sum_{i=1}^{N} \gamma_i$ is smaller than the threshold γ_T. The antenna replacement scheme can be viewed as a generalization of traditional SSC scheme, while the output SNR of each branch is equal to $\sum_{j=1}^{N} \gamma_j$. Therefore, we can write the CDF of the combined SNR as

$$F_{\gamma_c}(x) = \begin{cases} \Pr[\gamma_T \leq \sum_{j=1}^{N} \gamma_j < x] + \Pr\left[\sum_{j=1}^{N} \gamma_j < \gamma_T \ \& \ \sum_{j=1}^{N} \gamma_j < x\right], & x \geq \gamma_T, \\ \Pr\left[\sum_{j=1}^{N} \gamma_j < \gamma_T \ \& \ \sum_{j=1}^{N} \gamma_j < x\right], & 0 \leq x < \gamma_T. \end{cases} \quad (5.57)$$

Following the similar analytical approach for SSC, we can then obtain the closed-form expression of the PDF of γ_c under the i.i.d. Rayleigh fading assumption as

$$p_{\gamma_c}(x) = \begin{cases} \left(2 - e^{-\frac{\gamma_T}{\bar{\gamma}}} \sum_{k=0}^{L-1} \frac{1}{k!}\left(\frac{\gamma_T}{\bar{\gamma}}\right)^k\right) \frac{x^{L-1}}{(L-1)!\bar{\gamma}^L} e^{-\frac{x}{\bar{\gamma}}}, & x \geq \gamma_T, \\ \left(1 - e^{-\frac{\gamma_T}{\bar{\gamma}}} \sum_{k=0}^{L-1} \frac{1}{k!}\left(\frac{\gamma_T}{\bar{\gamma}}\right)^k\right) \frac{x^{L-1}}{(L-1)!\bar{\gamma}^L} e^{-\frac{x}{\bar{\gamma}}}, & 0 \leq x < \gamma_T. \end{cases} \quad (5.58)$$

We now consider the case that there are fewer than N antennas for replacement, i.e., $1 \leq L < N$. Let $\gamma_{i:N}$ denote the ith largest SNR out of N ones, $\gamma_{1:N} > \gamma_{2:N} > \ldots > \gamma_{N:N}$. According to the mode of operation of the transmit antenna replacement scheme, the decoder output SNR γ_c is given by

$$\gamma_c = \begin{cases} \sum_{i=1}^{N} \gamma_i, & \sum_{i=1}^{N} \gamma_i \geq \gamma_T, \\ \sum_{i=1}^{N-L} \gamma_{i:N} + \sum_{j=1}^{L} \gamma_j, & \sum_{i=1}^{N} \gamma_i < \gamma_T. \end{cases} \quad (5.59)$$

Let Γ_j denote the sum of the largest j SNRs among N ones, i.e., $\Gamma_j = \sum_{i=1}^{j} \gamma_{i:N}$. We can write the CDF of γ_c as

$$F_{\gamma_c}(x) = \begin{cases} \Pr[\gamma_T \le \Gamma_N < x] \\ \quad + \Pr\left[\Gamma_N < \gamma_T \,\&\, \Gamma_{N-L} + \sum_{j=1}^{L} \gamma_j < x\right], & x \ge \gamma_T, \\ \Pr\left[\Gamma_N < \gamma_T \,\&\, \Gamma_{N-L} + \sum_{j=1}^{L} \gamma_j < x\right], & 0 \le x < \gamma_T. \end{cases} \qquad (5.60)$$

Applying the Bayesian rule and the i.i.d. fading path assumption, we can rewrite the joint probability in (5.60) as

$$\Pr\left[\Gamma_N < \gamma_T \,\&\, \Gamma_{N-L} + \sum_{j=1}^{L} \gamma_j < x\right]$$

$$= \Pr\left[\Gamma_{N-L} + \sum_{j=1}^{L} \gamma_j < x \Big| \Gamma_N < \gamma_T\right] \Pr[\Gamma_N < \gamma_T]$$

$$= \Pr[\gamma_s < x] \Pr[\Gamma_N < \gamma_T], \qquad (5.61)$$

where we define a new random variable $\gamma_s \equiv \sum_{j=1}^{L} \gamma_j + (\Gamma_{N-L}|\Gamma_N < \gamma_T)$. The PDF of the combined SNR can be obtained by taking the derivative of (5.60) as

$$p_{\gamma_c}(x) = \begin{cases} p_{\Gamma_N}(x) + \Pr[\Gamma_N < \gamma_T] p_{\gamma_s}(x), & x \ge \gamma_T, \\ \Pr[\Gamma_N < \gamma_T] p_{\gamma_s}(x), & 0 \le x < \gamma_T, \end{cases} \qquad (5.62)$$

where $p_{\gamma_s}(x)$ denotes the PDF of γ_s. Since γ_s is the sum of the two random variables $\sum_{j=1}^{L} \gamma_j$ and $(\Gamma_{N-L}|\Gamma_N < \gamma_T)$, we can calculate $p_{\gamma_s}(x)$ using convolution operation as

$$p_{\gamma_s}(x) = \int_0^x p_{\sum_{j=1}^{L} \gamma_j}(x-z) p_{\Gamma_{N-L}|\Gamma_N < \gamma_T}(z) dz. \qquad (5.63)$$

Noting that the PDF $p_{\sum_{j=1}^{L} \gamma_j}(\cdot)$ is readily available in closed form, we focus in the following on the derivation of the conditional PDF $p_{\Gamma_{N-L}|\Gamma_N < \gamma_T}(x)$.

The CDF of $\Gamma_{N-L}|\Gamma_N < \gamma_T$ can be written as

$$F_{\Gamma_{N-L}|\Gamma_N < \gamma_T}(x) = \frac{1}{\Pr[\Gamma_N < \gamma_T]} \Pr[\Gamma_{N-L} < x, \Gamma_N < \gamma_T], \qquad (5.64)$$

where the joint probability can be calculated as

$$\Pr[\Gamma_{N-L} < x, \Gamma_N < \gamma_T] = \Pr[\Gamma_{N-L} < x, \Gamma_{N-L} + \gamma_{N-L+1:N} + z_l < \gamma_T] \qquad (5.65)$$

$$= \int_0^x \int_0^{\gamma_T - y} \int_0^{\gamma_T - y - \gamma} p_{\Gamma_{N-L}, \gamma_{N-L-1:N}, z_l}(y, \gamma, z) dz d\gamma dy.$$

In (5.65), z_l denotes the sum of the $L-1$ smallest SNRs among the total N ones, i.e., $z_l = \sum_{i=N-L+2}^{N} \gamma_{i:N}$ and $p_{\Gamma_{N-L}, \gamma_{N-L-1:N}, z_l}(y, \gamma, z)$ denotes the joint PDF of Γ_{N-L}, $\gamma_{N-L-1:N}$ and z_l, which is investigated in the Appendix. For the i.i.d. Rayleigh fading environment, the three-dimensional PDF is available in closed form as

$$p_{\Gamma_{N-L}, \gamma_{N-L-1:N}, z_l}(y, \gamma, z) = \frac{N!}{(L+1)!(N-L-2)!\bar{\gamma}^N} \frac{[y - (N-L-2)\gamma]^{N-L-3}}{(N-L-3)!(L-2)!}$$

$$\times e^{-\frac{y+\gamma+z}{\bar{\gamma}}} \mathcal{U}(y - (N-L-2)\gamma)$$

$$\times \sum_{i=0}^{L+1} \binom{L+1}{i} (-1)^i (z-i\gamma)^{L+2} \mathcal{U}(z-i\gamma),$$
$$\gamma > 0, y > (N-L-2)\gamma, z < (L+1)\gamma. \quad (5.66)$$

Consequently, the conditional PDF $p_{\Gamma_{N-L}|\Gamma_N < \gamma_T}(x)$ can be written as

$$p_{\Gamma_{N-L}|\Gamma_N < \gamma_T}(x) = \frac{d}{dx} \frac{\Pr[\Gamma_{N-L} < x, \Gamma_N < \gamma_T]}{\Pr[\Gamma_N < \gamma_T]}$$

$$= \frac{1}{\Pr[\Gamma_N < \gamma_T]} \int_0^{\min[\gamma_T - x, \frac{x}{N-L}]} \int_0^{\min[\gamma_T - x - \gamma, (L-1)\gamma]}$$
$$\times p_{\Gamma_{N-L}, \gamma_{N-L-1:N}, z_l}(y, \gamma, z) dz d\gamma, \ 0 \le x < \gamma_T. \quad (5.67)$$

Finally, after substituting (5.67) into (5.63) and then into (5.62), we arrive at the finite-integral form of the PDF of the combined SNR, γ_c, given by

$$p_{\gamma_c}(x) =$$
$$\begin{cases} p_{\Gamma_N}(x) + \int_0^{\gamma_T} p_{\sum_{l=N+1}^{N+L} \gamma_l}(x-y) \int_0^{\min[\gamma_T - y, \frac{y}{N-L}]} \int_0^{\min[\gamma_T - y - \gamma, (L-1)\gamma]} \\ \quad p_{\Gamma_{N-L}, \gamma_{N-L-1:N}, z_l}(y, \gamma, z) dz d\gamma dy, & x \ge \gamma_T, \\ \int_0^x p_{\sum_{l=N+1}^{N+L} \gamma_l}(x-y) \int_0^{\min[\gamma_T - y, \frac{y}{N-L}]} \int_0^{\min[\gamma_T - y - \gamma, (L-1)\gamma]} \\ \quad \times p_{\Gamma_{N-L}, \gamma_{N-L-1:N}, z_l}(y, \gamma, z) dz d\gamma dy, & 0 \le x < \gamma_T. \end{cases}$$
$$(5.68)$$

Using the obtained PDF of the received SNR, we can study the outage and average BER performance of the antenna replacement scheme over i.i.d. Rayleigh fading.

5.4.3 Numerical Results

Fig. 5.19 plots the outage probability of the OSTBC transmission with the transmit antenna replacement system. We vary the number of replacement antennas L while keeping four transmitting antennas and threshold γ_T fixed to 2 dB. Note that the $L = 0$ case corresponds to traditional open-loop transmit diversity with OSTBC. We note that when $L < N$, the outage performance of the proposed systems improves with increasing number of replacement antennas L, especially when the outage threshold is smaller than the switching threshold. However, if $L = N$, the outage performance actually degrades considerably for smaller values of the switching threshold.

Fig. 5.20 shows the average BER of the proposed transmit diversity system with fixed switching threshold $\gamma_T = 16$ dB. Again, we vary the number of replacement antennas from 0 to $N = 4$. It is interesting to see from this figure that as the number of antennas used for replacement increases, the error performance of the proposed scheme improves. When $L = N$, however, the error performance also degrades significantly, while still slightly better than no replacement antenna case.

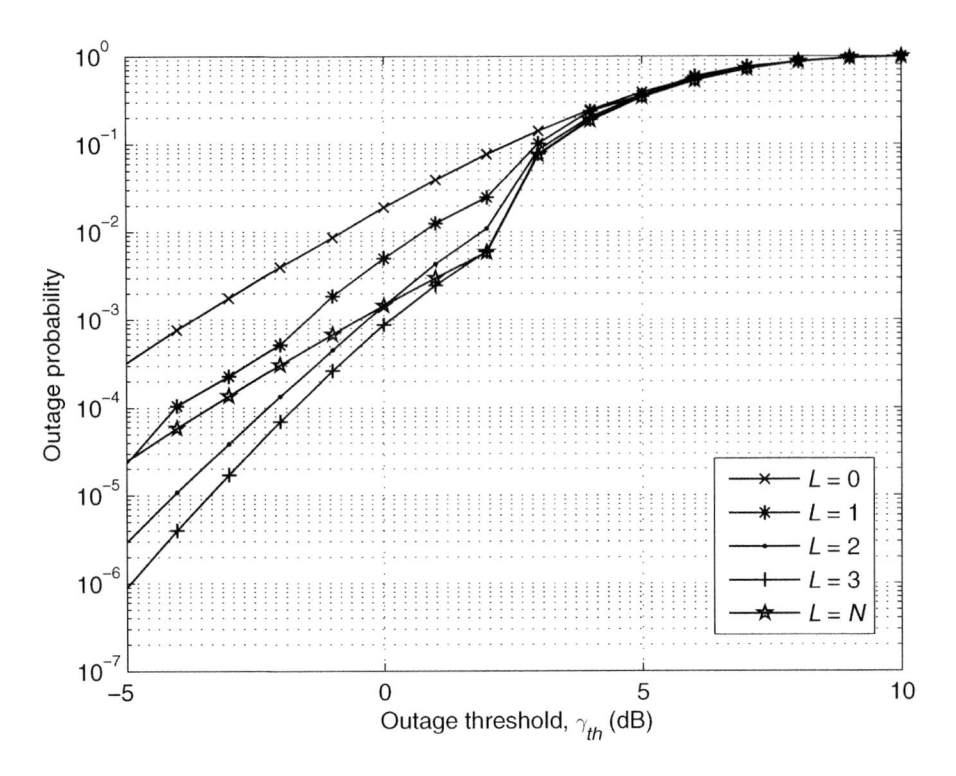

Figure 5.19 Outage probability of OSTBC transmission with antenna replacement and N-branch MRC ($N = 4$ and $\gamma_T = 2$). Reprint with permission from [18]. ©2008 IEICE.

In Fig. 5.21, we investigate the average BER performance of different transmit diversity configurations while fixing the total number of antennas at the transmitter $N + L$ to 6. The total transmitting power is also fixed irrespective of the number of antennas used for orthogonal STBC transmission. As we can see, as L increases and as long as $L < N$, the proposed transmit diversity system offers improved performance over the case when all six antennas are used for OSTBC transmission. In particular, the configuration with four transmitting antennas and two switching antennas offer about 1 dB gain over the six-antenna orthogonal STBC system and more than half a dB gain over the configuration with five transmitting antennas and one replacement antenna.

5.5 RAKE Finger Management during Soft Handover

Another application of adaptive combining is finger management for RAKE receiver operating in the handover region. RAKE receivers are commonly used in wideband wireless systems, such as WCDMA and UWB systems, to mitigate the effect of fading. RAKE receivers also facilitate the soft handover (SHO) in the handover region. Specifically, the mobile receiver can combine resolvable paths from both the serving base station (BS) and the target BS, which can greatly reduce the probability of

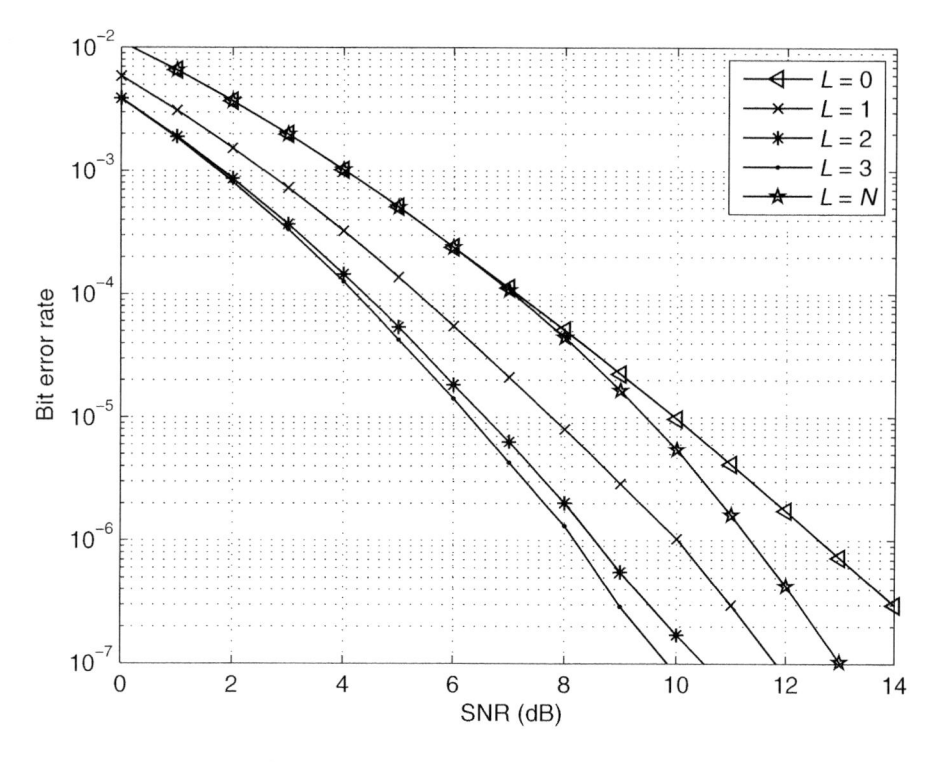

Figure 5.20 Average BER of BPSK of OSTBC transmission with antenna replacement and N-branch MRC over i.i.d. Rayleigh fading channels ($\gamma_T = 16$ dB). Reprint with permission from [18]. ©2008 IEICE.

dropped connections when the mobile receiver is near the cell boundary [19, Section 9.5.1]. On the other hand, serving the mobile user with more than one BS will incur a significant amount of system overhead, usually termed as SHO overhead. Note that user data signals need to be sent to and from all involved BSs to enjoy the diversity benefit.

In this section, we apply the concept of adaptive combining to develop several low-complexity RAKE finger management schemes and investigate their performance and complexity [20, 21]. The main idea is to minimize the SHO overhead by letting the receiver scan the additional resolvable paths from the target BS only if the received signal from the serving BS is of unsatisfactory quality. We will show that the resulting scheme can reduce the unnecessary path estimations and the SHO overhead compared to the conventional GSC-based finger management scheme [22, 23], which always uses the best resolved paths from both BSs.

5.5.1 Mode of Operation

We consider a mobile unit that is equipped with an L_c-finger RAKE receiver and is capable of despreading signals from different BSs with different fingers, and thus facilitating

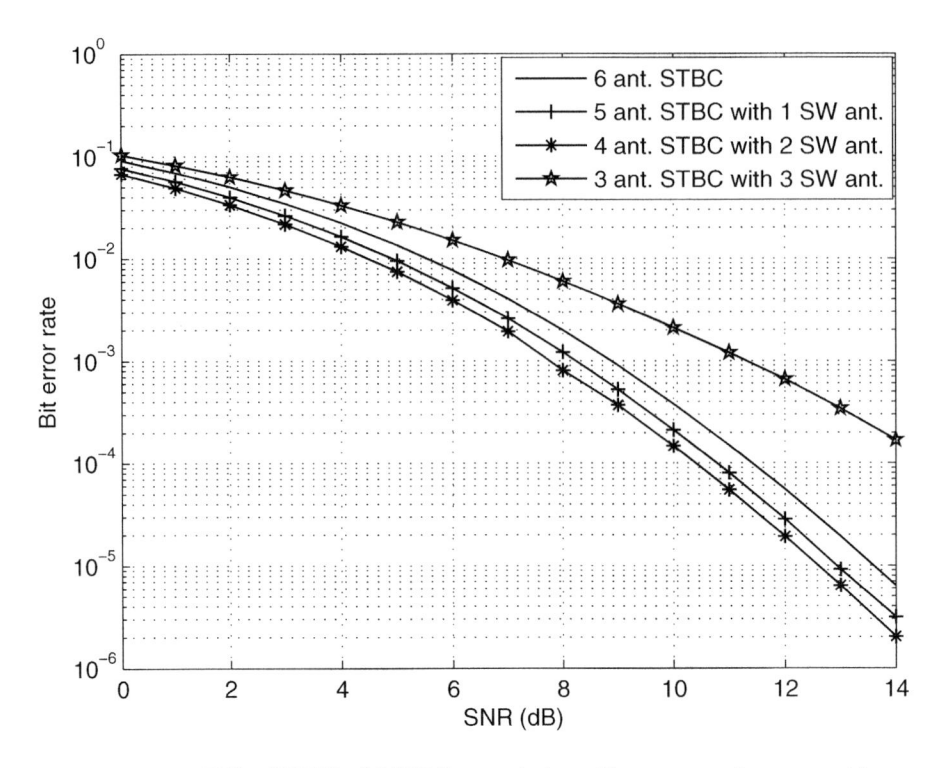

Figure 5.21 Average BER of BPSK of OSTBC transmission with antenna replacement with $N + L = 6$ transmit antennas over i.i.d. Rayleigh fading channels. Reprint with permission from [18]. ©2008 IEICE.

the SHO process. Without loss of generality, we assume that there are L resolvable paths from the serving BS and L_a paths from the target BS. We focus on the receiver operation when the mobile unit is moving from the coverage area of its serving BS to that of a target BS. As the mobile unit enters the SHO region, the RAKE receiver relies at first on the L resolvable paths from the serving BS and as such starts with L_c/L-GSC. If we let $\Gamma_{i:j}$ be the sum of the i largest SNRs among j ones, i.e., $\Gamma_{i:j} = \sum_{k=1}^{i} \gamma_{k:j}$ where $\gamma_{k:j}$ is the kth largest SNRs among j ones, then the total received SNR with GSC is given by $\Gamma_{L_c:L}$. At the beginning of every time slot, the receiver compares the received SNR, $\Gamma_{L_c:L}$, with a certain target SNR, denoted by γ_T. If $\Gamma_{L_c:L}$ is greater than or equal to γ_T, the mobile unit continues to rely solely on its current serving BS and SHO is not initiated. On the other hand, whenever $\Gamma_{L_c:L}$ falls below γ_T, the mobile unit will initiate SHO by estimating and combining resolvable paths from the target BS using the available figures.

We consider two finger update strategies in this work, full GSC [20] or block change [21]. With full GSC, the RAKE receiver reassigns its L_c fingers to the L_c strongest paths among the total $L + L_a$ available resolvable paths from both the serving and the target BSs (i.e., the RAKE receiver uses $(L + L_a)/L_c$-GSC). Now the total received SNR is

5.5 RAKE Finger Management during Soft Handover

given by $\Gamma_{L_c:L+L_a}$. Based on the above mode of operation, we can see that the final combined SNR, denoted by γ_c, with full GSC scheme, is mathematically given by

$$\gamma_c = \begin{cases} \Gamma_{L_c:L+L_a}, & \Gamma_{L_c:L} < \gamma_T; \\ \Gamma_{L_c:L}, & \Gamma_{L_c:L} \geq \gamma_T. \end{cases} \quad (5.69)$$

With the block change scheme, the RAKE receiver compares the sums of two groups of path SNRs: the sum of the L_s smallest paths among the L_c currently used paths from the serving BS (i.e., $\sum_{i=L_c-L_s+1}^{L_c} \gamma_{i:L}$); and the sum of the L_s strongest paths from the target BS (i.e., $\sum_{i=1}^{L_s} \gamma_{i:L_a}$). If $\sum_{i=L_c-L_s+1}^{L_c} \gamma_{i:L} < \sum_{i=1}^{L_s} \gamma_{i:L_a}$, then the receiver starts using the best paths from the target BS by replacing the L_s weakest paths from the serving BS. Otherwise, no path change occurs. Let

$$Y = \sum_{i=1}^{L_c-L_s} \gamma_{i:L}, \quad Z = \sum_{i=L_c-L_s+1}^{L_c} \gamma_{i:L},$$

$$\text{and} \quad W = \sum_{i=1}^{L_s} \gamma_{i:L_a}.$$

According to the above mode of operation, the final combined SNR, γ_c, with block change is given by

$$\gamma_c = \begin{cases} Y + \max\{Z, W\}, & Y + Z < \gamma_T; \\ Y + Z, & Y + Z \geq \gamma_T. \end{cases} \quad (5.70)$$

Note that the block change scheme only needs to compare the sum of two groups of path SNRs, and as such avoids reordering all the paths, which is required for the full GSC scheme. Therefore, the block change scheme leads to a reduction of both SNR comparison and SHO overhead. In the following, we analyze the performance of the block change scheme and compare its performance with the full GSC scheme through numerical examples.

5.5.2 Statistics of Output SNR

Based on the mode of operation of the block change scheme summarized in (5.70), the CDF of the combined SNR γ_c, $F_{\gamma_c}(x)$, can be written as

$$F_{\gamma_c}(x) = \begin{cases} \Pr[Y + \max\{Z, W\} < x], & 0 \leq x < \gamma_T; \\ \Pr[\gamma_T \leq Y + Z < x] \\ \quad + \Pr[Y + \max\{Z, W\} < x, \\ \quad Y + Z < \gamma_T], & x \geq \gamma_T \end{cases}$$

$$
= \begin{cases} \Pr[Z \geq W, Y + Z < x] \\ \quad + \Pr[Z < W, Y + W < x], & 0 \leq x < \gamma_T; \\ \Pr[\gamma_T \leq Y + Z < x] \\ \quad + \Pr[Z \geq W, Y + Z < \gamma_T] \\ \quad + \Pr[Z < W, Y + W < x, \\ \qquad Y + Z < \gamma_T], & x \geq \gamma_T. \end{cases} \quad (5.71)
$$

In (5.71), $\Pr[\gamma_T \leq Y + Z < x]$ can be easily calculated using the CDF of the combined SNR of L/L_c-GSC. Noting that W is independent of Y and Z, we can calculate the joint probabilities in (5.71) as

$$\Pr[Z \geq W, Y + Z < x] \tag{5.72}$$
$$= \int_0^x \int_0^{x-y} \int_w^{x-y} p_{Y,Z}(y,z) p_W(w) dz\, dw\, dy,$$
$$\Pr[Z < W, Y + W < x] \tag{5.73}$$
$$= \int_0^x \int_0^{x-y} \int_0^w p_{Y,Z}(y,z) p_W(w) dz\, dw\, dy,$$
$$\Pr[Z \geq W, Y + Z < \gamma_T] \tag{5.74}$$
$$= \int_0^{\gamma_T} \int_0^{\gamma_T - y} \int_w^{\gamma_T - y} p_{Y,Z}(y,z) p_W(w) dz\, dw\, dy,$$

and

$$\Pr[Z < W, Y + W < x, Y + Z < \gamma_T] \tag{5.75}$$
$$= \int_0^{\gamma_T} \int_0^{x-y} \int_0^{\min\{w, \gamma_T - y\}} p_{Y,Z}(y,z) p_W(w) dz\, dw\, dy.$$

We can see that $p_W(w)$ is simply the PDF of the combined SNR with L_a/L_s-GSC. The joint PDF of Y and Z, $p_{Y,Z}(y,z)$ is the joint PDF of two partial sums of ordered random variables. Based on the order statistics result in the Appendix, $p_{Y,Z}(y,z)$ can be calculated as

$$p_{Y,Z}(y,z) = \int_0^{\frac{z}{L_s}} \int_{\frac{z}{L_s}}^{\frac{y}{L_c - L_s}} p_{A, \gamma_{L_c - L_s:L}, B, \gamma_{L_c:L}}(y - \alpha, \alpha, z - \beta, \beta) d\alpha\, d\beta,$$
$$y > \frac{L_c - L_s}{L_s} z, \tag{5.76}$$

where $p_{A, \gamma_{l:L}, B, \gamma_{k:L}}(a, \alpha, b, \beta)$ is the joint PDF of four random variables defined as follows:

$$A = \sum_{i=1}^{L_c - L_s - 1} \gamma_{i:L} \qquad B = \sum_{i=L_c - L_s + 1}^{L_c - 1} \gamma_{i:L}$$
$$\underbrace{\gamma_{1:L}, \ldots, \gamma_{L_c - L_s - 1:L}, \gamma_{L_c - L_s:L}}_{Y = \sum_{i=1}^{L_c - L_s} \gamma_{i:L}}, \underbrace{\gamma_{L_c - L_s + 1:L}, \ldots, \gamma_{L_c - 1:L}, \gamma_{L_c:L}}_{Z = \sum_{i=L_c - L_s + 1}^{L_c} \gamma_{i:L}}, \gamma_{L_c + 1:L}, \ldots, \gamma_{L:L}. \tag{5.77}$$

5.5 RAKE Finger Management during Soft Handover

For i.i.d. Rayleigh fading channels, we can obtain the closed-form expression for the joint PDF, $p_{A,\gamma_{l:L},B,\gamma_{k:L}}(a,\alpha,b,\beta)$, as

$$p_{A,\gamma_{l:L},B,\gamma_{k:L}}(a,\alpha,b,\beta) \tag{5.78}$$
$$= \frac{L! e^{-(a+\alpha+b+\beta)/\bar{\gamma}}(1-e^{-\beta/\bar{\gamma}})^{L-k}[a-(l-1)\alpha]^{l-2}}{(L-k)!(k-l-1)!(k-l-2)!(l-1)!(l-2)!\bar{\gamma}^k}$$
$$\times \sum_{j=0}^{k-l-1}\binom{k-l-1}{j}(-1)^j[b-\beta(k-l-j-1)-\alpha j]^{k-l-2}$$
$$\times \mathcal{U}(\alpha)\mathcal{U}(\alpha-\beta)\mathcal{U}(a-(l-1)\alpha)$$
$$\times \mathcal{U}(b-\beta(k-l-j-1)-\alpha j),$$
$$0 < (k-l-1)\beta < b < (k-l-1)\alpha.$$

The joint PDF of Y and Z (5.76) involves only finite integrations of elementary functions and, as such, can be easily calculated with mathematical software, such as Mathematica. Also note that even though (5.76) is valid only when $l \geq 2$ and $k \geq l+2$, all other cases can be similarly solved by following the steps used to derive (5.78).

Finally, considering (5.76) together with (5.72)–(5.75), the CDF of γ_t, $F_{\gamma_c}(x)$, in (5.71) can be obtained. Differentiating (5.71) with respect to x, we can obtain, after some manipulations, the following generic expression for the PDF of the combined SNR, γ_c, as

$$p_{\gamma_c}(x) = \begin{cases} \int_0^x \left(p_{Y,Z}(y,x-y)F_W(x-y) \right. \\ \left. + p_W(x-y)\int_0^{x-y} p_{Y,Z}(y,z)dz \right)dy, & 0 \leq x < \gamma_T; \\ p_{Y+Z}(x) \\ + \int_0^{\gamma_T}\left(p_W(x-y)\int_0^{\gamma_T-y} p_{Y,Z}(y,z)dz \right)dy, & x \geq \gamma_T, \end{cases} \tag{5.79}$$

where $p_{Y,Z}(\cdot,\cdot)$ is defined in (5.76), $p_{Y+Z}(\cdot)$ is the PDF of the combined SNR with L/L_c-GSC, $p_W(\cdot)$ and $F_W(\cdot)$ are the PDF and CDF of the combined SNR with L_a/L_s-GSC, respectively, for both of which the closed-form expression for the i.i.d. Rayleigh fading environment is available (see (5.27) and (5.18)).

5.5.3 Complexity Analysis

We expect that the full GSC scheme provides a better performance than the block change scheme as it selects the strongest path during SHO. However, the block change scheme enjoys a lower complexity as it avoids the need for a full reordering process of all available paths from the serving and the target BSs. In this section, we investigate this complexity saving of block change scheme by quantifying the average number of path estimations, the average number of SNR comparisons, and the SHO overhead.

With the block change scheme, the RAKE receiver estimates L paths in the case of $\Gamma_{L_c/L} \geq \gamma_T$ or $L+L_a$ in the case of $\Gamma_{L_c/L} < \gamma_T$. Hence, the average number of path estimations of the block change scheme is the same as that of the full GSC scheme [20, Eq. (25)].

We now calculate the average number of required SNR comparisons. Noting that the average number of SNR comparisons for j/i-GSC, denoted by $C_{j/i-GSC}$, can be obtained as[1]

$$C_{j/i-GSC} = \sum_{k=1}^{\min[i,j-i]} (j-k), \qquad (5.80)$$

we can express the average number of SNR comparisons for the full GSC scheme and the block change scheme as

$$C_{Full} = \Pr\left[\Gamma_{L_c/L} \geq \gamma_T\right] C_{L/L_c-GSC} \qquad (5.81)$$
$$+ \Pr\left[\Gamma_{L_c/L} < \gamma_T\right] C_{(L+L_a)/L_c-GSC}$$

and

$$C_{Block} = C_{L/L_c-GSC} + \Pr\left[\Gamma_{L_c/L} < \gamma_T\right] \qquad (5.82)$$
$$\times \left(C_{L_c/L_s-GSC} + C_{L_a/L_s-GSC} + 1\right),$$

respectively.

Since the SHO is attempted whenever $\Gamma_{L_c/L}$ is below γ_T, the probability of the SHO attempt is the same as the outage probability of L_c/L-GSC evaluated at γ_T, i.e., $F_{\Gamma_{L_c/L}}(\gamma_T)$. The SHO overhead, denoted by β, is commonly used to quantify the SHO activity in a network and is defined as [24, Eq. (9.2)]

$$\beta = \sum_{n=1}^{N} nF_n - 1, \qquad (5.83)$$

where N is the number of active BSs and F_n is the average probability that the mobile unit uses n-way SHO. Based on the mode of operation of the block change scheme, we can express its SHO overhead, β, as

$$\beta = \begin{cases} F_{Y+Z}(\gamma_T) \Pr\left[Z < W | Y+Z < \gamma_T\right], & L_s < L_c; \\ 0, & L_s = L_c. \end{cases} \qquad (5.84)$$

Note that $F_{Y+Z}(\gamma_T)$ is the CDF of L_c/L-GSC output SNR evaluated at γ_T. Since W is independent of Z and Y, we can calculate the conditional probability, $\Pr[Z < W | Y+Z < \gamma_T]$, in (5.84) as

$$\Pr\left[Z < W | Y+Z < \gamma_T\right] = \int_0^\infty F_{Z|Y+Z<\gamma_T}(x) p_W(x) dx, \qquad (5.85)$$

---

[1] With the traditional sorting approach, we need $k-1$ comparisons to find the kth largest/smallest one after the previous $k-1$ largest/smallest ones have been found. We follow this traditional approach in order to perform an accurate complexity analysis while noting that with certain quick sorting algorithms, we just need $O(\log(n))$ comparisons for ordering n paths.

where the conditional CDF in (5.85) can be obtained as

$$F_{Z|Y+Z<\gamma_T}(x) = \frac{\Pr\left[Z < x, Y+Z < \gamma_T\right]}{\Pr\left[Y+Z < \gamma_T\right]} \quad (5.86)$$

$$= \frac{1}{F_{Y+Z}(\gamma_T)} \times \begin{cases} \int_0^x \int_{(L_c-L_s)z/L_s}^{\gamma_T-z} f_{Y,Z}(y,z)dydz, & 0 \le x < \frac{L_s}{L_c}\gamma_T; \\ \int_0^{L_s\gamma_T/L_c} \int_{(L_c-L_s)z/L_s}^{\gamma_T-z} f_{Y,Z}(y,z)dydz, & x \ge \frac{L_s}{L_c}\gamma_T. \end{cases}$$

After successive substitutions from (5.86) to (5.84), we can finally obtain the analytical expression of the SHO overhead.

5.5.4 Numerical Results

With the statistics of the combined SNR, we can evaluate the performance of the block change scheme and compare it with that of the full GSC scheme. In Fig. 5.22, we consider the effect of the switching threshold by plotting the average BER of the block change scheme and the full GSC scheme versus the average SNR per path, $\overline{\gamma}$, for various values of γ_T over i.i.d. Rayleigh fading channels when $L = 5, L_a = 5, L_c = 3$, and $L_s = 2$. From this figure, it is clear that the higher the threshold, the better the performance, as we expect intuitively. We can observe that in good channel conditions (i.e., $\overline{\gamma}$ is relatively large compared to γ_T), both schemes become insensitive to variations in γ_T. Since combining additional paths from target BS is not necessary. On the other hand, in the case of large threshold values, the full GSC scheme shows better performance since with the full GSC scheme, instead of comparing and replacing blocks, the L_c largest paths are selected among the $L + L_a$ ones.

In Fig. 5.23, we vary the block size, L_s, with two values of γ_T. We can see that for smaller threshold, the variations of the block size do not affect the performance since in this case no replacement is needed. However, when the threshold is set high, we can observe the performance variation with the value of L_s. For our chosen set of parameters, the best performance, of block change scheme, can be acquired when $L_s = 2$. This is because when $L_s = 1$, we achieve limited benefit from the additional path while when $L_s = 3$, we switch to three paths of target BS without selecting the strongest ones.

In Fig. 5.24, we plot (a) the average number of SNR comparisons, (b) the average BER, and (c) the SHO overhead versus the output threshold, γ_T, of the block change and the full GSC schemes for various values of L_s over i.i.d. Rayleigh fading channels when $L = 5, L_a = 5, L_c = 3$, and $\overline{\gamma} = 0$ dB. For comparison purposes, we also plot those for conventional $(L + L_a)/L_c$-GSC. Note that the full GSC scheme is acting as $(L + L_a)/L_c$-GSC when the output threshold becomes large. Hence, we can observe from all the subfigures that the full GSC scheme converges to GSC as γ_T increases.

Recall that the block change scheme has the same path estimation load as the full GSC scheme. However, from Fig. 5.24(a), we can see that the block change scheme leads to a great reduction of the SNR comparison load compared to the full GSC scheme. The maximum reduction occurs when $\gamma_T > 8$ dB and $L_s = 2$. However, from Fig. 5.24(b), we can observe that the block change scheme leads to slight performance loss compared to the full GSC as well as the conventional GSC schemes over the same SNR region.

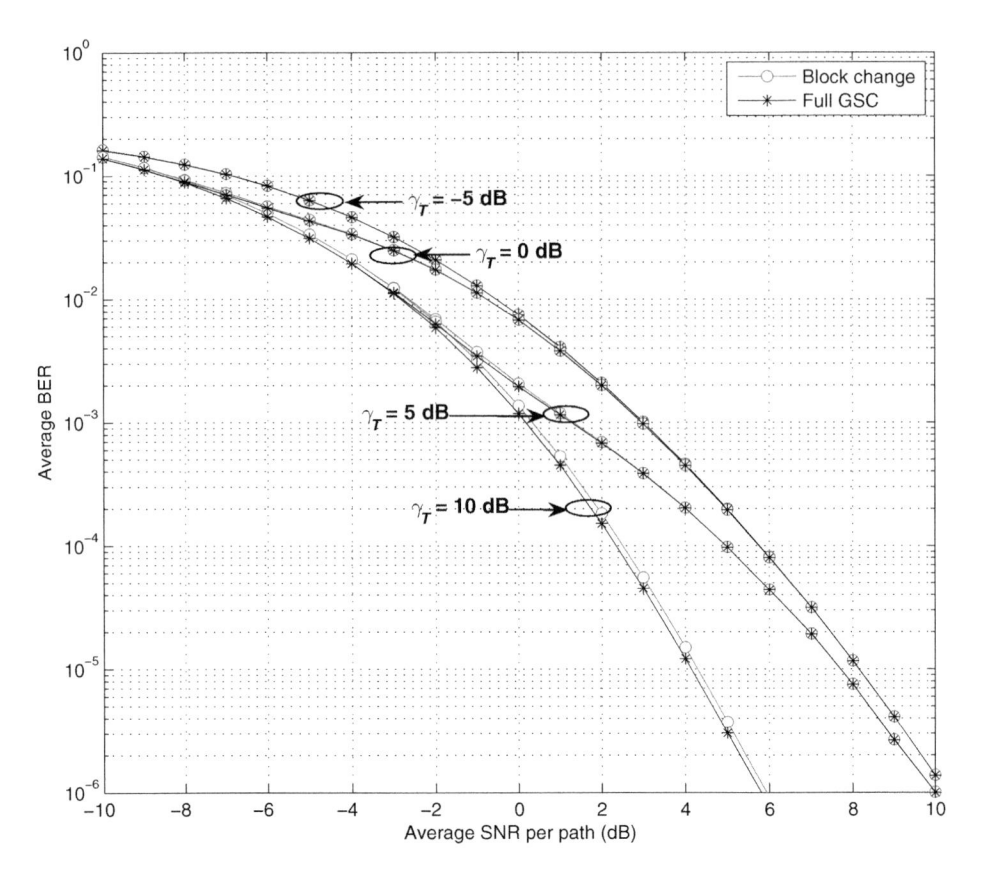

Figure 5.22 Average BER of BPSK versus the average SNR per path, $\overline{\gamma}$, of the block change and the full GSC schemes for various values of γ_T over i.i.d. Rayleigh fading channels when $L = 5, L_a = 5, L_c = 3$, and $L_s = 2$. Reprint with permission from [17]. ©2008 IEEE.

Fig. 5.24(c) presents the SHO overhead and corresponding simulation results to verify our analysis. It is clear from this figure that the receiver has a higher chance to use two-way SHO as L_s decreases. This is because as L_s decreases, the probability that the sum of the L_s smallest paths among the L_c currently used paths from the serving BS is less than the sum of the L_s strongest paths from the target BS is increasing and, as such, we have a higher chance to replace paths. The reduction of SHO overhead is again most significant when $\gamma_T > 8$ dB and $L_s = 2$. at the expense of a slight performance loss in comparison to the full GSC scheme.

5.6 Joint Adaptive Modulation and Diversity Combining

Adaptive transmission has been widely utilized to achieve efficient and reliable transmission over time-varying wireless channels. The basic idea of adaptive transmission is to vary the transmission parameters with the instantaneous fading channel

5.6 Joint Adaptive Modulation and Diversity Combining

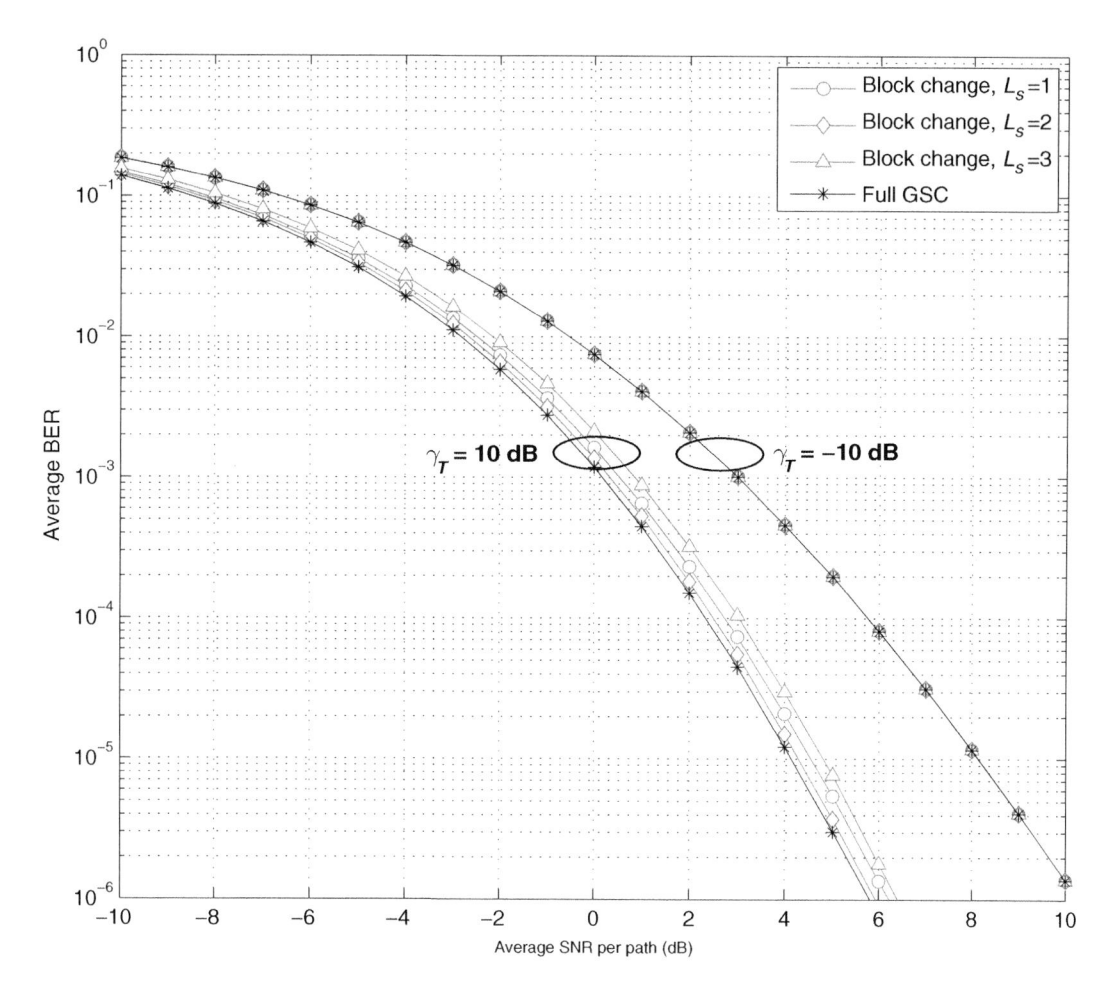

Figure 5.23 Average BER of BPSK versus the average SNR per path, $\bar{\gamma}$, of the block change and the full GSC schemes for various values of L_s and γ_T over i.i.d. Rayleigh fading channels when $L = 5, L_a = 5$, and $L_c = 3$. Reprint with permission from [17]. ©2008 IEEE.

conditions in order to either maximize the transmission rate or minimize power consumption while satisfying a certain instantaneous reliability requirement [25–27]. Both adaptive transmission and adaptive diversity combining schemes utilize some predetermined threshold in their operations. Motivated by this observation, we investigate the joint design of adaptive transmission and adaptive combining in this section.

Adaptive modulation is one of the most popular rate-adaptive transmission schemes. In this section, we present several joint adaptive modulation and diversity combining (AMDC) schemes. In these schemes, the receiver jointly determines the most appropriate modulation scheme and diversity combiner structure based on the current channel conditions and the desired BER requirements. The proposed AMDC systems can efficiently explore the bandwidth and power resource by transmitting at higher data rate

130 Adaptive Diversity Combining

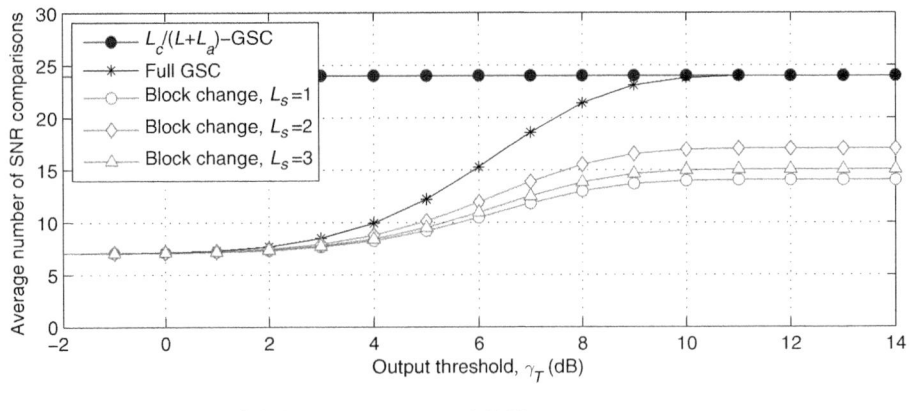

(a) Average number of SNR comparisons

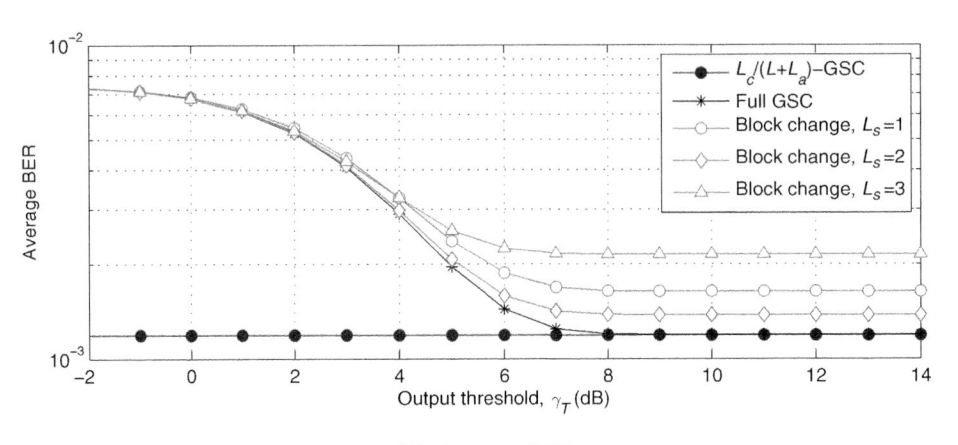

(b) Average BER

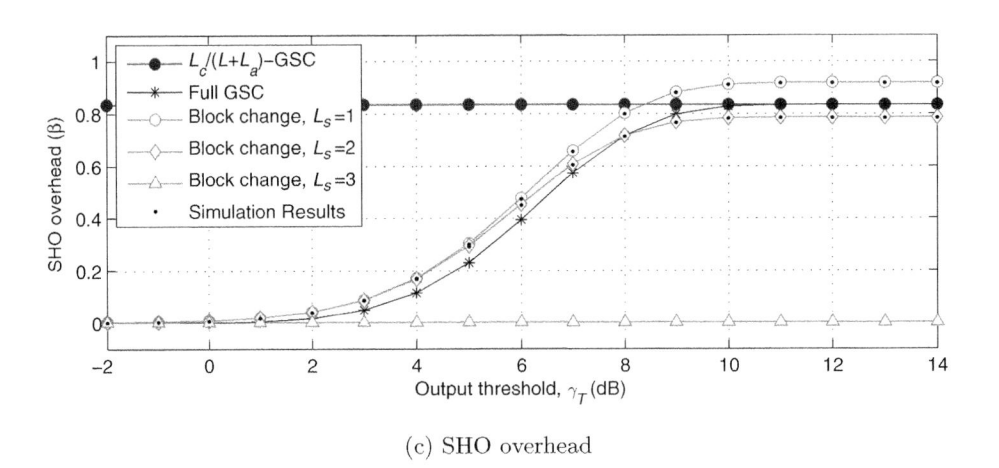

(c) SHO overhead

Figure 5.24 Complexity trade-off versus the output threshold, γ_T, of the block change and the full GSC schemes, and conventional GSC for various values of L_s over i.i.d. Rayleigh fading channels with $L = 5, L_a = 5, L_c = 3$, and $\overline{\gamma} = 0$ dB. Reprint with permission from [17]. ©2008 IEEE.

and/or combining the least number of diversity paths under favorable channel conditions. On the other hand, the AMDC system also responds to channel degradation with an increase in the number of combined paths and/or a reduction in the data rate. The system model of the joint design is illustrated in Fig. 5.25.

While the joint designs are applicable to different adaptive combining and adaptive modulation schemes, we focus on the MS-GSC scheme [2, 3, 5] and constant-power variable-rate uncoded M-ary QAM based adaptive modulation scheme studied in [26]. In particular, the value range of the instantaneous received SNR is divided into $N+1$ regions with thresholds γ_{T_n}, $n = 0, 1, \ldots, N$. The modulation scheme 2^n-QAM is used if the received SNR γ is in the nth region, i.e. $\gamma_{T_n} \leq \gamma < \gamma_{T_{n+1}}$. Depending on the primary objective of the joint design, we can arrive at an AMDC scheme with a high power efficiency (called power-efficient AMDC), an AMDC scheme with a high bandwidth efficiency (called bandwidth-efficient AMDC), and an AMDC scheme with a high bandwidth efficiency and improved power efficiency at the cost of a higher error rate than the bandwidth-efficient AMDC scheme (called bandwidth-efficient and power-greedy AMDC) [29], all of which satisfy the desired BER requirement. For all three AMDC schemes under consideration, we quantify through accurate analysis their power consumption (in terms of average number of combined diversity paths), spectral efficiency (in terms of average number of transmitted bits/s/Hz), and performance (in terms of average BER).

5.6.1 Power-Efficient AMDC Scheme

The primary objective of the power-efficient AMDC scheme is to minimize the processing power consumption of the diversity combiner, i.e., to minimize the average number of combined/active diversity paths during data reception. Once this primary objective is met, this scheme tries to achieve the largest possible spectral efficiency while meeting the target BER requirement. According to these objectives, the diversity combiner will perform just enough combining operations such that at least the lowest modulation mode, e.g., BPSK, will exhibit an instantaneous BER smaller than the

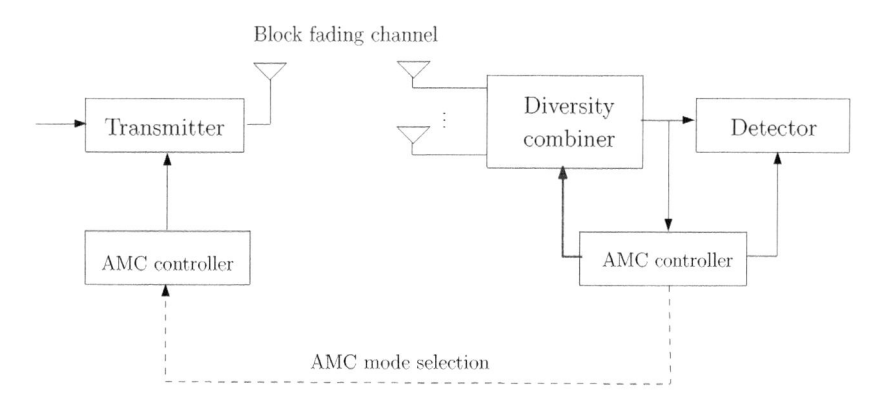

Figure 5.25 System model of joint adaptive modulation and diversity combining.

target value. In particular, the combiner tries to increase the output SNR γ_c above the threshold for BPSK, denoted by γ_{T_1}, by performing MS-GSC diversity. After estimating and ranking the L available diversity paths, the combiner first checks if the SNR of the strongest path is greater than γ_{T_1}. If so, the combiner just uses the received signal over the strongest path. If not, the combiner checks the combined SNR of the first two strongest paths. If the combined SNR is still less than γ_{T_1}, then the combiner checks the combined SNR of the first three strongest paths. This process is continued until either (1) the combined SNR become greater than γ_{T_1} or (2) all L available paths have been combined. In the first case, the receiver starts to determine the modulation mode to be used by checking which interval the resulting output SNR falls into. In particular, the receiver sequentially compares the output SNR with respect to the thresholds, $\gamma_{T_2}, \gamma_{T_3}, \ldots, \gamma_{T_N}$. Whenever the receiver finds that the output SNR is smaller than $\gamma_{T_{n+1}}$ but greater than γ_{T_n}, it selects the modulation mode n for the data reception and feeds back that particular modulation mode to the transmitter. If the combined SNR of all L available branches is still below γ_{T_1}, the receiver may ask the transmitter to either (1) transmit using the lowest modulation mode in violation of the target BER requirement (option 1) or (2) buffer the data and wait for more favorable channel conditions (option 2).

Statistics of Output SNR

According to the mode of operation described above, we can see that the combined SNR, γ_c, of the power-efficient AMDC scheme is the same as the combined SNR of MS-GSC with γ_{T_1} as the output threshold. In other words, the CDF of the received SNR, $F_{\gamma_c}(\cdot)$, of the power-efficient AMDC based on MS-GSC is given by

$$F_{\gamma_c}(\gamma) = \begin{cases} F_{\gamma_c}^{MSC(\gamma_{T_1})}(\gamma), & \text{for option 1;} \\ \begin{cases} F_{\gamma_c}^{MSC(\gamma_{T_1})}(\gamma), & \gamma > \gamma_{T_1}; \\ F_{\gamma_c}^{MSC(\gamma_{T_1})}(\gamma_{T_1}), & 0 < \gamma \leq \gamma_{T_1} \end{cases} & \text{for option 2,} \end{cases} \quad (5.87)$$

where $F_{\gamma_c}^{MSC(\gamma_{T_1})}(\cdot)$ denotes the CDF of the combined SNR with L-branch MS-GSC and using γ_{T_1} as the output threshold. The generic expression of $F_{\gamma_c}^{MSC(\gamma_{T_1})}(\cdot)$ was given in (5.21) with γ_T changed to γ_{T_1} and $L_c = L$, which is available in closed form for the i.i.d. Rayleigh fading environment in [5, Eq. (30)].

Correspondingly, the PDF of the received SNR, $p_{\gamma_c}(\cdot)$, is given by

$$p_{\gamma_c}(\gamma) = \begin{cases} p_{\gamma_c}^{MSC(\gamma_{T_1})}(\gamma), & \text{for option 1;} \\ p_{\gamma_c}^{MSC(\gamma_{T_1})}(\gamma)\mathcal{U}(\gamma - \gamma_{T_1}) + p_{\gamma_c}^{MSC(\gamma_{T_1})}(\gamma_{T_1})\delta(\gamma), & \text{for option 2,} \end{cases} \quad (5.88)$$

where $\delta(\cdot)$ is the delta function and $p_{\gamma_c}^{MSC(\gamma_{T_1})}(\cdot)$ denotes the PDF of the combined SNR with L-branch MS-GSC using γ_{T_1} as the output threshold, which is available in closed form for the i.i.d. Rayleigh fading environment in (5.23) with γ_T changed to γ_{T_1} and $L_c = L$.

5.6 Joint Adaptive Modulation and Diversity Combining

Processing Power and Performance Analysis

The power consumption for diversity combining can be quantified in terms of the average number of combined paths [5]. It can be shown that the average number of combined paths with the power-efficient AMDC is given for option 1 by

$$\overline{N}_C = 1 + \sum_{i=1}^{L-1} F_{\Gamma_i}(\gamma_{T_1}),$$

and for option 2 by

$$\overline{N}_C = 1 + \sum_{i=1}^{L-1} F_{\Gamma_i}(\gamma_{T_1}) - L F_{\gamma_c}^{L-MRC}(\gamma_{T_1}),$$

where $F_{\Gamma_i}(\cdot)$ is the CDF of the combined SNR with the L/i-GSC scheme (which is given in closed form for i.i.d. Rayleigh fading in (5.18)) and $F_{\gamma_c}^{L-MRC}(\cdot)$ is the CDF of the combined SNR with L-branch MRC scheme (which is given in closed form for i.i.d. Rayleigh fading in Table 5.1).

The average spectral efficiency of an adaptive modulation system can be calculated as [26, Eq. 33]

$$\eta = \sum_{n=1}^{N} n P_n,$$

where n is the number of bits carried by a modulated symbol with nth modulation scheme and P_n is the probability that the nth modulation scheme is used. For the power-efficient AMDC system under consideration, it can be shown that P_n is given by

$$P_n = \begin{cases} \begin{cases} F_{\gamma_c}^{MSC(\gamma_{T_1})}(\gamma_{T_{n+1}}) - F_{\gamma_c}^{MSC(\gamma_{T_1})}(\gamma_{T_n}), & n \geq 2; \\ F_{\gamma_c}^{MSC(\gamma_{T_1})}(\gamma_{T_2}), & n = 1 \end{cases} & \text{for option 1;} \\ F_{\gamma_c}^{MSC(\gamma_{T_1})}(\gamma_{T_{n+1}}) - F_{\gamma_c}^{MSC(\gamma_{T_1})}(\gamma_{T_n}), & \text{for option 2.} \end{cases} \quad (5.89)$$

Therefore, the average spectral efficiency of the power efficient AMDC is given by

$$\eta = \begin{cases} N - \sum_{n=2}^{N} F_{\gamma_c}^{MSC(\gamma_{T_1})}(\gamma_{T_n}), & \text{for option 1;} \\ N - \sum_{n=1}^{N} F_{\gamma_c}^{MSC(\gamma_{T_1})}(\gamma_{T_n}), & \text{for option 2.} \end{cases} \quad (5.90)$$

The average BER of adaptive modulation system can be calculated as [26, Eq. (35)]

$$<BER> = \frac{1}{\eta} \sum_{n=1}^{N} n \, \overline{BER}_n, \quad (5.91)$$

where $\overline{BER}_n$ is the average error rate for modulation scheme n, given by

$$\overline{BER}_n = \int_{\gamma_{T_n}}^{\gamma_{T_{n+1}}} BER_n(\gamma) p_{\gamma_c}^{MSC(\gamma_{T_1})}(\gamma) d\gamma, \quad (5.92)$$

where $BER_n(\gamma)$ is the BER of the modulation scheme n over the AWGN channel, with received SNR equal to γ, an approximate expression of which was given in (2.32). Therefore, the average BER of the power-efficient AMDC scheme can be calculated as

$$<BER> = \begin{cases} \frac{1}{\eta}\left(\int_0^{\gamma_{T_2}} BER_1(\gamma)\, p_{\gamma_c}^{MSC(\gamma_{T_1})}(\gamma)\, d\gamma \right. \\ \left. + \sum_{n=2}^{N} n \int_{\gamma_{T_n}}^{\gamma_{T_{n+1}}} BER_n(\gamma)\, p_{\gamma_c}^{MSC(\gamma_{T_1})}(\gamma)\, d\gamma \right), & \text{for option 1;} \\ \frac{1}{\eta}\sum_{n=1}^{N} n \int_{\gamma_{T_n}}^{\gamma_{T_{n+1}}} BER_n(\gamma)\, p_{\gamma_c}^{MSC(\gamma_{T_1})}(\gamma)\, d\gamma, & \text{for option 2.} \end{cases} \quad (5.93)$$

5.6.2 Bandwidth-Efficient AMDC Scheme

The primary objective of the bandwidth-efficient AMDC scheme is to maximize the spectral efficiency. As such, the receiver with this scheme performs the necessary combining operations so that the highest order modulation mode can be used while satisfying the instantaneous BER requirement. More specifically, the receiver tries first to increase the output SNR γ_c above the threshold of the highest modulation mode, i.e., γ_{T_N}, by employing the MS-GSC type of diversity. After estimating and ranking all the available diversity paths, the combiner sequentially checks the SNR of the strongest path, the combined SNR of the two strongest paths, the combined SNR of the three strongest paths, etc. Whenever the combined SNR is larger than γ_{T_N}, the receiver stops checking and informs the transmitter to use 2^N-QAM as the modulation mode. If the combined SNR of all available paths is still below γ_{T_N}, the receiver selects the modulation mode corresponding to the SNR interval in which the combined SNR falls. In particular, the receiver sequentially compares the output SNR with the thresholds, $\gamma_{T_{N-1}}, \gamma_{T_{N-2}}, \ldots,$ γ_{T_1}. Whenever the receiver finds that the output SNR is smaller than $\gamma_{T_{n+1}}$ but greater than γ_{T_n}, it selects the modulation mode n and feeds back this selected mode to the transmitter. If, in the worst case, the combined SNR of all the available paths ends up being below γ_{T_1}, the receiver has the same two termination options as for the power-efficient scheme (i.e., to transmit using the lowest order modulation mode (option 1) or to wait until channel condition improves (option 2)).

Statistics of Output SNR

Based on the mode of operation described above, we can see that the combined SNR, γ_c, of the bandwidth-efficient AMDC system is the same as the combined SNR of MS-GSC diversity with γ_{T_N} as the output threshold. Therefore, the CDF of the received SNR of the bandwidth-efficient AMDC scheme is given by

$$F_{\gamma_c}(\gamma) = \begin{cases} F_{\gamma_c}^{MSC(\gamma_{T_N})}(\gamma), & \text{for option 1;} \\ \begin{cases} F_{\gamma_c}^{MSC(\gamma_{T_N})}(\gamma), & \gamma > \gamma_{T_1}; \\ F_{\gamma_c}^{MSC(\gamma_{T_N})}(\gamma_{T_1}), & 0 < \gamma \le \gamma_{T_1} \end{cases} & \text{for option 2,} \end{cases} \quad (5.94)$$

where $F_{\gamma_c}^{MSC(\gamma_{T_N})}(\cdot)$ denotes the CDF of the combined SNR with L-branch MS-GSC using γ_{T_N} as the output threshold. Correspondingly, the PDF of the received SNR is given by

5.6 Joint Adaptive Modulation and Diversity Combining

$$p_{\gamma_c}(\gamma) = \begin{cases} p_{\gamma_c}^{MSC(\gamma_{T_N})}(\gamma), & \text{for option 1;} \\ p_{\gamma_c}^{MSC(\gamma_{T_N})}(\gamma)\mathcal{U}(\gamma - \gamma_{T_1}) + p_{\gamma_c}^{MSC(\gamma_{T_N})}(\gamma_{T_1})\delta(\gamma), & \text{for option 2,} \end{cases} \quad (5.95)$$

where $p_{\gamma_c}^{MSC(\gamma_{T_N})}(\cdot)$ denotes the PDF of the combined SNR with L-branch MS-GSC using γ_{T_N} as the output threshold.

Processing Power and Performance Analysis

With the statistics of the output SNR, we can study the processing power and the performance of the bandwidth-efficient AMDC scheme. For conciseness, we just list the analytical results below.

- Average number of combined branches

$$\overline{N}_C = \begin{cases} 1 + \sum_{i=1}^{L-1} F_{\Gamma_i}(\gamma_{T_N}), & \text{for option 1;} \\ 1 + \sum_{i=1}^{L-1} F_{\Gamma_i}(\gamma_{T_N}) - L F_{\gamma_c}^{L-MRC}(\gamma_{T_1}), & \text{for option 2.} \end{cases} \quad (5.96)$$

- Average spectral efficiency

$$\eta = \begin{cases} N - \sum_{n=2}^{N} F_{\gamma_c}^{MSC(\gamma_{T_N})}(\gamma_{T_n}), & \text{for option 1;} \\ N - \sum_{n=1}^{N} F_{\gamma_c}^{MSC(\gamma_{T_N})}(\gamma_{T_n}), & \text{for option 2.} \end{cases} \quad (5.97)$$

- Average BER

$$<BER> = \begin{cases} \frac{1}{\eta}\left(\int_0^{\gamma_{T_2}} BER_1(\gamma) p_{\gamma_c}^{MSC(\gamma_{T_N})}(\gamma)\, d\gamma \right. \\ \left. + \sum_{n=2}^{N} n \int_{\gamma_{T_n}}^{\gamma_{T_{n+1}}} BER_n(\gamma) p_{\gamma_c}^{MSC(\gamma_{T_N})}(\gamma)\, d\gamma \right), & \text{for option 1;} \\ \frac{1}{\eta} \sum_{n=1}^{N} n \int_{\gamma_{T_n}}^{\gamma_{T_{n+1}}} BER_n(\gamma) p_{\gamma_c}^{MSC(\gamma_{T_N})}(\gamma)\, d\gamma, & \text{for option 2.} \end{cases}$$
(5.98)

5.6.3 Bandwidth-Efficient and Power-Greedy AMDC Scheme

The bandwidth-efficient and power-greedy AMDC scheme can be viewed as a modified bandwidth-efficient scheme with better power efficiency at the cost of a slightly higher error rate. Basically, the receiver combines the smallest number of diversity branches such that the highest order modulation mode can be used while satisfying the instantaneous BER requirement. During this process, the receiver ensures that combining additional branches does not make a higher order modulation mode feasible.

More specifically, the receiver tries first to increase the output SNR γ_c above the threshold of 2^N-QAM, i.e., γ_{T_N}, by applying MS-GSC diversity. After estimating and ranking all the available diversity paths, the combiner sequentially checks the SNR of the strongest path, the combined SNR of the two strongest paths, the combined SNR of the three strongest paths, etc. Whenever the combined SNR is larger than γ_{T_N}, the receiver selects 2^N-QAM as the modulation mode and uses the current combiner structure for data reception during the subsequent data burst. If the combined SNR of all available paths is still below γ_{T_N}, the receiver determines the highest feasible modulation mode by checking which interval the combined SNR of these branches falls into. The receiver sequentially compares the output SNR with respect to the thresholds, $\gamma_{T_{N-1}}, \gamma_{T_{N-2}}, \ldots, \gamma_{T_1}$. Whenever the receiver finds that the output SNR is smaller than $\gamma_{T_{n+1}}$ but greater than γ_{T_n}, it selects the modulation mode n for data transmission. Before the data transmission, the receiver selects the minimum combiner structure (i.e., with the minimum number of active branches) such that the output SNR is still greater than γ_{T_n}. Specifically, the receiver sequentially turns off the weakest branches until a further branch turnoff will lead to an output SNR below γ_{T_n}. If, in the worst case, the combined SNR of all the available branches is below γ_{T_1}, the receiver has the same two terminating options as for previously presented schemes.

Statistics of Output SNR

Based on the mode of operation described above, we can see that the output SNR, γ_c, of the bandwidth-efficient and power-greedy AMDC scheme is the same as the combined SNR of MS-GSC with γ_{T_N} as the output threshold only for $\gamma_c > \gamma_{T_N}$ case, i.e.,

$$F_{\gamma_c}(x) = F_{\gamma_c}^{MSC(\gamma_{T_N})}(x), \quad x > \gamma_{T_N}. \qquad (5.99)$$

For the case of $\gamma_{T_1} \leq \Gamma < \gamma_{T_N}$, the statistics are more complicated to calculate. It can be shown that the CDF of the received SNR γ_c over the range of $[\gamma_{T_n}, \gamma_{T_{n+1}})$ is mathematically given by

$$F_{\gamma_c}(x) = F_{\gamma_c}^{L-MRC}(\gamma_{T_n}) + \Pr\left[\gamma_{T_n} \leq \gamma_{1:L} < x \ \& \ \sum_{k=1}^{L} \gamma_k < \gamma_{T_{n+1}}\right]$$
$$+ \sum_{l=2}^{L} \Pr\left[\sum_{j=1}^{l-1} \gamma_{j:L} < \gamma_{T_n} \leq \sum_{j=1}^{l} \gamma_{j:L} < x \ \& \ \sum_{k=1}^{L} \gamma_k < \gamma_{T_{n+1}}\right], \qquad (5.100)$$

where $F_{\gamma_c}^{L-MRC}(\cdot)$ denotes the CDF of the combined SNR with L-branch MRC and $\gamma_{j:L}$ denotes the jth largest path SNR among the L available ones. Let y_l denote the sum of the $l-1$ largest ordered path SNRs, i.e., $y_l = \sum_{j=1}^{l-1} \gamma_{j:L}$, and let z_l denote the sum of the $L-l$ smallest ordered path SNRs, i.e., $z_l = \sum_{j=l+1}^{L} \gamma_{j:L}$. The CDF can be

5.6 Joint Adaptive Modulation and Diversity Combining

rewritten as

$$F_{\gamma_c}(x) = F^{(L)}(\gamma_{T_n}) + \Pr\left[\gamma_{T_n} \leq \gamma_{1:L} < x \,\&\, \gamma_{1:L} + z_1 < \gamma_{T_{n+1}}\right]$$
$$+ \sum_{l=2}^{L-1} \Pr\left[y_l < \gamma_{T_n} \leq y_l + \gamma_{l:L} < x \,\&\, y_l + \gamma_{l:L} + z_l < \gamma_{T_{n+1}}\right]$$
$$+ \Pr\left[y_L < \gamma_{T_n} \leq y_L + \gamma_{L:L} < x\right], \tag{5.101}$$

which can be calculated in terms of the joint PDFs of y_l, $\gamma_{l:L}$, and z_l as

$$F_{\gamma_c}(x) = F_{\gamma_c}^{L-MRC}(\gamma_{T_n}) + \int_{\gamma_{T_n}}^{x}\int_{0}^{\gamma_{T_{n+1}}-\gamma} p_{\gamma_{1:L},z_1}(\gamma,z)dzd\gamma \tag{5.102}$$
$$+ \sum_{l=2}^{L-1} \int_{0}^{\gamma_{T_n}} \int_{\gamma_{T_n}-y}^{\min[y/(l-1),x-y]} \int_{0}^{\gamma_{T_{n+1}}-y-\gamma} p_{y_l,\gamma_{l:L},z_l}(y,\gamma,z)dzd\gamma dy$$
$$+ \int_{0}^{\gamma_{T_n}} \int_{\gamma_{T_n}-y}^{\min[y/(L-1),x-y]} p_{y_L,\gamma_{L:L}}(y,\gamma)d\gamma dy, \quad \gamma_{T_n} \leq x < \gamma_{T_{n+1}}.$$

For the case of $\gamma_c < \gamma_{T_1}$, it can be shown that the CDF of γ_c for the MS-GSC-based AMDC scheme is given by

$$F_{\gamma_c}(x) = \begin{cases} F_{\gamma_c}^{L-MRC}(x), & \text{for option 1;} \\ F_{\gamma_c}^{L-MRC}(\gamma_{T_1}), & \text{for option 2,} \end{cases} \quad 0 < x < \gamma_{T_1}. \tag{5.103}$$

Thus, we have obtained the generic expression of the CDF of the combined SNR for the whole SNR value range. After differentiating $F_{\gamma_c}(x)$ with respect to x, we obtain a generic formula for the PDF of the output SNR, γ_c, as

$$p_{\gamma_c}(x) = \begin{cases} p_{\gamma_c}^{MSC(\gamma_{T_N})}(x), & x > \gamma_{T_N}; \\ \int_{0}^{\gamma_{T_{n+1}}-x} p_{\gamma_{1:L},z_1}(x,z)dz \\ \quad + \sum_{l=2}^{L-1} \left(\int_{\frac{l-1}{l}x}^{\gamma_{T_n}} \int_{0}^{\gamma_{T_{n+1}}-x} p_{y_l,\gamma_{l:L},z_l}(y,x-y,z)dzdy \right. \\ \quad \times \left. \left(\mathcal{U}(x-\gamma_{T_n}) - \mathcal{U}(x - \frac{l}{l-1}\gamma_{T_n}) \right) \right) \\ \quad + \int_{\frac{L-1}{L}x}^{\gamma_{T_n}} p_{y_L,\gamma_{L:L}}(y,x-y)dy \\ \quad \times (\mathcal{U}(x-\gamma_{T_n}) - \mathcal{U}(x - \frac{L}{L-1}\gamma_{T_n})), & \gamma_{T_n} \leq x < \gamma_{T_{n+1}}; \\ \begin{cases} p_{\gamma_c}^{L-MRC}(x), & \text{for option 1;} \\ \delta(x)p_{\gamma_c}^{L-MRC}(\gamma_{T_1}), & \text{for option 2,} \end{cases} & 0 < x < \gamma_{T_1}. \end{cases} \tag{5.104}$$

Note that the CDF and the PDF of the combined SNR are given in terms of the joint PDFs of y_l, $\gamma_{l:L}$, and z_l, which is extensively studied in the Appendix. Specifically, it has been shown that these joint PDFs are available in closed form for the i.i.d. Rayleigh fading case. Specifically, it has been shown that $f_{y_l,\gamma_{l:L},z_l}(y,\gamma,z)$ is given by

$$p_{y_l,\gamma_{l:L},z_l}(y,\gamma,z) = p_{\gamma_{l:L}}(\gamma) \times p_{z_l|\gamma_{l:L}=\gamma}(z) \times p_{y_l|\gamma_{l:L}=\gamma,z_l=z}(y)$$

$$= \frac{L!}{(L-l)!(l-1)!\bar{\gamma}^L} \frac{[y-(l-1)\gamma]^{l-2}}{(l-2)!(L-l-1)!} e^{-\frac{y+\gamma+z}{\bar{\gamma}}} \mathcal{U}(y-(l-1)\gamma)$$

$$\times \sum_{i=0}^{L-l} \binom{L-l}{i} (-1)^i (z-i\gamma)^{L-l-1} \mathcal{U}(z-i\gamma)$$

$$\gamma > 0, y > (l-1)\gamma, z < (L-l)\gamma, \tag{5.105}$$

where $\mathcal{U}(\cdot)$ is the unit step function. The joint PDFs $p_{\gamma_{l:L},z_l}(\gamma,z)$ and $p_{y_L,\gamma_{L:L}}(y,\gamma)$ can be obtained as marginals of the joint PDF given in (5.105).

Processing Power and Performance Analysis

We now study the performance and processing power of the bandwidth-efficient and power-greedy AMDC scheme based on the statistics of the output SNR that we just derived. Note that the bandwidth-efficient and power-greedy AMDC scheme always uses the highest achievable modulation mode, and as such it will have the same average spectral efficiency as the bandwidth-efficient AMDC scheme, which was analyzed in the previous section. We focus in what follows on the average number of combined branches and the average BER.

Let $P_{l,n}$ denote the probability that modulation mode n is used with l combined branches. We can calculate the average number of combined branches with the bandwidth-efficient and power-greedy AMDC scheme by averaging over all possible values of l and n as

$$\overline{N}_c = \begin{cases} \sum_{l=1}^{L}\sum_{n=1}^{N} l\, P_{l,n} + L F_{\gamma_c}^{L-MRC}(\gamma_{T_1}), & \text{for option 1;} \\ \sum_{l=1}^{L}\sum_{n=1}^{N} l\, P_{l,n}, & \text{for option 2.} \end{cases} \tag{5.106}$$

Based on the mode of operation of the bandwidth-efficient and power-greedy AMDC scheme, it can be shown that $P_{l,n}$ can be calculated for $n=N$ as

$$P_{l,N} = \begin{cases} 1 - F_{\gamma_c}^{L/1-GSC}(\gamma_{T_N}), & l=1 \\ F_{\gamma_c}^{L/(l-1)-GSC}(\gamma_{T_N}) - F_{\gamma_c}^{L/l-GSC}(\gamma_{T_N}), & 1 < l \leq L, \end{cases} \tag{5.107}$$

and for $n < N$ as

$$P_{l,n} =$$

$$\begin{cases} \Pr\left[\gamma_{T_n} \leq \gamma_{1:L} \;\&\; \sum_{k=1}^{L} \gamma_{k:L} < \gamma_{T_{n+1}}\right], & l=1 \\ \Pr\left[\sum_{j=1}^{l-1} \gamma_{j:L} < \gamma_{T_n} \leq \sum_{j=1}^{l} \gamma_{j:L} \;\&\; \sum_{k=1}^{L} \gamma_{k:L} < \gamma_{T_{n+1}}\right], & 1 < l < L \\ \Pr\left[\sum_{j=1}^{L-1} \gamma_{j:L} < \gamma_{T_n} \leq \sum_{j=1}^{L} \gamma_{j:L} < \gamma_{T_{n+1}}\right], & l=L, \end{cases} \tag{5.108}$$

which can be calculated using the joint PDFs of y_l, $\gamma_{l:L}$, and z_l as

$$P_{l,n} = \begin{cases} \int_{\gamma T_n}^{\gamma T_{n+1}} \int_0^{\gamma T_{n+1}-\gamma} p_{\gamma_{1:L},z_1}(\gamma,z) dz d\gamma, & l=1; \\ \int_{(l-1)\gamma T_n/l}^{(l-1)\gamma T_{n+1}/l} \int_{\gamma T_n-y}^{y/(l-1)} \int_0^{\gamma T_{n+1}-y-\gamma} p_{y_l,\gamma_{l:L},z_l}(y,\gamma,z) dz d\gamma dy \\ \quad + \int_{(l-1)\gamma T_{n+1}/l}^{\gamma T_n} \int_{\gamma T_n-y}^{\gamma T_{n+1}-y} \int_0^{\gamma T_{n+1}-y-\gamma} p_{y_l,\gamma_{l:L},z_l}(y,\gamma,z) dz d\gamma dy, & 1 < l < L; \\ \int_{(L-1)\gamma T_n/L}^{(L-1)\gamma T_{n+1}/L} \int_{\gamma T_n-y}^{y/(L-1)} p_{y_L,\gamma_{L:L}}(y,\gamma) d\gamma dy \\ \quad + \int_{(L-1)\gamma T_{n+1}/L}^{\gamma T_n} \int_{\gamma T_n-y}^{\gamma T_{n+1}-y} p_{y_L,\gamma_{L:L}}(y,\gamma) d\gamma dy, & l=L. \end{cases}$$

(5.109)

Following the same approach as the power-efficient AMDC while applying the PDF of received SNR given in (5.104), the average BER of the bandwidth-efficient and power-greedy AMDC scheme can be calculated as

$$<BER> = \begin{cases} \dfrac{\int_0^{\gamma T_2} BER_1(\gamma) p_{\gamma_c}(\gamma) d\gamma + \sum_{n=2}^N n \int_{\gamma T_n}^{\gamma T_{n+1}} BER_n(\gamma) p_{\gamma_c}(\gamma) d\gamma}{N - \sum_{n=2}^N F_{\gamma_c}^{MSC(\gamma_{T_N})}(\gamma_{T_n})}, & \text{for option 1;} \\ \dfrac{\sum_{n=1}^N n \int_{\gamma T_n}^{\gamma T_{n+1}} BER_n(\gamma) p_{\gamma_c}(\gamma) d\gamma}{N - \sum_{n=1}^N F_{\gamma_c}^{MSC(\gamma_{T_N})}(\gamma_{T_n})}, & \text{for option 2.} \end{cases}$$

(5.110)

5.6.4 Numerical Results

We now illustrate the analytical results through several numerical examples. In particular, we compare the processing power consumption, spectral efficiency, and average BER performance of three AMDC schemes. In the following numerical examples, we set the number of available diversity branches $L = 5$, the number of adaptive modulation modes $N = 4$, and the SNR thresholds are set to satisfy the instantaneous BER requirement of $BER_0 = 10^{-3}$.

Note that the greater the average number of combined paths during data reception, the larger the average receiver processing power consumption. We investigate the power efficiency of the three AMDC schemes under consideration in Fig. 5.26 by plotting the average number of combined branches of these AMDC schemes with two terminating options as a function of the average path SNR $\bar{\gamma}$. For reference purpose, the average

140 **Adaptive Diversity Combining**

number of combined branches of adaptive modulation without diversity and with L-branch MRC cases are also plotted. We observe that, the power-efficient AMDC scheme always combines the lowest number of diversity branches on average, for any value of $\overline{\gamma}$, whereas the bandwidth-efficient AMDC scheme combines the most among the three schemes, but still less than adaptive modulation with the L-branch MRC case. We also notice that when the average path SNR is very small, three AMDC schemes consumes more processing power with option 1 than with option 2 by combining more diversity branches.

The average spectral efficiency of three AMDC schemes with two options are plotted as a function of the average path SNR $\overline{\gamma}$ in Fig. 5.27. The average spectral efficiency curves of adaptive modulation without diversity and with L-branch MRC cases are also included. We can see that the bandwidth-efficient AMDC and the bandwidth-efficient/power-greedy AMDC schemes have the same spectral efficiency

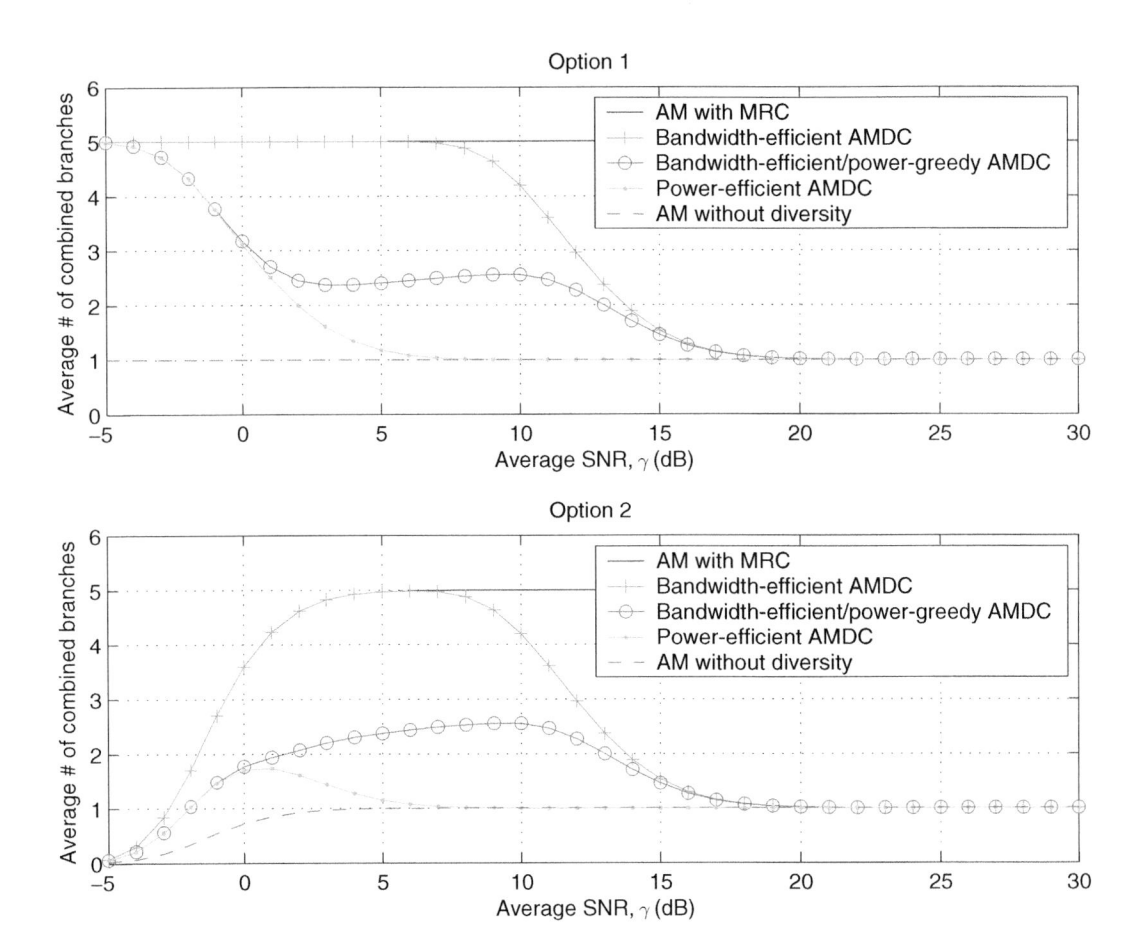

Figure 5.26 Average number of combined branches of three AMDC schemes with two options as a function of the average path SNR $\overline{\gamma}$ ($L = 5$, $N = 4$, and $\text{BER}_0 = 10^{-3}$). Reprint with permission from [29]. ©2007 IEEE.

5.6 Joint Adaptive Modulation and Diversity Combining

as adaptive modulation with L-branch MRC, which is much greater than that of the power-efficient AMDC scheme over the medium value range of $\overline{\gamma}$. Thus, by observing Figs. 5.26 and 5.27, there is a trade-off of spectral efficiency and power consumption between power-efficient and bandwidth-efficient AMDC schemes. We also observe that when the average SNR is very small, three AMDC schemes with option 1 exhibit higher spectral efficiency than with option 2, at the expense that the target BER is not met with option 1.[2]

Finally, we examine the BER performance of three AMDC schemes in Fig. 5.28. This figure confirms that when the average SNR is very small, the average BER of the three AMDC schemes with option 1 become larger than the target $BER_0 = 10^{-3}$, in

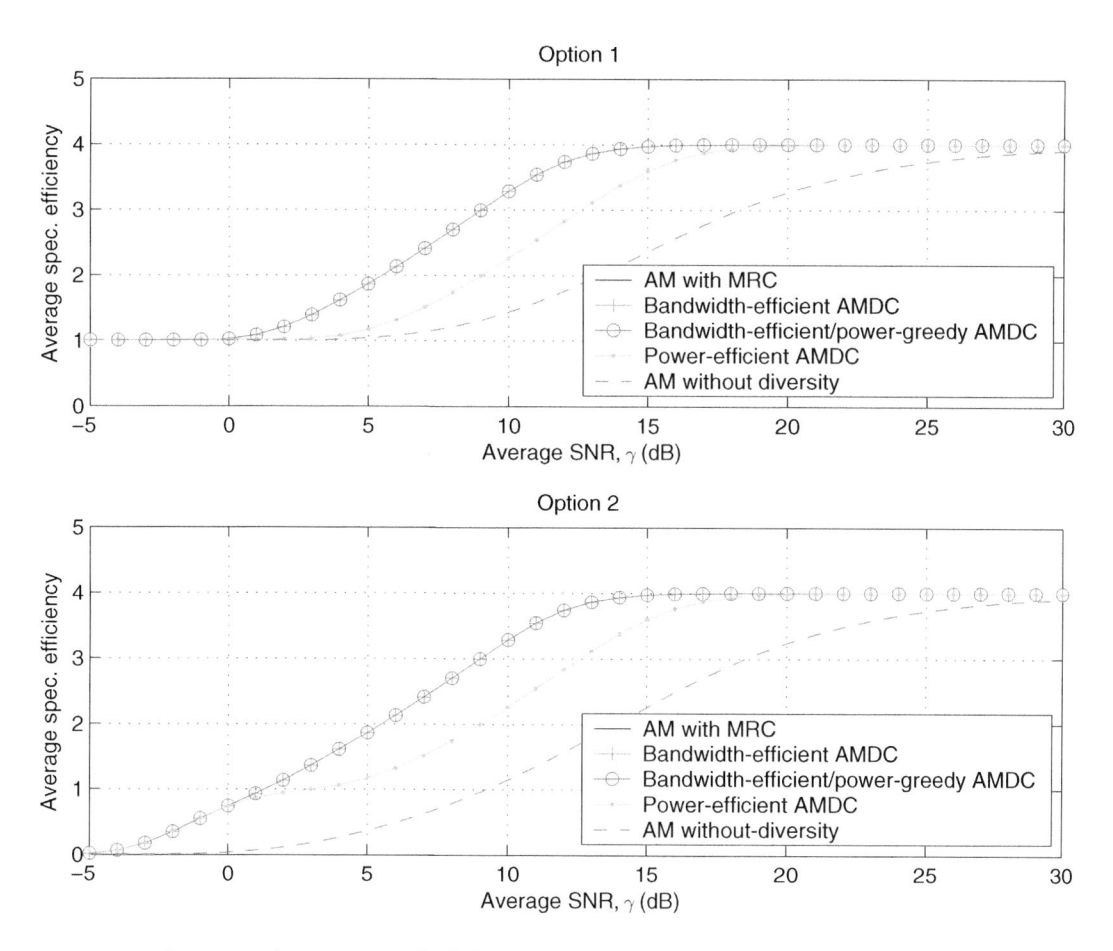

Figure 5.27 Average spectral efficiency of three AMDC schemes with two options as a function of the average path SNR $\overline{\gamma}$ ($L = 5, N = 4$, and $BER_0 = 10^{-3}$). Reprint with permission from [29]. ©2007 IEEE.

---

[2] Note that the average spectral efficiency is viewed as a valid performance measure only if the BER requirement is satisfied, then the results for option 1 are valid only for an average SNR greater than 1 dB.

violation of the instantaneous BER requirement. We can also see that the bandwidth-efficient/power-greedy AMDC scheme has the poorest error performance among three AMDC schemes, but still satisfies the BER requirement except over the low-SNR range with option 1. Comparing the error performance of the power-efficient AMDC and bandwidth-efficient AMDC, we find that over the low SNR range, the bandwidth-efficient scheme slightly outperforms the power-efficient scheme, whereas for the medium to high SNR range, the power-efficient scheme performs better. This is because when the SNR is low, the bandwidth-efficient scheme needs to combine more diversity branches, and when the SNR is higher the power-efficient scheme tends to settle on the lower adaptive modulation mode, which has better error protection.

Considering the three figures together, we can draw the following conclusions:

- There is a trade-off of power consumption versus spectral efficiency between the power-efficient AMDC scheme and the bandwidth-efficient AMDC scheme, which have comparable BER performance.

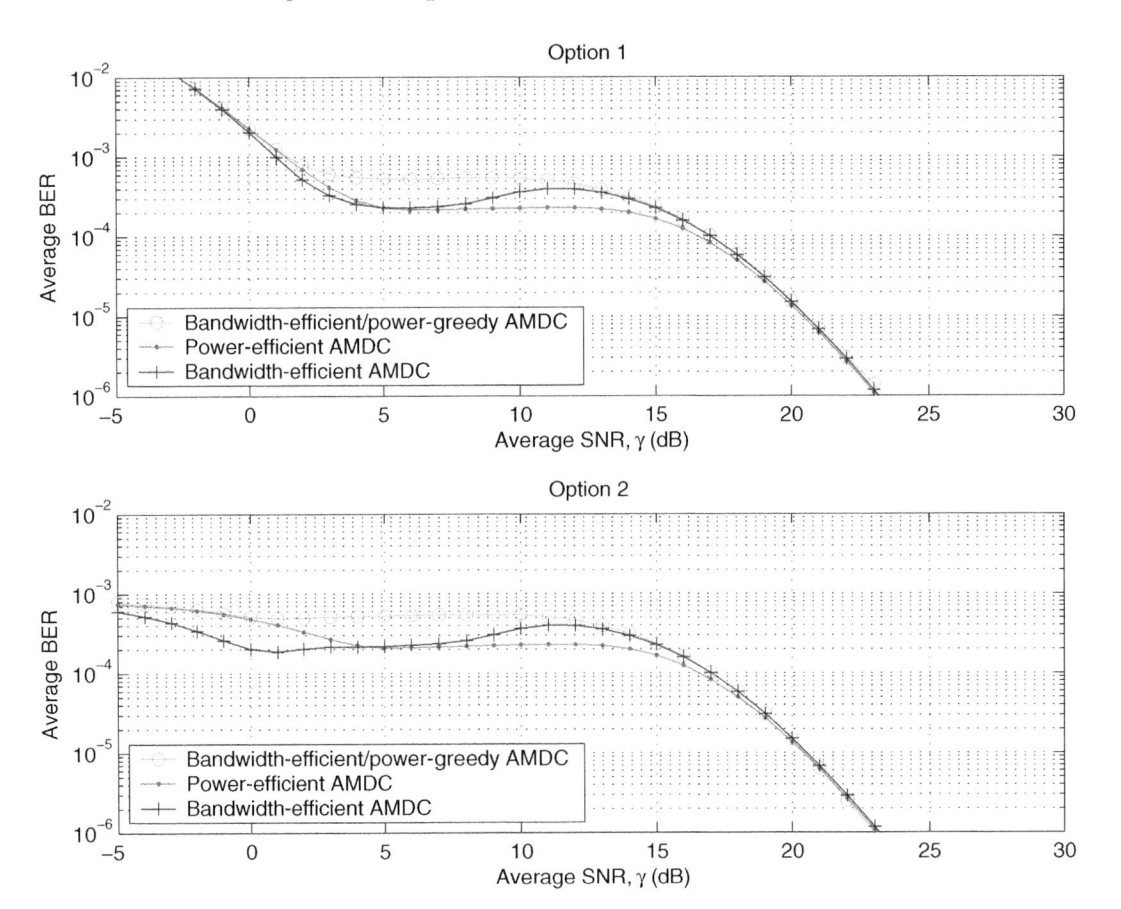

Figure 5.28 Average BER of three AMDC schemes with two options as a function of the average path SNR $\bar{\gamma}$ ($L = 5, N = 4$, and $BER_0 = 10^{-3}$). Reprint with permission from [29]. ©2007 IEEE.

- The bandwidth-efficient/power-greedy AMDC scheme offers better power efficiency than the bandwidth-efficient AMDC scheme at the cost of slightly poorer BER performance and with the same spectral efficiency.
- Option 1 for the worst-case scenario leads to a better spectral efficiency at the expense of higher power consumption and in violation of the BER requirement, whereas option 2 avoids the extra power consumption and BER requirement violation but causes a certain amount of transmission delay.

5.7 Summary

In this chapter, we introduced the idea of adaptive combining and its applications in various transmission scenarios, including traditional antenna reception diversity systems, transmit diversity systems, RAKE combining over soft handover region, and joint design of adaptive modulation and diversity combining. We carried out thorough trade-off analysis of performance versus complexity of the resulting systems. We demonstrated that adaptive combining can bring significant complexity and processing power savings to the wireless systems operating over fading environments, at a minimum or no loss in performance.

5.8 Further Reading

Minimum estimation and combining GSC (MEC-GSC) [30] extends the idea of MS-GSC by introducing a switching and examining combining (SEC) stage. Lioumpas et al. [31] propose an adaptive GSC scheme, which essentially applies the adaptive combining principle to the NT-GSC scheme. Bouida et al. combine adaptive combining based on MS-GSC with transmit power control in [32]. The joint design of adaptive modulation with OT-MRC scheme is presented in [33]. Lee and Ko extend the joint AMDC scheme to the multi-channel environment in [34]. Discrete transmit power control is considered together with the joint AMDC scheme in [35].

References

[1] H.-C. Yang and M.-S. Alouini, "MRC and GSC diversity combining with an output threshold," *IEEE Trans. Veh. Technol.*, 54, no. 3, pp. 1081–1090, 2005.

[2] S. W. Kim, D. S. Ha, and J. H. Reed, "Minimum selection GSC and adaptive low-power RAKE combining scheme," in *Proc. of IEEE Int. Symp. on Circuits and Systems. (ISCAS'03), Bangkok, Thailand*, vol. 4, May 2003, pp. 357–360.

[3] P. Gupta, N. Bansal, and R. K. Mallik, "Analysis of minimum selection H-S/MRC in Rayleigh fading," *IEEE Trans. Commun.*, 53, no. 5, pp. 780–784, 2005.

[4] R. K. Mallik, P. Gupta, and Q. T. Zhang, "Minimum selection GSC in independent Rayleigh fading," *IEEE Trans. Veh. Technol.*, 54, no. 3, pp. 1013–1021, 2005.

[5] H.-C. Yang, "New results on ordered statistics and analysis of minimum-selection generalized selection combining (GSC)," *IEEE Trans. Wireless Commun.*, 5, no. 7, 2006.

[6] M. K. Simon and M.-S. Alouini, *Digital Communications over Generalized Fading Channels*, 2nd ed., Wiley, 2004.

[7] I. S. Gradshteyn and I. M. Ryzhik, *Table of Integrals, Series, and Products*, 5th ed., Academic Press, 1994.

[8] H.-C. Yang and L. Yang, "Exact error analysis of output-threshold generalized selection combining (OT-GSC)," *IEEE Trans. Wireless Commun.*, 6, no. 9, pp. 3159–3162, 2007.

[9] V. Tarokh, N. Seshadri, and A. R. Calderbank, "Space-time codes for high data rate wireless communication: performance criterion and code construction," *IEEE Trans. Inf. Theory*, 44, no. 2, pp. 744–765, 1998.

[10] G. J. Foschini and M. J. Gans, "On limits of wireless communications in a fading environment when using multiple antennas," *Wirel. Pers. Commun.*, 6, no. 3, pp. 311–335, 1998.

[11] S. M. Alamouti, "A simple transmit diversity technique for wireless communications," *IEEE J. Sel. Areas Commun.*, 16, no. 8, pp. 1451–1458, 1998.

[12] D. A. Gore and A. J. Paulraj, "MIMO antenna subset selection with space-time coding," *IEEE Trans. Signal Processing*, 50, no. 10, pp. 2580–2588, 2002.

[13] A. F. Molisch, M. Z. Win, and J. H. Winters, "Reduced-complexity transmit/receive-diversity systems," *IEEE Trans. Signal Processing*, 51, no. 11, pp. 2729–2738, 2003.

[14] A. Ghrayeb and T. M. Duman, "Performance analysis of MIMO systems with antenna selection over quasi-static fading channels," *IEEE Trans. Veh. Technol.*, 52, no. 2, pp. 281–288, 2003.

[15] D. J. Love and R. W. Heath, "Diversity performance of precoded orthogonal space-time block codes using limited feedback," *IEEE Commun. Lett.*, 8, no. 5, pp. 305–307, 2004.

[16] S. Choi, Y.-C. Ko, and E. J. Powers, "Optimization of switched MIMO systems over Rayleigh fading channels," *IEEE Trans. Veh. Technol.*, 56, no. 1, pp. 103–114, 2007.

[17] S. Choi, H.-C. Yang, and Y.-C. Ko, "Performance analysis of transmit diversity systems with antenna replacement," *IEEE Trans. Veh. Technol.*, 57, no. 4, pp. 2588–2595, 2008.

[18] K.-H. Park, Y.-C. Ko, and H.-C. Yang, "Performance analysis of transmit diversity systems with multiple antenna replacement," *IEICE Trans. Commun.*, E91-B, no. 10, pp. 3281–3287, 2008.

[19] G. L. Stüber, *Principles of Mobile Communications*, 2nd ed., Kluwer Academic Publishers, 2000.

[20] S. Choi, M.-S. Alouini, K. A. Qaraqe, and H.-C. Yang, "Soft handover overhead reduction by RAKE reception with finger reassignment," *IEEE Trans. Commun.*, 56, no. 2, pp. 213–221, 2008.

[21] S. Choi, M.-S. Alouini, K. A. Qaraqe, and H.-C. Yang, "Fingers replacement method for RAKE receivers in the soft handover region," *IEEE Trans. Wireless Commun.*, 7, no. 4, pp. 1152–1156, 2008.

[22] M. Z. Win and J. H. Winters, "Analysis of hybrid selection/maximal-ratio combining in Rayleigh fading," *IEEE Trans. Commun.*, 47, no. 12, pp. 1773–1776, 1999.

[23] M.-S. Alouini and M. K. Simon, "An MGF-based performance analysis of generalized selective combining over Rayleigh fading channels," *IEEE Trans. Commun.*, 48, no. 3, pp. 401–415, 2000.

[24] H. Holma and A. Toskala, *WCDMA for UMTS*, revised ed., Wiley, 2001.

[25] A. J. Goldsmith and S.-G. Chua, "Adaptive coded modulation for fading channels," *IEEE Trans. Commun.*, 46, no. 5, pp. 595–602, 1998.

[26] M.-S. Alouini and A. J. Goldsmith, "Adaptive modulation over Nakagami fading channels," *Kluwer J. Wireless Commun.*, 13, no. 1–2, pp. 119–143, 2000.

[27] K. J. Hole, H. Holm, and G. E. Oien, "Adaptive multidimensional coded modulation over flat fading channels," *IEEE J. Select. Areas Commun.*, 18, no. 7, pp. 1153–1158, 2000.

[28] A. Goldsmith, *Wireless Communications*, Cambridge University Press, 2005.

[29] H.-C. Yang, N. Belhaj, and M.-S. Alouini, "Performance analysis of joint adaptive modulation and diversity combining over fading channels," *IEEE Trans. Commun.*, 55, no. 3, pp. 520–528, 2007.

[30] M.-S. Alouini and H.-C. Yang, "Minimum estimation and combining generalized selection combining (MEC-GSC)," *IEEE Trans. Wireless Commun.*, 6, no. 7, pp. 526–532, 2007.

[31] A. S. Lioumpas, G. K. Karagiannidis, and T. A. Tsiftsis, "Adaptive generalized selection combining (A-GSC) receivers," *IEEE Trans. Wireless Commun.*, 7, no. 12, pp. 5214–5219, 2008.

[32] Z. Bouida, N. Belhaj, M.-S. Alouini, and K. A. Qaraqe, "Minimum selection GSC with downlink power control," *IEEE Trans. Wireless Commun.*, 7, no. 7, pp. 2492–2501, 2008.

[33] Y.-C. Ko, H.-C. Yang, S.-S. Eom, and M.-S. Alouini, "Adaptive modulation and diversity combining based on output-threshold MRC," *IEEE Trans. Wireless Commun.*, 6, no. 10, pp. 3728–3737, 2007.

[34] S.-D. Lee and Y.-C. Ko, "Exact performance analysis of hybrid adaptive modulation schemes in multi-channel system," *IEEE Trans. Wireless Commun.*, 8, no. 6, pp. 3206–3215, 2009.

[35] A. Gjendemsjø, H.-C. Yang, G. E. Øien, and M.-S. Alouini, "Joint adaptive modulation and diversity combining with downlink power control," *IEEE Trans. Veh. Techno.*, 57, no. 4, pp. 2145–2152, 2008.

6 Multiuser Scheduling

Multiple antenna transmission techniques can improve the performance of wireless transmission by exploiting the diversity benefit. In certain practical scenarios, however, it might be challenging to implement multiple antennas at wireless terminals. If multiple users are present, we can extract a certain diversity benefit by exploring multiple antennas at different users [1, 2]. Since users are typically separately located, different user channels will most likely experience independent fading. At any given time instant, it is with high probability that at least one user channel will have favorable condition. The overall system performance will improve if channel access is always granted to the user with favorable channel condition. The resulting multiuser diversity transmission technique with user scheduling lead to an interesting new trade-off of performance gain versus user fairness.

In this chapter, we analyze the performance of several multiuser diversity transmission technologies and discuss their associated implementation complexity whenever appropriate. We first review the basics of multiuser diversity, including capacity benefit, fairness issue, and feedback load. We then analyze single-user scheduling schemes in terms of channel access statistics, such as channel access rate and access duration of an arbitrary user. After that, we extend the study to multiuser parallel scheduling. We present three different multiuser scheduling schemes and analyze their performance and efficiency. The chapter is concluded with a discussion on power allocation for parallel multiuser scheduling.

6.1 Multiuser Selection Diversity

We consider a generic cell with one base station (BS) and K active single-antenna users, as illustrated in Fig. 6.1. For the sake of clarity, we assume that the BS has only one antenna (the extension to a multiple-antenna BS will be considered in the next chapter) and the system operates in the time division multiple access (TDMA) fashion. During each TDMA time slot, the system will schedule the user with the favorable channel conditions for service. Assuming frequency flat fading, the quality of user channels can be characterized by the instantaneous received SNR, denoted by $\gamma_k, k = 1, 2, \ldots, K$.

To explore the diversity benefit over such a multiuser environment, a natural strategy is to schedule the user experiencing the largest instantaneous received signal-to-noise

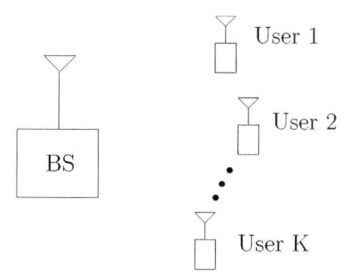

Figure 6.1 Sample multiuser system.

ratio (SNR) for data transmission during a particular time slot. Mathematically speaking, user k^*, where $k^* = \arg\max_k\{\gamma_k\}$, will be scheduled for transmission. It follows that the received SNR of the scheduled user is the largest one among K user SNRs, which are inherently independent as users are randomly distributed in the coverage area. The cumulative distribution function (CDF) of the received SNR over the time slot is given by [3]

$$F_{\gamma_{k^*}}(x) = \prod_{k=1}^{K} F_{\gamma_k}(x). \tag{6.1}$$

If users experience identical fading, which may be possible with a proper power-control mechanism, the probability density function (PDF) of the received SNR at the scheduled user becomes

$$p_{\gamma_{k^*}}(x) = K p_\gamma(x)[F_\gamma(x)]^{K-1}, \tag{6.2}$$

which is the PDF of combiner output SNR with K branch selection combining over i.i.d. diversity paths. Therefore, multiuser selection diversity can achieve the same diversity gain as selection-based antenna reception diversity systems.

Multiuser diversity technique improves the performance of wireless systems over fading channels by following a completely different approach from conventional diversity techniques. Conventional diversity techniques try to reduce deep channel fade by combining signal from multiple diversity paths together. Multiuser diversity, however, explores favorable fading channel conditions. As such, multiuser diversity exploits multipath fading rather than compensates for it. In certain scenarios, fading might be intentionally introduced with some random beamforming approach [4]. Multiuser diversity enjoys several inherent advantages over antenna diversity, including a simpler receiver structure, as a single antenna per receiver is sufficient, and naturally independent fading channels, since users are usually geographically separated. On the other hand, multiuser diversity requires additional system resources to collect users' channel state information, especially for non-reciprocal transmission scenarios. Furthermore, multiuser diversity may lead to severe fairness issue across users in the short term, while long-term fairness may be achieved through channel normalization and/or utilizing historical throughput information.

In real-world environment, the average channel gain corresponding to different users differs with the experienced path loss and shadowing processes. If the power-control mechanism is not available or not perfect, then the user with better average fading conditions might be scheduled much more frequently, which leads to the fairness issue among users. In this scenario, we can improve fairness among users by scheduling users based on their normalized SNR. Specifically, let $\overline{\gamma}_k$ denote the average received SNR of user k. We first calculated user k's normalized SNR, given by $\tilde{\gamma}_k = \frac{\gamma_k}{\overline{\gamma}_k}$ [5]. Then, we schedule User k^*, where

$$k^* = \arg\max_k \left\{ \frac{\gamma_k}{\overline{\gamma}_k} \right\}, \quad (6.3)$$

for transmission during a particular time slot. We can show that the PDF of the normalized SNR $\tilde{\gamma}_k$ under Rayleigh fading is given by

$$p_{\tilde{\gamma}_k}(x) = e^{-x}, \; k = 1, 2, \ldots, K. \quad (6.4)$$

The unfairness among users due to different path loss/shadowing effects is eliminated. It can be shown that the normalized SNR-based user scheduling scheme can achieve long-term fairness.

Alternatively, we can improve fairness by taking into account the historical throughput information of each user during user selection [4]. The basic principle is that those users who have received more channel access should be weighted less during the competition for channel access during upcoming time slots. A popular strategy to implement this principle is the so-called proportional fair scheduling, which was shown to be able to achieve long-term fairness. With proportional fair scheduling, the selected user over the ith time slot, denoted by $k^*(i)$, should have the maximum normalized instantaneous capacity, i.e.,

$$k^*(i) = \arg\max_k \left\{ \frac{C_k(i)}{R_k(i)} \right\}, \quad (6.5)$$

where $C_k(i)$ is the instantaneous rate of user k over slot i, given by

$$C_k(i) = \log_2(1 + \gamma_k(i)), \quad (6.6)$$

and $R_k(i)$ denotes the historical throughput of user k for up to the $i-1$th time slots, which can be updated with the following relationship:

$$R_k(i) = R_k(i-1), k \neq k^*, \quad (6.7)$$
$$R_{k^*}(i) = R_{k^*}(i-1) + C_{k^*}(i-1). \quad (6.8)$$

6.2 Performance Analysis of Multiuser Selection Diversity

In this section, we analyze the performance of multiuser selection diversity. We use system ergodic capacity as the performance metric to evaluate both absolute SNR based and normalized SNR based scheduling strategies. Meanwhile, the viability of these multiuser scheduling schemes in practical systems largely depends on the number of active users and the channel changing rate [1]. Specifically, while having more users increases the multiuser diversity gain, the average duration that each individual user is scheduled to communicate decreases. Clearly, a key measure to evaluate the viability of the multiuser scheduling algorithms is to determine the average channel access time based on the fading rate and the number of users. In this section, we also study the average access time (AAT) and the average access rate (AAR) of individual users in a multiuser environment [6]. AAT will be useful in setting scheduling interval since, if the interval is longer than AAT, the scheduler cannot track the channel variation fast enough and the scheduling gain will be seriously reduced. On the other hand, if the scheduling interval is too short compared to AAT, there will be too much unnecessary feedback. We also introduce another metric, termed as average waiting time (AWT), to characterize how long, on average, a user has to wait for the next access. AWT is important for configuring time-out timer for upper-layer protocols.

6.2.1 Absolute SNR-Based Scheduling

With the statistics of the received SNR of the scheduled user with absolute SNR-based scheduling given in (6.2), we can evaluate the system ergodic capacity of multiuser diversity transmission system [5]. Mathematically, the system ergodic capacity can be calculated by averaging the instantaneous capacity over the distribution of the received SNR, as

$$C_{\text{sys}} = \int_0^\infty \log_2(1+\gamma) p_{\gamma_{k^*}}(\gamma) d\gamma. \tag{6.9}$$

For i.i.d. Rayleigh fading, after appropriate substitution and carrying out integration, the capacity with multiuser diversity transmission specializes to

$$C_{\text{sys}} = K \log_2(e) \sum_{k=0}^{K-1} (-1)^k \frac{(K-1)!}{(k+1)!(K-k-2)!} e^{(k+1)/\overline{\gamma}} E_1\left(\frac{k+1}{\overline{\gamma}}\right), \tag{6.10}$$

where $E_1(\cdot)$ is the exponential integral function of the first order [7], defined by

$$E_1(x) = \int_1^\infty \frac{e^{-xt}}{t} dt, x \geq 0. \tag{6.11}$$

The multiuser diversity gain can also be demonstrated with the asymptotic behavior of γ_{k^*} when K approaches infinity. For the i.i.d. Rayleigh fading scenario, it can be shown that γ_{k^*}, which is the largest one of K i.i.d. exponential random variables, has a limiting distribution of the Gumbel type. More specifically, the limiting CDF of γ_{k^*} as K approaches infinity is given by

$$\lim_{K\to+\infty} F_{\gamma_{1:K}}(x) = \exp\left(-e^{-x+\log K}\right). \tag{6.12}$$

It follows that the average SNR of the selected user increases at the rate of $\log K$ as K approaches infinity [4].

With absolute SNR-based scheduling, the scheduler selects user i if and only if the instantaneous SNR of the user i is larger than that of every other user, i.e.,

$$\gamma_i \geq \gamma_j, j = 1, \ldots, L; j \neq i. \tag{6.13}$$

Equivalently, we can write

$$\gamma_i \geq \gamma_*, \tag{6.14}$$

where γ_* is the maximum SNR of other users, i.e.,

$$\gamma_* = \max_{j=1,\ldots,L, j \neq i} (\gamma_j). \tag{6.15}$$

Therefore, the AAR of user i is precisely the average number of times that the process $r = \sqrt{\gamma_i/\gamma_*} = \alpha_i/\alpha_*$ crosses level 1 per unit time, where α_i is the amplitude of the complex channel gain for user i and α_* is the largest amplitude of other users. It follows that the AAR of user i can be evaluated as the average level crossing rate (LCR) [8, Eq. 2.90] of the process r at level 1, which is given by

$$N_i = \int_0^\infty \dot{r} p_{r,\dot{r}}(1,\dot{r}) d\dot{r}, \tag{6.16}$$

where $\dot{r}$ is the time derivative of the process r and $p_{r,\dot{r}}(1,\dot{r})$ is the joint PDF of r and $\dot{r}$, which is given by [9, Eq. (9)][10, Eq. (15)]:

$$p_{r,\dot{r}}(1,\dot{r}) = \int_0^\infty \int_{-\infty}^\infty \alpha_*^2 p_{\alpha_i}(\alpha_* r) p_{\dot{\alpha}_i}(\dot{r}\alpha_* + \dot{\alpha}_* r) \tag{6.17}$$
$$\times p_{\alpha_*, \dot{\alpha}_*}(\alpha_*, \dot{\alpha}_*) d\dot{\alpha}_* d\alpha_*.$$

For Rayleigh fading user channels, the PDF of the channel amplitude α_i is given by

$$p_{\alpha_i}(\alpha) = \frac{2\alpha}{\Omega_i} \exp\left(-\frac{\alpha^2}{\Omega_i}\right), \quad \alpha \geq 0, \tag{6.18}$$

where Ω_i is the short-term average channel power gain of the ith user. The time derivative of signal amplitude process $\dot{\alpha}_i$ follows a normal distribution with zero mean and is independent of the signal amplitude α_i, with PDF given by [11, Eq. (1.3-34)]

$$p_{\dot{\alpha}_i}(\dot{\alpha}) = \frac{1}{\sqrt{2\pi}\sigma_i} \exp\left(-\frac{\dot{\alpha}^2}{2\sigma_i^2}\right), \tag{6.19}$$

where $\sigma_i^2 = \Omega_i \pi^2 f_i^2$ for isotropic scattering and f_i is the maximum Doppler frequency shift of the ith user. Finally, the joint PDF of α_* and $\dot{\alpha}_*$ is given by [12]

$$p_{\alpha_*, \dot{\alpha}_*}(x, \dot{x}) = \sum_{j=1, j \neq i}^L \frac{1}{\sqrt{2\pi}\sigma_j} \exp\left(-\frac{\dot{x}^2}{2\sigma_j^2}\right) \frac{2x}{\Omega_j} \exp\left(-\frac{x^2}{\Omega_j}\right) \tag{6.20}$$

$$\times \prod_{k=1, k \neq i, j}^L \left[1 - \exp\left(-\frac{x^2}{\Omega_k}\right)\right]$$

$$= \sum_{j=1, j \neq i}^L \sum_{\tau \in T_{ij}^L} \text{sign}(\tau) \frac{1}{\sqrt{2\pi}\sigma_j} \exp\left(-\frac{\dot{x}^2}{2\sigma_j^2}\right) \frac{2x}{\Omega_j} \exp\left(-\frac{x^2}{\Omega_j} - \tau x^2\right),$$

where T_{ij}^L is the set obtained by expanding the product $\prod_{k=1,k\neq i,j}^{L}\left[1-\exp\left(-\frac{x^2}{\Omega_k}\right)\right]$ then taking the natural logarithm of each term [13, 14], and $\text{sign}(\tau)$ is the corresponding sign of each term in the expansion. After proper substitution and carrying out integration, we can obtain the analytical expression of AAR for user i as

$$N_i = \sum_{j=1,j\neq i}^{L} \sum_{\tau \in T_{ij}^L} \text{sign}(\tau) \frac{\pi}{\sqrt{2}} \frac{\sqrt{\Omega_i f_i^2 + \Omega_j f_j^2}}{\Omega_i \Omega_j} \left(\frac{1}{\Omega_i} + \frac{1}{\Omega_j} + \tau\right)^{-3/2}. \quad (6.21)$$

The AAT of user i, denoted by T_i, is defined as the average time duration of user i's channel access. It can be shown that the AAT of user i can be calculated as

$$T_i = \frac{P_i}{N_i}, \quad (6.22)$$

where P_i denotes the probability that user i accesses the channel at any time instant, or equivalently, the average access probability (AAP) of user i. P_i can be calculated as

$$P_i = \Pr[\alpha_i \geq \alpha_*] = \int_0^\infty p_{\alpha_i}(x) \int_0^x p_{\alpha_*}(y) dy dx. \quad (6.23)$$

For the Rayleigh fading case under consideration, we can show after proper substitution and manipulation that the AAP of user i is given by

$$P_i = \int_0^\infty \frac{2x}{\Omega_i} \exp\left(-\frac{x^2}{\Omega_i}\right) \prod_{k=1,k\neq i}^{L} \left[1 - \exp\left(-\frac{x^2}{\Omega_k}\right)\right] dx \quad (6.24)$$

$$= \sum_{\tau' \in T_i^L} \text{sign}(\tau) \frac{1}{1 + \tau' \Omega_i}.$$

Similarly, the AWT of user i, which characterizes the average time duration a user has to wait for the next channel access, can be calculated as

$$W_i = \frac{1 - P_i}{N_i} = \frac{1}{N_i} - T_i. \quad (6.25)$$

6.2.2 Normalized SNR-Based Scheduling

The CDF of the scheduled user's SNR γ_k^* with the normalized SNR-based user scheduling can be calculated as

$$F_{\gamma_{k^*}}(\gamma) = \Pr[\gamma_{k^*} < \gamma] \quad (6.26)$$

$$= \sum_{i=1}^{K} \Pr[\gamma_{k^*} < \gamma; \gamma_{k^*} = \gamma_i]$$

$$= \sum_{i=1}^{K} \int_0^{\frac{\gamma}{\gamma_i}} p_{\tilde{\gamma}_k}(x) \prod_{k=1,k\neq i}^{K} F_{\tilde{\gamma}_k}(x) dx.$$

After taking derivatives with respect to γ, the PDF of the received SNR at the scheduled user is obtained, after some manipulations, as

$$p_{\gamma_{k^*}}(\gamma) = \sum_{i=1}^{K} \frac{1}{\bar{\gamma}_i} p_{\tilde{\gamma}_i}(\frac{\gamma}{\bar{\gamma}_i}) \prod_{k=1, k \neq i}^{K} F_{\tilde{\gamma}_k}(\frac{\gamma}{\bar{\gamma}_i}). \qquad (6.27)$$

For the Rayleigh fading model, after proper substitution and carrying out integration, the system capacity with normalized SNR-based scheduling is given by

$$C_{\text{sys}} = \log_2(e) \sum_{i=1}^{K} \sum_{k=0}^{K-1} \frac{(-1)^k}{1+k} \binom{K-1}{k} e^{(k+1)/\bar{\gamma}_i} E_1\left(\frac{k+1}{\bar{\gamma}_i}\right). \qquad (6.28)$$

With normalized SNR-based scheduling, the BS schedules the user with the largest normalized SNR, or equivalently, the user with largest normalized channel amplitude, defined as $\beta_k = \frac{\alpha_k}{\sqrt{\Omega_k}}$. The AAR and AAT of user i in this case can be calculated using the up-crossing rate of the process,

$$r = \beta_i / \beta_*, \qquad (6.29)$$

where

$$\beta_* = \max_{j=1,\ldots,L, j \neq i}(\beta_j), \qquad (6.30)$$

at level 1. The rate can be similarly calculated as the absolute SNR-based case but using the PDFs of β and $\dot{\beta}$. As the result of normalization, the PDF of β_k becomes

$$p_{\beta_k}(x) = 2x \exp(-x^2), \qquad (6.31)$$

the PDF of its time derivative $\dot{\beta}_k$ is

$$p_{\dot{\beta}_k}(\dot{x}) = \frac{1}{\sqrt{2\pi}\sigma_k} \exp\left(-\frac{\dot{x}^2}{2\sigma_k^2}\right), \qquad (6.32)$$

where $\sigma_k^2 = \pi^2 f_k^2$, and the joint PDF of β_* and $\dot{\beta}_*$ becomes

$$p_{\beta_*, \dot{\beta}_*}(x, \dot{x}) = \sum_{j=1, j \neq i}^{L} \frac{2x}{\sqrt{2\pi}\sigma_j} \exp\left(-\frac{\dot{x}^2}{2\sigma_j^2} - x^2\right)[1 - \exp(-x^2)]^{L-2}. \qquad (6.33)$$

Following the exact same steps as in the absolute SNR-based scheduling case, we get the AAR of user i, N_i, as

$$N_i = \sum_{j=1, j \neq i}^{L} \frac{\pi}{\sqrt{2}} \sqrt{f_i^2 + f_j^2} \sum_{n=0}^{L-2} \binom{L-2}{n} (-1)^n (n+2)^{-3/2}. \qquad (6.34)$$

With normalized SNR-based scheduling, the access probability of user i becomes $1/L$. It follows that the AAT of user i is given by

$$T_i = \frac{1}{N_i L}. \tag{6.35}$$

After proper substitution and some rearrangement, we can rewrite T_i as

$$T_i = \frac{C}{\sum_{j=1, j \neq i}^{L} \sqrt{f_i^2 + f_j^2}}, \tag{6.36}$$

where C denotes some constant independent of the Doppler frequency shift. If the network has a fast-moving user k among many other stationary users, we have

$$T_i = \begin{cases} \frac{C}{(L-1) f_k}, & i = k; \\ \frac{C}{f_k}, & i \neq k, \end{cases} \tag{6.37}$$

which implies that the AAT of the fast-moving user is roughly $1/(L-1)$ times that of other stationary users.

6.2.3 Numerical Results

Fig. 6.2 compares the system capacity of multiuser selection diversity systems over Rayleigh fading channels. In particular, the capacity over i.i.d. fading and i.n.d. fading with both absolute SNR-based scheduling and normalized SNR-based scheduling are plotted as the function of the number of users, K. We can see that the system capacity increases with K as the system enjoys more diversity gain. We also observe that the fairness achieved with normalized SNR-based scheduling comes at the cost of a capacity loss in comparison with the absolute SNR-based scheduling over i.n.d. fading. This is intuitively correct, since the absolute SNR-based scheduling tends to exploit the more frequent favorable channel conditions of strong (on average) users at the expense of a certain unfairness to relatively weak users. On the other hand, the normalized SNR-based scheduling tends to protect weak (on average) users while sacrificing a fraction of system capacity.

We study the AAT and AAR performance of the normalized SNR-based scheduling scheme in Figs. 6.3 and 6.4. Fig. 6.3 plots the AAR for user i versus its maximum Doppler frequency shift f_i in hertz when the number of users $K = 4$, 8, and 16. We set the maximum Doppler frequency shifts for users other than user i to 50 Hz. It can be seen that, when f_i increases, the AAR of user i increases as a result of more rapid channel fluctuation. On the other hand, over the same f_i range, the AAR of user i decreases when there are more users in the system.

Fig. 6.4 plots the AAT of each user versus the number of users K for different values of the common maximum frequency shifts of all users denoted by f_m. We can see that, when the number of users increases, the AAT for each user decreases. When the maximum Doppler shift increases, the AAT of each user also decreases as the result of faster channel variation. Note that in this figure, the AAT is in the range of milliseconds.

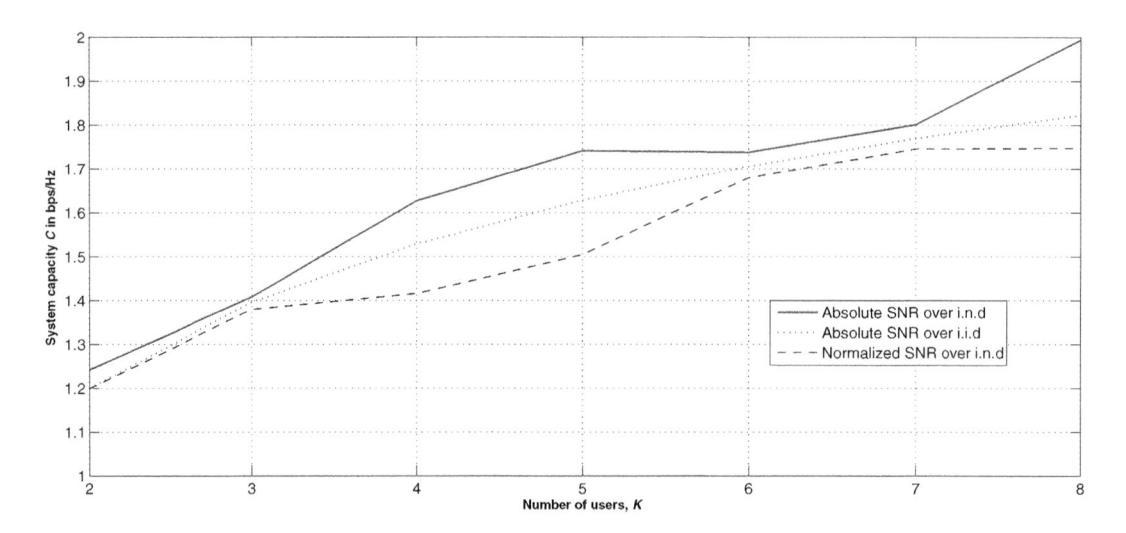

Figure 6.2 Capacity comparison of multiuser selection diversity schemes over fading channels.

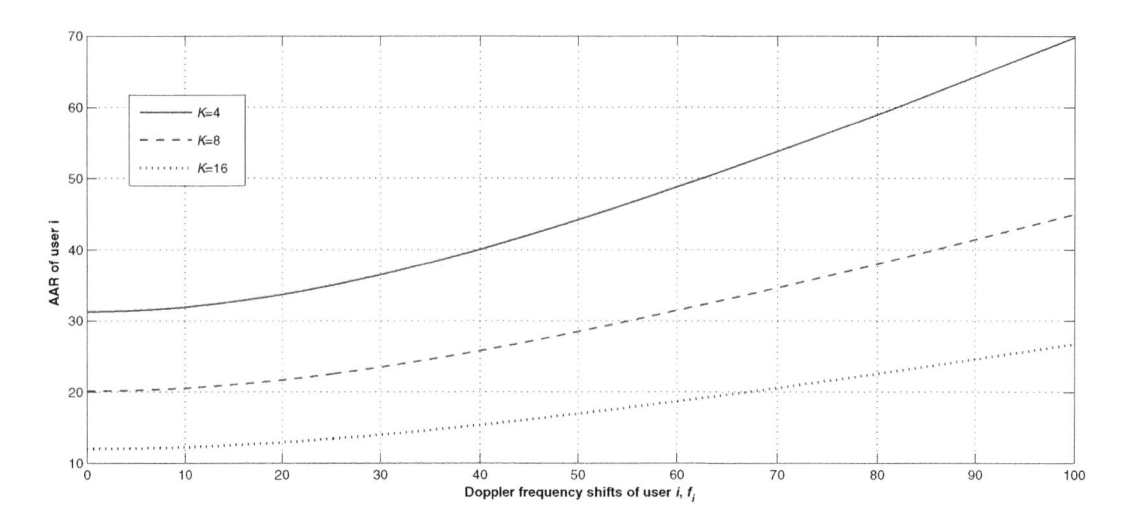

Figure 6.3 AAR of user i versus its maximum Doppler frequency shift, f_i, when the maximum Doppler frequency shifts of all other users are equal to 50 Hz.

6.3 Multiuser Diversity with Limited Feedback

The availability of users' instantaneous channel state information is essential to the implementation of multiuser diversity transmission. For the downlink transmission scenario, the BS needs to collect the channel quality information corresponding to all users in order to select the best user for data transmission. This will translate into a huge amount of channel probing and/or channel quality feedback, which consumes additional system resources. It is of great practical importance if we can reduce the feedback load

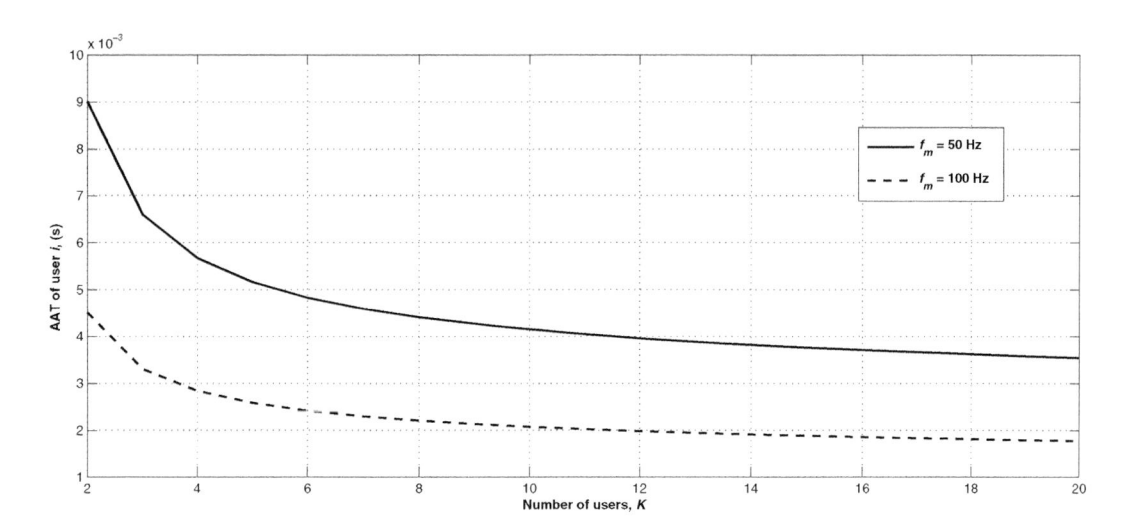

Figure 6.4 Normalized AAT versus the number of users K with different maximum Doppler frequency shifts.

while maintaining nearly the same multiuser diversity gain. In this context, the selective multiuser diversity scheduling scheme has demonstrated to be an attractive solution [5].

6.3.1 Selective Multiuser Diversity

The basic idea of selective multiuser diversity is to allow only those users whose channel qualities are good enough, i.e., with received SNR above a certain threshold γ_T, to feed back their channel state information. With multiuser diversity scheduling, the BS will select a single best user for transmission. As such, only users with a good enough channel have a chance to be selected. Furthermore, the BS only needs the channel quality of the scheduled user for capacity-achieving rate adaptation. Intuitively, we can expect the selective multiuser diversity approach to achieve the same diversity gain as conventional multiuser diversity if at least one user feeds back. On the other hand, selective multiuser scheduling may lead to scheduling outage when no user feeds back. It is not difficult to show that the probability of scheduling outage under the assumption of i.i.d. faded user channels is

$$P_{\text{out}} = [F_\gamma(\gamma_T)]^K, \tag{6.38}$$

where $F_\gamma(\cdot)$ denotes the common CDF of user SNRs.

An easy solution to avoid the scheduling outage is to randomly probe and select a user for data transmission when no user feeds back. The SNR of the scheduled user is given by

$$\gamma_{k^*} = \begin{cases} \max_k\{\gamma_k\}, & \text{if no outage;} \\ \text{rand}_k\{\gamma_k\}, & \text{if outage.} \end{cases} \tag{6.39}$$

It follows that the CDF of the received SNR γ_k^* over i.i.d. fading channels can be shown to be

$$F_{\gamma_{k^*}}(x) = \begin{cases} \sum_{k=1}^{K} \dfrac{K!}{k!(K-k)!} [F_\gamma(\gamma_T)]^{K-k} \\ \quad \times [F_\gamma(x) - F_\gamma(\gamma_T)]^k, & x \geq \gamma_T; \\ [F_\gamma(\gamma_T)]^{K-1} F_\gamma(x), & x < \gamma_T. \end{cases} \quad (6.40)$$

The PDF of γ_k^* can be routinely derived after taking the derivative with respect to x as

$$p_{\gamma_{k^*}}(x) = \begin{cases} \sum_{k=1}^{K} \dfrac{K!}{(k-1)!(K-k)!} [F_\gamma(\gamma_T)]^{K-k} \\ \quad \times [F_\gamma(x) - F_\gamma(\gamma_T)]^{k-1} p_\gamma(x), & x \geq \gamma_T; \\ [F_\gamma(\gamma_T)]^{K-1} p_\gamma(x), & x < \gamma_T, \end{cases} \quad (6.41)$$

which can be utilized to evaluate the system capacity. The average feedback load (AFL) in terms of the average number of SNR feedbacks per scheduling time slot can be determined as

$$\overline{N} = \sum_{k=1}^{K} k \Pr[k \text{ users feedback}] \quad (6.42)$$

$$= \sum_{k=1}^{K} \frac{K!}{(K-k)!(k-1)!} [F_\gamma(\gamma_T)]^{K-k} [1 - F_\gamma(\gamma_T)]^k.$$

When the user channels are independent but not identically distributed, we can utilize normalized SNR during feedback thresholding as well as the user selection process. In particular, user k will feed back if its normalized SNR $\tilde{\gamma}_k = \frac{\gamma_k}{\bar{\gamma}_k}$ is greater than a particular threshold $\tilde{\gamma}_T$. It follows that the probability of scheduling outage is given by

$$P_{\text{out}} = [F_{\tilde{\gamma}}(\tilde{\gamma}_T)]^K, \quad (6.43)$$

where $F_{\tilde{\gamma}}(\cdot)$ is the CDF of normalized user SNRs. The AFL in terms of the average number of SNR feedbacks per scheduling time slot can be determined as

$$\overline{N} = \sum_{k=1}^{K} \frac{K!}{(K-k)!(k-1)!} [F_{\tilde{\gamma}}(\tilde{\gamma}_T)]^{K-k} [1 - F_{\tilde{\gamma}}(\tilde{\gamma}_T)]^k. \quad (6.44)$$

The statistics of the received SNR at the scheduled user can also be.

6.3.2 Switched Multiuser Scheduling

Switched multiuser scheduling further simplifies the user selection procedure and reduces the feedback load by looking for an acceptable user, i.e., a user with channel quality above a predefined switching threshold, instead of the best user. In particular, the user scheduling process works as follows. The BS starts probing the users in a sequential fashion, requesting the SNR of a user and then comparing it to the switching threshold.

If the SNR is below the threshold, the BS moves on to another user. This user SNR probing/checking process continues until (1) either a user with SNR above the threshold is found (this user is selected for the subsequent transmission), (2) all user SNRs have been examined without finding an acceptable user, in which case either the last examined user is selected (for simplicity), or the best user among all the probed users is selected, or the transmission is held for a period longer than the channel coherence time before a new sequential search [15].

To analyze the performance of the switched multiuser diversity system, we need to derive the statistics of the scheduled user's SNR. We first consider the case that the last examined user is selected when no acceptable user is found. According to the mode of operation of switched multiuser scheduling, the SNR of the scheduled user is given by

$$\gamma_{k^*} = \begin{cases} \gamma_1, & \gamma_1 \geq \gamma_T; \\ \gamma_2, & \gamma_1 < \gamma_T, \gamma_2 \geq \gamma_T; \\ \gamma_3, & \gamma_1 < \gamma_T, \gamma_2 < \gamma_T, \gamma_3 \geq \gamma_T; \\ \vdots & \vdots \\ \gamma_K, & \gamma_i < \gamma_T, i = 1, 2, \ldots, K-1. \end{cases} \quad (6.45)$$

It follows that the CDF of the scheduled user's SNR over i.n.d. fading channels is given by

$$F_{\gamma_{k^*}}(x) = \begin{cases} \prod_{k=1}^{K-1} F_{\gamma_k}(\gamma_T) F_{\gamma_K}(x), & x < \gamma_T; \\ \sum_{k=1}^{K} \prod_{j=0}^{k-1} F_{\gamma_j}(\gamma_T)[F_{\gamma_k}(x) - (F_{\gamma_k}(\gamma_T)] \\ + \prod_{k=1}^{K} F_{\gamma_k}(\gamma_T), & x \geq \gamma_T. \end{cases} \quad (6.46)$$

The PDF of the scheduled user's SNR can then be obtained by taking the derivative as

$$p_{\gamma_{k^*}}(x) = \begin{cases} \prod_{k=1}^{K-1} F_{\gamma_k}(\gamma_T) p_{\gamma_K}(x), & x < \gamma_T; \\ \sum_{k=1}^{K} [\prod_{j=0}^{k-1} F_{\gamma_j}(\gamma_T)] p_{\gamma_k}(x), & x \geq \gamma_T. \end{cases} \quad (6.47)$$

If the best user among all the probed users is selected when no acceptable user can be found, the mode of operation can be summarized as

$$\gamma_{k^*} = \begin{cases} \gamma_1, & \gamma_1 \geq \gamma_T; \\ \gamma_2, & \gamma_1 < \gamma_T, \gamma_2 \geq \gamma_T; \\ \gamma_3, & \gamma_1 < \gamma_T, \gamma_2 < \gamma_T, \gamma_3 \geq \gamma_T; \\ \vdots & \vdots \\ \max_k\{\gamma_k\}, & \gamma_i < \gamma_T, i = 1, 2, \ldots, K-1. \end{cases} \quad (6.48)$$

The PDF of the scheduled user's SNR over i.n.d fading channels becomes

$$p_{\gamma_{k^*}}(x) = \begin{cases} \sum_{k=1}^{K} \prod_{j=1, j\neq k}^{K} F_{\gamma_j}(x) p_{\gamma_k}(x), & x < \gamma_T; \\ \sum_{k=1}^{K} [\prod_{j=0}^{k-1} F_{\gamma_j}(\gamma_T)] p_{\gamma_k}(x), & x \geq \gamma_T. \end{cases} \quad (6.49)$$

The feedback load of switched multiuser scheduling depends on the channel realization. In particular, the system will need k user feedback when the first $k-1$ users are found unacceptable but the kth user is acceptable. As such, the number of user feedback is a discrete random variable with probability mass function (PMF) given by

$$\Pr[N = k] = \begin{cases} \prod_{j=1}^{k-1} F_{\gamma_j}(\gamma_T)(1 - F_{\gamma_k}(\gamma_T)), & k = 1, 2, \ldots, K-1; \\ \prod_{j=1}^{K-1} F_{\gamma_j}(\gamma_T), & k = K. \end{cases} \quad (6.50)$$

For i.i.d. fading user channels, the AFL per scheduling interval is

$$N = \frac{1 - F_\gamma(\gamma_T)^K}{1 - F_\gamma(\gamma_T)}, \quad (6.51)$$

where $F_\gamma(\cdot)$ denotes the common CDF of user SNRs.

6.3.3 Performance and Complexity Comparison

We now compare the performance and feedback load requirements of selective multiuser scheduling and switched multiuser scheduling with reference to the multiuser selection diversity systems. We adopt the average spectrum efficiency (ASE) with constant-power variable-rate adaptive M-QAM transmission (see Chapter 2) as the performance measure. Specifically, the value range of the scheduled user's SNR is divided into $N+1$ regions, with threshold values denoted by $0 < \gamma_{T_1} < \gamma_{T_2} < \ldots < \gamma_{T_N} < \infty$. When the SNR falls into the nth region, i.e., $\gamma_{T_n} \le \gamma_{k^*} < \gamma_{T_{n+1}}$, the M-QAM scheme with constellation size 2^n is used for transmission. The threshold values γ_{T_n} for different constellation sizes are determined according to the instantaneous BER requirement. As such, the ASE of the system can be calculated as [16, Eq. 33]

$$\text{ASE} = \sum_{n=1}^{N} R_n P_n, \quad (6.52)$$

where P_n is the probability of using the nth constellation size and R_n is the corresponding spectral efficiency. P_n can be calculated using the distribution function of the scheduled user's SNR as

$$P_n = \int_{\gamma_{T_n}}^{\gamma_{T_{n+1}}} p_{\gamma_{k^*}}(x) dx. \quad (6.53)$$

The operations of both selective multiuser scheduling and switched multiuser scheduling require a certain threshold γ_T. The threshold setting for different schemes lead to different trade-offs between ASE and AFL. In particular, for selective multiuser scheduling, a larger γ_T value leads to lower AFL but more frequent scheduling outage, which in turn results in smaller ASE. For the switched multiuser scheduling with the best user selection in the worst case, larger γ_T will improve the ASE performance at the cost of higher AFL. We now illustrate these trade-offs with numerical examples.

In Fig. 6.5, the ASE of selective and switched multiuser scheduling schemes over i.i.d. Rayleigh fading are depicted for different average SNR levels. The results for switched

6.3 Multiuser Diversity with Limited Feedback

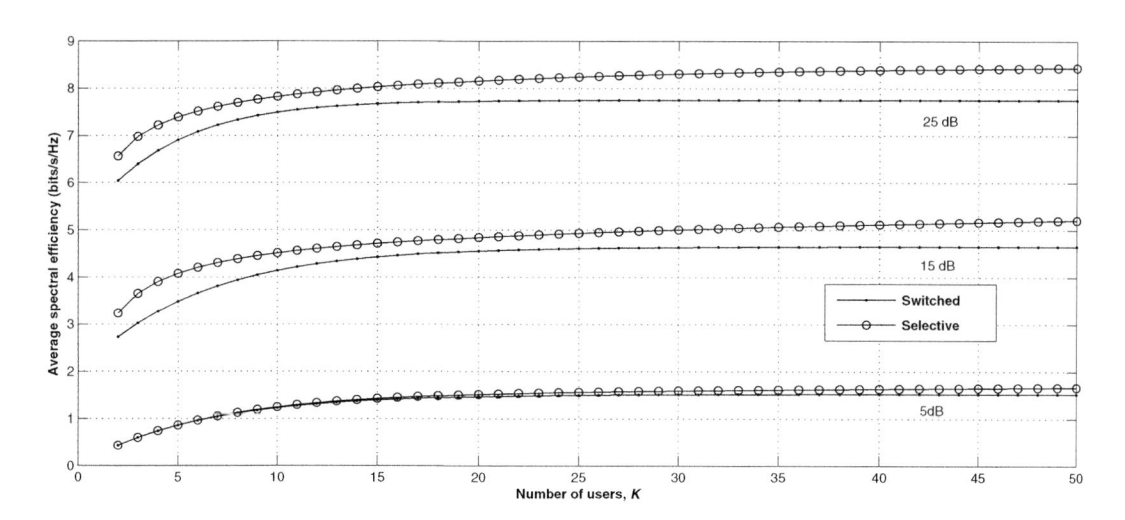

Figure 6.5 Average spectral efficiency of selective and switched multiuser scheduling schemes over i.i.d. Rayleigh fading with $\overline{\gamma} = 5$, 15, and 25.

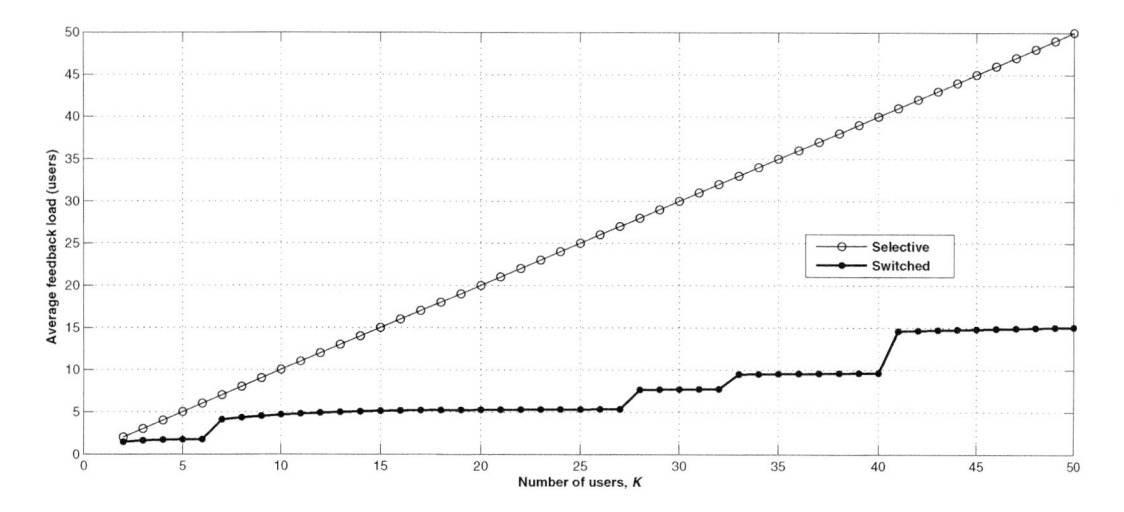

Figure 6.6 Average feedback load of selective and switched multiuser scheduling schemes over i.i.d. Rayleigh fading with $\overline{\gamma} = 15$.

scheduling are based on the optimal thresholds in a maximum ASE sense, subject to no restrictions on AFL. In Fig. 6.6, the AFL corresponding to the ASE results at $\overline{\gamma} = 15$ dB in Fig. 6.5 is depicted. From the results presented in Figs. 6.5 and 6.6, we can see that the reduced AFL of the switched scheduling schemes does not translate into a large performance loss in ASE compared to the selective scheduling scheme. Hence, the additional gain offered by the selective user scheme by always identifying the best user is limited.

6.4 Multiuser Parallel Scheduling

In many wireless systems, multiple users can be scheduled at the same time. This applies to, for example, the TDMA systems where multiple time slots within a channel coherence time T_c are to be allocated to users, or the FDMA systems where multiple parallel frequency subchannels with a channel coherence bandwidth B_c are to be allocated. Parallel channel access is also feasible in wideband CDMA systems or ultra-wideband (UWB) systems. In this scenario, we need efficient scheduling algorithms to carry out user selection. While there exist some optimization-based approaches in the literature, they often require the solution of a multidimensional nonlinear optimization problem, which makes them impractical in real-world environments [17]. In this section, we present several low-complexity multiuser parallel scheduling schemes and investigate their performance and complexity.

All user scheduling schemes can be implemented using either the absolute SNRs or the normalized SNRs of user channels. In this section, for the sake of clarity, we focus on the absolute SNR-based approach with the assumption that the received SNRs at different users experience independent and identical fading after proper power-control mechanisms. The generalization to nonidentical fading is also discussed. We also assume the multiuser channel is orthogonal, in either time, frequency, or code domains, and their mutual interference and external interference are negligible.

6.4.1 Generalized Selection Multiuser Scheduling (GSMuS)

GSMuS [18] originates from the conventional GSC diversity technique. With this scheme, the system schedules a fixed number of the best users among all K active users in each scheduling interval. Specifically, the scheduler ranks the instantaneous SNRs of all K users in descending order, denoted by $\gamma_{1:K} \geq \gamma_{2:K} \geq \ldots \geq \gamma_{K:K}$, where $\gamma_{k:K}$ is the kth largest user SNR. The scheduler will then choose K_s users with the largest SNRs for simultaneous transmission over available resource blocks. We can immediately see that, similar to the conventional multiuser diversity scheme, GSMuS requires the channel information feedback from all users. The analysis of the GSMuS scheme is simpler than that for GSC diversity systems as we only need the marginal statistics of the first K_s largest SNR, instead of their sum. In particular, when user channels experience i.i.d. fading, the PDF of the kth largest user SNR can be easily obtained from the classical order statistics result as

$$p_{\gamma_{k:K}}(x) = \frac{K!}{(K-k)!(k-1)!}[F_\gamma(x)]^{K-k}[1 - F_\gamma(x)]^{k-1} p_\gamma(x), \qquad (6.54)$$

where $F_\gamma(x)$ and $p_\gamma(x)$ are the common CDF and PDF of user SNRs, available for popular fading channel models. For the more general i.n.d. fading channel case, the PDF of the kth largest user SNR is given by

$$p_{\gamma_{k:K}}(s) = \sum_{\substack{n_1,\ldots,n_{k-1} \\ n_1<n_2<\ldots<n_{k-1}}} \sum_{n_k} p_{n_k}(x) \left[\prod_{l=1}^{k-1}[1-F_{n_l}(x)]\right]\left[\prod_{l'=k+1}^{K} F_{n_{l'}}(x)\right], \quad (6.55)$$

where $n_i \in \{1, 2, \ldots, K\}$, $i = 1, \ldots, K_s$, are the index of the ith best user, $p_{n_k}(x)$ is the PDF of the kth best user SNR, and $F_{n_l}(x)$ is the CDF of the lth largest user SNR. Note that summation $\sum_{\substack{n_1,\ldots,n_{k-1} \\ n_1<n_2<\ldots<n_{k-1}}}$ is carrying over all possible index sets of the largest $k-1$ user SNRs out of the total K users, $\sum_{n_k}$ over the possible indexes of the kth selected user and there are total $\frac{K!}{(K-k)!(k-1)!}$ terms in the double summations. It can be verified that (6.55) reduces to (6.54) with the i.i.d. assumption.

With the above PDFs of the scheduled users' SNRs, we can readily evaluate the performance of the GSMuS scheme. We again use the ASE of adaptive M-QAM as our performance metric. The ASE of the kth scheduled user with the GSMuS scheme can be calculated as [16, Eq. 33]

$$\text{ASE}_k = \sum_{n=1}^{N} R_n P_{n,k}, \quad (6.56)$$

where $\{R_n\}_{n=1}^{N}$ are the spectral efficiencies of N available M-QAM schemes and $P_{n,k}$ is the probability of using the nth scheme for kth scheduled user, which can be determined as

$$P_{n,k} = \int_{\gamma T_n}^{\gamma T_{n+1}} p_{\gamma_{k:K}}(\gamma) d\gamma. \quad (6.57)$$

The total system ASE can then be determined by summing the ASE of scheduled users, as

$$\text{ASE}_t = \sum_{k=1}^{K_s} \text{ASE}_k. \quad (6.58)$$

We can also derive the channel access statistics of an arbitrary user, including AAR and AAT defined in the previous section [18]. In particular, the AAR of user k can be calculated as the up-crossing rate at level 1 of the process $r_k = \alpha_k/\alpha_*$, where α_k is the channel amplitude of user k and $\alpha_* = \alpha_{K_s:K-1}$ is the K_sth largest channel amplitude among the other $K-1$ users (excluding user k). Note that with GSMuS, user k will be selected whenever its channel amplitude becomes one of the K_s largest ones among all users. Applying the result of LCR, the AAR of user k can given by

$$N_k = \int_0^{\infty} \dot{r} p_{r_k,\dot{r}_k}(1,\dot{r}) d\dot{r}, \quad (6.59)$$

where $\dot{r}_k$ is the time derivative of the process r_k and $f_{r_k,\dot{r}_k}(1,\dot{r})$ is the joint PDF of r_k and $\dot{r}_k$, which can be calculated in terms of the PDF of α_k, $\dot{\alpha}_k$, and the joint PDF of α_* and $\dot{\alpha}_* = \dot{\alpha}_{K_s:K-1}$, as in (6.17). Assuming that the $\alpha_{K_s:K-1}$ and $\dot{\alpha}_{K_s:K-1}$ are independent, it can be shown that the joint PDF of $f_{\alpha_*,\dot{\alpha}_*}(x,\dot{x})$ is given by

$$p_{\alpha_*,\dot{\alpha}_*}(x,\dot{x}) = \sum_{\substack{n_1,\ldots,n_{K_s-1} \\ n_1<n_2<\ldots<n_{K_s-1}}} \sum_{n_{K_s}} p_{\alpha_{n_{K_s}}}(x) \left[\prod_{l=1}^{K_s-1}[1-F_{\alpha_{n_l}}(x)]\right] \quad (6.60)$$

$$\times \left[\prod_{l'=K_s+1}^{K} F_{\alpha_{n_{l'}}}(x)\right] p_{\dot{\alpha}_{n_{K_s}}}(\dot{x}),$$

where $n_i \in \{1,\ldots,k-1,k+1,\ldots,K\}$ ($n_i \neq k$), $i=1,\ldots,K_s$, are the index of the ith best user SNR among $K-1$ users. Note that the summation $\sum_{\substack{n_1,\ldots,n_{K_s-1} \\ n_1<n_2<\ldots<n_{K_s-1}}}$ is carried over all possible index sets of the largest K_s-1 branch SNRs and $\sum_{n_{K_s}}$ over the possible indexes of the K_sth user, out of the total $K-1$ users, except for the kth user. For the i.n.d. Rayleigh fading scenario, after proper substitutions and carrying out the integration, the AAR of user k can be shown to be given by

$$N_k = \sum_{\substack{n_1,\ldots,n_{K_s-1} \\ n_1<n_2<\ldots<n_{K_s-1}}} \sum_{n_{K_s}} \sum_{\tau \in T_{1\ldots K_s}^{K-1}} \text{sign}(\tau)\frac{\pi}{\sqrt{2}} \quad (6.61)$$

$$\times \frac{\sqrt{\Omega_k f_k^2 + \Omega_{n_{K_s}} f_{n_{K_s}}^2}}{\Omega_k \Omega_{n_{K_s}}} \left(\frac{1}{\Omega_k} + \sum_{l=1}^{K_s}\frac{1}{\Omega_{n_l}} + \tau\right)^{-3/2},$$

where $T_{1\ldots K_s}^{K-1}$ denotes the set obtained by expanding the product $\prod_{l'=K_s+1}^{K-1}\left[1-\exp\left(-\frac{x^2}{\Omega_{n_{l'}}}\right)\right]$ then taking the natural logarithm of each term [13, 14], and sign(τ) is the corresponding sign of each term in the expansion.

Similarly, the AAP of user k can be calculated as the probability that its channel amplitude α_k becomes larger than the K_sth largest channel amplitude among the other $K-1$ users $\alpha_* = \alpha_{K_s:K-1}$. Mathematically speaking, we have

$$P_k = \int_0^\infty p_{\alpha_k}(x) \int_0^x p_{\alpha_{K_s:K-1}}(y)dy\,dx. \quad (6.62)$$

After substituting (6.18) and (6.55) and carrying out integration and some manipulation, we can obtain the closed-form expression of the AAP of user k for the Rayleigh fading case as

$$P_k = \sum_{\substack{n_1,\ldots,n_{K_s-1} \\ n_1<n_2<\ldots<n_{K_s-1}}} \sum_{n_{K_s}} \sum_{\tau' \in T_{1\ldots K_s}^{K-1}} \text{sign}(\tau')\frac{\Omega_k}{\Omega_{n_{K_s}}(1+(\tau'+\sum_{l=1}^{K_s-1}\frac{1}{\Omega_{n_l}})\Omega_k)}. \quad (6.63)$$

The AAT of user k then can be readily calculated as the ratio of AAP over AAR, i.e., $T_k = P_k/N_k$.

6.4.2 On–Off-Based Scheduling

The on–off-based scheduling (OOBS) scheme is inspired by the AT-GSC diversity combining scheme [19] and the on–off type of scheduling schemes discussed in [5, 20]. The difference is, however, that the OOBS scheme will process the selected users in parallel without combining [21]. The mode of operation of the OOBS scheme can be summarized as follows. At the beginning of each scheduling time interval, the BS sends a pilot signal to all users. User i will estimate its received SNR γ_i and compare it to a preselected threshold SNR, denoted by γ_T. The user will feed back its SNR information to the BS only if its SNR is above the threshold γ_T. Only these acceptable users are then scheduled by the BS for subsequent transmission. If no user has an acceptable SNR, the BS simply waits a time period, in the order of the channel coherence time, before starting a new round of scheduling. Note that with the GSMuS scheme, the BS needs always to collect the SNR of all K users to determine the K_s best users for scheduling. The OOBS scheme will lead to a time-varying feedback load and the number of scheduled users is also randomly varying.

We first evaluate the AFL of the OOBS scheme. As only users with acceptable channel conditions will feed back their channel SNRs, the AFL of the OOBS scheme over the i.i.d. fading environment is the same as that of selective multiuser scheduling scheme discussed in the previous section, which is given by

$$\text{AFL} = \sum_{k=0}^{K} k \binom{K}{k} \left[1 - F_\gamma(\gamma_T)\right]^k \left[F_\gamma(\gamma_T)\right]^{K-k}$$
$$= K \left[1 - F_\gamma(\gamma_T)\right]. \tag{6.64}$$

For the i.i.d. Rayleigh fading case, (6.64) specializes, after substituting in the common CDF of Rayleigh faded received SNR, to

$$\overline{N} = K \exp\left(-\frac{\gamma_T}{\bar{\gamma}}\right). \tag{6.65}$$

Therefore, the OOBS scheme requires, on average, less feedback load than the GSMuS scheme. Note that the average number of scheduled users with the OOBS scheme is equal to the number of acceptable users. In practical systems, the number of users that can be scheduled for simultaneous transmission may be limited by other constraints.

We now evaluate the ASE performance of the OOBS scheme over the i.i.d. fading environment. Based on the mode of operation of the OOBS scheme, the received SNR of scheduled users has a truncated PDF from left at γ_T. It can be shown that the PDF of the received SNR of a scheduled user is given by

$$p_{\gamma_s}(\gamma) = \begin{cases} \frac{p_\gamma(\gamma)}{1 - F_\gamma(\gamma_T)} & \gamma \geq \gamma_T; \\ 0 & \text{otherwise,} \end{cases} \tag{6.66}$$

where $p_\gamma(\cdot)$ and $F_\gamma(\cdot)$ are the common PDF and CDF of user SNRs, respectively. The ASE of a scheduled user with OOBS scheme can be calculated as [16, Eq. 33]

$$\text{ASE}_u = \sum_{n=1}^{N} R_n P_n, \tag{6.67}$$

where $\{R_n\}_{n=1}^{N}$ are the spectral efficiencies of N available modulation schemes and P_n is the probability of using the nth modulation scheme, which can be obtained as

$$P_n = \int_{\gamma_{T_n}}^{\gamma_{T_{n+1}}} \frac{p_\gamma(\gamma)}{1 - F_\gamma(\gamma_T)} U(\gamma - \gamma_T) d\gamma, \qquad (6.68)$$

where $U(\cdot)$ denotes the unit step function. The overall system ASE can then be determined by multiplying the average number of scheduled users to the ASE of a scheduled user, as

$$\text{ASE}_t = K \sum_{n=1}^{N} R_n \int_{\gamma_{T_n}}^{\gamma_{T_{n+1}}} p_\gamma(\gamma) U(\gamma - \gamma_T) d\gamma. \qquad (6.69)$$

For the Rayleigh fading environment, (6.69) specializes to

$$\text{ASE}_t = K \sum_{n=1}^{N} R_n P'_n, \qquad (6.70)$$

where

$$P'_n = \begin{cases} \exp\left(-\frac{\max(\gamma_T, \gamma_{T_n})}{\bar{\gamma}}\right) - \exp\left(-\frac{\gamma_{T_{n+1}}}{\bar{\gamma}}\right) & \text{if } \gamma_{T_{n+1}} \geq \gamma_T; \\ 0 & \text{if } \gamma_{T_{n+1}} < \gamma_T. \end{cases} \qquad (6.71)$$

6.4.3 Switching-Based Multiuser Scheduling

The switching-based multiuser scheduling (SBS) scheme is inspired by the switching-based diversity combining schemes [22, 23]. With the SBS scheme, the BS schedules a predetermined fixed total number of acceptable users, denoted by K_s, and if necessary, the best unacceptable users in a GSMuS fashion [21]. More specifically, at the beginning of each scheduling period, the BS sends a pilot signal sequentially to each user to request channel state information feedback. Each user estimates its SNR γ_i using the pilot signal and feeds it back to the BS. The BS will compare each received SNR with the preselected SNR threshold γ_T. If the user SNR γ_i is greater than γ_T, then the user channel is considered acceptable and the user will be scheduled for transmission. This process of estimation, feedback, and comparison is repeated until the BS finds K_s acceptable users, after which the data transmission will start without further probing. If the BS only finds $K_a < K_s$ acceptable users after probing all K user channels, the base station will rank those $K - K_a$ users with unacceptable channels and schedule the best $K_s - K_a$ unacceptable users along the already selected K_a acceptable users. As the SBS scheme may not always select the best users for transmission, its performance will be worse than the GSMuS scheme, especially when the threshold is relative small. On the other hand, the SBS scheme can approach the performance of the GSMuS scheme as γ_T increases. Note that when γ_T goes to infinity, the SBS scheme becomes equivalent to the GSMuS scheme. The advantage of the SBS scheme is lower feedback load. The GSMuS scheme always needs K user SNR feedback during each scheduling period, whereas the feedback with SBS scheme varies between K_s and K. In the following, we

6.4 Multiuser Parallel Scheduling

derive the analytical expressions of AFL and ASE with the SBS scheme over i.i.d. fading environment.

To calculate AFL, we consider two mutually exclusive cases depending on whether there are K_s acceptable users or not. If there are fewer than K_s acceptable users among the total K users, the SBS scheme needs to request SNR feedback from all K users. As a result, the amount of SNR feedback becomes K and the number of unacceptable users takes values from $(K - K_s + 1)$ to K and follows a binomial distribution. Since the diversity paths are i.i.d. faded, the probability that there are fewer than K_s acceptable users can be written as

$$P_B = \sum_{l=K-K_s+1}^{K} \binom{K}{l} [F_\gamma(\gamma_T)]^l [1 - F_\gamma(\gamma_T)]^{K-l}. \tag{6.72}$$

When there are at least K_s acceptable users among the total K users, the SBS scheme will terminate the process of SNR feedback as soon as K_s acceptable users are found. In this case, the number of SNR feedbacks takes values from K_s to K and follows a negative binomial (or Pascal) distribution, with the probability of K feedbacks given by

$$P_k = \binom{k-1}{k-K_s} [1 - F_\gamma(\gamma_T)]^{K_s} [F_\gamma(\gamma_T)]^{k-K_s}, \quad k = K_s, \ldots, K. \tag{6.73}$$

Finally, by combining these two mutually exclusive cases, we can calculate the overall AFL during a guard period as

$$\begin{aligned}
\text{AFL} &= \sum_{k=K_s}^{K} k\, P_k + K P_B \\
&= \sum_{k=K_s}^{K} k \binom{k-1}{k-K_s} [1 - F_\gamma(\gamma_T)]^{K_s} [F_\gamma(\gamma_T)]^{k-K_s} \\
&\quad + K \sum_{l=K-K_s+1}^{K} \binom{K}{l} [F_\gamma(\gamma_T)]^l [1 - F_\gamma(\gamma_T)]^{K-l}.
\end{aligned} \tag{6.74}$$

For the i.i.d. Rayleigh fading case, after replacing $F_\gamma(\gamma)$ with $1 - \exp\left(-\frac{\gamma}{\bar{\gamma}}\right)$, we obtain the following closed-form expression for AFL as

$$\begin{aligned}
\text{AFL} &= \sum_{k=K_s}^{K} k \binom{k-1}{k-K_s} \left[\exp\left(-\frac{\gamma_T}{\bar{\gamma}}\right)\right]^{K_s} \left[1 - \exp\left(-\frac{\gamma_T}{\bar{\gamma}}\right)\right]^{k-K_s} \\
&\quad + K \sum_{l=K-K_s+1}^{K} \binom{K}{l} \left[1 - \exp\left(-\frac{\gamma_T}{\bar{\gamma}}\right)\right]^l \left[\exp\left(-\frac{\gamma_T}{\bar{\gamma}}\right)\right]^{K-l}.
\end{aligned} \tag{6.75}$$

To calculate the total system ASE, we first derive the statistics of the received SNR of a scheduled user. Considering the cases that the number of acceptable users K_a is greater than or equal to K_s and that K_a is smaller than K_s separately, we can write the PDF of the received SNR of a scheduled user with SBS scheme as

$$p_{\gamma_s}(\gamma) = \Pr[K_a \geq K_s] p_{\gamma_{s_1}}(\gamma) + \Pr[K_a < K_s] p_{\gamma_{s_2}}(\gamma), \tag{6.76}$$

where $p_{\gamma_{s_1}}(\gamma)$ is the PDF of a scheduled user when $K_a \geq K_s$, the probability of which is given by

$$\Pr[K_a \geq K_s] = \sum_{K_a=K_s}^{K} \binom{K}{K_a} [1 - F_\gamma(\gamma_T)]^{K_a} [F_\gamma(\gamma_T)]^{K-K_a}, \quad (6.77)$$

and $p_{\gamma_{s_2}}(\gamma)$ is the PDF of a scheduled user in the case of $K_a < K_s$, the probability of which is equal to

$$\Pr[K_a < K_s] = \sum_{K_a=0}^{K_s-1} \binom{K}{K_a} [1 - F_\gamma(\gamma_T)]^{K_a} [F_\gamma(\gamma_T)]^{K-K_a}. \quad (6.78)$$

In the case of $K_a \geq K_s$, i.e., all scheduled users have acceptable SNR, the PDF of the received SNR at a scheduled user is the truncated PDF from below at γ_T, which is given by [22, Eq. (5)]

$$p_{\gamma_{s_1}}(\gamma) = \frac{p_\gamma(\gamma)}{1 - F_\gamma(\gamma_T)} U(\gamma - \gamma_T), \quad (6.79)$$

where $U(x)$ is the unit step function. For the case of $K_a < K_s$, we first note that with the SBS scheme, $K_a < K_s$ holds if and only if the K_sth largest user SNR is below the threshold and, as such, some unacceptable users are scheduled. Therefore, the PDF of a scheduled user's SNR is the PDF of any one of the K_s best user SNRs, given the fact that the K_sth best user SNR is below the threshold. Since the scheduled user has equal probability to be the first to the K_s best users, the PDF of a scheduled user can be written as

$$p_{\gamma_{s_2}}(\gamma) = \frac{1}{K_s} \left(\sum_{i=1}^{K_s-1} p_{\gamma_{i:K}|\gamma_{K_s:K}<\gamma_T}(\gamma) + \frac{p_{\gamma_{K_s:K}}(\gamma)}{F_{\gamma_{K_s:K}}(\gamma_T)} (1 - U(\gamma - \gamma_T)) \right), \quad (6.80)$$

where $p_{\gamma_{i:K}|\gamma_{K_s:K}<\gamma_T}(\gamma)$ is the conditional PDF of the ith ($i < K_s$) best user SNR given the K_sth best user SNR is below the threshold γ_T, which can be calculated using the joint PDF of $\gamma_{i:K}$ and $\gamma_{K_s:K}$ as

$$p_{\gamma_{i:K}|\gamma_{K_s:K}<\gamma_T}(x) = \frac{1}{\int_0^{\gamma_T} p_{\gamma_{K_s:K}}(\gamma) d\gamma} \int_0^{\min(\gamma_T,x)} p_{\gamma_{i:K},\gamma_{K_s:K}}(x,y) \, dy, \quad (6.81)$$

where $p_{\gamma_{i:K}}(\gamma)$ is the PDF of the ith largest user SNR and $p_{\gamma_{i:K},\gamma_{K_s:K}}(x,y)$ is the joint PDF of the ith and K_sth largest user SNRs, both of which have been discussed in earlier chapters but are reproduced here for convenience:

$$p_{\gamma_{i:K},\gamma_{K_s:K}}(x,y) = \frac{K!}{(K-K_s)!(K_s-i-1)!(i-1)!}$$
$$\times [F_\gamma(y)]^{K-K_s} [F_\gamma(x) - F_\gamma(y)]^{K_s-i-1}$$
$$\times [1 - F_\gamma(x)]^{i-1} p_\gamma(x) p_\gamma(y), \quad (6.82)$$
$$x > y \text{ and } 1 \leq i < K_s \leq K$$

$$p_{\gamma_{j:K}}(\gamma) = \binom{K}{j} j \cdot p_\gamma(\gamma) [1 - F_\gamma(\gamma)]^{j-1} [F_\gamma(\gamma)]^{K-j}. \quad (6.83)$$

Note that the second term in the parentheses of (6.80) corresponds to the case of $i = K_s$, i.e., the PDF of the K_sth largest user SNR given that it is smaller than the threshold.

Finally, after proper substitution, the PDF of the received SNR at a scheduled user with the SBS scheme can be obtained as

$$p_{\gamma_s}(\gamma) = \Pr[K_a < K_s] \frac{1}{K_s} \left(\sum_{i=1}^{K_s-1} p_{\gamma_{i:K_s} | \gamma_{K_s:K} < \gamma_T}(\gamma) \right.$$
$$\left. + \frac{p_{\gamma_{K_s:K}}(\gamma)}{F_{\gamma_{K_s:K}}(\gamma_T)} (1 - U(\gamma - \gamma_T)) \right)$$
$$+ \Pr[K_a \geq K_s] \frac{p_\gamma(\gamma)}{1 - F_\gamma(\gamma_T)} U(\gamma - \gamma_T), \quad (6.84)$$

which can be utilized to evaluate the ASE of the SBS scheme. Specifically, the ASE of a scheduled user can be calculated, after applying (6.84) to the definition in (6.67), as

$$\text{ASE}_u = \sum_{n=1}^{N} R_n \int_{\gamma_{T_n}}^{\gamma_{T_{n+1}}} p_{\gamma_s}(\gamma) \, d\gamma. \quad (6.85)$$

Note that the SBS scheme always schedules K_s users. Therefore, the total system ASE can be simply calculated as $\text{ASE}_t = K_s \text{ASE}_u$.

6.4.4 Numerical Results

In this subsection, we discuss the trade-off involved in three multiuser parallel scheduling schemes presented previously through selected numerical examples. Note that the number of scheduled users with the OOBS scheme is randomly varying, whereas both GSMuS and SBS schemes schedule K_s users. To allow a fair comparison between these scheduling schemes, we select the feedback threshold γ_T for the OOBS scheme such that the average number of scheduled users with the OOBS scheme is equal to K_s. For the i.i.d. Rayleigh fading scenario of interest, starting from (6.64), the threshold values satisfying this constraint can be calculated by solving the following equation for γ_T^*:

$$K_s = K \exp\left(-\frac{\gamma_T^*}{\bar{\gamma}}\right), \quad (6.86)$$

which leads to

$$\gamma_T^* = \bar{\gamma} \ln\left(\frac{K}{K_s}\right). \quad (6.87)$$

Note that γ_T^* varies with $\bar{\gamma}$.

In Fig. 6.7, we present the AFL of the OOBS scheme, the SBS scheme, and the GSMuS scheme over i.i.d. Rayleigh fading with $K = 5$ and $K_s = 3$. Specifically, we plot the AFL of three schemes with different threshold values as the function of the average SNR $\bar{\gamma}$. We first note that the AFL of the SBS scheme is decreasing from K and eventually converging to K_s as $\bar{\gamma}$ increases while that of the GSC-based scheme is always

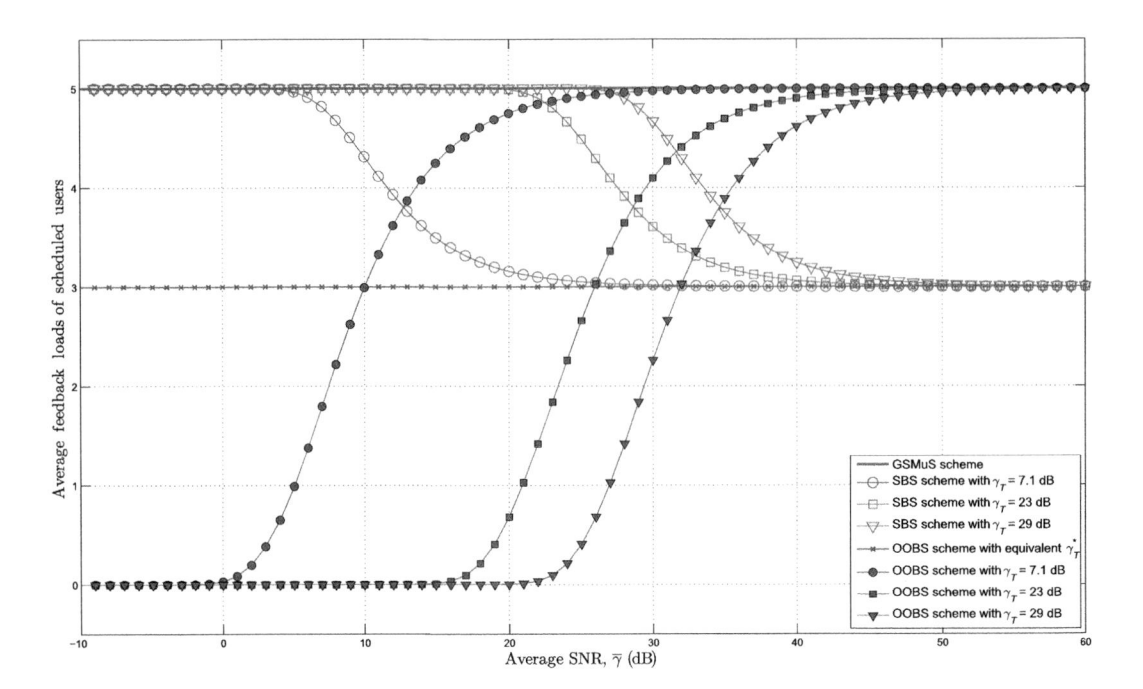

Figure 6.7 Average feedback loads for the (1) OOBS scheme, (2) SBS scheme, and (3) GSMuS scheme ($K = 5$ and $K_s = 3$). Reprint with permission from [21]. ©2009 IEEE.

equal to K. We can also see that the AFL of the SBS scheme increases as γ_T increases for a fixed $\bar{\gamma}$. On the other hand, for fixed γ_T, the AFL of the OOBS scheme is increasing from 0 to K, as $\bar{\gamma}$ increases. For fixed $\bar{\gamma}$, the AFL of the OOBS scheme decreases as γ_T increases because the feedback load is the same as the number of scheduled users in the OOBS scheme and the number of scheduled users decreases as γ_T increases. Finally, with the equivalent threshold γ_T^*, the AFL of the OOBS scheme remains constant.

Fig. 6.8 presents the ASE of the OOBS, SBS, and GSMuS schemes with adaptive coded M-QAM modulation as a function of average SNR $\bar{\gamma}$ over the i.i.d. Rayleigh fading with $K = 5$ and $K_s = 3$. We can see that, with a fixed γ_T, the ASE of the OOBS scheme is increasing and eventually converging to $K \cdot R_N$ as $\bar{\gamma}$ increases, because the number of scheduled users increases and eventually converges to K, and the total data rate depends on the number of scheduled users. We also notice that the ASE of the OOBS scheme with small γ_T is converging faster than that with large γ_T. When the channel condition is good, the ASE of the OOBS scheme is much larger than that of the GSMuS scheme because the number of scheduled users with the OOBS scheme is converging to K while the number of scheduled users with the GSMuS scheme is always K_s. By applying the equivalent threshold γ_T^* in the OOBS scheme, when the channel condition is poor, the OOBS scheme has almost the same ASE as the GSMuS scheme and the ASE of the OOBS scheme is slightly higher than that of the GSMuS

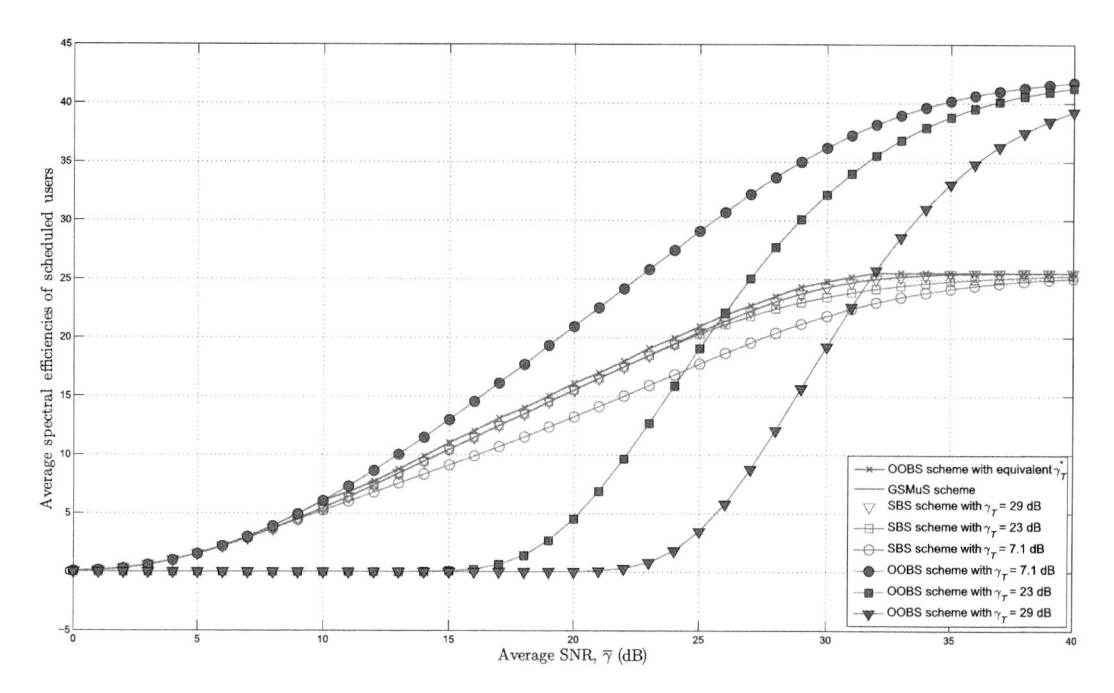

Figure 6.8 Average spectral efficiencies of adaptive coded M-QAM modulation for the (1) OOBS scheme, (2) SBS scheme, and (3) GSC-based scheduling scheme over i.i.d. Rayleigh fading conditions with $K = 5$ and $K_s = 3$. Reprint with permission from [21]. ©2009 IEEE.

scheme as the channel quality improves. For high average SNR (above 35 dB), the ASE of the OOBS scheme is reconverging to that of the GSMuS scheme, $K_s \cdot R_N$. Comparing the ASE between the SBS scheme and the GSMuS scheme, we can see that for a fixed γ_T, when the channel condition is poor, the SBS scheme has almost the same ASE as the GSMuS scheme and as $\bar{\gamma}$ increases, the GSMuS scheme has a slightly better ASE; eventually these two schemes have the same ASE again.

6.5 Power Allocation for Parallel Scheduling

So far, we have assumed that the transmit power is uniformly allocated to the scheduled users, irrespective of their instantaneous channel conditions. The performance of the multiuser parallel scheduling scheme studied in the previous section can be further improved with power allocation among scheduled users. Power allocation for the GSMuD scheme was investigated by Ma *et. al.* [24]. Specifically, optimal power allocation strategies based on the one-dimensional or two-dimensional water-filling solution derived there achieve better sum rate performance than equal power allocation. In this section, we focus on transmit power allocation for selected users with the SBS scheme [25].

According to the mode of operation of the SBS scheme, when there are not enough acceptable users, at least one user will have an SNR smaller than the threshold, which will limit the performance of the SBS scheme. One way to reduce or eliminate the number of scheduled unacceptable users is to redistribute the transmission power to different scheduled users while satisfying the total transmitting power constraint. Note that because of the discrete nature of adaptive modulation, such power redistribution among scheduled users may not affect the achieved ASE of those acceptable users, as the same modulation mode can be used for all values of the received SNRs in the same interval. With this motivation in mind, we study in this section several threshold-based power allocation algorithms for the SBS scheme.

6.5.1 Power Allocation Schemes

The main purpose of power reallocation for the scheme is to increase the number of scheduled acceptable users without consuming additional transmit power. This is achieved by reallocating the transmit power for the scheduled acceptable users to those scheduled unacceptable users. Based on this general principle, we present three specific power reallocation algorithms. For later reference, γ_i' denotes the SNR of the scheduled user after power allocation, K_a denotes the number of acceptable users before power allocation, and γ_T denotes the preselected SNR threshold.

Algorithm 1: Ranking Maintained Power Reallocation
This algorithm maintains the SNR ranking of scheduled acceptable users when extracting power from acceptable users. The algorithm extracts only the required transmitting power from the acceptable users and allocates it to the unacceptable users sequentially from the strongest user to the weakest user. The operation of Algorithm 1 is summarized as follows. After SBS-based user scheduling, the system calculates the required additional transmission power for the unacceptable users in dB scale as $\gamma_{req} = \sum_{i=K_a+1}^{K_s} (\gamma_T - \gamma_i)$. Then, the system tries to extract the same amount of excess power from K_a acceptable users while ensuring the acceptance of user SNR, as:

1. If $\left(\gamma_i - \frac{\gamma_{req}}{K_a}\right) \geq \gamma_T$, then we reduce the transmitting power for user i in the amount of $\frac{\gamma_{req}}{K_a}$. Then the resulting received SNR of user i becomes $\gamma_i' = \gamma_i - \frac{\gamma_{req}}{K_a}$.
2. If $\left(\gamma_i - \frac{\gamma_{req}}{K_a}\right) < \gamma_{TP}$, then we reduce the transmitting power for user i in the amount of $(\gamma_i - \gamma_{TP})$. As such, the modified SNR of the acceptable user is $\gamma_i' = \gamma_T$.

The total excess transmission power from the acceptable users are

$$\gamma_{excess} = \sum_{i=1}^{K_a} \left(\gamma_i - \gamma_i'\right), \qquad (6.88)$$

which will be less or equal to γ_{req}. The system will then allocate γ_{excess} to the unacceptable users. More specifically, if $\gamma_{excess} = \gamma_{req}$, the received SNRs of all scheduled unacceptable users will be increased to γ_T and as such become acceptable with power reallocation. If $\gamma_{excess} < \gamma_{req}$, then γ_{excess} is sequentially allocated to the unacceptable users, starting from the one with the strongest SNR. With this approach, the number of remaining unacceptable users will be the smallest after power reallocation.

Algorithm 2: Rate Maintained Power Reallocation

This algorithm extracts transmit power from scheduled acceptable users while ensuring that their supported transmission rates remain unchanged. Similar to Algorithm 1, the extracted power is allocated to unacceptable users sequentially starting from the strongest user. Specifically, excess SNR of the ith acceptable user is calculated as $\gamma_{excess,i} = \gamma_i - \gamma_{T_j}$, where γ_{T_j} is the lower threshold for modulation mode that ith user's SNR can support, i.e., $\gamma_{T_j} \leq \gamma_i \leq \gamma_{T_{j+1}}$. The total excess transmission power from all acceptable users becomes

$$\gamma_{excess} = \sum_{i=1}^{K_a} \gamma_{excess,i}, \tag{6.89}$$

which will be reallocated to those unacceptable users, such that the number of scheduled unacceptable users is minimized. In particular, this excess transmission power is then allocated to the unacceptable users in a similar fashion as in Algorithm 1. The only difference is that in this case, some excess power may remain after power allocation to all unacceptable users, in which case the system equally allocates it to all the scheduled users. Note that Algorithm 2 has lower complexity than Algorithm 1 as the system does not need to calculate γ_{req}.

Algorithm 3: Equal SNR Power Reallocation

With Algorithm 3, the received SNR of all acceptable users after power reallocation will be the same. Specifically, the transmission power will be redistributed among the scheduled users with the best channel conditions such that their received SNRs are equal and above the threshold γ_T. The mode of operation of Algorithm 3 is as follows. When some unacceptable users are scheduled, i.e., $K_a < K_s$, the BS will first determine the number of acceptable users after power reallocation, denoted by K'_a. In particular, if $\sum_{k=1}^{i} \gamma_{k:K} \geq i\gamma_T$ but $\sum_{k=1}^{i+1} \gamma_{k:K} < (i+1)\gamma_T$, then we conclude that $K'_a = i$ and the received SNR of all K'_a acceptable users after power reallocation is equal to $\frac{1}{i}\sum_{k=1}^{i} \gamma_{k:K}$. Note that if $K_a \leq K'_a < K_s$, then there will still remain $K_s - K'_a$ scheduled unacceptable users after power reallocation, whose SNR will not be changed with power reallocation. This algorithm has the lowest implementation complexity among the three algorithms.

6.5.2 Performance Analysis

In this subsection, we analyze the performance of power allocation algorithm 3 for SBS scheme.

We first calculate the average numbers of scheduled acceptable users before and after power reallocation based on Algorithm 3. The average number of scheduled acceptable users before power reallocation can be calculated as

$$\overline{K}_a = \sum_{i=1}^{K_s-1} i \Pr[i \text{ acceptable users in total}] + K_s \Pr[\text{at least } K_s \text{ acceptable users}]. \quad (6.90)$$

Using the joint PDF and marginal PDF of ordered statistics, the probabilities involved can be calculated as

$$\Pr[i \text{ acceptable users in total}] = \int_{\gamma_T}^{\infty} \int_{0}^{\gamma_T} p_{\gamma_{i:K}, \gamma_{i+1:K}}(x, y) dy dx \quad (6.91)$$

and

$$\Pr[\text{at least } K_s \text{ acceptable users}] = \int_{\gamma_T}^{\infty} p_{\gamma_{K_s:K}}(z) dz, \quad (6.92)$$

where $p_{\gamma_{i:K}, \gamma_{i+1:K}}(x, y)$ is the joint PDF of the ith largest and the $i+1$ largest received SNRs, which is given by

$$
\begin{aligned}
p_{\gamma_{i:K}, \gamma_{j:K}}(x, y) &= \frac{K!}{(K-j)!(j-i-1)!(i-1)!} \\
&\quad \times [F_\gamma(y)]^{K-j} [F_\gamma(x) - F_\gamma(y)]^{j-i-1} [1 - F_\gamma(x)]^{i-1} \\
&\quad \times p_\gamma(x) p_\gamma(y), \\
&\quad x > y \ \& \ 1 \le i < j \le K,
\end{aligned} \quad (6.93)
$$

and $p_{\gamma_{K_s:K}}(z)$ is the PDF of the K_sth largest received SNR, given by

$$p_{\gamma_{(K_s:K)}}(\gamma) = \binom{K}{K_s} K_s \cdot p_\gamma(\gamma) [1 - F_\gamma(\gamma)]^{K_s-1} [F_\gamma(\gamma)]^{K-K_s}. \quad (6.94)$$

Based on the mode of operation of Algorithm 3, the average number of scheduled acceptable users after power reallocation can be calculated as

$$\overline{K}'_a = \sum_{i=1}^{K_s-1} i \Pr\left[\sum_{k=1}^{i} \gamma_{k:K} \ge i\gamma_T \ \& \ \sum_{k=1}^{i+1} \gamma_{k:K} < (i+1)\gamma_T\right] \quad (6.95)$$

$$+ K_s \Pr\left[\sum_{k=1}^{K_s} \gamma_{k:K} \ge K_s \gamma_T\right].$$

In this case, the probabilities involved in the summation can be calculated using the joint PDF of $\sum_{k=1}^{i} \gamma_{k:K}$ and $\gamma_{i+1:K}$ and the marginal PDF of $\sum_{k=1}^{K_s} \gamma_{i:K}$, respectively, as

$$\Pr\left[\sum_{k=1}^{i}\gamma_{k:K}\geq i\gamma_T\ \&\ \sum_{k=1}^{i+1}\gamma_{k:K}<(i+1)\gamma_T\right] \qquad (6.96)$$

$$=\int_{0}^{\gamma_T}\int_{i\gamma_T}^{(i+1)\gamma_T-x}p_{\sum_{k=1}^{i}\gamma_{k:K},\gamma_{i+1:K}}(x,y)dydx,$$

and

$$\Pr\left[\sum_{k=1}^{K_s}\gamma_{k:K}\geq K_s\gamma_T\right]=\int_{K_s\gamma_T}^{\infty}p_{\sum_{k=1}^{K_s}\gamma_{k:K}}(z)dz. \qquad (6.97)$$

These joint statistics of the partial sums of order statistics were investigated in the Appendix. For example, the joint PDF of $\sum_{i=1}^{j}\gamma_{i:K}$ and $\gamma_{j+1:K}$ can be shown to be given by

$$p_{\gamma_{j+1:K},\sum_{i=1}^{j}\gamma_{i:K}}(x,y)=\frac{K!}{(K-j-1)!(j)!}[F_\gamma(x)]^{K-j-1}[1-F_\gamma(x)]^j$$

$$\times p_\gamma(x)p_{\sum_{i=1}^{j}\gamma_i^+}(y),$$

$$x\geq 0,\ y\geq jx, \qquad (6.98)$$

where $p_{\sum_{i=1}^{j}\gamma_i^+}(\gamma)$ denotes the PDF of the sum of j i.i.d. random variables γ_i^+, whose PDF is a truncated version of the PDF of γ_i on the left at x.

6.5.3 Numerical Results

The average number of scheduled acceptable users, the ASE, and the average BER with three power allocation algorithms are investigated through Monte Carlo computer simulations. In our simulation, we assume the coded adaptive modulation of [26] with $N=8$ modes and i.i.d. Rayleigh fading conditions.

Fig. 6.9 presents the average number of scheduled acceptable users with three power allocation algorithms over i.i.d. Rayleigh fading with $K=5$, $K_s=3$, and $\gamma_T=7.1$ dB. As we can see, after power allocation, the average number of scheduled acceptable users increases significantly for all three algorithms. Among the three algorithms, the third algorithm has the best performance and the second algorithm has a better performance than the first one.

Fig. 6.10 presents the ASE of SBS with three power allocation algorithms over i.i.d. Rayleigh fading with $K=5$, $K_s=3$, and $\gamma_T=7.1$ dB. We can see that when the average SNR is close to the SNR threshold, all three algorithm lead to some improvement to the ASE. On the other hand, when the average SNR is much larger than the SNR threshold, the first algorithm will not increase the ASE, as most scheduled user will have acceptable SNR. Meanwhile, the other two algorithms can still achieve certain performance improvement, with the third algorithm slightly outperforming the second algorithm.

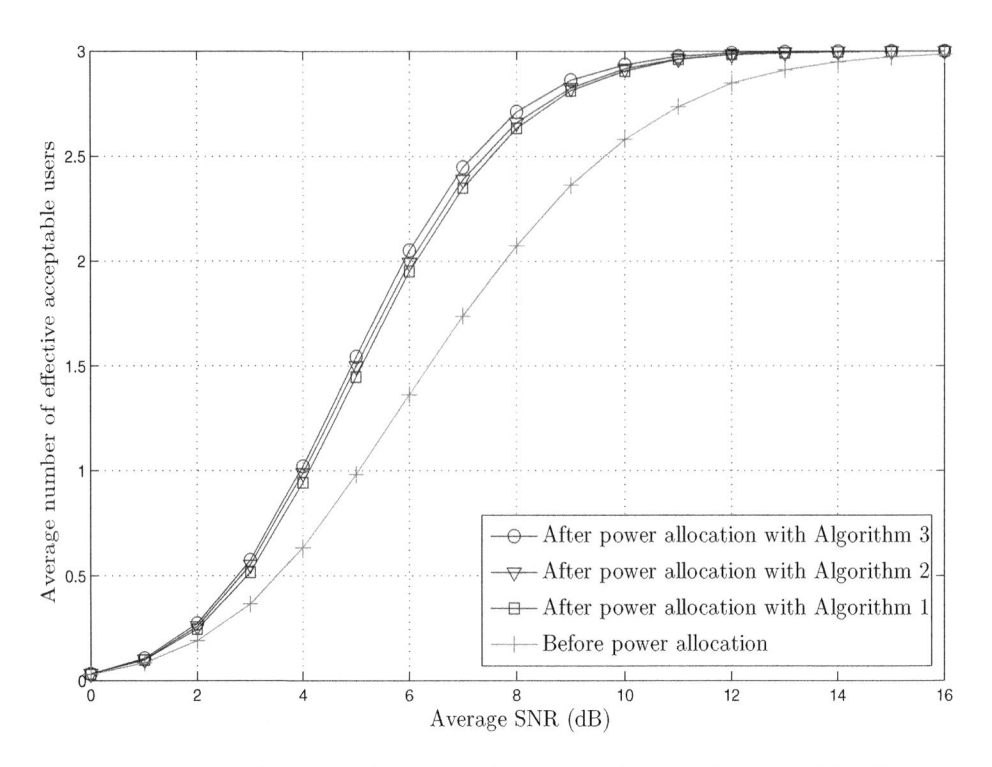

Figure 6.9 Average number of effective acceptable users with SBS over i.i.d. Rayleigh fading conditions with $K = 5$, $K_s = 3$, and $\gamma_T = 7.1$ dB. Reprint with permission from [25]. ©2009 IEEE.

Fig. 6.11 presents the average BER of SBS with three power allocation algorithms over i.i.d. Rayleigh fading with $K = 5$, $K_s = 3$, and $\gamma_T = 7.1$ dB. After power allocation with Algorithm 1, the average BER performance is degraded only when the average SNR is close to the SNR threshold because the excess SNRs are extracted from the acceptable users. In the region of high $\overline{\gamma}$, the extracted SNR from the acceptable users is decreased and then the average BER approaches the average BER prior to the application of the power allocation. Similar to the ASE results, the average BER of Algorithm 2 acts like the ASE of Algorithm 2 when the average SNR is close to the SNR threshold for power allocation. Considering the three algorithms together, it is clear that Algorithm 3 provides the best performance and the least complexity.

6.6 Summary

In this chapter, we considered the design and analysis of user scheduling schemes for wireless communication systems. We focused on those schemes that have low implementation complexity. Both single-user scheduling and multiple-user scheduling scenarios were considered. We quantified the performance of single-user scheduling schemes from both the capacity gain and channel access time/rate perspectives. Three

6.6 Summary

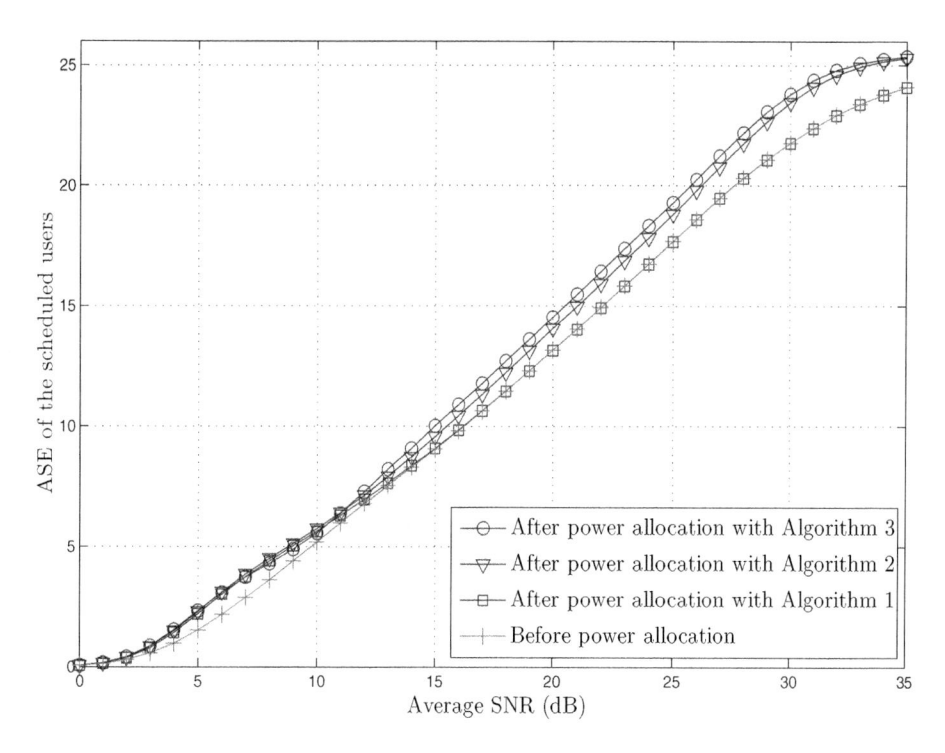

Figure 6.10 ASE (bits/s/Hz) with SBS over i.i.d. Rayleigh fading conditions with $K = 5$, $K_s = 3$, and $\gamma_T = 7.1$ dB. Reprint with permission from [25]. ©2009 IEEE.

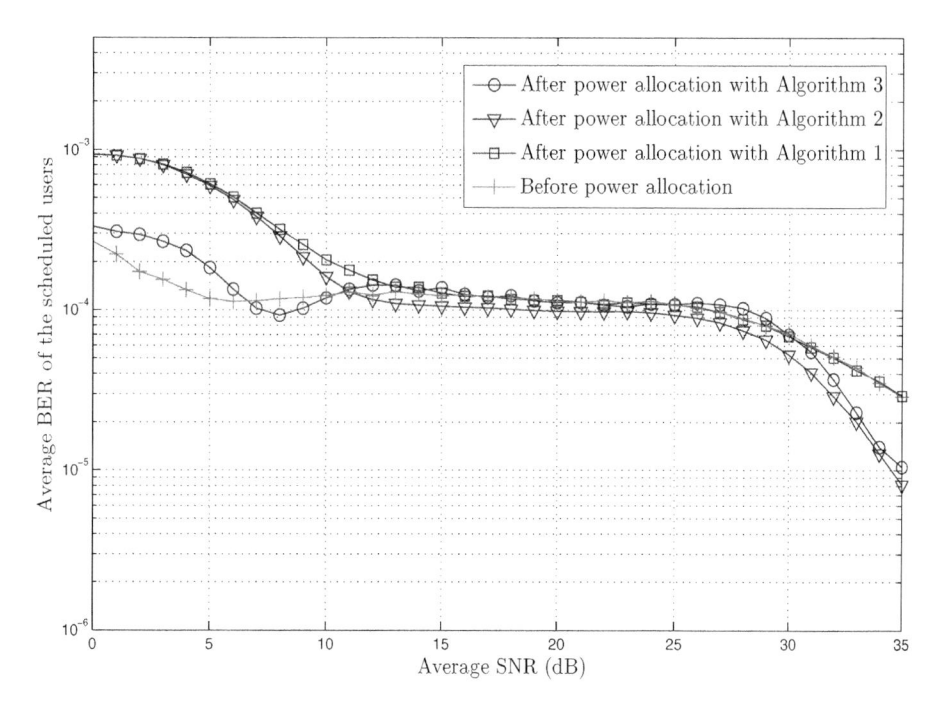

Figure 6.11 Average BER with SBS over i.i.d. Rayleigh fading conditions with $K = 5$, $K_s = 3$, and $\gamma_T = 7.1$ dB. Reprint with permission from [25]. ©2009 IEEE.

representative multiuser scheduling schemes were analyzed to quantify their different performance versus complexity trade-offs. Finally, we studied power allocation strategies for multiuser parallel scheduling.

6.7 Further Reading

Multiuser scheduling schemes based on switched diversity were studied in [15, 27, 28]. The performance of multiuser scheduling in multi-antenna scenarios with different diversity combining schemes was analyzed in [29]. Feedback load reduction strategies for multiuser diversity systems were considered in [30–33]. Ma and Tepedelenlioglu quantified the effect of outdated feedback on the performance of multiuser diversity systems in [34]. The joint design of multiuser scheduling schemes with adaptive diversity combining strategy was considered in [35, 36].

References

[1] R. Knopp and P. Humblet, "Information capacity and power control in single-cell multiuser communications," in *Proc. IEEE Int. Conf. Commun. (ICC'95)*, Seattle, WA, vol. 1, 1995, pp. 331–335.

[2] D. N. C. Tse, "Optimal power allocation over parallel Gaussian channels," in *Proc. Int. Symp. Inform. Theory (ISIT'97)*, Ulm, Germany, 1997, p. 27.

[3] H. A. David, *Order Statistics*, Wiley, 1981.

[4] P. Viswanath, D. Tse, and R. Laroia,"Opportunistic beamforming using dumb antennas," *IEEE Trans. Inform. Theory*, 48, pp. 1277–1294, 2002.

[5] D. Gesbert and M.-S. Alouini, "How much feedback is multi-user diversity really worth?" in *Proc. IEEE Int. Conf. Commun. (ICC'04)*, Paris, France, 2004, pp. 234–238.

[6] L. Yang and M.-S. Alouini, "Performance analysis of multiuser selection diversity," *IEEE Trans. Veh. Technol.*, 55, pp. 1003–1018, 2006.

[7] M. Abramowitz and I. A. Stegun, *Handbook of Mathematical Functions with Formulas, Graphs, and Mathematical Tables*, 9th ed., Dover, 1970.

[8] G. L. Stüber, *Principles of Mobile Communication*, 2nd ed., Kluwer, 2000.

[9] G. W. Lank and L. S. Reed, "Average time to loss of lock for an automatic frequency control loop with two fading signals and a related probability density function," *IEEE Trans. Inf. Theory*, 12, no. 1, pp. 73–75, 1966.

[10] J.-P. M. G. Linnartz and R. Prasad, "Threshold crossing rate and average non-fade duration in a Rayleigh-fading channel with multiple interferers," *Archiv Fur Elektronik und Ubertragungstechnik Electronics and Communication*, 43, no. 6, pp. 345–349, 1989.

[11] W. C. Jakes, *Microwave Mobile Communications*, Wiley, 1974.

[12] X. Dong and N. C. Beaulieu, "Average level crossing rate and average fade duration of selection diversity," *IEEE Commun. Lett.*, 5, no. 10, pp. 396–398, 2001.

[13] T. Eng, N. Kong, and L. B. Milstein, "Comparison of diversity combining techniques for Rayleigh-fading channels," *IEEE Trans. Commun.*, 44, no. 9, pp. 1117–1129, 1996.

[14] T. Eng, N. Kong, and L. B. Milstein, "Correction to 'Comparison of diversity combining techniques for Rayleigh-fading channels'," *IEEE Trans. Commun.*, 46, no. 9, p. 1111, 1998.

[15] B. Holter, M.-S. Alouini, G. E. Øien, and H.-C. Yang, "Multiuser switched diversity transmission," in *Proc. IEEE Vehicular Technology Conf. (VTC'04-Fall)*, Los Angeles, CA, 2004, pp. 2038–2043.

[16] M.-S. Alouini and A. J. Goldsmith, "Adaptive modulation over Nakagami fading channels," *Kluwer J. Wireless Commun.*, 13, no. 1–2, pp. 119–143, 2000.

[17] R. Kwan and C. Leung, "Downlink scheduling optimization in CDMA networks," *IEEE Commun. Letters*, 8, pp. 611–613, 2004.

[18] Y. Ma, J. Jin, and D. Zhang, "Throughput and channel access statistics of generalized selection multiuser scheduling," *IEEE Trans. Wireless Commun.*, 7, no. 8, pp. 2975–2987, 2008.

[19] M. K. Simon and M.-S. Alouini, "Performance analysis of generalized selection combining with threshold test per branch (T-GSC)," *IEEE Trans. Veh. Technol.*, 51, no. 5, pp. 1018–1029, 2002.

[20] J. Hömäläinen and R. Wichman, "Capacities of physical layer scheduling strategies on a shared link," *Wireless Personal Commun.*, 39, no. 1, pp. 115–134, 2006.

[21] S. Nam, M.-S. Alouini, H.-C. Yang, and K. A. Qaraqe, "Threshold-based parallel multiuser scheduling," *IEEE Trans. Wireless Commun.*, 8, no. 4, pp. 2150–2159, 2009.

[22] H.-C. Yang and M. -S. Alouini, "Generalized switch and examine combining (GSEC): A low-complexity combining scheme for diversity rich environments," *IEEE Trans. Commun.*, 52, no. 10, pp. 1711–1721, 2004.

[23] H.-C. Yang and L. Yang, "Tradeoff analysis of performance and complexity on GSECps diversity combining scheme," *IEEE Trans. Wireless Commun.*, 7, no. 1, pp. 32–36, 2008.

[24] Y. Ma, D. Zhang, and R. Schober, "Capacity-maximizing multiuser scheduling for parallel channel access," *IEEE Signal Processing Lett.*, 14, pp. 441–444, 2007.

[25] S. Nam, H.-C. Yang, M.-S. Alouini, and K. A. Qaraqe, "Performance evaluation of threshold-based power allocation algorithms for down-link switched-based parallel scheduling," *IEEE Trans. Wireless Commun.*, 8, no. 4, pp. 1744–1753, 2009.

[26] K. J. Hole, H. Holm, and G. E. Oien, "Adaptive multidimensional coded modulation over flat fading channels," *IEEE J. Select. Areas Commun.*, 18, no. 7, pp. 1153–1158, 2000.

[27] Y. S. Al-Harthi, A. H. Tewfik, and M.-S. Alouini, "Multiuser diversity with quantized feedback," *IEEE Trans. Wireless Commun.*, 6, no. 1, pp. 330–337, 2007.

[28] H. Nam and M.-S. Alouini, "Multiuser switched diversity scheduling systems with per-user threshold," *IEEE Trans. on Commun.*, 58, no. 5, pp. 1321–1326, 2010.

[29] C.-J. Chen and L.-C. Wang, "A unified capacity analysis for wireless systems with joint multiuser scheduling and antenna diversity in Nakagami fading channels," *IEEE Trans. Commun.*, 54, no. 3, pp. 469–478, 2006.

[30] T. Tang and R. W. Heath, "Opportunistic feedback for downlink multiuser diversity," *IEEE Commun. Lett.*, 9, no. 10, pp. 948–950, 2005.

[31] S. Sanayei and A. Nosratinia, "Opportunistic downlink transmission with limited feedback," *IEEE Trans. Inform. Theory*, 53, no. 11, pp. 4363–4372, 2007.

[32] Y. Xue and T. Kaiser, "Exploiting multiuser diversity with imperfect one-bit channel state feedback," *IEEE Trans. Veh. Technol.*, 56, no. 1, pp. 183–193, 2007.

[33] J. So and J. M. Cioffi, "Feedback reduction scheme for downlink multiuser diversity," *IEEE Trans. Wireless Commun.*, 8, no. 2, pp. 668–672, 2009.

[34] Q. Ma and C. Tepedelenlioglu, "Practical multiuser diversity with outdated channel feedback," *IEEE Trans. Veh. Technol.*, 54, no. 4, pp. 1334–1345, 2005.

[35] K.-H. Park, Y.-C. Ko, and M.-S. Alouini, "Joint adaptive combining and multiuser downlink scheduling," *IEEE Trans. Veh. Technol.*, 57, no. 5, pp. 2958–2968, 2008.

[36] S. B. Halima, M.-S. Alouini, and K. A. Qaraqe, "Joint MS-GSC combining and down-link multiuser diversity scheduling," *IEEE Trans. Wireless Commun.*, 8, no. 7, pp. 3536–3545, 2009.

7 Multiuser MIMO Transmissions

MIMO transmission can significantly increase the capacity of wireless transmission systems through spatial multiplexing. Implementing the same number of antennas at user devices as base station (BS) is, however, very challenging, due to size/cost constraints. As such, the scenario where a multiple-antenna BS serves multiple users, each with a single antenna, as illustrated in Fig. 7.1, is of practical interest. In such a scenario, the potential spatial multiplexing gain offered by multiple antennas at the BS can still be explored through the simultaneous transmission to multiple users. If we consider the antennas at different users jointly, a virtual MIMO transmission system is formed, with multiple co-located antennas at the BS side and multiple distributed antennas at the user side. The resulting multiuser MIMO systems have attracted a significant amount of research and practical interest.

In this chapter, we carry out the design and analysis of transmission schemes for multiuser MIMO systems. We focus on low-complexity linear transmission techniques and study their performance through accurate mathematical analysis. We first introduce the zero-forcing beamforming (ZFBF) transmission scheme and analyze its performance with two different user scheduling strategies. We then investigate the random unitary beamforming (RUB) transmission scheme. Several user scheduling strategies for RUB are presented and analyzed. Both signal antenna per user and multiple antenna per user cases are studied. Special emphasis is put on the quantification of the trade-off between sum-capacity performance and complexity/feedback load.

7.1 Introduction

To explore the spatial multiplexing gain of multiuser MIMO systems, the BS will serve multiple users simultaneously over the same frequency band. Interuser interference naturally occurs and needs to be effectively mitigated. Generally, the downlink transmission from BS to users is more challenging, since each user typically can only rely on its own received signal. Note that the BS can jointly process the received signal from all users for uplink transmission. To control the interuser interference for downlink transmission, the BS needs to apply proper preprocessing to the user signal before transmission, which is typically referred to as precoding.

Various precoding schemes for downlink multiuser MIMO transmission, also known as MIMO broadcast channels, have been proposed in the literature. These schemes are in

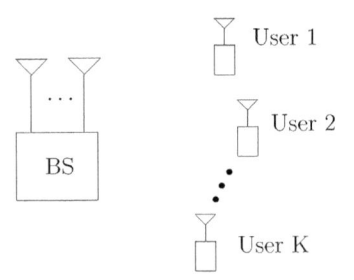

Figure 7.1 Multiuser MIMO system.

general classified into two categories: nonlinear precoding schemes and linear precoding schemes. Typically, nonlinear schemes achieve better performance, which is characterized by the total capacity of all users, also referred to as sum-rate capacity, at the cost of higher implementation complexity. For example, the dirty paper coding (DPC) technique [1] is a nonlinear precoding scheme that precancels interuser interference before transmission with the full knowledge of channel state information. It has been shown that DPC achieves the optimal sum-rate performance over MIMO broadcast channels [2–4]. The sum-rate capacity of DPC scales at a rate of $M \log \log K$, where M is the number of transmit antennas and K is the total number of receive antennas at different users. Other nonlinear precoding schemes include Tomlinson–Harashima precoding and vector perturbation techniques.

Nonlinear precoding schemes have high computational complexity. For example, the successive encoding process required by DPC has prohibitively high computational complexity. DPC also requires the complete channel state information at the transmitter (CSIT) side. Therefore, it is of great practical interest to design multiuser MIMO transmission schemes that can achieve high sum-rate capacity with low complexity and a minimum CSIT requirement. Linear precoding schemes, on the other hand, have much lower implementation complexity. Zero-forcing beamforming [5, 6] and RUB [7–9] are two of the most popular examples. It has been shown that together with proper user scheduling schemes these linear beamforming schemes can achieve the optimal sum-rate performance in an asymptotic sense. With ZFBF, the precoding matrix is designed as the pseudo-inverse of the channel matrix of the selected users. With RUB, the precoding matrix consists of M random orthonormal beamforming vectors and users are selected based on their channel quality on different beamforming directions. In the following sections, we will present and analyze several practical user selection strategies for multiuser MIMO systems employing linear precoding schemes (also known as linear beamforming transmission) and focus specifically on their exact sum-rate performance evaluation.

To establish a common context for the discussion, we now present the general signal model for linear beamforming transmission in multiuser MIMO systems.

We consider the downlink transmission of a single-cell wireless system, as illustrated in Fig. 7.1. The BS is equipped with M antennas and as such can serve as many as M

different selected users with linear beamforming. The transmitted signal vector from M antennas over one symbol period can be written as

$$\mathbf{x} = \sum_{m=1}^{M} \mathbf{b}_m s_m, \qquad (7.1)$$

where s_m is the information symbol, destined for the mth selected user, and $\mathbf{b}_m$ is the beamforming vector for the mth selected user. The transmitted signal vector $\mathbf{x}$ has an average power constraint of $tr\{\mathbf{E}[\mathbf{x}\mathbf{x}^H]\} \leq P$, where P is the maximum average transmission power, $\mathbf{E}[\cdot]$ denotes the statistical expectation, and $(\cdot)^H$ the Hermitian transpose. We assume that there are a total of K active users in the cell, where $K \geq M$. Each user is equipped with a single antenna. We further assume that with a certain slow power control mechanism, the users experience homogeneous flat Rayleigh fading. In particular, the channel gain from the ith antenna to the kth mobile users, denoted by h_{ik}, is assumed to be an independent zero mean complex Gaussian random variable with unitary variance, i.e., $h_{ik} \sim \mathcal{CN}(0,1)$. As such, the instantaneous multiple-input single-output (MISO) channel from the BS to the kth mobile user can be characterized by a zero mean complex Gaussian channel vector, denoted by $\mathbf{h}_k = \{h_{1k}, h_{2k}, \cdots, h_{Mk}\}^T$.

During each symbol period, the BS simultaneously serves a subset of M selected users using all M available beams. The received symbol of the ith selected user, when the jth beam is assigned to it, can be written as

$$y_i = \mathbf{h}_i^T \mathbf{x} + n_i = \mathbf{h}_i^T \mathbf{b}_j s_j + \sum_{m=1, m \neq j}^{M} \mathbf{h}_i^T \mathbf{b}_m s_m + n_i, \qquad (7.2)$$

where n_i are independent zero mean additive Gaussian noise with unit variance. It is easy to see that the received signal will experience interuser interference as well as additive noise. The metric that will determine the quality of detection becomes the SINR (signal to interference plus noise ratio), which is given for the ith user receiving on the jth beamforming direction, assuming uniform power allocation to different users, by

$$\gamma_{i,j} = \frac{\frac{P}{M} |\mathbf{h}_i^T \mathbf{b}_j|^2}{\frac{P}{M} \sum_{m=1, m \neq j}^{M} |\mathbf{h}_i^T \mathbf{b}_m|^2 + 1}, \quad j = 1, 2, \ldots, M. \qquad (7.3)$$

The interuser interference will be controlled through beamforming vector design and/or user selection, as will be detailed in the following sections.

Finally, the total sum rate of the multiuser MIMO system with linear beamforming is given by

$$R = \mathbf{E}\left[\sum_{m=1}^{M} \log_2(1 + \gamma_m^B) \right], \qquad (7.4)$$

where γ_m^B is the received SINR on the mth beamforming direction.

7.2 Zero-forcing Beamforming Transmission

Zero-forcing beamforming is a practical multiuser transmission strategy for multiuser MIMO systems [5, 6]. With ZFBF, one user's beamforming vector is designed to be orthogonal to other selected users' channel vectors, i.e., $|\mathbf{h}_i^T \mathbf{b}_m| = 0$ for $i \neq m$ in (7.3). As such, ZFBF can completely eliminate multiuser interference. A proper user selection scheme is essential for ZFBF-based multiuser MIMO systems to fully benefit from multiuser diversity gain. The desired properties of selected user channels are (1) large channel power gain, $|h_i|^2$, and (2) near orthogonality to one another, $|h_i^T h_j| \approx 0$ for $i \neq j$. The optimum user selection strategy for ZFBF involves the exhaustive search of all possible user subsets, which becomes prohibitive to implement when the number of users is large. Several low-complexity user selection strategies have been proposed and studied in the literature [6, 10–12]. Among them, two greedy search algorithms, the successive projection ZFBF (SUP-ZFBF) and greedy weight clique ZFBF (GWC-ZFBF), are attractive due to their implementation simplicity and ability to achieve the same sum-rate scaling rate as the DPC scheme. We will carry out accurate statistical analyses on these two schemes for the important two-transmit-antenna special case [13]. These analytical results will not only facilitate the trade-off study between them but also apply to the parameter optimization for each scheme.

7.2.1 Dual-User ZFBF Transmission

We consider the multiuser MIMO system as shown in Fig. 7.1. Specifically, the BS is equipped with two transmit antennas. There are total K users in the system, where $K \geq 2$, and each user has one receive antenna. We assume a flat homogeneous Rayleigh fading channel model. The channel gain from the ith transmit antenna to the kth user, denoted by h_{ki}, is assumed to be i.i.d. zero mean complex Gaussian random variable with unitary variance. As such, the instantaneous MISO channel from the BS to the kth mobile user can be characterized by a zero mean complex Gaussian channel vector, denoted by $\mathbf{h}_k = [h_{k1} \; h_{k2}]^T$, $k = 1, 2, \ldots, K$. As such, the norm square of the kth user's channel vector $\|\mathbf{h}_k\|^2$ (termed *channel power gain*) is a Chi-square distributed random variable with four degrees of freedom, with probability density function (PDF) and cumulative distribution function (CDF) given by

$$p_{\|\mathbf{h}\|^2}(x) = xe^{-x}, \; x \geq 0, \tag{7.5}$$

and

$$F_{\|\mathbf{h}\|^2}(x) = \gamma(2, x), \; x \geq 0, \tag{7.6}$$

respectively, where $\gamma(2, x) = \int_0^x t e^{-t} dt$ is the lower incomplete gamma function. We assume that the BS knows the downlink channel vector for every user either through feedback mechanism or the reciprocity of the channel in time division duplexing (TDD) mode.

Let $\pi(1)$ and $\pi(2)$ denote the indexes of the selected users and $\mathbf{b}_{\pi(1)}$ and $\mathbf{b}_{\pi(2)}$ denote the beamforming vectors for the two selected users. With ZFBF, the beamforming vector

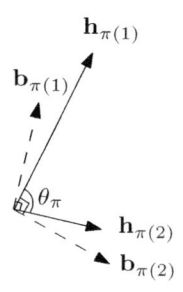

Figure 7.2 Zero-forcing beamforming for two-user case.

of one user is chosen to be orthogonal to the channel vector of the other user, as shown in Fig. 7.2. The transmitted signal vector from two antennas over one symbol period can then be written as

$$\mathbf{x} = \sum_{i=1}^{2} \sqrt{P_{\pi(i)}} \mathbf{b}_{\pi(i)} s_{\pi(i)}, \tag{7.7}$$

where $s_{\pi(i)}$ and $P_{\pi(i)}$ are the data symbol and transmit power scaling factor, respectively, for user $\pi(i)$. The power constraint imposed on the transmitted signal is $E[\|\mathbf{x}\|^2] \leq P$. Because of the orthogonality between beamforming vectors and channel vectors, the received symbol at user $\pi(i)$ is consequently given by

$$r_{\pi(i)} = \mathbf{h}_{\pi(i)}^T \mathbf{x} + n_{\pi(i)} = \sqrt{P_{\pi(i)}} \mathbf{h}_{\pi(i)}^T \mathbf{b}_{\pi(i)} s_{\pi(i)} + n_{\pi(i)}, \tag{7.8}$$

where $n_{\pi(i)}$ is the additive zero mean Gaussian noise with variance N_0. Note that the interference between two user transmissions is completely eliminated, whereas the effective channel power gain for user $\pi(i)$ becomes the norm square of the projection of its channel vector $\mathbf{h}_{\pi(i)}$ onto the corresponding beamforming vector $\mathbf{b}_{\pi(i)}$, which is termed the *projection power* here. Let θ_π, as illustrated in Fig. 7.2, denote the angle between $\mathbf{h}_{\pi(1)}$ and $\mathbf{h}_{\pi(2)}$. The effective power gain $\gamma_{\pi(i)}$ can be written as

$$\gamma_{\pi(i)} = \|\mathbf{h}_{\pi(i)}\|^2 |\sin(\theta_\pi)|^2. \tag{7.9}$$

Note that $|\sin(\theta_\pi)|^2$ can be viewed as the projection power loss factor due to ZFBF. Assuming the total transmit power P is equally divided between two selected users, the instantaneous sum rate of the system is given by

$$R = \sum_{i=1}^{2} \log(1 + \frac{\rho}{2} \gamma_{\pi(i)}), \tag{7.10}$$

where $\rho = \frac{P}{N_0}$ denotes the total transmit SNR.

7.2.2 User Selection Strategies

The BS will select two users to serve simultaneously using either the SUP-ZFBF or the GWC-ZFBF scheme. Basically, both schemes perform user selection sequentially from the user subset that contains users whose channel vectors are near-orthogonal to the channel vectors of already selected users. Here, two channel vectors are claimed to be near-orthogonal if the angle between the two channel vectors, θ, satisfies $|\sin(\theta)|^2 \geq \lambda_d$, where $0 < \lambda_d < 1$ is a constant value. With GWC-ZFBF, the next selected user is the one with the largest channel power gain, whereas with SUP-ZFBF, the next user is the one with the largest projection power of its channel vector onto the complement space of the space spanned by the channel vectors of already selected users. The mode of operations of GSW-ZFBF and SUP-ZFBF are summarized below.

GWC-ZFBF

1. Order all K users based on their channel power gain $\|\mathbf{h}_k\|^2$ as $\|\mathbf{h}_{(1)}\|^2 > \|\mathbf{h}_{(2)}\|^2 > \ldots > \|\mathbf{h}_{(K)}\|^2$.
2. Select the one with the largest channel power gain as the first user, i.e., $\pi(1) = (1)$.
3. Examine sequentially the angle between $\mathbf{h}_{\pi(1)}$ and $\mathbf{h}_{(k)}$, $k = 2, 3, \ldots, K$ and select the first user whose channel vector is near-orthogonal to $\mathbf{h}_{\pi(1)}$. Equivalently, the second selected user will be the one with the largest channel power gain among all users in the set

$$\mathcal{U} = \{i | |\sin(\theta_i)|^2 = 1 - \frac{|<\mathbf{h}_i, \mathbf{h}_{\pi(1)}>|^2}{\|\mathbf{h}_i\|^2 \|\mathbf{h}_{\pi(1)}\|^2} \geq \lambda_d\}, \quad (7.11)$$

where θ_i is the angle between $\mathbf{h}_i$ and $\mathbf{h}_{\pi(1)}$.

SUP-ZFBF

1. Select the user with the largest channel power gain as the first user.
2. Calculate and rank the projection power onto the complementary space of $\mathbf{h}_{\pi(1)}$ for all users in $\mathcal{U}$, which is defined in (7.11).
3. Select the user with the largest projection power as the second user, i.e.,

$$\pi(2) = \max_{i \in \mathcal{U}} \{\|\mathbf{h}_i\|^2 |\sin(\theta_i)|^2\}. \quad (7.12)$$

As SUP-ZFBF needs to calculate the projection power of all users in $\mathcal{U}$ while GWC-ZFBF stops searching when it encounters the first user that is near-orthogonal to $\mathbf{h}_{\pi(1)}$, SUP-ZFBF exhibits higher computational complexity than GWC-ZFBF.

7.2.3 Sum-Rate Analysis

In this subsection, we analyze the ergodic sum rate of ZFBF-based multiuser MIMO systems with SUP-ZFBF and GWC-ZFBF strategies. For that purpose, we first derive the statistics of channel power gains of selected users $\|\mathbf{h}_{\pi(i)}\|^2$, $i = 1, 2$ and projection power loss factor $|\sin(\theta_\pi)|^2$.

Common Analysis for Both Schemes

For both user selection schemes, the first selected user has the largest channel power gain among K users. Since the power gains of K users are i.i.d. random variables, the PDF of $\|\mathbf{h}_{\pi(1)}\|^2$ for both schemes is expressed as [14]

$$p_{\|\mathbf{h}_{\pi(1)}\|^2}(x) = K p_{\|\mathbf{h}\|^2}(x) (F_{\|\mathbf{h}\|^2}(x))^{K-1}. \quad (7.13)$$

If no user's channel vector is near-orthogonal to the channel vector of the first selected user, i.e., $\mathcal{U} = \emptyset$, the BS transmits to the first selected user using the traditional beamforming approach. In this case, the sum rate for both user selection schemes is expressed as

$$\mathbf{E}[R|\mathcal{U} = \emptyset] = \int_0^\infty \log(1 + \frac{\rho}{2}x) p_{\|\mathbf{h}_{\pi(1)}\|^2}(x) dx. \tag{7.14}$$

It is shown in [6] that $|\sin(\theta)|^2$, where θ is the angle between a random channel vector and $\mathbf{h}_{\pi(1)}$, follows the standard uniform distribution over the interval [0, 1]. Therefore, after the selection of the first user, the probability that a remaining user belongs to $\mathcal{U}$ is $\Pr[|\sin(\theta)|^2 \geq \lambda_d] = 1 - \lambda_d$. It follows that

$$\Pr[\mathcal{U} = \emptyset] = \lambda_d^{K-1} \quad \text{and} \quad \Pr[\mathcal{U} \neq \emptyset] = 1 - \lambda_d^{K-1}, \tag{7.15}$$

where $\Pr[\mathcal{U} \neq \emptyset]$ is the probability of $\mathcal{U} \neq \emptyset$ and as such the second user can be selected. Considering these two mutually exclusive cases of $\mathcal{U} = \emptyset$ and $\mathcal{U} \neq \emptyset$, the ergodic sum rate of both schemes can be calculated as

$$\mathbf{E}[R] = \Pr[\mathcal{U} = \emptyset]\mathbf{E}[R|\mathcal{U} = \emptyset] + \Pr[\mathcal{U} \neq \emptyset]\mathbf{E}[R|\mathcal{U} \neq \emptyset]. \tag{7.16}$$

We now need to determine the average sum rate when $\mathcal{U}$ is not empty. This analysis requires the PDF of $\|\mathbf{h}_{\pi(2)}\|^2$ and $|\sin(\theta_\pi)|^2$, which will be derived separately for GWC-ZFBF and SUP-ZFBF in the following.

GWC-ZFBF-Specific Analysis

With GWC-ZFBF, the second selected user is the one with the ith largest channel power gain, i.e., $\pi(2) = (i)$, if and only if user (i) belongs to $\mathcal{U}$ and none of the users from (2) to $(i-1)$ does. Therefore, the probability of $\pi(2) = (i)$, given that $\mathcal{U}$ is not empty, can be calculated as

$$\Pr\left[\pi(2) = (i)|\mathcal{U} \neq \emptyset\right] = \frac{(1 - \lambda_d)}{1 - \lambda_d^{K-1}} \lambda_d^{i-2}, \quad i = 2, 3, \ldots, K. \tag{7.17}$$

By applying the total probability theorem, the PDF of the channel power gain of the second selected user is obtained as

$$p_{\|\mathbf{h}_{\pi(2)}\|^2}(x|\mathcal{U} \neq \emptyset) = \sum_{i=2}^{K} \Pr\left[\pi(2) = (i)|\mathcal{U} \neq \emptyset\right] \cdot p_{\|\mathbf{h}_{(i)}\|^2}(x), \tag{7.18}$$

where $p_{\|\mathbf{h}_{(i)}\|^2}(\cdot)$ is the PDF of the ith largest channel power gain among K users, given by [14]

$$p_{\|\mathbf{h}_{(i)}\|^2}(x) = K\binom{K-1}{i-1} p_{\|\mathbf{h}\|^2}(x)\left(1 - F_{\|\mathbf{h}\|^2}(x)\right)^{i-1} F_{\|\mathbf{h}\|^2}^{K-i}(x). \tag{7.19}$$

After proper substitution of (7.7) and (7.9) and some mathematical manipulations, (7.18) can be rewritten as

$$p_{\|\mathbf{h}_{\pi(2)}\|^2}(x|\mathcal{U} \neq \emptyset) = \qquad (7.20)$$
$$\frac{K(1-\lambda_d)}{\lambda_d(1-\lambda_d^{K-1})}p_{\|\mathbf{h}\|^2}(x)\left(\left(\lambda_d + (1-\lambda_d)F_{\|\mathbf{h}\|^2}(x)\right)^{K-1} - F_{\|\mathbf{h}\|^2}^{K-1}(x)\right).$$

As the second user is always selected from set $\mathcal{U}$, the projection power loss factor, $|\sin(\theta_\pi)|^2$, follows a uniform distribution over the interval $[\lambda_d, 1]$. Therefore, the ergodic sum rate for the GWC-ZFBF scheme conditioned on $\mathcal{U} \neq \emptyset$ can be calculated, by averaging (7.10) over the distribution of $\|\mathbf{h}_{\pi(i)}\|^2$, $i = 1, 2$ and $|\sin(\theta_\pi)|^2$ as

$$\mathbb{E}[R|\mathcal{U} \neq \emptyset] = \sum_{i=1}^{2} \int_0^\infty \int_{\lambda_d}^1 \log(1 + \frac{\rho}{2}xy) \qquad (7.21)$$
$$\cdot p_{|\sin(\theta_\pi)|^2}(y|\mathcal{U} \neq \emptyset)dy\, p_{\|\mathbf{h}_{\pi(i)}\|^2}(x|\mathcal{U} \neq \emptyset)dx,$$

which can be further simplified to

$$\mathbb{E}[R|\mathcal{U} \neq \emptyset] = \sum_{i=1}^{2} \int_0^\infty \frac{g(x)}{1-\lambda_d} p_{\|\mathbf{h}_{\pi(i)}\|^2}(x|\mathcal{U} \neq \emptyset)dx, \qquad (7.22)$$

where

$$g(x) = \frac{2}{\rho x}[(1+\frac{\rho x}{2})\log(1+\frac{\rho x}{2}) - (1+\frac{\lambda_d \rho x}{2})\log(1+\frac{\lambda_d \rho x}{2})] - \frac{1}{\ln(2)}, \qquad (7.23)$$

and $p_{\|\mathbf{h}_{\pi(i)}\|^2}(\cdot)$ for $i = 1$ and 2 are given in (7.13) and (7.20), respectively.

Finally, the final expression of sum-rate capacity of the multiuser MIMO system with the GWC-ZEBF strategy can be obtained by substituting (7.14), (7.15), and (7.22) into (7.16) as

$$\mathbb{E}[R] = \lambda_d^{K-1} \int_0^\infty \log(1+\frac{\rho}{2}x)p_{\|\mathbf{h}_{\pi(1)}\|^2}(x)dx \qquad (7.24)$$
$$+\frac{1-\lambda_d^{K-1}}{1-\lambda_d}\sum_{i=1}^{2}\int_0^\infty g(x)p_{\|\mathbf{h}_{\pi(i)}\|^2}(x|\mathcal{U} \neq \emptyset)dx.$$

SUP-ZFBF-Specific Analysis

With SUP-ZFBF, the second selected user is the user in $\mathcal{U}$ that has the largest projection power onto the complementary space of the first selected user's channel vector. From (7.9), the second selected user's projection power is also its effective power gain. As such, the ergodic capacity of the second selected user can be obtained using the PDF of the largest projection power among all users in $\mathcal{U}$.

We first need to find the PDF of the channel power gain for a user in $\mathcal{U}$, which is denoted as $\|\tilde{\mathbf{h}}\|^2$. Noting that users in $\mathcal{U}$ cannot have the largest channel power gain, the PDF of $\|\tilde{\mathbf{h}}\|^2$ can be obtained as

$$p_{\|\tilde{\mathbf{h}}\|^2}(x) = \frac{1}{K-1}\sum_{i=2}^{K}f_{\|\mathbf{h}_{(i)}\|^2}(x) \qquad (7.25)$$
$$= \frac{K}{K-1}p_{\|\mathbf{h}\|^2}(x)(1-F_{\|\mathbf{h}\|^2}^{K-1}(x)),$$

7.2 Zero-forcing Beamforming Transmission

where $p_{\|\mathbf{h}_{(i)}\|^2}(x)$ is the PDF of the ith largest channel power gain. Letting $\|\tilde{\mathbf{h}}_p\|^2$ represent the projection power of a user in $\mathcal{U}$, we have $\|\tilde{\mathbf{h}}_p\|^2 = \|\tilde{\mathbf{h}}\|^2 |\sin(\tilde{\theta})|^2$, where $\tilde{\theta}$ is the angle between channel vectors of that user and the first selected user. Since $|\sin(\tilde{\theta})|^2$ follows the uniform distribution over the interval $[\lambda_d, 1]$, the CDF of $\|\tilde{\mathbf{h}}_p\|^2$ can be obtained as

$$F_{\|\tilde{\mathbf{h}}_p\|^2}(z) = \Pr[\|\tilde{\mathbf{h}}\|^2 |\sin(\tilde{\theta})|^2 < z] = \frac{1}{1-\lambda_d} \int_{\lambda_d}^{1} \int_{0}^{\frac{z}{y}} p_{\|\tilde{\mathbf{h}}\|^2}(x) dx dy. \quad (7.26)$$

After taking the derivative with respect to z, the corresponding PDF is given by

$$p_{\|\tilde{\mathbf{h}}_p\|^2}(z) = \frac{1}{1-\lambda_d} \int_{\lambda_d}^{1} \frac{1}{y} p_{\|\tilde{\mathbf{h}}\|^2}(\frac{z}{y}) dy. \quad (7.27)$$

Denote the cardinality of the set $\mathcal{U}$, i.e., the number of users in $\mathcal{U}$, as $|\mathcal{U}|$. As the channel directions of users are independent, the distribution of $|\mathcal{U}|$ conditioned on $\mathcal{U} \neq \emptyset$ can be shown to be given by

$$\Pr[|\mathcal{U}| = k | \mathcal{U} \neq \emptyset] \quad (7.28)$$
$$= \frac{1}{1-\lambda_d^{K-1}} \binom{K-1}{k} (1-\lambda_d)^k \lambda_d^{K-1-k}, \quad k = 1, \ldots, K-1.$$

When $|\mathcal{U}| = k$, the effective channel gain of the second selected user, $\gamma_{\pi(2)}$, is equal to the largest projection power among k users and its PDF is thus given by

$$p_{\gamma_{\pi(2)}}(g \mid |\mathcal{U}| = k) = k p_{\|\tilde{\mathbf{h}}_p\|^2}(g) F_{\|\tilde{\mathbf{h}}_p\|^2}^{k-1}(g). \quad (7.29)$$

Combining (7.28) and (7.29) and averaging over $|\mathcal{U}|$, the PDF of $\gamma_{\pi(2)}$, given that $\mathcal{U} \neq \emptyset$, can be obtained as

$$p_{\gamma_{\pi(2)}}(g | \mathcal{U} \neq \emptyset) = \sum_{k=1}^{K-1} \Pr(|U| = k | \mathcal{U} \neq \emptyset) p_{\gamma_{\pi(2)}}(g \mid |U| = k) \quad (7.30)$$
$$= \frac{(K-1)(1-\lambda_d)\lambda_d^{K-2}}{1-\lambda_d^{K-1}} f_{\|\tilde{\mathbf{h}}_p\|^2}(g) \Big(1 + \frac{1-\lambda_d}{\lambda_d} F_{\|\tilde{\mathbf{h}}_p\|^2}(g)\Big)^{K-2}.$$

To derive the distribution of $|\sin(\theta_\pi)|^2$, we start with the joint PDF of a remaining user's channel power, $\|\tilde{\mathbf{h}}\|^2$, and its projection power loss factor, $|\sin(\tilde{\theta})|^2$,

$$p_{\|\tilde{\mathbf{h}}\|^2, |\sin(\tilde{\theta})|^2}(x, y) = \frac{1}{1-\lambda_d} p_{\|\tilde{\mathbf{h}}\|^2}(x), \quad x \geq 0, \lambda_d \leq y \leq 1. \quad (7.31)$$

With a change of variables, the joint PDF of projection power $\|\tilde{\mathbf{h}}_p\|^2$ and $|\sin(\tilde{\theta})|^2$ of a remaining user is obtained from (7.31) as

$$p_{\|\tilde{\mathbf{h}}_p\|^2, |\sin(\tilde{\theta})|^2}(z, y) = \frac{1}{(1-\lambda_d)y} p_{\|\tilde{\mathbf{h}}\|^2}(\frac{z}{y}), \quad z \geq 0, \lambda_d \leq y \leq 1. \quad (7.32)$$

It follows that the conditional PDF of $|\sin(\tilde{\theta})|^2$ given that $\|\tilde{\mathbf{h}}_p\|^2 = z$ is given by

$$p_{|\sin(\tilde{\theta})|^2}(y \mid \|\tilde{\mathbf{h}}_p\|^2 = z) = \frac{1}{(1-\lambda_d)y} p_{\|\tilde{\mathbf{h}}\|^2}(\frac{z}{y}) \Big/ p_{\|\tilde{\mathbf{h}}_p\|^2}(z), \quad z \geq 0, \lambda_d \leq y \leq 1. \quad (7.33)$$

Since the projection power of the second selected user is given by (7.30), we can obtain the PDF of $|\sin(\theta_\pi)|^2$ conditioned on $\mathcal{U} \neq \emptyset$ by averaging (7.33) over (7.30) as

$$p_{|\sin(\theta_\pi)|^2}(y|\mathcal{U} \neq \emptyset) = \int_0^\infty p_{|\sin(\tilde{\theta})|^2}(y|\|\tilde{\mathbf{h}}_p\|^2 = z) p_{\gamma_{\pi(2)}}(z|\mathcal{U} \neq \emptyset) dz \qquad (7.34)$$

$$= \frac{(K-1)\lambda_d^{K-2}}{(1-\lambda_d^{K-1})y} \int_0^\infty p_{\|\tilde{\mathbf{h}}\|^2}\left(\frac{z}{y}\right)\left(1 + \frac{1-\lambda_d}{\lambda_d} F_{\|\tilde{\mathbf{h}}_p\|^2}(z)\right)^{K-2} dz, \lambda_d \leq y \leq 1.$$

Finally, $E[R|\mathcal{U} \neq \emptyset]$ for the SUP-ZFBF scheme as

$$E[R|\mathcal{U} \neq \emptyset] = \int_0^\infty \log(1 + \frac{\rho}{2}g) f_{\gamma_{\pi(2)}}(g|\mathcal{U} \neq \emptyset) dg + \qquad (7.35)$$

$$\int_0^\infty \int_{\lambda_d}^1 \log(1 + \frac{\rho}{2}xy) p_{\|\mathbf{h}_{\pi(1)}\|^2}(x) p_{|\sin(\theta_\pi)|^2}(y|\mathcal{U} \neq \emptyset) dy dx.$$

Consequently, the expression of the ergodic sum rate of the multiuser MIMO system with the SUP-ZEBF strategy can be obtained by substituting (7.14), (7.15), and (7.35) into (7.16).

7.2.4 Numerical Results

Fig. 7.3 illustrates the PDF of the projection power loss factor $|\sin(\theta_\pi)|^2$ for both GWC-ZFBF and SUP-ZFBF schemes with different λ_d and/or K values. As we can see, $|\sin(\theta_\pi)|^2$ with the GWC-ZFBF scheme always follows a uniform distribution over the interval $[\lambda_d, 1]$. On the other hand, $|\sin(\theta_\pi)|^2$ with the SUP-ZFBF scheme is not uniformly distributed. This is because its user selection is based on projection power, which is correlated to channel direction. We can also observe from Fig. 7.3 that the PDF of $|\sin(\theta_\pi)|^2$ with SUP-ZFBF is a monotonically increasing function and the mass of the PDF shifts to the right as K increases. We can conclude that SUP-ZFBF can explore more multiuser directional diversity gain than GWC-ZFBF by selecting "more orthogonal" users.

In Fig. 7.4, we plot the ergodic sum rate of both the GWC-ZFBF and SUP-ZFBF schemes as a function of channel direction constraint λ_d. As we can see, while the ergodic sum rate with GWC-ZFBF is a unimodal function of λ_d, that of SUP-ZFBF remains roughly constant for the small to medium values of λ_d and decreases when λ_d becomes very close to 1. Note that when λ_d is small, the GWC-ZFBF scheme may select less orthogonal users, which incurs larger projection power loss, whereas the SUP-ZFBF scheme tends to select "more orthogonal" users. On the other hand, when λ_d is very large, the number of candidate users for selection decreases, which leads to sum-rate performance degradation for both schemes. With the analytical sum-rate expression derived here, we can easily find the optimal value of λ_d by using one-dimensional optimization algorithms. We can also observe that the optimal value of λ_d increases with the number of users K, which indicates that when more users are in the system we should trade channel power diversity for directional diversity gain.

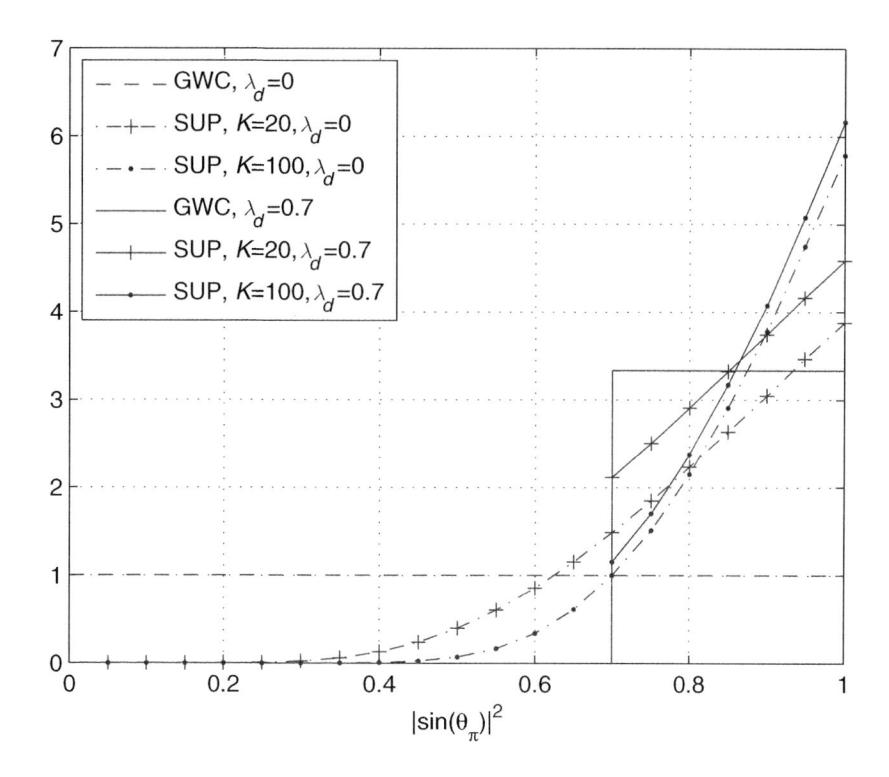

Figure 7.3 The PDF of $|\sin(\theta_\pi)|^2$ for GWC-ZFBF and SUP-ZFBF ($\rho = 10$ dB). Reprint with permission from [13]. ©2009 IEEE.

As an additional example, Fig. 7.5 plots the ergodic sum rate with both GWC-ZFBF and SUP-ZFBF schemes as a function of the number of users K. The sum-rate performance of a more complex scheme proposed in [12], dubbed a D-S scheme, is also plotted for comparison purposes. As shown in this figure, the performance of the GWC-ZFBF scheme with optimal λ_d values for different K approaches that of SUP-ZFBF, especially for a larger value of K. In addition, the performance gap between the SUP-ZFBF and D-S schemes is very small if λ_d for SUP-ZFBF is properly chosen. Note that both GWC-ZFBF and SUP-ZFBF enjoy much lower computational complexity than the D-S scheme.

7.3 Random Unitary Beamforming (RUB) Transmission

RUB, also known as random beamforming, is a low-complexity solution for exploring spatial multiplexing gain over multiuser MIMO broadcast channels [7–9]. With RUB, the BS generates M random orthonormal beams to communicate with as many as M selected users simultaneously. To explore multiuser diversity gain, each user will feed back some quality information, usually in terms of SINR, of each beam, based on which the BS carries out proper user selection. The major complexity saving of RUB

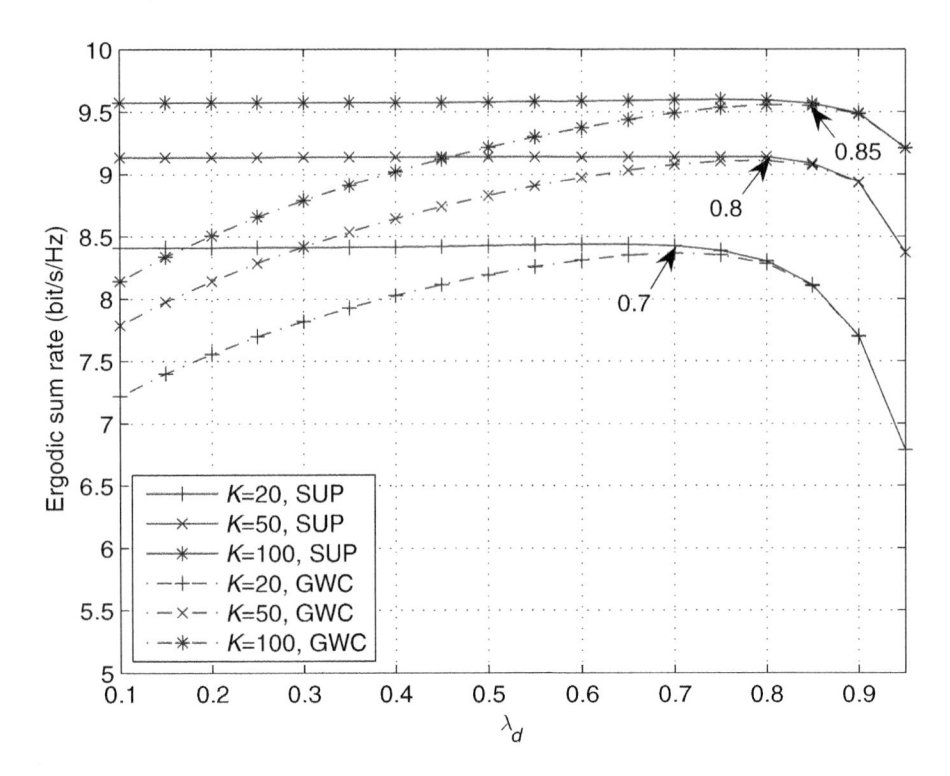

Figure 7.4 Effects of λ_d on the ergodic sum rate of the GWC-ZFBF and SUP-ZFBF schemes ($\rho = 10$ dB). Reprint with permission from [13]. ©2009 IEEE.

over ZFBF is the lower feedback load, as users will no longer need to feed back their complete channel vectors. It has been proved [9] that if each user just feeds back its maximum beam SINR and the index of the corresponding beam, RUB can achieve the same sum-rate scaling law as that of DPC (i.e., $M \log \log K$) when K approaches infinity. Note that the resulting feedback load is only a real number and an integer per user.

7.3.1 Asymptotic Sum-Rate Analysis of RUB Systems

We consider the downlink transmission of a single-cell wireless system. The M-antenna BS employs M random orthonormal beams, generated from an isotropic distribution [7], to serve M selected users simultaneously over the same frequency channel. Let $\mathcal{U} = \{\mathbf{u}_1, \mathbf{u}_2, \ldots, \mathbf{u}_M\}$ denote the set of beamforming vectors, assumed to be known to both BS and users. The transmitted signal vector from M antennas over one symbol period can be written as

$$\mathbf{x} = \sum_{m=1}^{M} \mathbf{u}_m s_m, \qquad (7.36)$$

where s_m is the information symbol, destined for the mth selected user.

7.3 Random Unitary Beamforming (RUB) Transmission

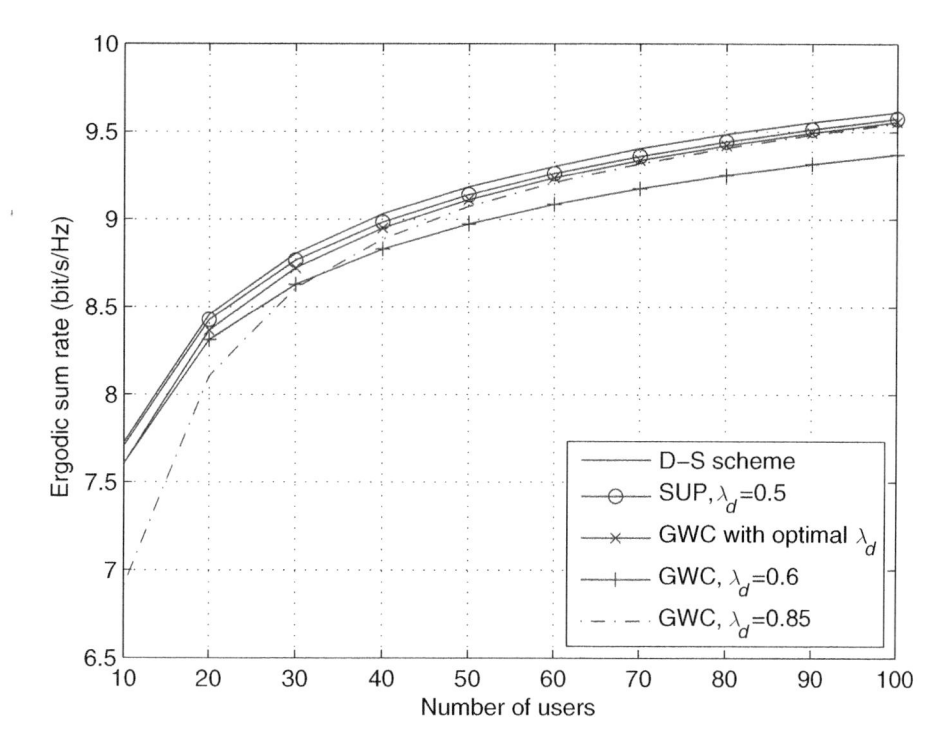

Figure 7.5 Ergodic sum rate of the GWC-ZFBF and SUP-ZFBF schemes as a function of the number of users ($\rho = 10$ dB). Reprint with permission from [13]. ©2009 IEEE.

We assume that there are a total of K active users in the cell, where $K \geq M$. Each user is equipped with a single antenna. We further assume that with a certain slow power control mechanism, the users experience homogeneous Rayleigh fading. In particular, the channel gain from the ith antenna to the kth mobile users, denoted by h_{ik}, is assumed to be an independent zero mean complex Gaussian random variable with unitary variance, i.e., $h_{ik} \sim \mathcal{CN}(0, 1)$. Consequently, the SINR of user i while using the jth beam is given by

$$\gamma_{i,j} = \frac{\frac{P}{M}|\mathbf{h}_i^T \mathbf{u}_j|^2}{\frac{P}{M} \sum_{m=1, m \neq j}^{M} |\mathbf{h}_i^T \mathbf{u}_m|^2 + 1}, \quad j = 1, 2, \ldots, M. \tag{7.37}$$

We assume that the users can accurately estimate their own channel vectors and determine the instantaneous SINR corresponding to different beams using Eq. (7.37) with the knowledge of beamforming vector set $\mathcal{U}$.

The most straightforward feedback strategy is to have every user feed back their received SNR on all M beamforming directions, which involves M real number feedback per user. The BS then selects for each beam the user with the highest SINR. If K is large enough, the probability that a user is the best user on two different beams and the probability that a beam is not requested by any user will be very small. As a result, the SINR of the selected user on a beam will be the largest one of all user SINRs on this beam. Mathematically, we have $\gamma_j^B \cong \max\{\gamma_{1,j}, \gamma_{2,j}, \ldots, \gamma_{K,j}\} \doteq \gamma_{1:K,j}$. To investigate the statistics of beam SINR $\gamma_{1:K,j}$, we first derive the statistics of the received SINR of a user on an arbitrary beam $\gamma_{i,j}$, as defined in (7.37). We first rewrite $\gamma_{i,j}$ as

$$\gamma_{i,j} = \frac{|\mathbf{h}_i^T \mathbf{u}_j|^2}{\sum_{m=1, m \neq j}^{M} |\mathbf{h}_i^T \mathbf{u}_m|^2 + \rho}, \tag{7.38}$$

where $\rho = M/P$. Since $\mathbf{u}_j$ are orthonormal beamforming vectors, it can be shown that $|\mathbf{h}_i^T \mathbf{u}_j|^2$, $j = 1, 2, \ldots, M$, are i.i.d. Chi-square random variables with two degrees of freedom. It follows that the summation term in the denominator follows a Chi-square distribution with $2M - 2$ degrees of freedom and independent of the numerator. As such, the PDF of the SINR $\gamma_{i,j}$ can be written as

$$p_{\gamma_{i,j}}(x) = \frac{e^{-\rho x}}{(1+x)^M} \left(\rho(1+x) + M - 1 \right). \tag{7.39}$$

The CDF can be obtained after carrying out integration as

$$F_{\gamma_{i,j}}(x) = 1 - \frac{e^{-\rho x}}{(1+x)^{M-1}}. \tag{7.40}$$

Since the user channels are assumed to be independent, $\gamma_{i,j}$, $i = 1, 2, \ldots, K$ are i.i.d. random variables. As such, the CDF of their maximum $\max\{\gamma_{1,j}, \gamma_{2,j}, \ldots, \gamma_{K,j}\}$ is simply $(F_{\gamma_{i,j}}(x))^M$, which can be applied to the sum-rate performance analysis of RUB systems.

To investigate the asymptotic sum-rate performance of RUB as K approaches infinity, we now examine the limiting distribution of $\gamma_{1:K,j}$. It can be shown that

$$\lim_{x \to \infty} \frac{1 - F_{\gamma_{i,j}}(x)}{p_{\gamma_{i,j}}(x)} = \lim_{x \to \infty} \left(\frac{1}{\rho} - \frac{(M-1)/\rho}{\rho(1+x) + M - 1} \right) \tag{7.41}$$

$$= \frac{1}{\rho}.$$

Therefore, the limiting distribution of $\gamma_{1:K,j}$ is of the Gumbel type. Solving the equation $1 - F_{\gamma_{i,j}}(b_K) = 1/K$ for b_K, we can show that b_K is approximately given by [9]

$$b_K = \frac{1}{\rho} \log K - \frac{M-1}{\rho} \log \log K. \tag{7.42}$$

The limiting CDF of $\gamma_{1:K,j}$ as K approaches infinity is then given by

$$\lim_{K \to +\infty} F_{\gamma_{1:K} - b_K}(x) = \exp\left(-e^{-x/a_K}\right). \tag{7.43}$$

We can conclude that $\gamma_{1:K,j}$ increases at the rate $\frac{1}{\rho} \log K$ when K approaches infinity. As such, the sum rate of RUB systems will scale at rate $M \log \log K$ asymptotically.

7.3.2 Low-Complexity User Selection Strategies

We now consider the user selection strategies for RUB with reduced feedback load. In particular, we investigate the best beam SINR and index (BBSI)-based [9], quantized best beam SINR and index (QBBSI)-based, and best index (BBI)-based RUB schemes [15, 16]. The modes of operation of these strategies are summarized in the following.

Best Beam SINR and Index Strategy

With the BBSI strategy, each active user in the system feeds back its best beam index and the corresponding SINR value to the BS. For example, if the mth beam leads to the largest SINR for the kth user, i.e., $\gamma_{k,m} = \gamma_{k,1:M}$, then the kth user will feed back the beam index m and the corresponding SINR value $\gamma_{k,m}$. Note that with the BBSI strategy, each user needs to feed back a real number for the SINR value and a finite integer for the index of its best beam.

Based on the feedback information, the BS assigns a beam to the user with the largest SINR value among all users who feed back the index of that beam. Specifically, the BS ranks all K feedback best beam SINRs. If $\gamma_{k,m}$ is the largest one among all K SINRs, then the BS selects the kth user for the mth beam. After that, the BS will rank the feedback SINRs for the remaining beams. If now $\gamma_{n,l}$ is the largest one, where $l \neq m$ and $n \neq k$, then the BS assigns the lth beam to the nth user. This process is continued until either all beams have been assigned to selected users or there are some unrequested beams remaining. In the latter case, the BS will randomly select users for the remaining beams.

Quantized Best Beam SINR and Index Strategy

In practical systems, only a quantized value of the best beam SINR can be feed back, which leads to the QBBSI strategy. Similar to the BBSI strategy, the users with the QBBSI strategy will first calculate the instantaneous SINRs for all M beams using its channel vector and determine the largest one after some comparison. Then, the users feed back the quantized SINR value of their best beam in the following fashion. In particular, the value range of SINR is divided into N intervals, with boundary values given as $0 = \alpha_0 < \alpha_1 < \alpha_2 < \ldots < \alpha_N = \infty$. If the SINR value of the best beam for the kth user, $\gamma_{k,1:M}$, is in the ith interval, $i = 1, 2, \ldots N$, i.e., $\alpha_{i-1} < \gamma_{k,1:M} < \alpha_i$, then the kth user will feedback the index of that interval, i, together with the index of its best beam. The BS allocates available beams to selected users based on the feedback information. Specifically, a beam will be assigned to a user who has the largest quantized SINR value, i.e., who feeds back the largest SINR interval index, among all the users requesting that beam. Note that a tie may exist in this case, which will be resolved using random selection. For those beams that no user requests, the BS will assign them to randomly selected users. Based on the model of operation of the QBBSI strategy, the amount of feedback per user is equal to $\log_2 MN$ bits.

Best Beam Index Strategy

With the BBI strategy, each user only feeds back the index of its best beam. After receiving the best beam information of all users, the BS allocates a beam to a randomly selected user among all users requesting that beam, i.e., all users who feed back the corresponding beam index. It may happen that no user requests one or several beams. In this case, the BS will assign that beam to a randomly selected user. Note that with the BBI strategy, each user only needs to feed back $\log_2 M$ bits of information.

We can see that the user with these strategies will only feed back the SINR of its best beam or even only the index of its best beam. The major difficulty in the sum-rate analysis for these schemes resides in the determination of the statistics of ordered beam SINRs for a particular user. Note that while the beam SINRs of different users can be assumed to be independent random variables [9], those corresponding to the same users are correlated to each other as they involve the same channel vector. As such, the SINR of the best beam of a user is the largest one among some correlated random variables. With the application of the order statistics results in the Appendix [17], we derive the accurate statistical characterization of ordered beam SINRs for a user, in terms of the PDF and the CDF, which are then applied to obtain the exact analytical expressions of the ergodic sum rate of different RUB schemes. These accurate analytical results greatly facilitate the tradeoff investigation of different RUB schemes.

7.3.3 Statistics of Ordered Beam SINRs

We now derive the statistics of the ordered beam SINRs for user i. Note that the beam SINRs for the same user, i.e., $\gamma_{i,j}$ in (7.37) with the same i but different j, are correlated random variables as they involve the same channel vector $\mathbf{h}_i$.

For notation conciseness, we use α_j to denote the norm square of the projection of user i's channel vector onto the jth beam direction, i.e., $\alpha_j = |\mathbf{h}_i^T \mathbf{u}_j|^2$, $j = 1, 2, \ldots, M$. The M beam SINRs for user i can be rewritten as

$$\gamma_{i,j} = \frac{\alpha_j}{\sum_{m=1, m \neq j}^{M} \alpha_m + \rho}, \quad j = 1, 2, \ldots, M. \tag{7.44}$$

We are interested in the statistics of the ordered version of these M correlated random variables, i.e., $\gamma_{i,1:M} \geq \gamma_{i,2:M} \geq \ldots \geq \gamma_{i,M:M}$. We first note that all M beam SINRs for the ith user are calculated using the same set of projection powers $\{\alpha_j\}_{j=1}^{M}$ and, as such, they are correlated random variables. On the other hand, we also note from (7.44) that the interference terms for signal on a particular beam are always the project power of the channel vector onto the remaining $M - 1$ beams. Therefore, we conclude that $\gamma_{i,j}$ is the largest beam SINR for user i if and only if α_j is the largest one among M projection norm squares for this user, i.e., $\alpha_j = \alpha_{1:M}$. Based on this observation, the largest SINR of user i, denoted by γ_i^*, can be written in terms of the ordered norm squares $\{\alpha_{m:M}\}_{m=1}^{M}$ as

$$\gamma_i^* = \frac{\alpha_{1:M}}{z_1 + \rho}, \tag{7.45}$$

where z_1 is the sum of the remaining $M - 1$ norm squares, i.e., $z_1 = \sum_{m=1}^{M} \alpha_m - \alpha_{1:M}$. Consequently, the CDF of $\gamma_{i,1:M}$ can be calculated in terms of the joint PDF of $\alpha_{1:M}$ and z_1, denoted by $f_{\alpha_{1:M}, z_1}(y, z)$, as

$$F_{\gamma_{i,1:M}}(x) = \int_0^\infty \int_0^{x(\rho+z)} p_{\alpha_{1:M}, z_1}(y, z) dy dz. \tag{7.46}$$

It follows that the PDF of γ_i^* can be calculated using the joint PDF of $\alpha_{1:M}$ and z_1 as

$$p_{\gamma_i^*}(x) = \int_0^\infty (z + \rho) \cdot p_{\alpha_{1:M}, z_1}((z + \rho)x, z) dz. \tag{7.47}$$

7.3 Random Unitary Beamforming (RUB) Transmission

From (7.44), we can also conclude that $\gamma_{i,j}$ is the lth largest SINR for user i if and only if α_j is the lth largest one among all M norm squares, i.e., $\gamma_{i,l:M} = \gamma_{i,j}$ if and only if $\alpha_j = \alpha_{l:M}$. Consequently, $\gamma_{i,l:M}$ can be rewritten as

$$\gamma_{i,l:M} = \frac{\alpha_{l:M}}{(w_l + z_l) + \rho}, \tag{7.48}$$

where $w_l = \sum_{m=1}^{l-1} \alpha_{m:M}$ and $z_l = \sum_{m=l+1}^{M} \alpha_{m:M}$. It can be shown that the PDF of $\gamma_{i,l:M}$ can then be calculated in terms of the joint PDF of w_l, $\alpha_{l:M}$, and z_l, denoted by $f_{w_l, \alpha_{l:M}, z_l}(w, y, z)$, as

$$p_{\gamma_{i,l:M}}(x) = \int_0^\infty \int_0^{\frac{(M-l)w}{l-1}} (\rho + z + w) \cdot p_{w_l, \alpha_{l:M}, z_l}(w, x(\rho + z + w), z) dz dw, \tag{7.49}$$

where the integration limits are set using the fact $w/(l-1) > \alpha_{l:M} > z/(M-l)$.

The PDFs $f_{\alpha_{1:M}, z_1}(\cdot, \cdot)$ and $f_{w_l, \alpha_{l:M}, z_l}(\cdot, \cdot, \cdot)$ are joint PDFs of partial sums of ordered random variables, the general expressions of which have been obtained in previous chapters. As such, the statistical results of the ordered SINR are quite general and independent of the distribution α_j. For the specific Rayleigh channel model under consideration, it can be shown, after proper substitution and carrying out integration, that we can obtain the following closed-form expression for the PDF of the largest beam SINR of a user as

$$p_{\gamma_i^*}(x) = \sum_{j=0}^{M-1} \frac{(-1)^j M!}{(M-j-1)! j!} \exp\left(-\frac{(1+j)\rho x}{jx-1}\right) \tag{7.50}$$

$$\frac{(M + \rho - 1 + (\rho - jM + j)x)(1 - jx)^{M-3}}{(1+x)^M}.$$

7.3.4 Sum-Rate Analysis

BBSI Strategy

Based on the mode of operation of the BBSI strategy with ordered beam assignment, we can see that the SINR of the first beam is always the largest one of all K best beam SINRs, i.e., $\gamma_1^B = \gamma_{1:K}^*$, where $\gamma_{i:K}^*$ denotes the ith largest best beam SINRs among all K ones. Noting that the largest SINRs of different users are independent, the PDF of γ_1^B can be easily obtained as

$$p_{\gamma_1^B}(x) = K F_{\gamma_i^*}(x)^{K-1} p_{\gamma_i^*}(x), \tag{7.51}$$

where $F_{\gamma_i^*}(\cdot)$ and $p_{\gamma_i^*}(\cdot)$ are the CDF and PDF of the best beam SINR for a particular user, given in (7.46) and (7.47) respectively.

We now consider the SINR of the second beam. Based on the mode of operation of the BBSI strategy, the SINR of the second beam may be the ith largest one among all K best beam SINRs, $i = 2, 3, \ldots, K$, if the previous $i - 2$ largest ones are for the first beam. In particular, if, for example, the second and third largest best beam SINRs are also for the first beam, then the SINR of the second beam will be equal to the fourth largest best beam SINR. Since the user SINRs for different beams are i.i.d., each beam has the same probability to lead to the largest SINR for a particular user. It can be shown that the probability that the second beam is assigned to a user with the ith largest best beam SINRs among all K ones, i.e., $\gamma_2^B = \gamma_{i:K}^*, i = 2, 3, \ldots, K$, is given by

$$P_i^2 = \left(\frac{1}{M}\right)^{i-2}\left(1 - \frac{1}{M}\right), \tag{7.52}$$

where $\frac{1}{M}$ gives the probability that the best SINR of a user is for a particular beam. In the worst case, where all users request the same beam, the SINR of the second beam is equal to the SINR of a randomly chosen user.[1] Note that the SINR of a randomly assigned beam has the same probability to be the jth largest one, $j = 2, 3, \ldots, M$, among M beam SINRs. Correspondingly, the PDF of the SINR of a randomly assigned beam can be obtained as

$$p_{\gamma^R}(x) = \frac{1}{M-1}\sum_{j=2}^{M} p_{\gamma_{i,j:M}}(x), \tag{7.53}$$

where $f_{\gamma_{i,j:M}}(\cdot)$ is the PDF of the jth largest beam SINR for a user, given in (7.49). The probability that the second beam is assigned to a random user is equal to $(\frac{1}{M})^{K-1}$. Consequently, the PDF of the SINR for the second beam γ_2^B can be shown to be given by

$$p_{\gamma_2^B}(x) = \sum_{i=2}^{K} P_i^2 p_{\gamma_{i:K}^*}(x) + \left(\frac{1}{M}\right)^{K-1} p_{\gamma^R}(x), \tag{7.54}$$

where $f_{\gamma_{i:K}^*}(x)$ is the PDF of the ith largest best beam SINR among all K ones, given by [14]

$$p_{\gamma_{i:K}^*}(x) = \frac{K!}{(K-i)!(i-1)!}F_{\gamma_i^*}(x)^{(K-i)} \tag{7.55}$$
$$\times [1 - F_{\gamma_i^*}(x)]^{i-1} p_{\gamma_i^*}(x).$$

Similarly, the SINR of the third beam may be the ith largest best beam SINRs among all K ones, $i = 3, 4, \ldots, K$, if one of the previous $i-2$ largest best beam SINR is assigned to the second beam and the remaining $i - 3$ largest ones are for the first and second beams. It can be shown that the probability that the third beam is assigned to a user with the ith largest best beam SINRs among all K users, i.e., $\gamma_3^B = \gamma_{i:K}^*, i = 3, 4, \ldots, K$, is given by

$$P_i^3 = \sum_{j=0}^{i-3}\left(\frac{1}{M}\right)^j\left(1 - \frac{1}{M}\right)\left(\frac{2}{M}\right)^{i-3-j}\left(1 - \frac{2}{M}\right), \tag{7.56}$$

where $\frac{2}{M}$ gives the probability that the largest SINR for a particular user is for the first and second assigned beams. Note that in the worst case, where all best beam SINRs belong to the first and second beams, whose probability is $\left(\frac{2}{M}\right)^{K-1}$, the SINR of the

---

[1] This approach is to maintain (7.37) as valid. Alternatively, we may turn off those beams to reduce mutual interference.

7.3 Random Unitary Beamforming (RUB) Transmission

Table 7.1 Sample scenario for the event that the mth beam is assigned to the user with the ith largest best beam SINR among all K users, i.e., $\gamma_m^B = \gamma_{i:K}^*$.

Indexes of ordered best beam SINRs	Associated beams	# of users
1	First beam	1
2 to $n_1 + 1$	First beam	n_1
$n_1 + 2$	Second beam	1
$n_1 + 3$ to $n_1 + n_2 + 2$	First two beams	n_2
$\vdots$	$\vdots$	$\vdots$
$\sum_{k=1}^{j-1} n_k + j$	jth beam	1
$\sum_{k=1}^{j-1} n_k + j + 1$ to $\sum_{k=1}^{j-1} n_k + n_j + j$	First j beams	n_j
$\vdots$	$\vdots$	$\vdots$
$\sum_{j=1}^{m-2} n_j + m$ to $\sum_{j=1}^{m-2} n_j + n_{m-1} + m - 1$	First $m-1$ beams	n_{m-1}
$i = \sum_{j=1}^{m-1} n_j + 1$	mth beam	1

third assigned beam is equal to the SINRs of a randomly chosen user, whose PDF was given in (7.53). Consequently, the PDF of the SINR for the third beam with the BBSI strategy can be shown to be given by

$$p_{\gamma_3^B}(x) = \sum_{i=3}^{K} P_i^3 p_{\gamma_{i:K}^*}(x) + \left(\frac{2}{M}\right)^{K-1} p_{\gamma^R}(x). \tag{7.57}$$

In general, the SINR of the mth beam may be the ith largest best beam SINRs, where $i = m, m+1, \ldots, K$. More specifically, $\gamma_m^B = \gamma_{i:K}^*$ when a scenario shown in Table 7.1 occurs, where n_j denotes the number of users whose best beam is one of the first j assigned beams. After determining the probability of the scenario in Table 7.1 for a particular vector $\{n_j\}_{j=1}^{m-1}$ and summing over all possible vectors $\{n_j\}_{j=1}^{m-1}$, while noting that $\{n_j\}_{j=1}^{m-1}$ needs to satisfy $n_j \in \{0, 1, \ldots, i-m\}$ and $\sum_{j=1}^{m-1} n_j = i - m$, we obtain the probability that the mth beam is assigned to the user with the ith largest best beam SINR, i.e., $\gamma_m^B = \gamma_{i:K}^*$, $i = m, m+1, \ldots, K$ as

$$P_i^m = \sum_{\substack{\sum_{j=1}^{m-1} n_j = i-m \\ n_j \in \{0,1,\cdots,i-m\}}} \left[\prod_{j=1}^{m-1} \left(\frac{j}{M}\right)^{n_j} \left(1 - \frac{j}{M}\right)\right] \tag{7.58}$$

$$i = m, m+1, \ldots, K.$$

Consequently, after noting that it is with a probability of $\left(\frac{m-1}{M}\right)^{K-1}$ that the mth beam is assigned to a random selected user, we can obtain the PDF of the SINR on the mth beam γ_m^B as

$$p_{\gamma_m^B}(x) = \sum_{i=m}^{K} P_i^m \, p_{\gamma_{i:K}^*}(x) + \left(\frac{m-1}{M}\right)^{K-1} p_{\gamma^R}(x), \tag{7.59}$$

noting that additional SINR feedback from the randomly selected users is required to achieve the data rate. When K is not too small, the extra feedback load can be neglected since the probability of beams not being requested is small.

Finally, after applying (7.51) and (7.59) in (7.4), the exact sum-rate expression for the BBSI strategy can be calculated as

$$R = \int_0^\infty \log_2(1+x) K F_{\gamma_i^*}(x)^{K-1} p_{\gamma_i^*}(x) dx \tag{7.60}$$
$$+ \sum_{m=2}^{M} \left(\sum_{i=m}^{K} P_i^m \int_0^\infty \log_2(1+x) \, p_{\gamma_{i:K}^*}(x) dx \right.$$
$$+ \left. \left(\frac{m-1}{M}\right)^{K-1} \int_0^\infty \log_2(1+x) p_{\gamma^R}(x) dx \right),$$

where P_i^m and $p_{\gamma_{i:K}^*}(x)$ are given in (7.58) and (7.55), respectively.

Quantized Best Beam SINR and Index Strategy

Let us consider one of the M available beams. It can be shown that the probability that exactly j users among the total K users request this particular beam is equal to

$$\Pr[N_r = j] = \frac{K!}{(K-j)!j!} \left(\frac{1}{M}\right)^j \left(\frac{M-1}{M}\right)^{K-j}, \tag{7.61}$$
$$j = 0, 1, \ldots, K.$$

If $N_r = 0$, then that beam will be assigned to a randomly selected user whose best beam is a different one. In this case, the PDF of the SINR of this beam is the same as the one given in (7.53). If $N_r = j > 0$, then that beam will be assigned to a user for whom this beam is the best one. It can be shown that the probability that the beam is assigned to a user with best beam SINR in the ith interval is given by

$$\Pr[\gamma^B \in (\alpha_{i-1}, \alpha_i) | N_r = j > 0] \tag{7.62}$$
$$= F_{\gamma_{i,1:M}}(\alpha_i)^j - F_{\gamma_{i,1:M}}(\alpha_{i-1})^j, \, 1 \le i \le N,$$

where $F_{\gamma_{i,1:M}}(\cdot)$ is the CDF of the SINR of the best beam of a user, as given in (7.46). The corresponding SINR PDF of the beam when it is assigned to a user with best beam SINR in the ith interval can then be obtained as

$$p_{\gamma^B | \gamma^B \in (\alpha_{i-1}, \alpha_i)}(x) \tag{7.63}$$
$$= \frac{1}{F_{\gamma_{i,1:M}}(\alpha_i) - F_{\gamma_{i,1:M}}(\alpha_{i-1})} p_{\gamma_{i,1:M}}(x), \alpha_{i-1} \le x \le \alpha_i.$$

After successively removing the conditioning, the SINR PDF for the beam can be obtained as

$$p_{\gamma^B}(x) = \sum_{j=1}^{K} \frac{K!}{(K-j)!j!} \left(\frac{1}{M}\right)^j \left(\frac{M-1}{M}\right)^{K-j} \qquad (7.64)$$

$$\times \sum_{i=1}^{N} \frac{F_{\gamma_{i,1:M}}(\alpha_i)^j - F_{\gamma_{i,1:M}}(\alpha_{i-1})^j}{F_{\gamma_{i,1:M}}(\alpha_i) - F_{\gamma_{i,1:M}}(\alpha_{i-1})} p_{\gamma_{i,1:M}}(x)$$

$$\times (\mathcal{U}(x - \alpha_{i-1}) - \mathcal{U}(x - \alpha_i)) + \left(\frac{M-1}{M}\right)^K p_{\gamma^R}(x),$$

where $p_{\gamma^R}(\cdot)$ was given in (7.53).

Finally, the sum rate of an RUB-based multiuser MIMO system with QBBSI-based user selection can be calculated as

$$R^{QBBSI} = M \int_0^\infty \log_2(1+x) p_{\gamma_m^B}(x) dx \qquad (7.65)$$

$$= M \left(\sum_{j=1}^{K} \frac{(K)!}{(K-j)!j!} \left(\frac{1}{M}\right)^j \left(\frac{M-1}{M}\right)^{K-j} \right.$$

$$\times \left(\sum_{i=1}^{N} \frac{F_{\gamma_{i,1:M}}(\alpha_i)^j - F_{\gamma_{i,1:M}}(\alpha_{i-1})^j}{F_{\gamma_{i,1:M}}(\alpha_i) - F_{\gamma_{i,1:M}}(\alpha_{i-1})} \int_{\alpha_{i-1}}^{\alpha_i} \log_2(1+x) p_{\gamma_{i,1:M}}(x) dx \right.$$

$$\left. + \left(\frac{M-1}{M}\right)^k \frac{1}{M-1} \sum_{j=2}^{M} \int_0^\infty \log_2(1+x) p_{\gamma_{i,j:M}}(x) dx \right).$$

BBI Strategy

According to the mode of operation of the BBI strategy, the SINR of the user assigned to the mth beam, γ_m^B in (7.4), may follow two different types of distribution. More specifically, if there is at least one user requesting the mth beam, then this beam will be the best beam of the selected user. Therefore, γ_m^B is the largest one among M beam SINRs for a user, i.e., $\gamma_m^B = \gamma_i^*$, whose PDF is given in (7.47). On the other hand, when no user requests the mth beam and that beam is assigned to a randomly chosen user, γ_m can be any one of M beam SINRs, except for the largest one, for the selected user. Therefore, the PDF of γ_m^B for this case was given in (7.53), in terms of the PDF of the jth largest beam SINR for a user $f_{\gamma_{i,j:M}}(\cdot)$.

Let N_a denote the number of beams that are requested by at least one user. Clearly, N_a is a discrete random variable, taking values $1, 2, \ldots, M$. We can easily see that the probability that only one beam is active, i.e., $N_a = 1$ is equal to the probability that all k users feed back the same beam index, which can be calculated as

$$\Pr[N_a = 1] = \binom{M}{1} \left(\frac{1}{M}\right)^K. \qquad (7.66)$$

The probability of exactly two beams being active can be calculated by subtracting from the probability that at most two beams are active, which is given by $\binom{M}{2}\left(\frac{2}{M}\right)^K$, the probability of only one of those two beams being active, which is equal to $\binom{M}{2}\binom{2}{1}\left(\frac{1}{M}\right)^K$. Therefore, we have

$$\Pr[N_a = 2] = \binom{M}{2}\left[\left(\frac{2}{M}\right)^K - \binom{2}{1}\left(\frac{1}{M}\right)^K\right]. \tag{7.67}$$

Similarly, the probability that exactly three beams are active can be calculated by subtracting from the probability that at most three beams are active, the probabilities that not all three beams are active. Hence, the probability of $N_a = 3$ is given by

$$\Pr[N_a = 3] = \binom{M}{3}\left\{\left(\frac{3}{M}\right)^K - \binom{3}{2}\right.$$

$$\left[\left(\frac{2}{M}\right)^K - \binom{2}{1}\left(\frac{1}{M}\right)^K\right] - \binom{3}{1}\left(\frac{1}{M}\right)^K\right\}. \tag{7.68}$$

In general, the probability that exactly m beams out of M available beams are active given that k users feed back their best beam indexes can be recursively calculated as

$$\Pr[N_a = m] = \binom{M}{m}\left\{\left(\frac{m}{M}\right)^K\right.$$

$$\left. - \sum_{j=1}^{m-1}\binom{m}{j}\frac{\Pr[N_a = j]}{\binom{M}{j}}\right\}, \; m = 1, 2, \ldots, M. \tag{7.69}$$

We can also prove by induction that (7.69) can be simplified to the following compact expression

$$\Pr[N_a = m] = \frac{1}{M^K}\binom{M}{m}\cdot\sum_{q=1}^{m}(-1)^{m-q}\binom{m}{q}q^K, \; m = 1, 2, \ldots, M. \tag{7.70}$$

It is easy to verify that (7.70) holds for the $i = 1$ case. Assume that (7.70) holds for all $1 \leq i \leq m - 1$, then we have

$$\Pr[N_a = i] = \frac{1}{M^K}\binom{M}{i}\cdot \tag{7.71}$$

$$\sum_{q=1}^{i}(-1)^{i-q}\binom{i}{q}q^K, \; i = 1, 2, \ldots, m-1.$$

We now consider the case of $i = m$. From (7.69), we have

$$\Pr[N_a = m] = \binom{M}{m}\left\{\left(\frac{m}{M}\right)^k - \right. \tag{7.72}$$

$$\left. \sum_{j=1}^{m-1}\binom{m}{j}\frac{\Pr[M_a = j|N_f = k]}{\binom{M}{j}}\right\}.$$

Substituting (7.71) into (7.72), we have

$$\Pr[M_a = m | N_f = k]$$

$$= \frac{1}{M^k}\binom{M}{m}\left\{m^k - \sum_{j=1}^{m-1}\binom{m}{j}\sum_{q=1}^{j}(-1)^{j-q}\binom{j}{q}q^k\right\}$$

$$= \frac{1}{M^k}\binom{M}{m}\left\{\sum_{q=1}^{m}(-1)^{m-q}\binom{m}{q}q^k\right\}.$$

Therefore, (7.70) holds in general.

Finally, the sum rate of an RUB-based multiuser MIMO system with the BBI strategy can be obtained, by conditioning on the number of beams that have been requested by at least one user, as

$$R^{BBI} = \sum_{i=1}^{M}\Pr[N_a = i]\left(i\int_{0}^{\infty}\log_2(1+x)p_{\gamma_i^*}(x)dx\right.$$

$$\left. + \frac{M-i}{M-1}\sum_{j=2}^{M}\int_{0}^{\infty}\log_2(1+x)p_{\gamma_{i,j:M}}(x)dx\right), \quad (7.73)$$

where $\Pr[N_a = i]$ is given in (7.70).

7.3.5 Numerical Results

In Fig. 7.6, we plot the exact sum rate of a MIMO broadcast channel with the BBSI-based RUB as given in (7.60) as the function of the number of users under consideration. On the same figure, we also plot the upper and lower bounds derived in [9] for the asymptotic sum-rate analysis. As we can see, the upper bound is much closer to the exact sum rate, especially when the number of users is large. We also notice that when the number of users is small, there is a noticeable gap between the upper bound and the exact sum rate. As such, the analytical sum-rate expression provides a valuable tool to predict the system performance when the number of users in the system is limited.

In Fig. 7.7, we plot the sum-rate performance of a MIMO broadcast channel with BBSI-based RUB as given in (7.60) as the function of the average SNR for different numbers of active users K and different numbers of antennas (or, equivalently, the number of beams) M. First of all, we can observe a common trend that, when SNR increases, the sum rate of RUB systems increases over the low SNR (signal-to-noise ratio) region but saturates in the high SNR region. If we compare the curves with the same K value but different M values, we can see that while using more antennas, i.e., transmitting to more users simultaneously, helps slightly increase the total sum rate over the low SNR region, activating fewer beams leads to a much better sum-rate performance in the high SNR region. These behaviors can be explained as follows. Over the high SNR region, the multiuser interference becomes a dominant factor in terms of limiting the system sum rate. Note that a user for the $M = 2$ case will experience only one-third of the

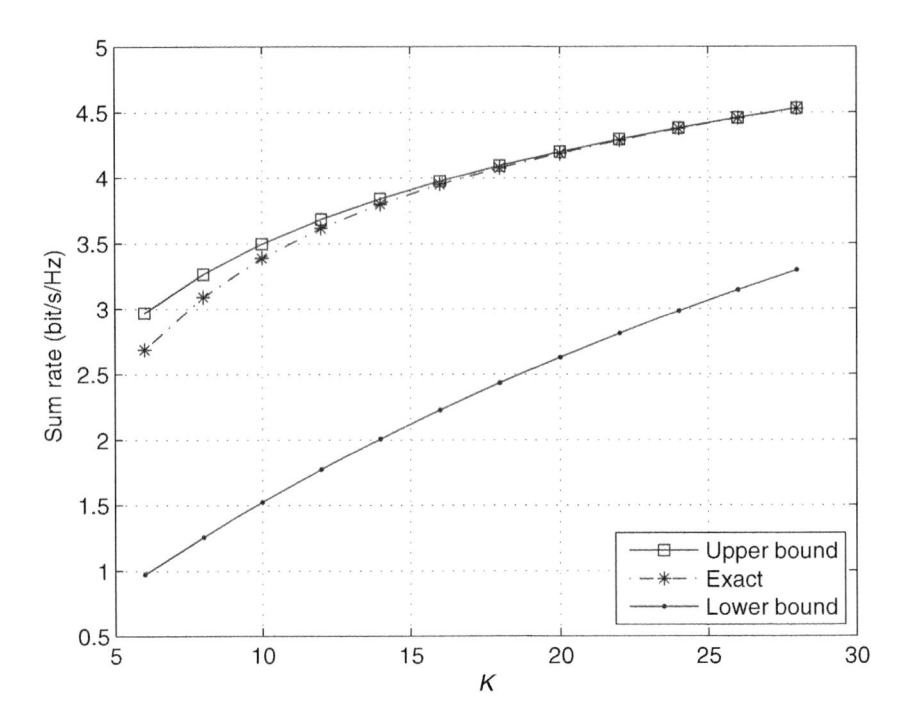

Figure 7.6 Sum-rate capacity in comparison with upper and lower bounds ($M = 4, P = 5$ dB). Reprint with permission from [18]. ©2009 IEEE.

amount of multiuser interference experienced by users for the $M = 4$ cases. In addition, with the same total transmitting power constraint, the users in the $M = 2$ case enjoy twice the transmitting power as the users in the $M = 4$ cases, which contributes further to the sum-rate advantage of the $M = 2$ case over the high SNR region. On the other hand, noise is the major limiting factor over the low SNR region, whereas the multiuser interference can be negligible. As such, transmitting to more users translates to approximately a linear increase in the system sum rate and can overcome the power loss due to the total power constraint.

In Fig. 7.8 we compare the sum-rate performance of different RUB schemes considered in this work for the case of four transmit antennas and a total of 20 users. The sum rates of the BBSI schemes with quantized SINR feedback (QBBSI), which leads to lower feedback load, are presented. Note that if $N = 2^r$ SINR intervals are used, then each user needs to feed back r additional bits for its best beam SINR feedback. As we can see, the sum-rate performance of all three schemes improves as average SNR increases and saturates in the high SNR region, as expected. On the other hand, the sum rate of the BBI-based RUB scheme converges to a much smaller value than other schemes. It is worth noting that the QBBSI-based scheme offers much larger sum-rate capacity than the BBI scheme over high SNR, even with $N = 2$ SINR intervals, i.e., one additional bit feedback for quantized best beam SINR. We also notice that when the number of quantization levels increases, the sum rate of the QBBSI scheme improves

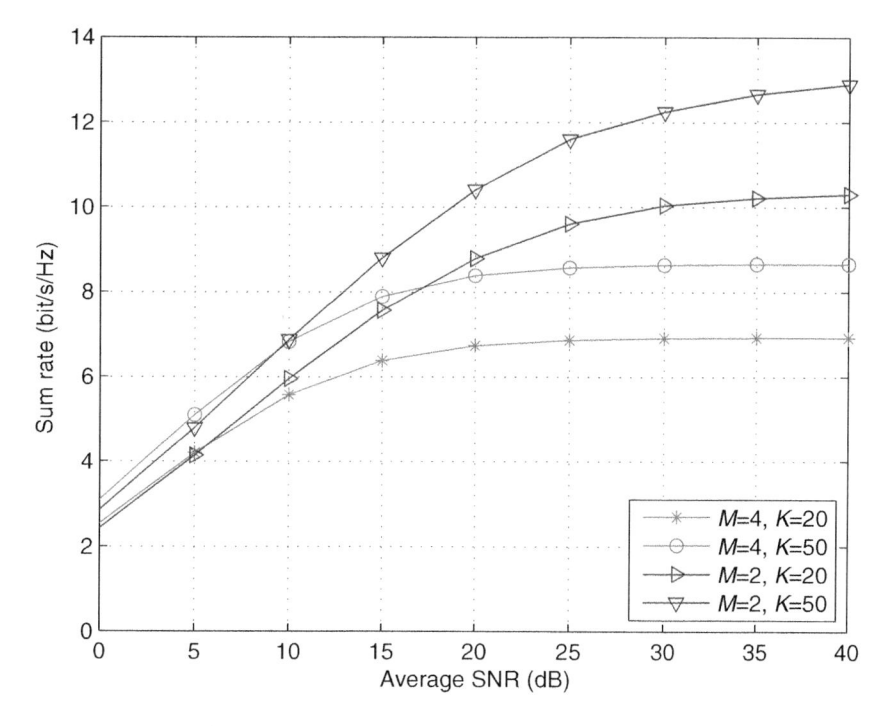

Figure 7.7 Sum-rate performance for different numbers of transmit antennas and/or numbers of users. Reprint with permission from [18]. ©2009 IEEE.

and quickly approaches that of the BBSI scheme. Note that the sum rate of QBBSI with $N = 8$ intervals is within 0.3 bps/Hz of that of the BBSI scheme.

As a final numerical example, we plot the sum rate of different RUB schemes as a function of the number of active users in the system in Fig. 7.9. We assume that the total transmitting power budget is fixed to 10 dB with four transmit antennas. The SINR threshold values for QBBSI schemes are arbitrarily selected to be the same as the ones used in Fig. 7.8. As we can clearly observe, while the sum rate of BBSI and QBBSI-based RUB schemes increases as the number of active users increases, that of the BBI scheme quickly saturates at a small value (≈ 3.8 bps/Hz). This is because without any knowledge of SINR values, the BBI scheme cannot fully explore the multiuser diversity gain. We also see from Fig. 7.9 that the sum rate of QBBSI increases at a higher rate as K increases when the number of quantization levels becomes larger.

7.4 RUB with Conditional Best Beam Index Feedback

In this section, we present another low-feedback user scheduling strategy for an RUB-based multiuser MIMO system, termed adaptive beam activation with conditional best beam index feedback (ABA-CBBI) [19]. In particular, we apply feedback thresholding to reduce the feedback load [15, 20]. With the ABS-CBBI scheme, only those users

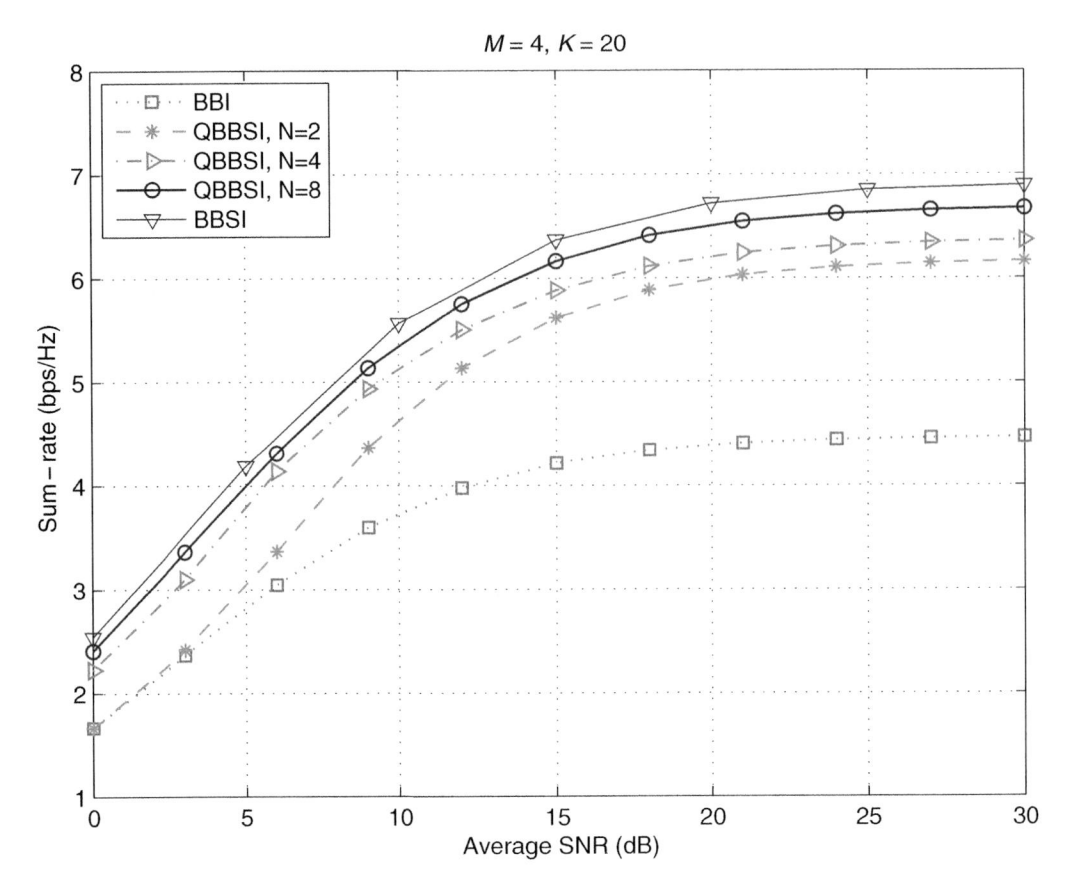

Figure 7.8 Sum-rate performance of different RUB schemes as the function of average SNR (SINR thresholds for QBBSI: 3 dB for $N = 2$; $-5, 0$ and 5 dB for $N = 4$; $-6, -3, 0, 3, 6, 9$, and 12 dB for $N = 8$).

with their largest SINR greater than a threshold will feed back their best beam indexes. When the SINR threshold is high and/or the number of users is moderate, it may happen that a beam is not requested by any user. Noting that randomly selecting a user for such a beam will introduce unnecessary interuser interference and therefore cannot guarantee high SINR for selected users, we can activate only those beams that are requested by at least one user. We study the sum-rate performance of the resulting ABA-CBBI strategy through accurate statistical analysis. In particular, we derive the distribution of the number of active beams and the distribution of their resulting SINR, which are then applied to obtain the exact analytical expression of the overall system sum rate. We also accurately quantify the average feedback load of the proposed ABA-CBBI scheme. Based on these analytical results, we study the effect of the SINR threshold on the sum-rate performance and its optimization. We show that the resulting ABA-CBBI scheme can effectively explore multiuser diversity gain and control interuser interference for sum-rate performance benefit.

7.4 RUB with Conditional Best Beam Index Feedback

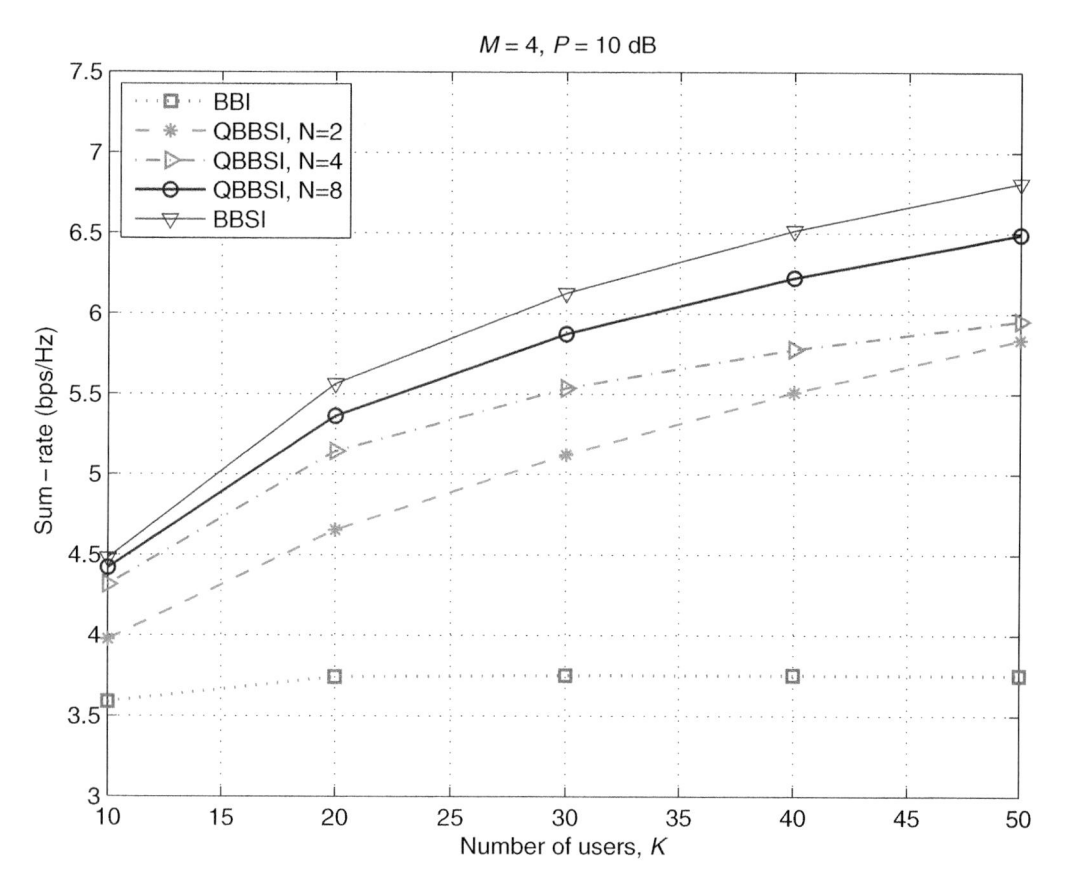

$$M = 4, P = 10 \text{ dB}$$

Figure 7.9 Sum-rate performance of different RUB schemes as the function of the number of active users (SINR thresholds for QBBSI: 3 dB for $N = 2$; $-5, 0$ and 5 dB for $N = 4$; $-6, -3, 0, 3, 6, 9$, and 12 dB for $N = 8$).

7.4.1 Mode of Operation and Complexity Analysis

At the beginning of the scheduling period, each user will determine the instantaneous SINR on the available beams while assuming that all M beams will be active. Specifically, with the knowledge of beamforming vector set $\mathcal{U}$ and the estimated channel vector $\mathbf{h}_k$, the SINR of user k on the jth beam can be calculated as

$$\gamma_{k,j} = \frac{\frac{P}{M} |\mathbf{h}_k^T \mathbf{u}_j|^2}{\frac{P}{M} \sum_{m=1, m \neq j}^{M} |\mathbf{h}_k^T \mathbf{u}_m|^2 + 1}, \quad k = 1, 2, \ldots, K, \, j = 1, 2, \ldots, M. \quad (7.74)$$

The users will then perform some comparison to determine their best beams, which achieve the largest SINR among all beams, namely the beam with index $j^* = \arg\max_j(\gamma_{k,j})$. The corresponding SINR, denoted as γ_{k,j^*}, is called the best beam SINR of user k. Each user compares its best beam SINR with an SINR threshold, denoted by β. Only when its best beam SINR is greater than β will the user feed back its best beam index, which results in a feedback of $\log_2(M)$ bits per user. After collecting the best beam indexes feedback from qualified users, the BS will assign a beam randomly to a user requesting it. If a beam is not requested by any user, the BS will simply turn off that beam and redistribute the power to other beams. In this case, the number of active beams (or selected users) M_a will be less than the number of available beams M. Since a beam is randomly assigned to one of the users requesting it, this scheduling strategy has lower operational complexity than the BBSI scheme.

Average Feedback Load

The feedback load of the ABA-CBBI strategy varies over time as the number of qualified users changes with the channel condition. We can quantify the average feedback load (AFL) of this strategy as

$$F = KP_f \log_2 M, \qquad (7.75)$$

where P_f is the probability that a user is qualified to feed back. According to the mode of operation above, P_f is equal to the probability that the best beam SINR of a user while assuming all M beams are active, γ_{k,j^*}, is greater than the threshold β, i.e., $P_f = \Pr[\gamma_{k,j^*} > \beta]$. With the homogeneous Rayleigh fading assumption, it can be shown that γ_{k,j^*} are identically distributed with common PDF given by

$$p_{\gamma_{j^*}^{(M)}}(x) = \frac{M}{(M-2)!} \int_0^\infty (M/P + z)e^{-x(\frac{M}{P}+z)-z} \qquad (7.76)$$

$$\times \sum_{j=0}^{M-1} \binom{M-1}{j}(-1)^j \left(z - jx(\frac{M}{P} + z)\right)^{M-2} U\left(z - jx(\frac{M}{P} + z)\right) dz,$$

where $U(\cdot)$ denotes the unit step function. As such, the AFL with the proposed ABA-CBBI strategy is given by[2]

$$F = K \log_2 M \int_\beta^\infty p_{\gamma_{j^*}^{(M)}}(x) dx. \qquad (7.77)$$

Distribution of the Number of Active Beams

A unique feature of the proposed ABA-CBBI strategy is no random user selection for unrequested beams. As a result, the number of selected users/active beams, M_a, becomes a discrete random variable taking values in $[1, 2, \ldots, M]$. Note that if every user feeds back, the probability of a beam not being requested by any user will be small, even negligible, as long as the number of users is large enough. But with the conditional feedback strategies, this probability becomes larger as the number of users that feed back can be small due to the feedback thresholding. In the following, we derive the probability mass function (PMF) of M_a, which will be applied to the sum-rate analysis of the proposed scheme in the next section.

---

[2] This result corrects an error in [15, Eq. (13)] as the SINRs of a user for different beams are correlated random variables.

7.4 RUB with Conditional Best Beam Index Feedback

With the ABA-CBBI strategy, the number of users that are qualified to feed back, denoted by N_f, is random. The probability of $N_f = k$ is given by

$$\Pr[N_f = k] = \binom{K}{k} P_f^k (1-P_f)^{K-k}, \quad k = 0, 1, \ldots, K. \quad (7.78)$$

Given that k users feed back their best beam indexes, the probability that only one beam is active, i.e., $M_a = m$, can be calculated by following the similar derivation in the previous section as

$$\Pr[M_a = m | N_f = k] = \binom{M}{m} \left\{ \left(\frac{m}{M}\right)^k \right. \quad (7.79)$$

$$\left. - \sum_{j=1}^{m-1} \binom{m}{j} \frac{\Pr[M_a = j | N_f = k]}{\binom{M}{j}} \right\}, \quad m = 1, 2, \ldots, M,$$

which can be simplified to the following compact expression:

$$\Pr[M_a = m | N_f = k] = \frac{1}{M^k} \binom{M}{m} \cdot \sum_{q=1}^{m} (-1)^{m-q} \binom{m}{q} q^k, \quad m = 1, 2, \ldots, M. \quad (7.80)$$

Finally, combining (7.78) and (7.80), the probability that exact m beams are active can be obtained by applying the total probability theorem as

$$\Pr[M_a = m] = \sum_{k=1}^{K} \Pr[N_f = k] \Pr[M_a = m | N_f = k] \quad (7.81)$$

$$= \binom{M}{m} \sum_{q=1}^{m} (-1)^{m+q} \binom{m}{q} \left[\left(1 - \frac{M-q}{M} P_f\right)^K - (1-P_f)^K \right],$$

where the feedback probability P_f is given in (7.77).

7.4.2 Sum-Rate Analysis

In this section, we investigate the performance of the proposed ABA-CBBI strategy by analytically deriving the exact sum-rate expression for the resulting multiuser MIMO systems. Conditioning on the number of active beams, the average sum rate with ABA-CBBI strategy can be calculated as

$$E[R] = \sum_{m=1}^{M} \Pr[M_a = m] E\left[\sum_{i=1}^{m} \log_2(1 + \gamma_i)\right], \quad (7.82)$$

where $\Pr[M_a = m]$ was given in (7.81) and γ_i is the SINR of the ith selected user. According to the mode of operation of the ABA-CBBI strategy, the SINRs of users on different active beams are identically distributed. As such, we can rewrite the sum rate as

$$E[R] = \sum_{m=1}^{M} \Pr[M_a = m] m \int_0^{\infty} \log_2(1+x) p_{\gamma_B^{(m)}}(x) dx, \quad (7.83)$$

where $f_{\gamma_B^{(m)}}(\cdot)$ denotes the common PDF of the SINR of selected users given that exactly m beams are active.

We now proceed to derive the PDF $f_{\gamma_B^{(m)}}(\cdot)$. Note that $\gamma_B^{(m)}$ is also the best beam SINR of a feedback user when m beams are active. Meanwhile, the best beam SINR of such a user when assuming all M beams are active, $\gamma_B^{(M)}$, must be greater than the threshold β. Therefore, the SINR distribution for active beams with ABA-CBBI strategy is the conditional distribution of $\gamma_B^{(m)}$ given $\gamma_B^{(M)} > \beta$. It follows that the CDF of the SINR of the selected users when m beams are active can be determined as

$$F_{\gamma_B^{(m)}}(x) = \frac{\Pr[\gamma_B^{(m)} < x, \gamma_B^{(M)} > \beta]}{\Pr[\gamma_B^{(M)} > \beta]}. \tag{7.84}$$

To proceed further, we note that a beam becomes the best beam for a particular user if and only if the user's channel vector has the largest projection norm square ($|\mathbf{h}_k^T \mathbf{u}_j|^2$) on that beam direction. Therefore, we can rewrite $\gamma_B^{(m)}$ as

$$\gamma_B^{(m)} = \begin{cases} \frac{\alpha_{1:M}}{z_{m-1}+m/P}, & 1 < m \leq M, \\ P\alpha_{1:M}, & m = 1, \end{cases} \tag{7.85}$$

where $\alpha_{1:M} = \max\{|\mathbf{h}_i^T \mathbf{u}_j|^2, j = 1, 2, \ldots, M\}$ is the largest projection norm squares onto all M possible beam directions and z_{m-1} is the sum of arbitrary $m-1$ projection norm squares out of the remaining $M-1$ ones. As such, the joint probability in (7.84) can be rewritten as

$$\Pr[\gamma_B^{(m)} < x, \gamma_B^{(M)} > \beta] \tag{7.86}$$

$$= \begin{cases} \Pr[P\alpha_{1:M} < x, \frac{\alpha_{1:M}}{z_{M-1}+M/P} > \beta], & m = 1; \\ \Pr[\frac{\alpha_{1:M}}{z_{m-1}+m/P} < x, \frac{\alpha_{1:M}}{z_{m-1}+z_{M-m}+M/P} > \beta], & 1 < m < M; \\ \Pr[\beta < \frac{\alpha_{1:M}}{z_{M-1}+M/P} < x], & m = M, \end{cases}$$

where we define $z_{M-m} = z_{M-1} - z_{m-1}$ for the case of $1 < m < M$. Consequently, the joint probability of the case of $m = 1$ can be calculated using the joint PDF of $\alpha_{1:M}$ and z_{M-1} as

$$\Pr[\gamma_B^{(m)} < x, \gamma_B^{(M)} > \beta] \tag{7.87}$$

$$= \int_0^{x/P} dy \int_0^{\min\{y/\beta - M/P, (M-1)y\}} du \, p_{\alpha_{1:M}, z_{M-1}}(y, u),$$

where the joint PDF $p_{\alpha_{1:M}, z_{M-1}}(y, u)$ is given by

$$p_{\alpha_{1:M}, z_{M-1}}(y, u) = p_{\alpha_{1:M}}(y) p_{z_{M-1}|\alpha_{1:M}=y}(u) \tag{7.88}$$

$$= M e^{-y-u} \sum_{j=0}^{M-1} \binom{M-1}{j} \frac{(-1)^j (u-jy)^{M-2}}{(M-2)!} U(u-jy),$$

$$y > 0, u < (M-1)y.$$

7.4 RUB with Conditional Best Beam Index Feedback

The joint probability of the cases of $1 < m < M$ can be calculated using the joint PDF of $\alpha_{1:M}$, z_{m-1}, and z_{M-m} as

$$\Pr[\gamma_B^{(m)} < x, \gamma_B^{(M)} > \beta] = \int_0^\infty dy \int_{y/x-m/P}^{y/\beta-M/P} du \qquad (7.89)$$
$$\int_0^{y/\beta-u-M/P} dv\, p_{\alpha_{1:M}, z_{m-1}, z_{M-m}}(y, u, v).$$

The joint PDF of $\alpha_{1:M}$, z_{m-1}, and z_{M-m} can be determined by following the Bayesian approach. It has been shown that the unordered projection norm squares $|\mathbf{h}_i^T \mathbf{u}_j|^2$, $j = 1, 2, \ldots, M$, are i.i.d. Chi-square random variables with two degrees of freedom [7]. Due to the ordering process, $\alpha_{1:M}$, z_{m-1}, and z_{M-m} are correlated random variables. On the other hand, it can be shown that given $\alpha_{1:M} = y$, z_{m-1} and z_{M-m} are the sum of $m-1$ and $M-m$ i.i.d. random variables with truncated distribution on the right at y, respectively, and as such are independent. Specifically, the joint PDF of $\alpha_{1:M}$, z_{m-1}, and z_{M-m} can be written as

$$p_{\alpha_{1:M}, z_{m-1}, z_{M-m}}(y, u, v) = p_{\alpha_{1:M}}(y) p_{z_{m-1}|\alpha_{1:M}=y}(u) p_{z_{M-m}|\alpha_{1:M}=y}(v) \qquad (7.90)$$
$$= p_{\alpha_{1:M}}(y) p_{\sum_{j=1}^{m-1} \alpha_j^-}(u) p_{\sum_{j=m+1}^{M} \alpha_j^-}(v),$$

where α_j^- are i.i.d. random variables with the PDF for the Rayleigh fading scenario under consideration given by

$$p_{\alpha_j^-}(x) = \frac{e^{-x}}{1 - e^{-y}}, \quad 0 < x < y. \qquad (7.91)$$

After some mathematical manipulations, we can obtain the closed-form expression for the joint PDF of $\alpha_{1:M}$, z_{m-1}, and z_{M-m} as

$$p_{\alpha_{1:M}, z_{m-1}, z_{M-m}}(y, u, v) = M e^{-y-u-v} \sum_{j=0}^{m-1} \binom{m-1}{j} \frac{(-1)^j (u-jy)^{m-2}}{(m-2)!} U(u-jy)$$
$$\times \sum_{k=0}^{M-m} \binom{M-m}{k} \frac{(-1)^k (v-ky)^{M-m-1}}{(M-m-1)!} U(v-ky),$$
$$y > 0, u < (m-1)y, v < (M-m)y. \qquad (7.92)$$

The joint probability for the case of $m = M$ can be calculated using the PDF of $\gamma_B^{(M)}$, given in (7.76) as

$$\Pr[\gamma_B^{(m)} < x, \gamma_B^{(M)} > \beta] = \int_\beta^x p_{\gamma_B^{(M)}}(y) dy. \qquad (7.93)$$

After proper substitutions and taking the derivative, we obtain the PDF of the active beam SINR $\gamma_B^{(m)}$ as

Table 7.2 Features of different strategies

	SINR thresholding	SINR value feedback	Random selection
BBSI	No	Yes	Yes
CBBI	Yes	No	Yes
ABA-CBBI	Yes	No	No
BBI	No	No	Yes

$$p_{\gamma_B^{(m)}}(x) = \qquad (7.94)$$

$$\frac{1}{P_f} \begin{cases} \int_0^{\frac{x}{P\beta} - \frac{M}{P}} \frac{1}{P} p_{\alpha_{1:M}, z_{M-1}}(\frac{x}{P}, u) du, & m = 1; \\ \int_0^\infty \int_0^{\frac{y}{\beta} - \frac{y}{x} - \frac{M-m}{P}} \frac{y}{x^2} p_{\alpha_{1:M}, z_{m-1}, z_{M-m}}(y, \frac{y}{x} - \frac{m}{P}, v) dv dy, & 1 < m < M; \\ f_{\gamma_B^{(M)}}(x) U(x - \beta), & m = M. \end{cases}$$

Finally, the exact sum-rate expression for an RUB-based multiuser MIMO system with the ABA-CBBI strategy can be obtained by substituting (7.81) and (7.94) into (7.83). The final expression can be readily evaluated using mathematical software such as Maple or Mathematica.

7.4.3 Numerical Results

We now study the sum-rate performance of the proposed ABA-CBBI scheduling strategy through selected numerical examples. In particular, we compare its performance and complexity with some well-known user scheduling strategies for RUB-based multiuser MIMO systems. Table 7.2 summarizes the key features of different user scheduling strategies under consideration. Fig. 7.10 illustrates the effects of random user selection for unrequested beams on the sum-rate performance. In particular, we plot the sum rate of ABA-CBBI and conventional CBBI as a function of SNR. As we can see, with the same threshold value, the sum rate of ABA-CBBI is always greater than that of conventional CBBI, which shows that the RUB-based system is not benefiting from random user selection. We also notice from Fig. 7.10 that ABA-CBBI with different threshold values has better sum-rate performance over different SNR ranges. Therefore, to achieve the best sum-rate performance, we should use the optimal value of the feedback threshold β.

In Fig. 7.11, we plot the sum rate of the proposed ABA-CBBI strategy as a function of the feedback threshold β. For a fixed SNR, there exists an optimal value of β maximizing the sum rate, which is marked in the figure. Intuitively, when β is very small, almost all users will feed back their best beam indexes to the BS. Since the BS randomly assigns a beam to one of the users requesting it, that beam may be assigned to a user that does not achieve high SINR, which leads to relatively poorer sum-rate performance. On the other hand, when β is very large, few users will be qualified to feed back and some beams may not be requested. As such, not enough spatial multiplexing gain will be explored, which also causes poorer sum-rate performance.

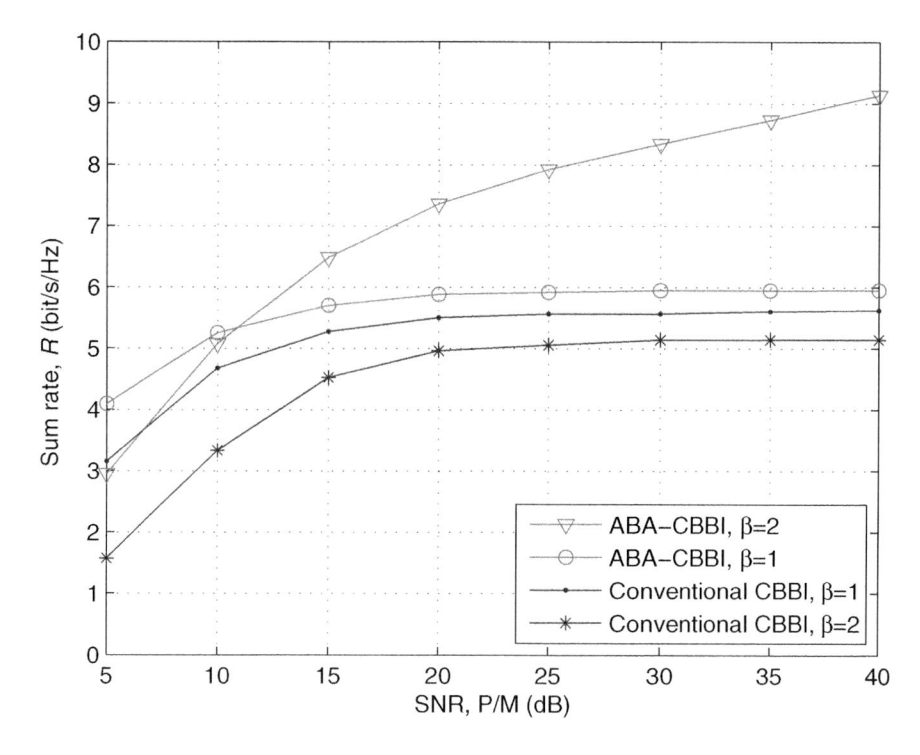

Figure 7.10 Effect of random user selection on sum-rate performance ($K = 20$). Reprint with permission from [19]. ©2009 IEEE.

In Fig. 7.12, we compare the sum-rate performance of the proposed ABA-CBBI strategy with two strategies without thresholding, BBSI and BBI. As we can see, the sum rate of ABA-CBBI with optimal threshold values approaches that of BBSI over the low SNR region, and as SNR increases, ABA-CBBI considerably outperforms BBSI. This somewhat surprising behavior can be explained as follows. While the sum rate of the BBSI strategy saturates over the high SNR region due to multiuser interference, ABA-CBBI manages to reduce multiuser interference by increasing the threshold value and then selecting fewer users to serve. As a result, the selected users will experience less multiuser interference, a bottleneck to achieving better sum-rate performance at high SNR for conventional RUB-based user scheduling schemes. By comparison, BBI has the worst performance due to its complete lack of a mechanism to guarantee the high SINR of selected users.

7.5 RUB Transmission with Receiver Linear Combining

Diversity combining can apply to enhance the performance of RUB-based multiuser MIMO systems when users are equipped with more than one antennas. Specifically, we assume each user has N ($N > 1$) antennas. A conventional approach to exploit multiple receive antennas is to treat each antenna independently and let each user feed back

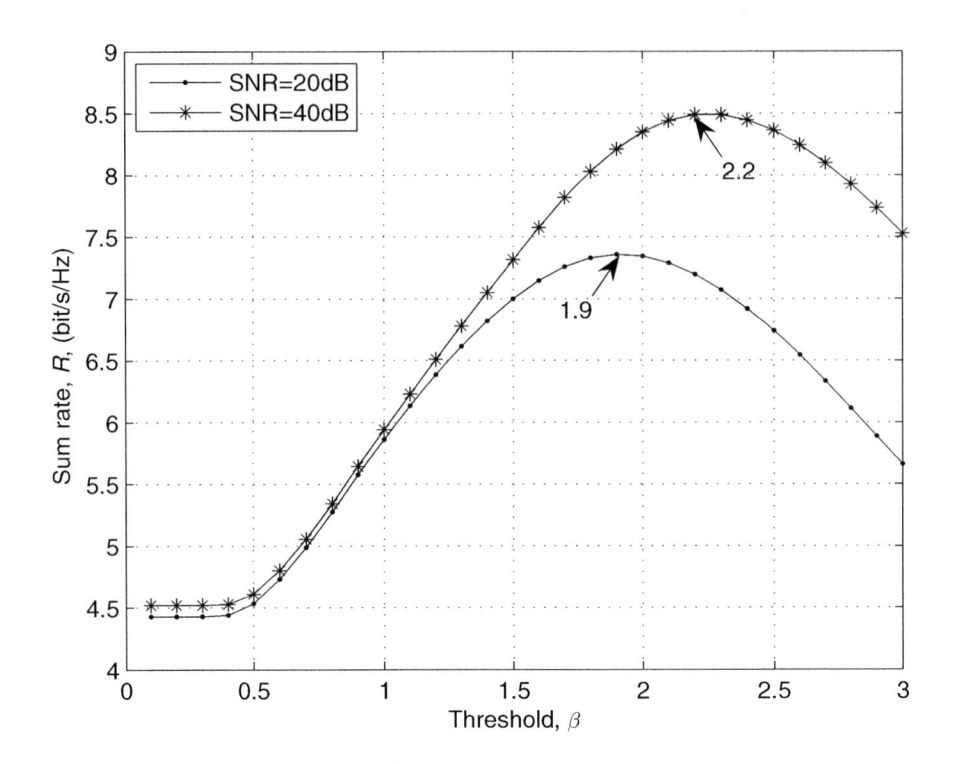

Figure 7.11 Sum rate of ABA-CBBI as a function of SINR threshold for different SNR values ($K = 20$). Reprint with permission from [19]. ©2009 IEEE.

the best beam indexes and corresponding SINR values for all its receive antennas [9]. Intuitively, this approach would improve the sum-rate performance since it is equivalent to increasing the number of single-antenna users in the system to NK. The main disadvantage with this approach is, however, that the selected receiver will use only one designated antenna for reception and the received signals at other receive antennas are simply discarded, which leads to an inefficient utilization of multiple receive antennas.

Alternatively, we can apply diversity combining technique to fully explore the multiple receive antennas at each mobile [21, 22]. In this section, we present and investigate RUB-based multiuser MIMO systems with selection combining (SC) [23] and optimum combining (OC) [24–26] schemes using two difference user scheduling strategies.[3]

7.5.1 System and Channel Model

We consider the downlink transmission of a single-cell wireless system. The BS is equipped with M antennas and each user is equipped with N antennas, where $N<M$ due to the size and cost limitations of user terminals. We assume that there are a total of K active users in the cell where $K \gg M$. With RUB, the BS utilizes M random orthonormal vectors generated from an isotropic distribution [7] to transmit to as many

---

[3] We omit the maximum ratio combining (MRC) scheme because MRC is optimal only in the noise-limited environment.

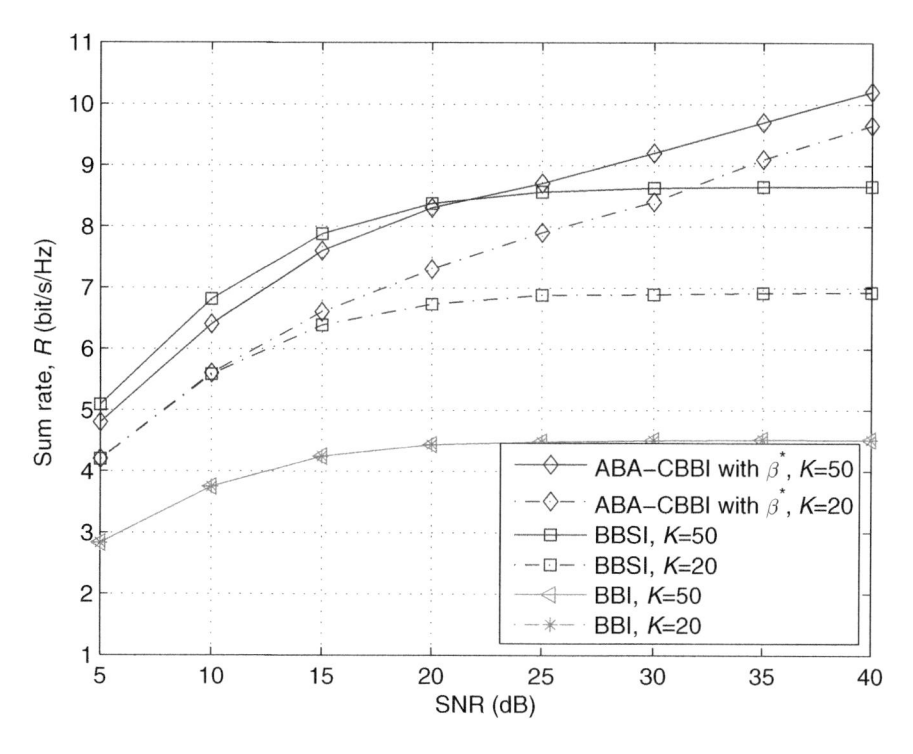

7.5 RUB Transmission with Receiver Linear Combining

Figure 7.12 Sum-rate comparison between ABA-CBBI, BBSI, and BBI strategies. Reprint with permission from [19]. ©2009 IEEE.

as M selected users simultaneously. We denote the vector set as $\mathcal{U} = \{\mathbf{u}_1, \mathbf{u}_2, \ldots, \mathbf{u}_M\}$. The transmitted signal vector from M antennas over one symbol period can be written as

$$\mathbf{x} = \sum_{m=1}^{M} \mathbf{u}_m s_m, \tag{7.95}$$

where s_m is the information symbol for the mth selected user. The transmitted signal vector $\mathbf{x}$ has an average power constraint of $\mathbf{E}(\mathbf{x}^H \mathbf{x}) \leq P$, where P is the maximum average transmitting power, $\mathbf{E}(\cdot)$ denotes the statistical expectation, and $(\cdot)^H$ stands for the Hermitian transpose.

We adopt a homogeneous flat fading channel model for analytical tractability. Specifically, the channel gains from each transmit antenna to each receive antenna of users follow i.i.d. zero mean complex Gaussian distribution with unit variance. The channel gain from the mth transmit antenna to the nth receive antenna of the kth mobile users is denoted by $h_{nm}^{(k)}$. As such, the $N \times M$ channel matrix for the kth user is constructed as

$$\mathbf{H}_k = \begin{pmatrix} h_{11}^{(k)} & h_{12}^{(k)} & \cdots & h_{1M}^{(k)} \\ h_{21}^{(k)} & h_{22}^{(k)} & \cdots & h_{2M}^{(k)} \\ \vdots & \vdots & \ddots & \vdots \\ h_{N1}^{(k)} & h_{N2}^{(k)} & \cdots & h_{NM}^{(k)} \end{pmatrix}, \tag{7.96}$$

with the nth row, denoted by $\mathbf{h}_n^{(k)}$, being the instantaneous MISO channel for the nth receive antenna of the kth user.

The received signal vector over a symbol period at the ith selected user, when the jth beam is assigned to it, can be written as

$$\mathbf{y}_j^{(i)} = \mathbf{H}_i \mathbf{x} + \mathbf{n}_i = \mathbf{H}_i \mathbf{u}_j s_j + \sum_{m=1, m \neq j}^{M} \mathbf{H}_i \mathbf{u}_m s_m + \mathbf{n}_i, \qquad (7.97)$$

where $\mathbf{n}_i$ is the $N \times 1$ additive Gaussian noise vector, whose entries follow independent complex Gaussian distribution with zero mean and variance σ_n^2. Note that $\mathbf{H}_i \mathbf{u}_j s_j$ is the desired signal for user i while $\sum_{m=1, m \neq j}^{M} \mathbf{H}_i \mathbf{u}_m s_m$ acts as interference.

The users apply diversity combining techniques, such as SC and OC, to combine the signal received from N different receive antennas. When beam j is assigned to the ith user, the combined signal at user i can be written as

$$z_j^{(i)} = \mathbf{w}_{ij}^H \mathbf{y}_j^{(i)} = \mathbf{w}_{ij}^H \mathbf{H}_i \mathbf{u}_j s_j + \mathbf{w}_{ij}^H \sum_{m=1, m \neq j}^{M} \mathbf{H}_i \mathbf{u}_m s_m + \mathbf{w}_{ij}^H \mathbf{n}_i, \qquad (7.98)$$

where $\mathbf{w}_{ij}$ denotes the weighting vector of user i for the jth beam. Assuming that the transmit power is equally allocated among M selected users, the SINR of the combined signal at the user i when the jth beam is assigned to it can be shown to be given by

$$\gamma_j^{(i)} = \frac{|\mathbf{w}_{ij}^H \mathbf{H}_i \mathbf{u}_j|^2}{\sum_{m=1, m \neq j}^{M} |\mathbf{w}_{ij}^H \mathbf{H}_i \mathbf{u}_m|^2 + \rho \|\mathbf{w}_{ij}\|^2}, \qquad (7.99)$$

where $\rho = \frac{M \sigma_n^2}{P}$ is the normalized transmit SNR. We assume that users have perfect knowledge of their downlink channel matrix through the channel estimation process. Applying (7.99), user i calculates the SINRs after beam-specific combining for all M available beams, $\gamma_j^{(i)}, j = 1, 2, \ldots, M$.

7.5.2 All Beam Feedback Strategy

With this strategy, each mobile evaluates the resulting SINR for each beam at the beginning of each time slot, after combining the received signals on all antennas with either SC or OC. Then, the mobile will feed back M effective beam SINRs to the BS. The incurred feedback load with this strategy is M real numbers per users. Note that OC performs active interference suppression by exploiting the interference structure, whereas SC simply selects the beam with the best SINR. After receiving the effective SINR information from all users, the BS schedules and starts data transmission to the users that reported the largest effective SINRs on different beams.

We analyze the asymptotic performance of this feedback strategy using the limiting distribution of the beam SINR [21]. It is worth noting that the probability of awarding multiple beams to the same user is rather small when the number of users K is large. As such, the SINR of the jth beam after user selection is equal to $\gamma_j^B =$

$\max\{\gamma_j^{(1)}, \gamma_j^{(2)}, \ldots, \gamma_j^{(i)}\}$, where $\gamma_j^{(i)}$, $i = 1, 2, \ldots, K$, are the resulting SINR at user i for the beam j after applying diversity combining. With the homogeneous fading assumption, we can show that the CDF of γ_j^B is given by

$$F_{\gamma_j^B}(x) = [F_{\gamma_j^{(i)}}(x)]^K. \tag{7.100}$$

In the following, we develop the limiting distribution of γ_j^B and use it to examine the asymptotic throughput and sum-rate scaling laws of different combining schemes. To make the analysis easy to follow, we focus on the case of four transmit antennas $M = 4$ and two receive antennas per user $N = 2$.

Selection Combining

For SC, the CDF of the resulting SINR after beam-specific combining is given by

$$F_{\gamma_j^{(i)}}(x) = \left(1 - \frac{e^{-\rho x}}{(1+x)^3}\right)^2. \tag{7.101}$$

It follows that the PDF of $\gamma_j^{(i)}$ is given by

$$p_{\gamma_j^{(i)}}(x) = 2\left(1 - \frac{e^{-\rho x}}{(1+x)^3}\right) \frac{e^{-\rho x}}{(1+x)^4} (\rho(1+x) + 3). \tag{7.102}$$

We can then verify that the following condition is satisfied:

$$\lim_{x \to +\infty} \frac{1 - F_{\gamma_j^{(i)}}(x)}{p_{\gamma_j^{(i)}}(x)} = 1/\rho > 0, \tag{7.103}$$

which implies that the limiting distribution of γ_j^B exists and follows the Gumbel type. More specifically, the limiting distribution of γ_j^B when K approaches infinity is given by

$$\lim_{K \to +\infty} F_{\gamma_j^B}(x) = \exp\left(-e^{-\frac{x - b_K}{a_K}}\right), \tag{7.104}$$

where the parameters a_K and b_K can be determined by numerically solving the equations $1 - F_{\gamma_j^{(i)}}(b_K) = 1/K$ and $1 - F_{\gamma_j^{(i)}}(a_K + b_K) = 1/(eK)$.

It can be shown that the sum-rate of the multiuser MIMO system with SC scales as [21]

$$\lim_{K \to +\infty} \frac{C^{SC}}{4 \log b_K} = 1. \tag{7.105}$$

When $\rho = 1$, we can approximately determine b_K as

$$b_K \approx \log 2K - 2\log(1 + \log 2K), \tag{7.106}$$

which implies that the sum-rate scales at the rate of $\log \log 2K$ as K approaches infinity.

Optimal Combining

The resulting SINR after applying OC for a particular beam is equivalent to the combined SINR with OC in the presence of $M - 1$ interfering sources. The CDF of the resulting SINR for the four transmit antenna and two receive antenna case under consideration is given by

$$F_{\gamma_j^{(i)}}(x) = 1 - \frac{e^{-\rho x}}{(1+x)^3}(1 - 3x - \rho x). \qquad (7.107)$$

It follows that the PDF of $\gamma_j^{(i)}$ is determined as

$$p_{\gamma_j^{(i)}}(x) = \frac{xe^{-\rho x}}{(1+x)^4}[(3\rho - \rho^2)x + (6 + 6\rho + \rho^2)]. \qquad (7.108)$$

Based on these results, it can be again shown that the limiting distribution of γ_j^B is of the Gumbel type. Following the same procedure as the SC case, we can show that the sum-rate scaling law for the OC scheme is given by [21]

$$\lim_{K \to +\infty} \frac{C^{OC}}{4 \log b_K} = 1. \qquad (7.109)$$

When $\rho = 1$, we can approximately determine b_K as

$$b_K \approx \log 4K - 2 \log \log K. \qquad (7.110)$$

As such, the sum-rate scales at the rate of $\log \log 4K$.

7.5.3 Best Beam Feedback Strategy

In this subsection, we investigate the sum-rate performance of RUB systems with diversity combining when each user only feeds back the best beam information. Specifically, the user first calculates the output SINR after beam-specific combining for each of the M available beams. Then, the user selects its best beam among M beams, i.e., the one that achieves the largest combiner output SINR. Let b_i denote the index of the best beam for user i. Mathematically, b_i is given by $b_i = \arg\max_j \gamma_j^{(i)}$. The index of the best beam b_i and its corresponding output SINR $\gamma_{b_i}^{(i)}$ are fed back to the BS. Note that the feedback load with receiver combining techniques is one integer (for best beam index) and one real number (for best beam output SINR) per user, which is much lower than that of the M beam feedback strategy discussed in the previous subsection.

The BS assigns a beam to the user that feeds back the largest SINR value among all users requesting that beam. Specifically, let $\mathcal{K}_m$ denote the user index set containing all users who feedback beam index m, i.e., $b_k = m$ if and only if $k \in \mathcal{K}_m$. If $\gamma_m^{(\hat{k})}$ is the largest one among all $\gamma_m^{(k)}$, $k \in \mathcal{K}_m$, then the mth beam will be assigned to user $\hat{k}$ and the SINR of the mth beam γ_m^B will be equal to $\gamma_m^{(\hat{k})}$. If a particular beam is not requested by any user, i.e., the set $\mathcal{K}_m$ is empty for such a beam, then the BS knows that this beam is not the best one for any user and will allocate it to a randomly chosen user.

In what follows, we analyze the sum-rate performance of the resulting system through accurate statistical analysis. Unlike the limiting distribution approach adopted in the previous subsection, which applies only when the number of users is very large, we derive the exact statistics of the feedback SINR for the SC scheme, based on which we arrive at the exact expression of the resulting sum rate for arbitrary number of users. By following a geometric approach, we also develop the statistics of output signal-to-interference ratio (SIR) over the high SNR region with the OC scheme, which is used to obtain an upper sum-rate bound. The upper bound is shown to be tight for a moderate number of users, as long as it is at least five times the number of transmit antennas.

Selection Combining
With the SC scheme, the combined SINR at user i for the jth beam is equal to the largest SINR among N ones corresponding to different receive antennas for this beam. Specifically, the SINR on user i's nth receive antenna, while assuming beam j is assigned to it, can be written as

$$\gamma_{n,j}^{(i)} = \frac{|(\mathbf{h}_n^{(i)})^T \mathbf{u}_j|^2}{\sum_{m=1, m \neq j}^{M} |(\mathbf{h}_n^{(i)})^T \mathbf{u}_m|^2 + \rho}, \quad 1 \leq n \leq N, \ 1 \leq j \leq M, \tag{7.111}$$

where $\mathbf{h}_n^{(i)}$ is the $M \times 1$ channel vector from M transmit antennas to the nth receive antenna of user i. The combined SINR for the jth beam is then mathematically given by $\gamma_j^{(i)} = \max\{\gamma_{1,j}^{(i)}, \gamma_{2,j}^{(i)}, \ldots, \gamma_{N,j}^{(i)}\}$. If a user is selected for data reception on a certain beam, the user will use only the best receive antenna for this beam during reception, which constitutes the main complexity advantage of the SC scheme.

The sum-rate performance of the resulting multiuser MIMO system can be analyzed by following a similar approach as in the previous section while treating each user as a single antenna receiver. On the other hand, because of the antenna selection process at the mobile users, the statistics of the feedback SINR are different from the single receive antenna case. In particular, $\gamma_{b_i}^{(i)}$ is the largest one of M i.i.d. combined SINRs, i.e., $\gamma_{b_i}^{(i)} = \max\{\gamma_1^{(i)}, \gamma_2^{(i)}, \ldots, \gamma_M^{(i)}\}$. Equivalently, the feedback SINR $\gamma_{b_i}^{(i)}$ can be viewed as the largest one among N best beam SINRs, i.e., $\gamma_{b_i}^{(i)} = \max\{\gamma_{1,j^*}^{(i)}, \gamma_{2,j^*}^{(i)}, \ldots, \gamma_{N,j^*}^{(i)}\}$, where $\gamma_{n,j^*}^{(i)} = \max\{\gamma_{n,1}^{(i)}, \gamma_{n,2}^{(i)}, \ldots, \gamma_{n,M}^{(i)}\}$ denotes the best beam SINR for the nth antenna at user i, whose PDF was obtained as

$$p_{\gamma_{n,j^*}^{(i)}}(x) = \sum_{j=0}^{M-1} \frac{(-1)^j M(M-1)}{(M-j-1)! j!} \int_0^\infty (\rho + z) \tag{7.112}$$

$$\times ((1-jx)z - jx\rho)^{M-2} e^{-(1+x)z - x\rho} U((1-jx)z - jx\rho) dz.$$

Therefore, the common PDF of the feedback SINRs $\gamma_{b_i}^{(i)}$ can be obtained as

$$p_{\gamma_{b_i}^{(i)}}(x) = N F_{\gamma_{n,j^*}^{(i)}}(x)^{N-1} p_{\gamma_{n,j^*}^{(i)}}(x), \tag{7.113}$$

where $F_{\gamma_{n,j^*}^{(i)}}(x) = \int_0^x p_{\gamma_{n,j^*}^{(i)}}(z) dz$ is the CDF of $\gamma_{n,j^*}^{(i)}$.

Based on the mode of operation for user selection, a certain beam will be assigned to a randomly selected user if no user feeds back the index of such a beam. In this case, the user will choose the best antenna, i.e., the receive antenna that leads to the largest SINR on this beam, for data reception. As such, the beam SINR for such a beam will be equal to the largest SINR among N i.i.d. ones, whose PDF is given by

$$f_{\gamma^R}(x) = N\left(F_{\gamma_{n,j}^{(i)}}(x)\right)^{N-1} p_{\gamma_{n,j}^{(i)}}(x), \qquad (7.114)$$

where $f_{\gamma_{n,j}^{(i)}}(x)$ and $F_{\gamma_{n,j}^{(i)}}(x)$ denote the PDF and CDF of $\gamma_{n,j}^{(i)}$ in (7.111), respectively. The closed-form expression of $f_{\gamma_{n,j}^{(i)}}(x)$ is given by [9, Eq. (14)]

$$p_{\gamma_{n,j}^{(i)}}(x) = \frac{e^{-x/\rho}}{(1+x)^M} \left(\frac{x+1}{\rho} + M - 1\right). \qquad (7.115)$$

Finally, following a similar analytical approach to the previous section, while noting that the PDF of user feedback SINR is given in (7.112) and the PDF of randomly assigned beam SINR in (7.115), the sum rate of the RUB system with SC at receivers can be calculated as

$$R^{SC} = \int_0^\infty \log_2(1+x) p_{\gamma_{b_i}^{(1:K)}}(x) dx \qquad (7.116)$$
$$+ \sum_{m=2}^{M} \sum_{i=m}^{K} \left(P_i^m \int_0^\infty \log_2(1+x) p_{\gamma_{b_i}^{(i:K)}}(x) dx \right.$$
$$+ \left. \left(\frac{m-1}{M}\right)^{K-1} \int_0^\infty \log_2(1+x) p_{\gamma^R}(x) dx \right),$$

where P_i^m denotes the probability that the mth beam is assigned to the user with the ith largest feedback SINR, i.e., $\gamma_m^B = \gamma_{b_i}^{(i:K)}, i = m, m+1, \ldots, K$, which is given by,

$$P_i^m = \sum_{\substack{\sum_{j=1}^{m-1} n_j = i-m \\ n_j \in \{0,1,\ldots,i-m\}}} \left[\prod_{j=1}^{m-1} \left(\frac{j}{M}\right)^{n_j} \left(1 - \frac{j}{M}\right) \right], \quad i = m, m+1, \ldots, K, \qquad (7.117)$$

and $p_{\gamma_{b_i}^{(i:K)}}(x)$ represents the PDF of the ith user feedback SINR, given in terms of $f_{\gamma_{b_i}^{(i)}}(x)$ as

$$p_{\gamma_{b_i}^{(i:K)}}(x) = \frac{K!}{(K-i)!(i-1)!} F_{\gamma_{b_i}^{(i)}}(x)^{K-i} [1 - F_{\gamma_{b_i}^{(i)}}(x)^M]^{i-1} p_{\gamma_{b_i}^{(i)}}(x). \qquad (7.118)$$

Optimal Combining
With the OC scheme [23, 24], it can be shown that the optimal weighting vector of user i for the jth beam is given by

$$\mathbf{w}_{ij}^* = \left(\sum_{m=1, m \neq j}^{M} \mathbf{H}_i \mathbf{u}_m (\mathbf{H}_i \mathbf{u}_m)^H + \rho \mathbf{I} \right)^{-1} \mathbf{H}_i \mathbf{u}_j, \qquad (7.119)$$

7.5 RUB Transmission with Receiver Linear Combining

and the corresponding maximum SINR of the combined signal is given by

$$\gamma_j^{(i)} = (\mathbf{H}_i\mathbf{u}_j)^H \mathbf{w}_{ij}^* = (\mathbf{H}_i\mathbf{u}_j)^H \left(\sum_{m=1, m\neq j}^{M} \mathbf{H}_i\mathbf{u}_m(\mathbf{H}_i\mathbf{u}_m)^H + \rho \mathbf{I} \right)^{-1} \mathbf{H}_i\mathbf{u}_j, \quad (7.120)$$

where $\mathbf{I}$ denotes the identity matrix. After proper comparison, user i will feedback the best beam index b_i and the corresponding combined SINR, $\gamma_{b_i}^{(i)}$, to the BS for user selection. If the user is selected for data reception on the b_ith beam, the user will combine the signal from different antennas using weights $\mathbf{w}_{ib_i}^*$ calculated using (7.119) for data detection.

After some mathematical manipulation, the optimal weight vector and the corresponding output SINR are simplified to

$$\mathbf{w}_{ij}^* = \frac{(\mathbf{H}_i\mathbf{H}_i^H + \frac{1}{\rho}\mathbf{I})^{-1}\mathbf{H}_i\mathbf{u}_j}{1 - (\mathbf{H}_i\mathbf{u}_j)^H(\mathbf{H}_i\mathbf{H}_i^H + \frac{1}{\rho}\mathbf{I})^{-1}\mathbf{H}_i\mathbf{u}_j}, \quad (7.121)$$

and

$$\gamma_j^{(i)} = \frac{(\mathbf{H}_i\mathbf{u}_j)^H(\mathbf{H}_i\mathbf{H}_i^H + \frac{1}{\rho}\mathbf{I})^{-1}\mathbf{H}_i\mathbf{u}_j}{1 - (\mathbf{H}_i\mathbf{u}_j)^H(\mathbf{H}_i\mathbf{H}_i^H + \frac{1}{\rho}\mathbf{I})^{-1}\mathbf{H}_i\mathbf{u}_j}, \quad (7.122)$$

respectively. Note that each user needs only to calculate the inversion of a single matrix $(\mathbf{H}_i\mathbf{H}_i^H + \frac{1}{\rho}\mathbf{I})$ now. Although the statistics of the combined SINR for a single beam can be obtained [27], because of the dependence between SINRs for different beams, it is very difficult, if not impossible, to derive the statistics of the feedback SINR with the OC scheme. Therefore, we focus on the sum-rate upper bound analysis over the high SNR region in what follows.

When the SNR is very high and the system becomes interference limited, the optimal weighting vector can be approximately given by, after setting ρ in (7.121) to infinity,

$$\mathbf{w}_{ij}^* = \frac{(\mathbf{H}_i\mathbf{H}_i^H)^{-1}\mathbf{H}_i\mathbf{u}_j}{1 - (\mathbf{H}_i\mathbf{u}_j)^H(\mathbf{H}_i\mathbf{H}_i^H)^{-1}\mathbf{H}_i\mathbf{u}_j}. \quad (7.123)$$

It follows that $\mathbf{w}_{ij}^*$ is proportional to the least square solution of $\mathbf{H}_i^H\mathbf{w} = \mathbf{u}_j$ [27, 28]. Therefore, with the optimal weighting vector, the angle between the effective MISO channel for user i, $\mathbf{H}_i^H\mathbf{w}_{ij}^*$, and the jth beam, $\mathbf{u}_j$ is minimized, which is equal to the angle between $\mathbf{u}_j$, and the range space of $\mathbf{H}_i^H$, denoted by $\mathcal{R}(\mathbf{H}_i^H)$.

LEMMA 7.1 *Over the high SNR region, the maximum output SINR at user i after performing optimal combining with respect to the jth beam is given by*

$$\gamma_{i,j}^* = |\cot(\theta_{ij})|^2, \quad (7.124)$$

where θ_{ij} is the angle between $\mathbf{u}_j$ and $\mathcal{R}(\mathbf{H}_i^H)$.

Proof. Let ρ in (7.122) go to infinity, the maximum output SINR at user i over the high SNR region becomes

$$\gamma_{i,j}^* = \frac{(\mathbf{H}_i\mathbf{u}_j)^H(\mathbf{H}_i\mathbf{H}_i^H)^{-1}\mathbf{H}_i\mathbf{u}_j}{1 - (\mathbf{H}_i\mathbf{u}_j)^H(\mathbf{H}_i\mathbf{H}_i^H)^{-1}\mathbf{H}_i\mathbf{u}_j}. \quad (7.125)$$

Since $\mathbf{H}_i^H(\mathbf{H}_i\mathbf{H}_i^H)\mathbf{H}_i$ is the projection matrix onto $\mathcal{R}(\mathbf{H}_i^H)$, we have

$$|\cos(\theta_{ij})|^2 = (\mathbf{H}_i\mathbf{u}_j)^H(\mathbf{H}_i\mathbf{H}_i^H)^{-1}\mathbf{H}_i\mathbf{u}_j. \qquad (7.126)$$

The lemma is proved after substituting (7.126) into (7.125). □

Since $\cot(\cdot)$ is a monotonically decreasing function, each user can feed back the index and SINR value of the beam that forms the smallest angle with $\mathcal{R}(\mathbf{H}_i^H)$.

It can be shown that the random variable $|\cos(\theta_{ij})|^2$, where θ_{ij} is the angle between a random vector of length M and a space spanned by N independent random vectors of the same length, follows the beta distribution parameterized by N and $M - N$, with PDF and CDF given by [6]

$$f_{|\cos(\theta)|^2}(x) = (M-N)\binom{M-1}{N-1}x^{N-1}\cdot(1-x)^{M-N-1}, \quad 0 \leq x \leq 1 \qquad (7.127)$$

and

$$F_{|\cos(\theta)|^2}(x) = \mathrm{I}(x; N, M), \quad 0 \leq x \leq 1, \qquad (7.128)$$

respectively. To derive an upper bound of the system sum rate, we assume that every beam of the BS is requested by at least one user. In this case, since a user can only be assigned to one beam, the SINR of the jth assigned beam can be shown to be equal to the largest one among $K - j + 1$ combined SINRs on that beam for the remaining $K - j + 1$ users ($j - 1$ users have been previously assigned). Mathematically speaking, the SINR of the jth assigned beam is equal to $|\cot(\theta_j^*)|^2$, where θ_j^* is the smallest angle among $K - j + 1$ ones. Correspondingly, $|\cos(\theta_j^*)|^2$ will be the largest one among $K - j + 1$ independent beta distributed random variables. The PDF of $|\cos(\theta_j^*)|^2$ is thus given by [14]

$$f_{|\cos(\theta_j^*)|^2}(x) = (K-j+1)\left(F_{|\cos(\theta)|^2}(x)\right)^{K-j}f_{|\cos(\theta)|^2}(x), \quad 0 \leq x \leq 1. \qquad (7.129)$$

Noting that $\log_2(1 + |\cot(\theta_j^*)|^2) = -\log_2(1 - |\cos(\theta_j^*)|^2)$, we obtain a sum-rate upper bound for the proposed scheme over the high SNR region as

$$R^{OC} = \sum_{j=1}^{M} -(K-j+1)\int_0^1 \log_2(1-x)\left(F_{|\cos(\theta)|^2}(x)\right)^{K-j}f_{|\cos(\theta)|^2}(x)dx. \qquad (7.130)$$

As this upper bound is derived based on the assumption that all beams are requested, it will be tight as long as the probability of all beams being requested is close to 1. In [19], we showed that the probability that each of M available beams is requested by at least one user is given by

$$\Pr[\text{every beam requested}] = \sum_{q=1}^{M}(-1)^{M-q}\binom{M}{q}\left(\frac{q}{M}\right)^K. \qquad (7.131)$$

This probability is plotted as a function of the number of users K and the number of beams M in Fig. 7.13. As we can see, when K is more than five times M, which is usually the case in practice, every beam will be requested with probability very close to 1.

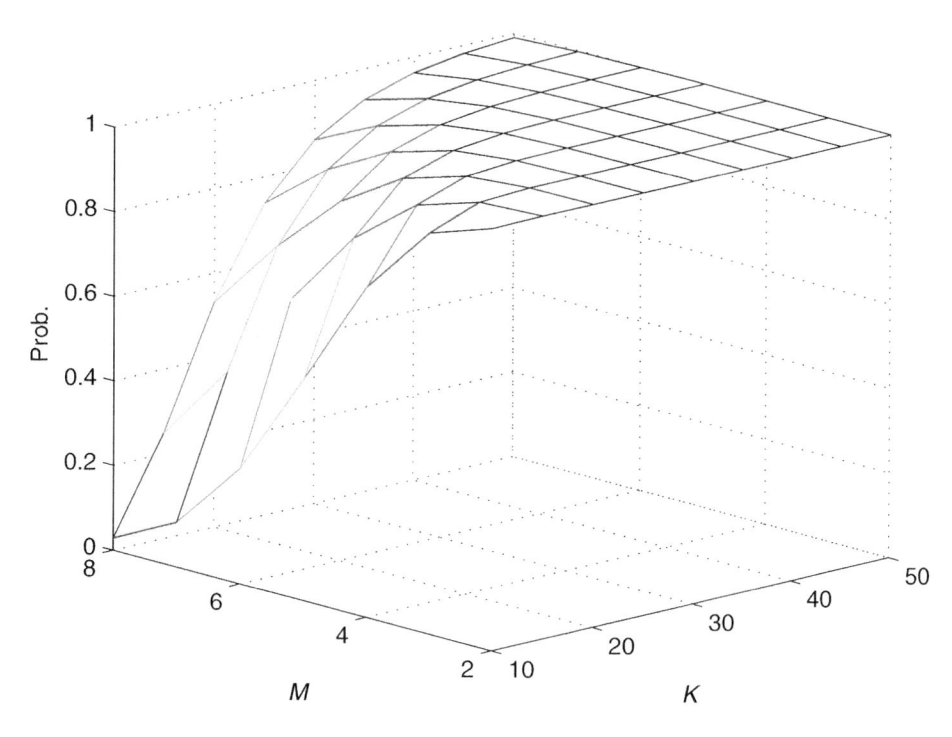

Figure 7.13 Probability of all beams being requested. Reprint with permission from [22]. ©2010 IEEE.

7.5.4 Numerical Results

In Fig. 7.14, we plot the sum rate of SC-based systems as a function of average SNR for different number of users. We set $M = 4$ and $N = 2$ here. The sum rate of the conventional approach, which treats a user's receive antennas independently, is also plotted for comparison. Though requiring only $1/N$ of the feedback load of the conventional approach, SC is shown to achieve nearly the same sum-rate performance. In Fig. 7.14, we also compare our analytical results given by (7.117) to simulation results. The perfect match verifies the analytical approach.

In Fig. 7.15, the sum rate of the OC-based system is plotted as a function of average SNR for different number of users. The sum rate of the conventional approach, M beam feedback, and the analytical upper bound given by (7.130) are also plotted. As expected, the analytical upper bound developed for the OC-based system is shown to be tight over the high SNR region. Compared to the conventional approach, OC achieves noticeably better sum-rate performance while requiring less feedback load. We further observe that the OC-based scheme offers nearly the same performance as the M beam feedback scheme, which mandates M real numbers per user feedback load. It is worthwhile to mention that the advantages of OC come at the cost of additional processing complexity of channel matrix inversion when calculating the combining weights and output SINR.

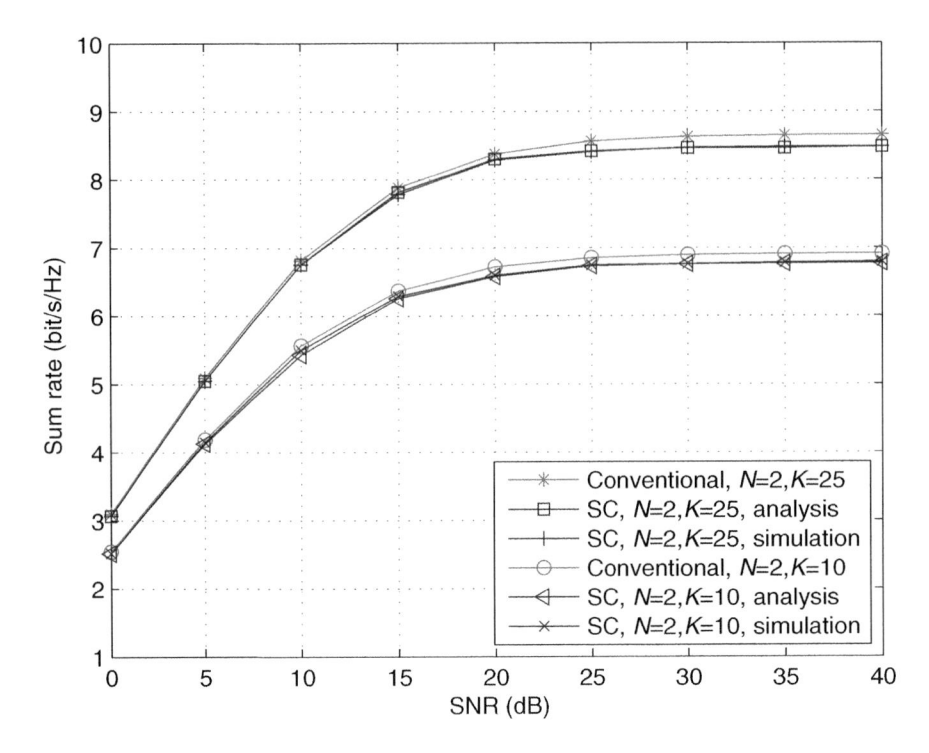

Figure 7.14 Sum-rate performance of RUB systems with the SC scheme ($M = 4$). Reprint with permission from [22]. ©2010 IEEE.

7.6 Summary

In this chapter, we studied low-complexity beamforming and user scheduling schemes for multiuser MIMO systems. While ZFBF can completely eliminate the interuser interference when operating over multiuser MIMO channels, RUB schemes incur much lower feedback load as the users only feed back their best beam SINR information. We derived the exact statistics of users' best beam SINR, which facilitates the analysis of different user scheduling schemes. Finally, we showed that the receiver diversity combining can offer significant performance gain for the multiple antennas per user case.

7.7 Further Reading

The capacity of multiuser MIMO systems and its achievement with DPC was investigated in [2, 3]. The optimality of ZFBF schemes over the multiuser MIMO channel was established examined in [6], whereas that for the RUB scheme can be found in [9]. [29, 30] present several additional user scheduling schemes with limited feedback for multiuser MIMO systems. A beamforming scheme with limited feedback for multiple access channels was considered in [31]. Finally, more details for optimal combining can be found in [23].

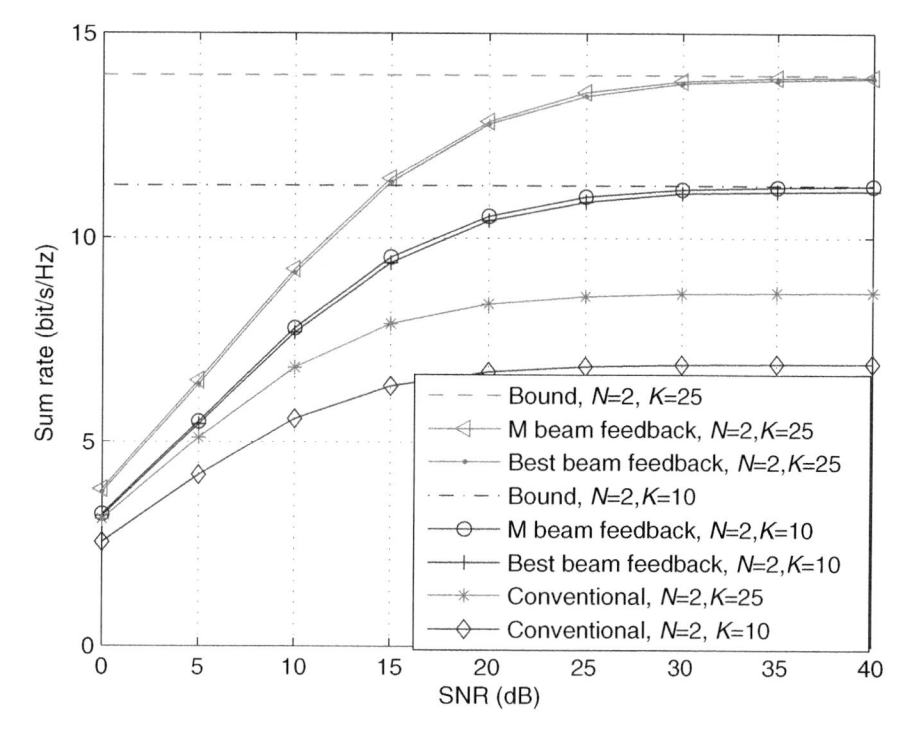

Figure 7.15 Sum-rate performance of RUB systems with OC scheme ($M = 4$). Reprint with permission from [22]. ©2010 IEEE.

References

[1] M. Costa, "Writing on dirty paper," *IEEE Trans. Inform. Theory*, 29, no. 5, pp. 439–441, 1983.

[2] G. Caire and S. Shamai, "On the achievable throughput of a multiantenna Gaussian broadcast channel," *IEEE Trans. Inform. Theory*, 49, no. 7, pp. 1691–1706, 2003.

[3] P. Viswanath and D. Tse, "Sum capacity of the vector Gaussian broadcast channel and uplink–downlink duality," *IEEE Trans. Inform. Theory*, 49, no. 8, pp. 1912–1921, 2003.

[4] N. Jindal and A. Goldsmith, "Dirty-paper coding versus TDMA for MIMO broadcast channels," *IEEE Trans. Inform. Theory*, 51, no. 5, pp. 1783–1794, 2005.

[5] W. Rhee, W. Yu, and J. M. Cioffi, "The optimality of beamforming in uplink multiuser wireless systems," *IEEE Trans. Wireless Commun.*, 3, no. 1, pp. 86–96, 2004.

[6] T. Yoo and A. Goldsmith, "On the optimality of multi-antenna broadcast scheduling using zero-forcing beamforming," *IEEE J. Select. Areas Commun.*, 24, no. 3, pp. 528–541, 2006.

[7] B. Hassibi and T. L. Marzetta, "Multiple-antennas and isotropically random unitary inputs: the received signal density in closed form," *IEEE Trans. Inform. Theory*, 48, no. 6, pp. 1473–1484, 2002.

[8] K. K. J. Chung, C.-S. Hwang, and Y. K. Kim, "A random beamforming technique in MIMO systems exploiting multiuser diversity," *IEEE J. Select. Areas Commun.*, 21, no. 5, pp. 848–855, 2003.

[9] M. Sharif, and B. Hassibi, "On the capacity of MIMO broadcast channels with partial side information," *IEEE Trans. Inform. Theory*, 51, no. 2, pp. 506–522, 2005.

[10] J. Kim, S. Park, J. H. Lee, J. Lee, and H. Jung, "A scheduling algorithm combined with zero-forcing beamforming for a multiuser MIMO wireless system," in *Proc. IEEE Semiannual Veh. Tech. Conf. (VTC' 2005)*, Dallas, Texas, vol. 1, September 2005, pp. 211–215.

[11] A. M. Toukebri, S. Aissa, and M. Maier, "Resource allocation and scheduling for multiuser MIMO systems: a beamforming-based strategy," in *Proc. IEEE Global Telecomm. Conf. (Globecom' 2006)*, San Francisco, CA, November 2006.

[12] G. Dimic and N. D. Sidiropoulos, "On downlink beamforming with greedy user selection: performance analysis and a simple new algorithm," *IEEE Trans. Sig. Pro.*, 53, no. 10, pp. 3857–3868, 2005.

[13] P. Lu and H.-C. Yang, "Sum-rate analysis of multiuser MIMO system with zero-forcing transmit beamforming," *IEEE Trans. Commun.*, 57, no. 9, pp. 2585–2589, 2009.

[14] H. A. David, *Order Statistics*, Wiley, 1981.

[15] J. Diaz, O. Simeone, O. Somekh, and Y. bar-Ness, "Scaling law of the sum-rate for multiantenna broadcast channels with deterministic or selective binary feedback," in *Proc. Inform. Theory Workshop (ITW'2006)*, Punta del Este, Uruguay, March 2006, pp. 298–301.

[16] J. Diaz, O. Simeone, and Y. bar-Ness, "Sum-rate of MIMO broadcast channels with one bit feedback," in *Proc. Int. Symp. Inform. Theory (ISIT'2006)*, Seattle, Washington, July 2006, pp. 1944–1948.

[17] Y.-C. Ko, H.-C. Yang, S.-S. Eom, and M. -S. Alouini, "Adaptive modulation with diversity combining based on output-threshold MRC," *IEEE Trans. Wireless Commun.*, 6, no. 10, pp. 3728–3737, 2007.

[18] H.-C. Yang, P. Lu, H.-K. Sung, and Y.-C. Ko, "Exact sum-rate analysis of MIMO broadcast channels with random unitary beamforming," *IEEE Trans. Commun.*, 59, no. 11, pp. 2982–2986, 2011.

[19] P. Lu and H.-C. Yang, "A simple and efficient user scheduling strategy for RUB-based multiuser MIMO systems and its sum rate analysis," *IEEE Trans. Veh. Tech.*, 58, no. 9, pp. 4860–4867, 2009.

[20] D. Gesbert and M.-S. Alouini, "How much feedback is multi-user diversity really worth?" in *Proc. IEEE Int. Conf. on Commun. (ICC'2004)*, Paris, France, June 2004, pp. 234–238.

[21] M. O. Pun, V. Koivunen, and H. V. Poor, "SINR analysis of opportunistic MIMO–SDMA downlink systems with linear combining," in *Proc. IEEE Int. Conf. Commun. (ICC'2008)*, Beijing, China, May 2008.

[22] P. Lu and H.-C. Yang, "Performance analysis for RUB-based multiuser MIMO systems with antenna diversity techniques," *IEEE Trans. Veh. Tech.*, 59, no. 1, pp. 490–494, 2010.

[23] M. K. Simon and M.-S. Alouini, *Digital Communication over Fading Channels*, 2nd ed., Wiley, 2004.

[24] J. H. Winters, "Optimum combining in digital mobile radio with cochannel interference," *IEEE Trans. Veh. Tech.*, 33, no. 3, pp. 144–155, 1984.

[25] A. Shah and A. M. Haimovich, "Performance analysis of optimum combining in wireless communications with Rayleigh fading and cochannel interference," *IEEE Trans. Commun.*, 46, no. 4, pp. 473–479, 1998.

[26] M. Chiani, M. Z. Win, A. Zanella, R. K. Mallik, and J. H. Winters, "Bounds and approximations for optimum combining of signals in the presence of multiple cochannel interferers and thermal noise," *IEEE Trans. Commun.*, 51, no. 2, pp. 296–307, 2003.

[27] H. S. Gao, P. J. Smith, and M. V. Clark, "Theoretical reliability of MMSE linear diversity combining in Rayleigh-fading additive interference channels," *IEEE Trans. Commun.*, 46, no. 5, pp. 666–672, 1998.

[28] S. Haykin, "Adaptive filter theory," in *Information and System Sciences*, Prentice Hall, 1986.

[29] J. So and J. M. Cioffi, "Multiuser diversity in a MIMO system with opportunistic feedback," *IEEE Trans. Veh. Tech.*, 58, no. 9, pp. 4909–4918, 2009.

[30] M. Pugh and B. D. Rao, "Reduced feedback schemes using random beamforming in MIMO broadcast channels," *IEEE Trans. Signal Process.*, 58, no. 3, pp. 1821–1832, March 2010.

[31] W. Dai, B. C. Rider, and Y. E. Liu, "Joint beamforming for multiaccess MIMO systems with finite rate feedback," *IEEE Trans. Wireless Commun.*, 8, no. 5, pp. 2618–2628, 2009.

8 Relay Transmission

Relay transmission has been widely used in satellite communication systems and terrestrial microwave communication systems. The primary objective of introducing relays in these systems was to extend the transmission distance. Relay transmission can also improve the coverage of cellular wireless systems. In particular, it serves as an efficient solution to enhance the quality of service for mobile users that experience severe path loss/shadowing effects. Furthermore, cooperative relay transmission can achieve diversity gains by exploring multiple antennas at different users [1–3].

In this chapter, we present several relay transmission schemes and mathematically characterize their performance. We first introduce basic relaying strategies for single relay systems, including nonregenerative amplify-and-forward and regenerative decode-and-forward systems. Then, we consider relay selection in nonregenerative multiple relay transmission systems. After that, we introduce the notion of cooperative relaying in the regenerative relay transmission setting. Finally, the effect of incremental relaying on regenerative relay transmission is investigated. The objective of this chapter is to establish a solid understanding of essential design options for relay transmission systems.

8.1 Basic Relaying Strategies

The basic structure of a relay transmission system is shown in Fig. 8.1. Specifically, the source node is sending information to the destination node with the help of the relay node. The participation of the relay node is especially valuable when the direct source–destination link is of poor quality. Given the broadcasting nature of most wireless transmission, the relay node can overhear the transmitted signal from the source and forward its received copy to the destination after proper processing. Amplify and forward (AF) and decode and forward (DF) are two of the most popular relaying strategies.

8.1.1 Amplify-and-Forward Relaying

Let us assume that the source node transmits signal $s(t)$ with symbol energy E_s. The received signal at the relay node is given, assuming a flat fading environment, by

$$r_1(t) = h_1 s(t) + n_1(t), \tag{8.1}$$

8.1 Basic Relaying Strategies

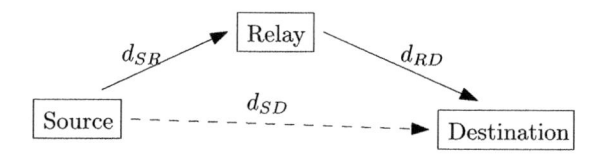

Figure 8.1 Relay transmission system.

where h_1 is the complex channel gain of the source–relay link and $n_1(t)$ is the additive white Gaussian noise (AWGN) at the relay node with power spectrum density $N_0/2$. With AF relaying, also known as nonregenerative relaying, the relay node will amplify the received signal with a gain G and then forward it to the destination. The received signal at the destination node is given by

$$r_2(t) = h_2 G(h_1 s(t) + n_1(t)) + n_2(t) = h_2 G h_1 s(t) + h_2 G n_1(t) + n_2(t), \qquad (8.2)$$

where h_2 is the complex channel gain of the relay–destination link and $n_2(t)$ is the AWGN at the destination node also with power spectrum density $N_0/2$. The received signal-to-noise ratio (SNR) at the destination node can be determined as

$$\gamma^{\text{AF}} = \frac{|h_1|^2 |h_2|^2 G^2 E_s}{|h_2|^2 G^2 N_0 + N_0}. \qquad (8.3)$$

While larger power amplifying gain G leads to larger γ^{AF}, the output power constraint of the relay node limits the value of G. Specifically, the output power of the relay node, given by $G^2 |h_1|^2 E_s + G^2 N_0$, should be no more than a certain maximum value. Let us assume that G is chosen such that the output symbol energy at the relay is also equal to E_s. Then, G can be determined as

$$G^2 = \frac{E_s}{|h_1|^2 E_s + N_0}. \qquad (8.4)$$

The resulting received SNR at the destination becomes

$$\gamma^{\text{AF}} = \frac{|h_1|^2 E_s/N_0 \cdot |h_2|^2 E_s/N_0}{|h_1|^2 E_s/N_0 + |h_2|^2 E_s/N_0 + 1} = \frac{\gamma_1 \gamma_2}{\gamma_1 + \gamma_2 + 1}, \qquad (8.5)$$

where γ_1 and γ_2 are the received SNR of the first and second hops, respectively. The performance of relay transmission with AF relaying can be analyzed with the statistics of γ^{AF}.

Over the high SNR region, where $|h_1|^2 E_s \gg N_0$, the amplifying gain G can be set as $\frac{1}{|h_1|^2}$. The corresponding end-to-end SNR with AF relaying can be approximated by [4]

$$\gamma^{\text{AF}} \approx \frac{\gamma_1 \gamma_2}{\gamma_1 + \gamma_2} = \tilde{\gamma}^{\text{AF}}, \qquad (8.6)$$

which serves as a lower bound of the end-to-end SNR and simplifies the corresponding performance analysis. Specifically, the SNR lower bound is closely related to the harmonic mean of two random variables, defined as

$$\mu_H(X_1, X_2) = \frac{2}{\frac{1}{X_1} + \frac{1}{X_2}} = 2 \frac{X_1 X_2}{X_1 + X_2}. \qquad (8.7)$$

It can be shown that the PDF of $\tilde{\gamma}^{\mathrm{AF}}$ over i.n.d. Rayleigh fading

$$p_{\tilde{\gamma}^{\mathrm{AF}}}(x) = \frac{2\gamma}{\overline{\gamma}_1 \overline{\gamma}_2} \exp\left(-\frac{\gamma}{\overline{\gamma}_1} - \frac{\gamma}{\overline{\gamma}_2}\right) \left[\frac{\overline{\gamma}_1 + \overline{\gamma}_2}{\sqrt{\overline{\gamma}_1 \overline{\gamma}_2}} K_1\left(\frac{2\gamma}{\sqrt{\overline{\gamma}_1 \overline{\gamma}_2}}\right) + 2K_0\left(\frac{2\gamma}{\sqrt{\overline{\gamma}_1 \overline{\gamma}_2}}\right)\right], \tag{8.8}$$

where $K_i(\cdot)$ is the ith-order modified Bessel function of the second kind [6], $\overline{\gamma}_1$ and $\overline{\gamma}_2$ are the average SNR of the first and the second hop, respectively. We can also derive the probability density function (PDF) of $\tilde{\gamma}^{\mathrm{AF}}$ over i.i.d. Nakagami fading as

$$p_{\tilde{\gamma}^{\mathrm{AF}}}(x) = \frac{2\sqrt{\pi}}{\Gamma^2(m)} \frac{m^m \gamma^{m-1}}{\overline{\gamma}^m} \exp\left(-\frac{4\gamma m}{\overline{\gamma}}\right) \Phi(\frac{1}{2} - m, 1 - m, \frac{4\gamma m}{\overline{\gamma}}), \tag{8.9}$$

where $\overline{\gamma}$ is the common average SNR of both hops, m is the common Nakagami fading parameter, and $\Phi(\cdot, \cdot, \cdot)$ is the confluent hypergeometric function [6, Eq. (13.2.5)].

These distribution functions can be readily applied to analyze the performance of AF relaying systems. In particular, the outage probability of AF relaying transmission over i.i.d. Nakagami fading is calculated as [7]

$$P_{\mathrm{out}} = \Pr[\tilde{\gamma}^{\mathrm{AF}} \leq \gamma_{\mathrm{th}}] = \frac{\sqrt{\pi} m \gamma_{\mathrm{th}}}{2^{2m-3} \Gamma^2(m) \overline{\gamma}} G_{23}^{21}\left(4\frac{m\gamma_{\mathrm{th}}}{\overline{\gamma}} \Big| \begin{array}{c} 0, m-\frac{1}{2} \\ m-1, 2m-1, -1 \end{array}\right), \tag{8.10}$$

where $G_{mn}^{pq}(\cdot)$ is the Meijers G-function defined in [6, Eq. (9.301)]. The average error rate analysis can be facilitated with the following MGF of $\tilde{\gamma}^{\mathrm{AF}}$ over i.i.d. Nakagami fading

$$M_{\tilde{\gamma}^{\mathrm{AF}}}(s) = {}_2F_1\left(m, 2m; m+\frac{1}{2}; \frac{\overline{\gamma}}{4m}s\right), \tag{8.11}$$

where ${}_2F_1(\cdot, \cdot; \cdot; \cdot)$ is the Gauss hypergeometric function defined in [6, Eq. (15.1.1)]. For example, the average bit error rate (BER) of binary differential phase-shift keying (DPSK) is given by $P_e^{\mathrm{AF}} = M_{\tilde{\gamma}^{\mathrm{AF}}}(-1)/2$.

8.1.2 Decode-and-Forward Relaying

With DF relaying, or equivalently regenerative relaying, the relay node first demodulates and decodes the source signal. Then, the relay will forward a re-encoded copy of the source signal to the destination. The relay transmission link can be viewed as a cascade connection of two independent wireless links. As such, the performance of DF relay transmission is limited by the weaker hop. The quality indicator of the DF relay link becomes the minimum of the two hop SNRs, i.e., $\min\{\gamma_1, \gamma_2\}$.

To further illustrate, let us consider the outage probability of a relaying transmission system with DF. Outage of such system occurs when either the source–relay hop or the relay–destination hop has very low received SNR, i.e., below the outage threshold γ_{th}. As such, the outage probability is calculated as

$$P_{\mathrm{out}} = 1 - \Pr[\text{no hop in outage}] = 1 - \Pr[\gamma_1 > \gamma_{\mathrm{th}}, \gamma_2 > \gamma_{\mathrm{th}}], \tag{8.12}$$

which can be shown to be equal to

$$P_{\text{out}} = \Pr[\min\{\gamma_1, \gamma_2\} < \gamma_{\text{th}}]. \tag{8.13}$$

For the Rayleigh fading environment, the PDF of $\min\{\gamma_1, \gamma_2\}$, given by

$$p_{\gamma_{\text{eq}}^{\text{DF}}}(\gamma) = \frac{1}{\overline{\gamma}_{\min}} e^{-\gamma/\overline{\gamma}_{\min}}, \tag{8.14}$$

where $\overline{\gamma}_{\min} = \frac{\overline{\gamma}_1 \overline{\gamma}_2}{\overline{\gamma}_1 + \overline{\gamma}_2}$, can be readily applied to evaluate the outage probability. For i.i.d. Nakagami fading hops, the outage probability can be evaluated as

$$P_{\text{out}} = 1 - \left(\frac{\Gamma(m, \frac{m\gamma_{\text{th}}}{\overline{\gamma}})}{\Gamma(m)} \right)^2. \tag{8.15}$$

Compared to AF relaying, DF relaying incurs higher complexity at the relay for detection and encoding/modulation processes. Such additional operations avoid the forwarding of the noise signal with AF relaying. Meanwhile, the relay node with DF relaying may forward undetected errors to the destination and cause error propagation, especially when the source–relay link is of poor quality. The overall probability of error for given received SNRs over both hops is given by

$$P_e^{\text{DF}}(\gamma_1, \gamma_2) = P_e(\gamma_1) + P_e(\gamma_2) - 2P_e(\gamma_1)P_e(\gamma_2), \tag{8.16}$$

where $P_e(\gamma_i)$ denotes the instantaneous BER of the ith hop, $i = 1, 2$. After averaging over the distribution functions of received SNRs, the average BER of DF relaying transmission over i.i.d. fading hops is given by

$$P_e^{\text{DF}} = 2P_e - 2P_e^2, \tag{8.17}$$

where P_e is the common average BER of a single hop, specializing to $P_e = (m/(m + \overline{\gamma}))^m/2$ for DPSK over Nakagami fading.

8.1.3 Numerical Results

Fig. 8.2 compares outage probability of regenerative (DF) and nonregenerative (AF) relaying systems over Nakagami fading channels. It is clear that regeneration improves the performance at low average SNR at the cost of an increased complexity. At high average SNR, the two systems are equivalent from the outage probability perspective. Note also that the difference between these two relaying schemes grows as the Nakagami fading parameter m increases.

Fig. 8.3 compares the average BER performance of DPSK with regenerative relaying and nonregenerative relaying systems. Similar to the outage probability results, it is clear from the figure that the regenerative scheme outperforms the non-regenerative scheme at low average SNR and the two schemes become equivalent at high average SNR. The difference between the two relaying schemes over low average SNR grows as m increases. Note that this is in contrast to the AWGN scenario, in which there is always (over the whole SNR range) a 3 dB performance gain of regenerative systems over nonregenerative systems when the hops are balanced.

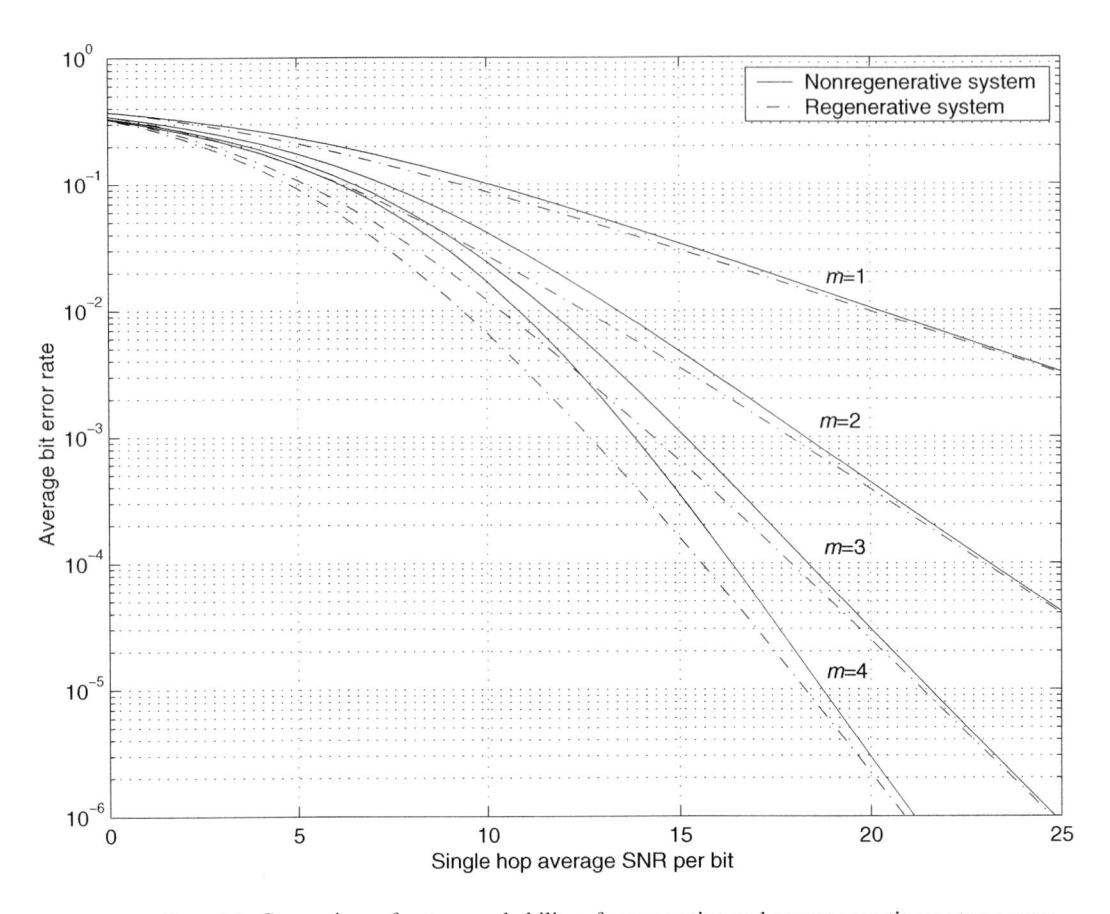

Figure 8.2 Comparison of outage probability of regenerative and nonregenerative systems over Nakagami fading channels. Reprint with permission from [7]. ©2004 IEEE.

8.2 Opportunistic Nonregenerative Relaying

In many wireless transmission scenarios, multiple relay nodes may be available to help the source–destination communication. These relay nodes could be idle users around the source and destination nodes. The antennas at these idle users can form a virtual antenna array to further enhance the performance of relay transmission. Various strategies can be applied to explore the diversity benefit of such an antenna array, including distributed beamforming and cooperative space-time transmission. While achieving the maximum diversity benefit, these schemes incur high implementation complexity for relay node coordination, synchronization, and interference mitigation. Opportunistic relaying (OR), where the best relay from a set of available relays is selected to cooperate [8–10], serves as an effective low-complexity solution.

In this section, we present and study an OR system with nonregenerative AF strategy [11]. With AF relaying, the best relay selection strategy is to select the relay that leads to the largest end-to-end SNR. Alternatively, the relay selection can be performed

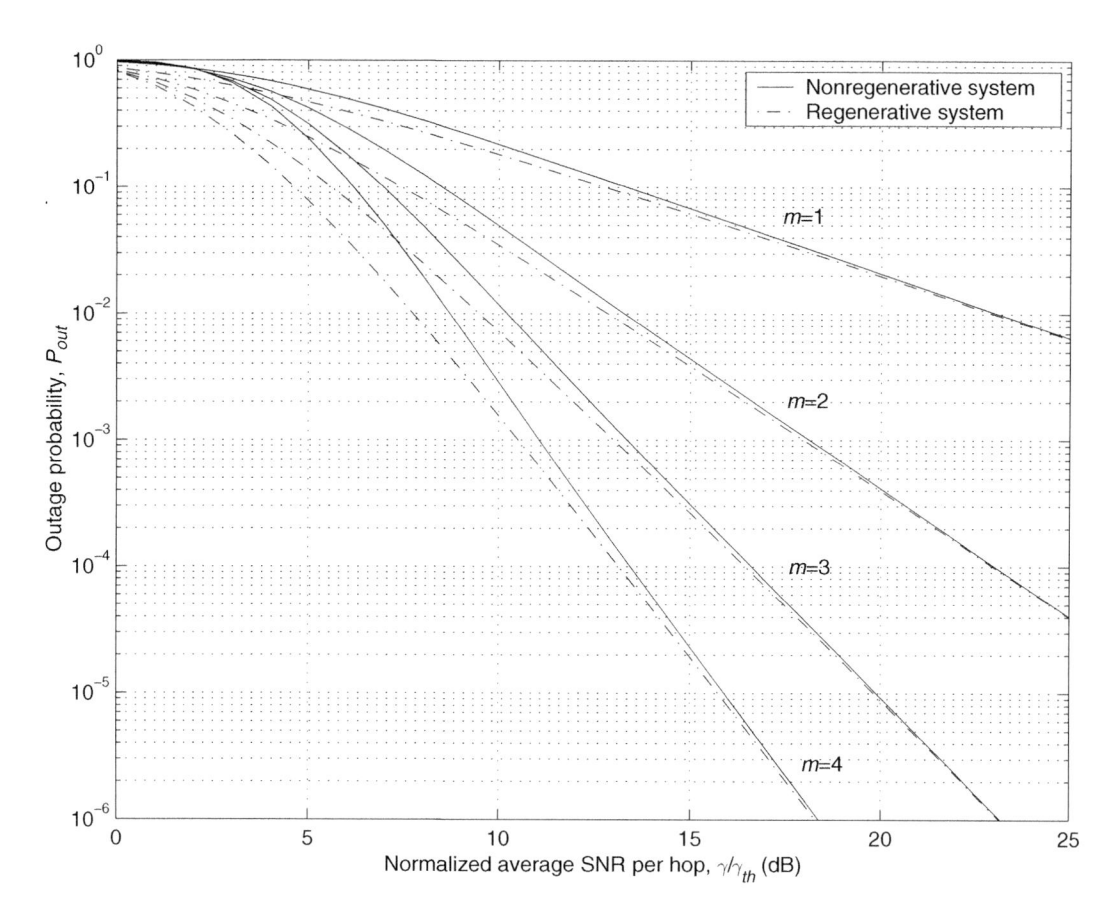

Figure 8.3 Average BER comparison between regenerative and nonregenerative systems. Reprint with permission from [7]. ©2004 IEEE.

using source–relay and/or relay–destination link qualities. Here, we present and study a max–min selection criterion for AF relaying strategy and compare its performance and complexity with the best relay selection strategy.

8.2.1 System Model

We assume that the source cannot communicate directly with the destination and must rely on the help of multiple relays, which are close to each other and form a cluster. We note that only a selected relay from the cluster will cooperate. The source, the destination, and the relays are denoted by s, d, and r_k, where $k \in \{1, \ldots, K\}$, respectively, each equipped with one antenna. We denote the flat fading channel gain between the source and the kth relay and the kth relay and the destination by h_{sr_k} and $h_{r_k d}$, respectively.

We consider half-duplex relays and as such the transmission process involves two phases. During the first phase, the source broadcasts its signal $s(t)$. The received signal by relay r_k is given by

$$y_{r_k}(t) = \sqrt{E_s} h_{sr_k} s(t) + n_{r_k}(t), \tag{8.18}$$

where $n_{r_k}(t)$ is the additive noise at the kth relay with noise spectrum density N_0. We denote the instantaneous received SNR at the kth relay by $\gamma_{sr_k} = E_s |h_{sr_k}|^2 / N_0$, which follows an exponential distribution over Rayleigh fading. We assume that relays are clustered together and as such different relays experience the same average received SNR, denoted by $\bar{\gamma}_{sr}$.

During the second phase, one selected relay, say r_k, will forward its received signal to the destination using the AF protocol. The received signal at the destination node is given by

$$y_d(t) = G h_{r_k d} y_{r_k}(t) + n_d(t), \tag{8.19}$$

where $n_d(t)$ is the additive-noise symbols at the destination, with the same noise density N_0, and G is the gain factor at the relay. To maintain the same transmitted symbol energy E_s, G should be set to

$$G = \sqrt{\frac{E_s}{E_s |h_{sr_k}|^2 + N_0}}. \tag{8.20}$$

The resulting end-to-end SNR through relay r_k is given by

$$\gamma^{AF} = \frac{\gamma_{sr_k} \gamma_{r_k d}}{\gamma_{sr_k} + \gamma_{r_k d} + 1}, \tag{8.21}$$

where $\gamma_{r_k d} = E_s |h_{r_k d}|^2 / N_0$ is the instantaneous received SNR over the link from relay r_k to the destination. We again assume that $\gamma_{r_k d}$ are i.i.d. random variables with the same average $\bar{\gamma}_{rd}$ as the relays are clustered together.

8.2.2 Relay Selection Strategies

The best relay selection strategy for OR system with AF relaying is to select the relay that leads to the largest end-to-end SNR, termed centralized OR here. The resulting end-to-end SNR is given by

$$\gamma_{C-OR} = \max_k \left(\frac{\gamma_{sr_k} \gamma_{r_k d}}{\gamma_{sr_k} + \gamma_{r_k d} + 1} \right). \tag{8.22}$$

The implementation of centralized OR, however, entails the centralized monitoring of end-to-end SNR corresponding to each relay, typically at the destination. The destination node also needs to inform the relays of the selection result. The exact error rate analysis of a centralized OR scheme is challenging. Most analytical results are approximate, with the application of the upper bound of end-to-end SNR, given by

$$\frac{\gamma_{sr_k} \gamma_{r_k d}}{\gamma_{sr_k} + \gamma_{r_k d} + 1} \leq \gamma_{up,k} = \min \left(\gamma_{sr_k}, \gamma_{r_k d} \right), \tag{8.23}$$

where $\gamma_{up,k}$ follows an exponential distribution for the Rayleigh fading environment with cumulative distribution function (CDF) given by

$$F_{\gamma_{up,k}}(z) = 1 - e^{-z/\bar{\gamma}}, \tag{8.24}$$

where $\bar{\gamma} = \frac{\bar{\gamma}_{sr} \bar{\gamma}_{rd}}{\bar{\gamma}_{sr} + \bar{\gamma}_{rd}}$.

8.2 Opportunistic Nonregenerative Relaying

The relay selection for an OR system with AF relaying can also be performed in a distributed manner, termed as distributed OR here. Note that every relay can obtain the (local) channel state information (CSI) from the request-to-send (RTS) and clear-to-send (CTS) signals coming from the source and the destination, respectively. By setting a timer proportional to $1/\min(\gamma_{sr_k}, \gamma_{r_k d})$, the relay with the smallest $\min(\gamma_{sr_k}, \gamma_{r_k d})$ can start relaying first, which suppresses other relays. The resulting distributed OR scheme effectively selects the relay leading to the largest minimum SNR over two hopes, i.e.,

$$r_* = \arg\max_k \min(\gamma_{sr_k}, \gamma_{r_k d}). \quad (8.25)$$

Note that the above selection criterion is often used for opportunistic DF relay networks. It follows that the end-to-end SNR of the OR system can be expressed as

$$\gamma_{D-OR} = \frac{\gamma_{sr_*} \gamma_{r_* d}}{\gamma_{sr_*} + \gamma_{r_* d} + 1}. \quad (8.26)$$

8.2.3 Outage Analysis

In this subsection, we study the outage performance of opportunistic relaying transmission with AF by deriving the CDF of the end-to-end SNR.

Centralized OR Scheme

With the centralized OR scheme, the CDF of the end-to-end SNR γ_{C-OR} is given by

$$F_{\gamma_{C-OR}}(z) \triangleq \Pr[\gamma_{C-OR} \leq z]$$
$$= \prod_{k=1}^{K} \Pr\left[\frac{\gamma_{sr_k} \gamma_{r_k d}}{\gamma_{sr_k} + \gamma_{r_k d} + 1} \leq z\right], \quad (8.27)$$

where we applied the assumption of independent fading across relays. Conditioning on $\gamma_{r_k d}$, $F_{\gamma_{C-OR}}(z)$ can be calculated as

$$F_{\gamma_{C-OR}}(z) = \prod_{k=1}^{K} \left(F_{\gamma_{r_k d}}(z) + \int_z^\infty \Pr\left[\gamma_{sr_k} \leq \frac{z(y+1)}{y-z} \Big| y\right] p_{\gamma_{r_k d}}(y) dy\right), \quad (8.28)$$

where $F_{\gamma_{r_k d}}(.)$ and $p_{\gamma_{r_k d}}(.)$ are the CDF and PDF of $\gamma_{r_k d}$, respectively. Using the identity [6, Eq. 3.471.9] after carrying out the integration in (8.28) and some manipulations, $F_{\gamma_{C-OR}}(.)$ for the i.i.d. Rayleigh fading environment can be determined as

$$F_{\gamma_{C-OR}}(z) = \left(1 - 2e^{-z/\bar{\gamma}} \sqrt{\frac{z(z+1)}{\bar{\gamma}_{sr} \bar{\gamma}_{rd}}} K_1\left(2\sqrt{\frac{z(z+1)}{\bar{\gamma}_{sr} \bar{\gamma}_{rd}}}\right)\right)^K, \quad (8.29)$$

where $K_1(\cdot)$ is the first-order modified Bessel function of the second kind. The outage probability of the OR system with the centralized OR can then be evaluated.

Over a high SNR regime, where $\bar{\gamma}_{sr}$ and $\bar{\gamma}_{rd}$ are very large, we can derive an approximate expression of $F_{\gamma_{C-OR}}(\cdot)$. Noting that $K_1(z) \approx 1/z$ when $z \to 0$ [6, Eq. 9.6.9], the CDF $F_{\gamma_{C-OR}}(\cdot)$ can be approximated in the high-SNR regime as

$$F_{\gamma_{C-OR}}(z) \approx (1 - e^{-z/\bar{\gamma}})^K. \tag{8.30}$$

We can see that Eq. (8.30) is also the CDF of the largest amount K upper bound $\gamma_{up,k}$.

Distributed OR Scheme

The CDF of γ_{D-OR} can be calculated as

$$\begin{aligned}F_{\gamma_{D-OR}}(z) &\triangleq \Pr\left[\frac{\gamma_{sr_*}\gamma_{r_*d}}{\gamma_{sr_*} + \gamma_{r_*d} + 1} < z\right] \\ &= F_{\gamma_{r_*d}}(z) + \int_z^\infty F_{\gamma_{sr_*}}\left(\frac{z(y+1)}{y-z}\right) p_{\gamma_{r_*d}}(y)dy,\end{aligned} \tag{8.31}$$

where $p_{\gamma_{r_*d}}(.)$ is the PDF of γ_{r_*d}, $F_{\gamma_{sr_*}}(.)$ and $F_{\gamma_{r_*d}}(.)$ are the CDFs of γ_{sr_*} and γ_{r_*d}, respectively. Note that although γ_{sr_k} and $\gamma_{r_k d}$ are independent for all $k \in \{1, \ldots, K\}$, γ_{sr_*} and γ_{r_*d} are not independent since r_* is the selected relay that enjoys the largest minimum hop SNRs.

We first derive the PDF $p_{\gamma_{r_*d}}(.)$ by rewriting it as

$$p_{\gamma_{r_*d}}(x) = \int_0^\infty p_{\gamma_{r_id}|Z_i=z}(x) \cdot p_{\max(Z_i)}(z)dz, \tag{8.32}$$

where $Z_i = \min(\gamma_{sr_i}, \gamma_{r_id})$. Applying the Bayesian rule, the conditional PDF can be expressed as

$$p_{\gamma_{r_id}|Z_i=z}(x) = \frac{p_{\gamma_{r_id},Z_i}(x,z)}{p_{Z_i}(z)}. \tag{8.33}$$

The PDF of Z_i can be determined from its CDF, given in (8.24), as

$$p_{Z_i}(z) = \frac{1}{\bar{\gamma}}e^{-z/\bar{\gamma}}, \tag{8.34}$$

where $\bar{\gamma} = \frac{\bar{\gamma}_{sr}\bar{\gamma}_{rd}}{\bar{\gamma}_{sr}+\bar{\gamma}_{rd}}$. The joint CDF $F_{\gamma_{r_id},Z_i}(x,z)$ can be expressed as

$$F_{\gamma_{r_id},Z_i}(x,z) = \Pr[\gamma_{r_id} < x, \min(\gamma_{sr_i}, \gamma_{r_id}) < z] \tag{8.35}$$

$$= \begin{cases} F_{\gamma_{r_id}}(x) - \left(F_{\gamma_{r_id}}(x) - F_{\gamma_{r_id}}(z)\right)\left(1 - F_{\gamma_{sr_i}}(z)\right), & x \geq z \\ F_{\gamma_{r_id}}(x), & x < z \end{cases}$$

We note that the joint CDF $F_{\gamma_{r_id},Z_i}(x,z)$ is not continuous along the x-axis. After taking the derivative, the joint PDF $p_{\gamma_{r_id},Z_i}(x,z)$ involves an impulse at the position $x = z$, given by

$$p_{\gamma_{r_id}}(z)\left(1 - F_{\gamma_{sr_i}}(z)\right)\delta(x-z). \tag{8.36}$$

Therefore, the PDF $p_{\gamma_{r_*d}}(.)$ can be rewritten as

$$\begin{aligned}p_{\gamma_{r_*d}}(x) &= \int_0^x \frac{p_{\gamma_{r_id}}(x)p_{\gamma_{sr_i}}(z)}{p_{Z_i}(z)} p_{\max(Z_i)}(z)dz \\ &\quad + \frac{p_{\gamma_{r_id}}(x)\left(1 - F_{\gamma_{sr_i}}(x)\right)}{p_{Z_i}(x)} p_{\max(Z_i)}(x),\end{aligned} \tag{8.37}$$

where the sifting property of the impulse function is applied.

8.2 Opportunistic Nonregenerative Relaying

It can be shown that the CDF and PDF of max(Z_i) can be determined as

$$F_{\max(Z_i)}(z) = \left(1 - e^{-z/\bar{\gamma}}\right)^K, \tag{8.38}$$

and

$$p_{\max(Z_i)}(z) = \sum_{i=1}^{K} \binom{K}{i} (-1)^{i-1} \frac{i}{\bar{\gamma}} e^{-iz/\bar{\gamma}}, \tag{8.39}$$

respectively. Substituting (8.34) and (8.39) in (8.32) and performing integration, we obtain the PDF of γ_{r_*d} as

$$p_{\gamma_{r_*d}}(x) = \sum_{i=1}^{K} \binom{K}{i} \frac{(-1)^{i-1}}{\bar{\gamma}_{sr}} \frac{i\bar{\gamma}}{i\bar{\gamma}_{rd} - \bar{\gamma}} \left(e^{-x/\bar{\gamma}_{rd}} - e^{-ix/\bar{\gamma}}\right) \tag{8.40}$$

$$+ \sum_{i=1}^{K} \binom{K}{i} \frac{(-1)^{i-1}}{\bar{\gamma}_{rd}} i\, e^{-ix/\bar{\gamma}}.$$

It follows that the CDF of γ_{r_*d} and γ_{sr_*} can be obtained as

$$F_{\gamma_{r_*d}}(x) = \sum_{i=1}^{K} \binom{K}{i} \frac{(-1)^{i-1}}{\bar{\gamma}_{sr}} \frac{i\bar{\gamma}}{i\bar{\gamma}_{rd} - \bar{\gamma}} \left[\Psi_{\frac{1}{\bar{\gamma}_{rd}}}(x) - \Psi_{\frac{i}{\bar{\gamma}}}(x)\right] \tag{8.41}$$

$$+ \sum_{i=1}^{K} \binom{K}{i} \frac{(-1)^{i-1}i}{\bar{\gamma}_{rd}} \Psi_{\frac{i}{\bar{\gamma}}}(x),$$

and

$$F_{\gamma_{sr_*}}(x) = \sum_{i=1}^{K} \binom{K}{i} \frac{(-1)^{i-1}}{\bar{\gamma}_{rd}} \frac{i\bar{\gamma}}{i\bar{\gamma}_{sr} - \bar{\gamma}} \left[\Psi_{\frac{1}{\bar{\gamma}_{sr}}}(x) - \Psi_{\frac{i}{\bar{\gamma}}}(x)\right] \tag{8.42}$$

$$+ \sum_{i=1}^{K} \binom{K}{i} \frac{(-1)^{i-1}i}{\bar{\gamma}_{sr}} \Psi_{\frac{i}{\bar{\gamma}}}(x),$$

respectively, where $\Psi_a(.)$ is given by

$$\Psi_a(x) = \frac{1}{a}\left(1 - e^{-ax}\right). \tag{8.43}$$

Hence, the second term in (8.31) can be expressed as

$$\int_z^\infty F_{\gamma_{sr_*}}\left(\frac{z(y+1)}{y-z}\right) p_{\gamma_{r_*d}}(y)\, dy = S_1 + S_2 + S_3 + S_4, \tag{8.44}$$

where S_1, S_2, S_3, and S_4 are given by

$$S_1 = \sum_{i=1}^{K} \binom{K}{i} \left(\frac{(-1)^{i-1}}{\bar{\gamma}_{rd}} \frac{i\bar{\gamma}}{i\bar{\gamma}_{sr} - \bar{\gamma}}\right) \sum_{j=1}^{K} \binom{K}{j} \left(\frac{(-1)^{j-1}}{\bar{\gamma}_{sr}} \frac{j\bar{\gamma}}{j\bar{\gamma}_{rd} - \bar{\gamma}}\right)$$

$$\times \left[\Lambda_{\frac{1}{\bar{\gamma}_{sr}},\frac{1}{\bar{\gamma}_{rd}}}(z) - \Lambda_{\frac{1}{\bar{\gamma}_{sr}},\frac{j}{\bar{\gamma}}}(z) - \Lambda_{\frac{i}{\bar{\gamma}},\frac{1}{\bar{\gamma}_{rd}}}(z) + \Lambda_{\frac{i}{\bar{\gamma}},\frac{j}{\bar{\gamma}}}(z)\right],$$

Relay Transmission

$$S_2 = \sum_{i=1}^{K} \binom{K}{i} \left(\frac{(-1)^{i-1}}{\bar{\gamma}_{rd}} \frac{i\bar{\gamma}}{i\bar{\gamma}_{sr} - \bar{\gamma}} \right) \sum_{j=1}^{K} \binom{K}{j} \frac{(-1)^{j-1}j}{\bar{\gamma}_{rd}} \left[\Lambda_{\frac{1}{\bar{\gamma}_{sr}},\frac{j}{\bar{\gamma}}}(z) - \Lambda_{\frac{i}{\bar{\gamma}},\frac{j}{\bar{\gamma}}}(z) \right],$$

$$S_3 = \sum_{i=1}^{K} \binom{K}{i} \frac{(-1)^{i-1}i}{\bar{\gamma}_{sr}} \left(\frac{(-1)^{j-1}}{\bar{\gamma}_{sr}} \frac{j\bar{\gamma}}{j\bar{\gamma}_{rd} - \bar{\gamma}} \right) \sum_{j=1}^{K} \binom{K}{j} \left[\Lambda_{\frac{i}{\bar{\gamma}},\frac{1}{\bar{\gamma}_{rd}}}(z) - \Lambda_{\frac{i}{\bar{\gamma}},\frac{j}{\bar{\gamma}}}(z) \right],$$

$$S_4 = \sum_{i=1}^{K} \binom{K}{i} \left(\frac{(-1)^{i-1}}{\bar{\gamma}_{sr}} i \right) \sum_{j=1}^{K} \binom{K}{j} \left(\frac{(-1)^{j-1}}{\bar{\gamma}_{rd}} j \right) \Lambda_{\frac{i}{\bar{\gamma}},\frac{j}{\bar{\gamma}}}(z). \tag{8.45}$$

Here, the function $\Lambda_{a,b}(.)$ is defined as

$$\Lambda_{a,b}(z) \triangleq \int_{z}^{\infty} \Psi_a \left(\frac{z(y+1)}{y-z} \right) e^{-by} dy, \tag{8.46}$$

and calculated using [6, Eq. 3.471.9] as

$$\Lambda_{a,b}(z) = \frac{1}{ab} e^{-bz} - 2\sqrt{\frac{z(z+1)}{ab}} e^{-(a+b)z} K_1 \left(2\sqrt{abz(z+1)} \right). \tag{8.47}$$

In the case of high SNR, the end-to-end SNR with distributed OR can be approximated by

$$\gamma_{D-OR} = \frac{\gamma_{sr_*} \gamma_{r_*d}}{\gamma_{sr_*} + \gamma_{r_*d}}. \tag{8.48}$$

This end-to-end SNR expression also applies when the noise variance at the selected relay r_* is unavailable and the relay sets the amplifying gain G to $|h_{sr_*}|^{-1}$. Following similar reasoning, we can derive the CDF of γ_{D-OR} as

$$F_{\gamma_{D-OR}}(z) \triangleq \Pr \left[\frac{\gamma_{sr_*} \gamma_{r_*d}}{\gamma_{sr_*} + \gamma_{r_*d}} < z \right] = F_{\gamma_{r_*d}}(z) + \int_{z}^{\infty} F_{\gamma_{sr_*}} \left(\frac{zy}{y-z} \right) p_{\gamma_{r_*d}}(y) dy, \tag{8.49}$$

where the second term can be evaluated as in (8.45) by replacing the function $\Lambda_{a,b}(.)$ by the function $\Xi_{a,b}(.)$, given by

$$\Xi_{a,b}(z) = \frac{1}{ab} e^{-bz} - 2\frac{z}{\sqrt{ab}} e^{-(a+b)z} K_1 \left(2z\sqrt{ab} \right). \tag{8.50}$$

The outage probability for the distributed OR system with AF relaying can be evaluated with these statistical results.

8.2.4 Average Error Rate Analysis

In this subsection, we analyze the average error rate of the OR systems using the CDF of end-to-end SNR derived in the previous subsection. We can invoke a general result on symbol error rate analysis using the CDF of the SNR as follows:

$$P_{s,sys} = -\int_{0}^{\infty} \frac{d}{d\gamma} P_e(x) F_{\gamma}(x) dx, \tag{8.51}$$

8.2 Opportunistic Nonregenerative Relaying

where $P_e(.)$ is the conditional error probability of the modulation schemes for given SNR value and $F_\gamma(.)$ is the CDF of the end-to-end SNR. The conditional error probability of several modulation schemes is commonly given by

$$P_e(x) = \alpha Q\left(\sqrt{\beta x}\right), \tag{8.52}$$

where α and β are the modulation-specific constants, and $Q(.)$ is the Gaussian function. For instance, $(\alpha, \beta) = (1, 2)$ for binary phase-shift keying (BPSK), $(\alpha, \beta) = (1, 1)$ for coherent binary frequency-shift keying (BFSK), and $(\alpha, \beta) = (2(M-1)/M, 6\log_2(M)/(M^2-1))$ for M-ary pulse amplitude modulation (PAM).

Centralized OR Scheme
We first analyze the error rate performance of the centralized OR scheme. Substituting $F_{\gamma_{C-OR}}(.)$ into (8.51), we get

$$P_{s,CSI} = \frac{\alpha}{2}\sqrt{\frac{\beta}{2\pi}} \int_0^\infty \frac{e^{-\frac{\beta}{2}z}}{\sqrt{z}} F_{\gamma_{C-OR}}(z)dz. \tag{8.53}$$

Unfortunately, using the exact expression of $F_{\gamma_{C-OR}}(.)$ given in (8.29) will not lead to a convenient closed-form result for (8.53). However, we can obtain an approximate result by applying the approximate expression of $F_{\gamma_{OAF}}(.)$ given by (8.30). Specifically, after applying the binomial expansion, we arrive at the following expression for the average error rate:

$$P_{s,CSI}^a = \frac{\alpha}{2} \sum_{i=0}^{K} \binom{K}{i}(-1)^i \sqrt{\frac{\beta\bar{\gamma}}{2i+\beta\bar{\gamma}}}. \tag{8.54}$$

Distributed OR Scheme
Likewise, using the exact expression of $F_{\gamma_{D-OR}}(.)$, which leads to

$$P_s = \frac{\alpha}{2}\sqrt{\frac{\beta}{2\pi}} \int_0^\infty \frac{e^{-\frac{\beta}{2}z}}{\sqrt{z}} F_{\gamma_{D-OR}}(z)dz, \tag{8.55}$$

may not lead to a closed-form expression for the average error rate of the distributed OR scheme. As such, we adopt an alternative approach by applying the CDF of γ_{D-OR} for the high SNR regime given in (8.49). This will allow us to derive a very accurate approximate closed-form expression as shown below.

Let us define $\Theta_{a,b}^{\alpha,\beta}$ as

$$\Theta_{a,b}^{\alpha,\beta} = \frac{\alpha}{2}\sqrt{\frac{\beta}{2\pi}} \int_0^\infty \frac{e^{-\frac{\beta}{2}z}}{\sqrt{z}} \Xi_{a,b}(z)dz. \tag{8.56}$$

Using [12, Eq. 2.16.6], we can calculate (8.56) in closed-form expression as

$$\Theta_{a,b}^{\alpha,\beta} = \frac{\alpha}{2ab}\sqrt{\frac{\beta}{2b+\beta}} - \frac{\alpha\sqrt{\beta}\Gamma(1/2)\Gamma(5/2)}{8ab\sqrt{a+b+\beta/2}}\,_2F_1\left(\frac{1}{4},\frac{3}{4};2;1-\frac{4ab}{(a+b+\beta/2)^2}\right), \tag{8.57}$$

where $_2F_1(.,.;.;.)$ is the hypergeometric function, which is given by [6, Eq. 9.14.2]. Similarly, we define and calculate $\Theta_{a,.}^{\alpha,\beta}$ as

$$\Theta_{a,.}^{\alpha,\beta} = \frac{\alpha}{2}\sqrt{\frac{\beta}{2\pi}}\int_0^\infty \frac{e^{-\frac{\beta}{2}z}}{\sqrt{z}}\Psi_a(z)dz = \frac{\alpha}{2a}\left(1 - \sqrt{\frac{\beta}{2a+\beta}}\right). \qquad (8.58)$$

Finally, applying (8.57) and (8.58), we determine an accurate approximation for the end-to-end SER of the relay transmission system with distributed OR scheme, given by

$$P_s^a = \sum_{i=0}^{4} s_i, \qquad (8.59)$$

where s_i are detailed in the following as

$$s_0 = \sum_{j=1}^{K}\binom{K}{j}\frac{(-1)^{j-1}}{\bar{\gamma}_{sr}}\frac{j\bar{\gamma}}{j\bar{\gamma}_{rd}-\bar{\gamma}}\left[\Theta_{\frac{1}{\bar{\gamma}_{rd}},.}^{\alpha,\beta} - \Theta_{\frac{j}{\bar{\gamma}},.}^{\alpha,\beta}\right] + \sum_{j=1}^{K}\binom{K}{j}\frac{(-1)^{j-1}j}{\bar{\gamma}_{rd}}\Theta_{\frac{j}{\bar{\gamma}},.}^{\alpha,\beta},$$

$$s_1 = \sum_{i=1}^{K}\binom{K}{i}\left(\frac{(-1)^{i-1}}{\bar{\gamma}_{rd}}\frac{i\bar{\gamma}}{i\bar{\gamma}_{sr}-\bar{\gamma}}\right)\sum_{j=1}^{K}\binom{K}{j}\left(\frac{(-1)^{j-1}}{\bar{\gamma}_{sr}}\frac{j\bar{\gamma}}{j\bar{\gamma}_{rd}-\bar{\gamma}}\right)$$
$$\times \left[\Theta_{\frac{1}{\bar{\gamma}_{sr}},\frac{1}{\bar{\gamma}_{rd}}}^{\alpha,\beta}(z) - \Theta_{\frac{1}{\bar{\gamma}_{sr}},\frac{j}{\bar{\gamma}}}^{\alpha,\beta}(z) - \Theta_{\frac{i}{\bar{\gamma}},\frac{1}{\bar{\gamma}_{rd}}}^{\alpha,\beta}(z) + \Theta_{\frac{i}{\bar{\gamma}},\frac{j}{\bar{\gamma}}}^{\alpha,\beta}(z)\right],$$

$$s_2 = \sum_{i=1}^{K}\binom{K}{i}\left(\frac{(-1)^{i-1}}{\bar{\gamma}_{rd}}\frac{i\bar{\gamma}}{i\bar{\gamma}_{sr}-\bar{\gamma}}\right)\sum_{j=1}^{K}\binom{K}{j}\frac{(-1)^{j-1}j}{\bar{\gamma}_{rd}}\left[\Theta_{\frac{1}{\bar{\gamma}_{sr}},\frac{j}{\bar{\gamma}}}^{\alpha,\beta}(z) - \Theta_{\frac{i}{\bar{\gamma}},\frac{j}{\bar{\gamma}}}^{\alpha,\beta}(z)\right],$$

$$s_3 = \sum_{i=1}^{K}\binom{K}{i}\frac{(-1)^{i-1}i}{\bar{\gamma}_{sr}}\sum_{j=1}^{K}\binom{K}{j}\left(\frac{(-1)^{j-1}}{\bar{\gamma}_{sr}}\frac{j\bar{\gamma}}{j\bar{\gamma}_{rd}-\bar{\gamma}}\right)\left[\Theta_{\frac{i}{\bar{\gamma}},\frac{1}{\bar{\gamma}_{rd}}}^{\alpha,\beta}(z) - \Theta_{\frac{i}{\bar{\gamma}},\frac{j}{\bar{\gamma}}}^{\alpha,\beta}(z)\right],$$

$$s_4 = \sum_{i=1}^{K}\sum_{j=1}^{K}\binom{K}{i}\binom{K}{j}\frac{(-1)^{i-1}i}{\bar{\gamma}_{sr}}\frac{(-1)^{j-1}j}{\bar{\gamma}_{rd}}\Theta_{\frac{i}{\bar{\gamma}},\frac{j}{\bar{\gamma}}}^{\alpha,\beta}(z). \qquad (8.60)$$

8.2.5 Numerical Results

In this subsection, we investigate the performance of an opportunistic relaying system through numerical examples. Both centralized OR and distributed OR schemes are studied in terms of the outage probability and average error rate metrics. To examine the effect of the location of the relay cluster, we assume a normalized linear network, where the source is located at (0, 0), the destination at (1, 0), and the relays at $(d, 0)$. We also assume a log-distance path loss model and ignore the shadowing effect. In particular, the mean channel power gains are related to relay cluster position d as

$$\sigma_{sd}^2 = 1,\ \sigma_{sr_k}^2 = d^{-\nu},\ \sigma_{r_k d}^2 = (1-d)^{-\nu}, \qquad (8.61)$$

where ν is the path loss exponent.

8.2 Opportunistic Nonregenerative Relaying

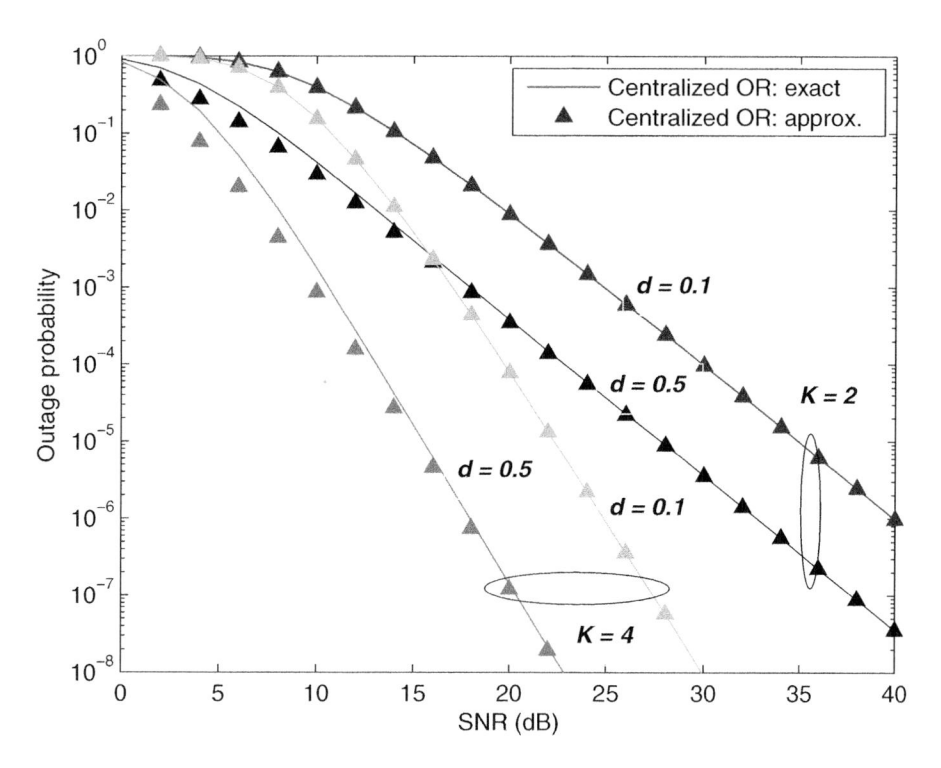

Figure 8.4 End-to-end outage probability of the opportunistic AF relaying system with centralized OR scheme. Reprint with permission from [11]. ©2013 Wiley.

Fig. 8.4 depicts the end-to-end outage performance of a centralized OR scheme in a linear network. We compare the analytical and simulation results where a cluster of K relays are located at different distances d from the source. We can see that the approximation given by (8.30) is tight over the high-SNR regime. We can also see that the outage performance improves with the increasing number of candidate relays. The outage performance is considerably better when the relay cluster is halfway between source and destination nodes than when the relay cluster is closer to the source. Fig. 8.5 presents the end-to-end outage performance of a distributed OR scheme over a linear network. We observe a similar effect of the number of relays and the relay cluster location on the outage performance of the distributed OR scheme. We can also observe that the approximate expression given in (8.49) matches with the exact and simulation results very well. These observations motivate us to perform average error rate analysis using the approximate expression of $F_{\gamma_{D-OR}}(.)$, which leads to convenient closed-form expressions. As an additional numerical example, we compare the average error rate performance of centralized OR and distributed OR schemes in Fig. 8.6. We can see that distributed OR can achieve nearly the same performance as the centralized OR scheme. The relay cluster location and the number of candidate relays have the same effect on the average error rate as on the outage probability. Note also that the derived SER expression in (8.59) offers a tight bound over all the SNR value range.

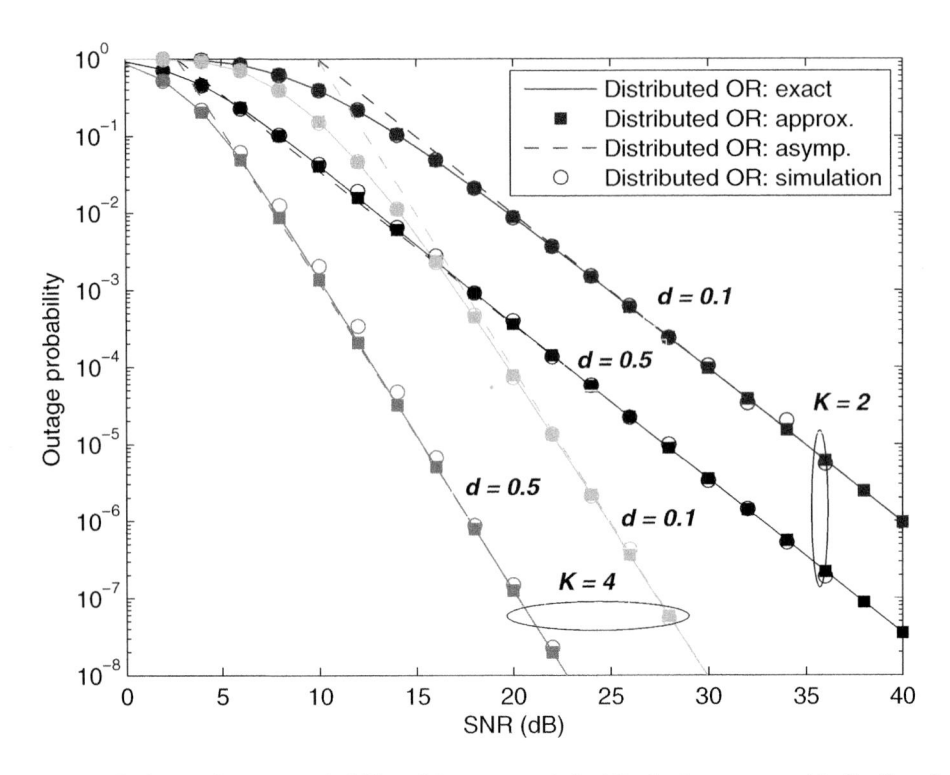

Figure 8.5 End-to-end outage probability of the opportunistic AF relaying system with distributed OR scheme. Reprint with permission from [11]. ©2013 Wiley.

8.3 Cooperative Opportunistic Regenerative Relaying

Transmission through relay nodes leads to alternative signaling paths. If the destination node can receive information over both the direct source–destination link and the relay link, then the resulting cooperative relay transmission system can enjoy additional diversity benefit. Typically, cooperative relay transmission is carried out over two consecutive time slots. During the first time slot, the source node broadcasts information signal to the relay and destination nodes. Over the second time slot, the relay node transmits to the destination node, whereas the destination applies diversity combining on the received signal copies over two time slots. In this section, we illustrate the benefit of cooperative relay transmission by considering an opportunistic regenerative relaying system [13–15].

8.3.1 Mode of Operation

We consider the same system model as the previous section, where one relay out of a cluster of relays r_k, $k = 1, 2, \ldots, K$, is selected to help the transmission from source s to destination d. All nodes are equipped with a single antenna. Unlike the previous section, we assume that the destination can receive the transmitted signal from the source

8.3 Cooperative Opportunistic Regenerative Relaying

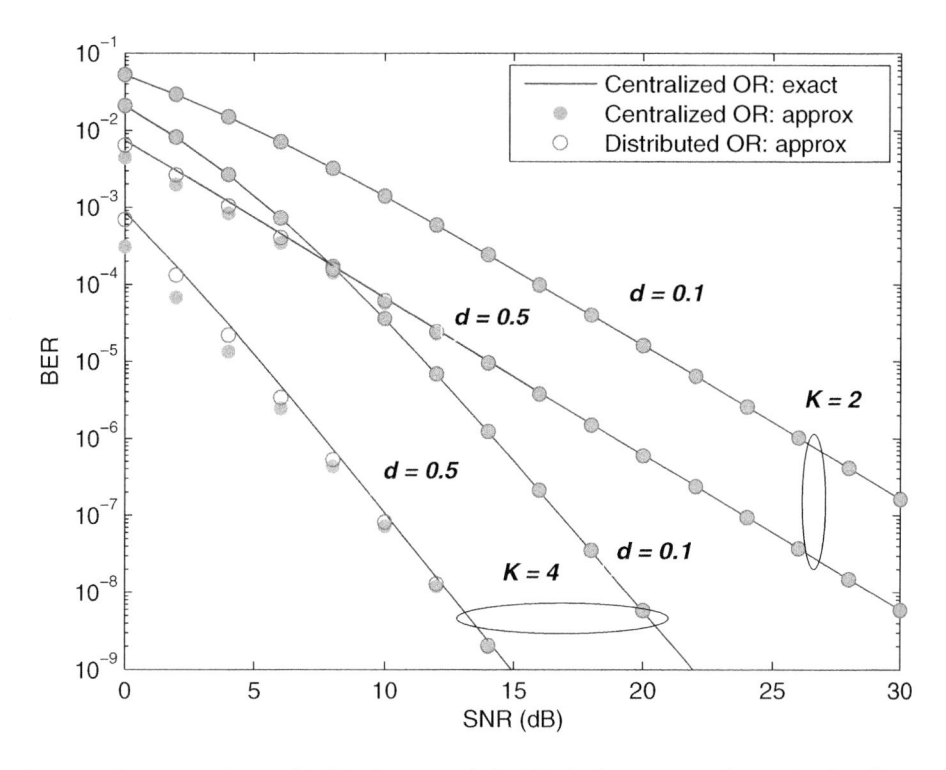

Figure 8.6 Average end-to-end BER of opportunistic AF relaying system using BPSK in a linear network. Reprint with permission from [11]. ©2013 Wiley.

directly. The received SNR of the direct source–destination link is denoted by γ_{sd}, which is an exponential random variable over Rayleigh fading.

We also assume the selected relay applies a regenerative DF relaying strategy. As such, the quality of the relay link is limited by the quality of the poorer hop. The optimal relay selection strategy is the max–min strategy, which selects the relay based on the following rule [16]:

$$r_* = \arg\max_k \min\left(\gamma_{sr_k}, \gamma_{r_k d}\right), \tag{8.62}$$

where γ_{sr_k} and $\gamma_{r_k d}$ are the received SNRs of the source to relay k link and the relay k to destination link, respectively. After the cooperative transmission from the selected relay, the destination eventually combines the received signals from direct transmission and relay transmission using a maximum ratio combining (MRC) receiver. As such, the combined SNR at the destination is equal to

$$\gamma_c = \gamma_{sd} + \gamma_{r_* d}. \tag{8.63}$$

8.3.2 Outage Analysis

In this subsection, we analyze the outage performance of the cooperative DF opportunistic relaying system. The exact statistics of γ_{sr_*} and $\gamma_{r_* d}$ derived in the previous section will be utilized in the analysis.

The outage probability of the cooperative opportunistic relaying transmission can be calculated by conditioning on whether the source to selected relay link is in outage or not, as

$$P_{\text{out}} = \Pr\left[\gamma_{sd} < \gamma_{\text{th}}\right] \Pr\left[\gamma_{sr_*} < \gamma_{\text{th}}\right] \\ + \Pr\left[\gamma_{sr_*} > \gamma_{\text{th}}, \gamma_{sd} + \gamma_{r_*d} < \gamma_{\text{th}}\right], \tag{8.64}$$

where γ_{th} is the outage threshold. Here, $\Pr\left[\gamma_{sd} < \gamma_{\text{th}}\right]$ and $\Pr\left[\gamma_{sr_*} < \gamma_{\text{th}}\right]$ are the probabilities that the direct link and the source to selected relay link are in outage and can be calculated, respectively, using the CDF of γ_{sd} and γ_{sr_*}, as

$$\Pr\left[\gamma_{sd} < \gamma_{\text{th}}\right] = 1 - e^{-\gamma_{\text{th}}/\bar{\gamma}_{sd}}, \tag{8.65}$$

and

$$\Pr\left[\gamma_{sr_*} < \gamma_{\text{th}}\right] \triangleq F_{\gamma_{sr_*}}(\gamma_{\text{th}}), \tag{8.66}$$

where $F_{\gamma_{sr_*}}(\cdot)$ is given by (8.42). Although γ_{sr_k} and γ_{r_kd} are independent for all relays, γ_{sr_*} and γ_{r_*d} cannot be treated as independent random variables, as they correspond to the selected relay. Conditioning on the SNR of the direct link, the third probability in (8.64) can be calculated as

$$\Pr\left[\gamma_{sr_*} > \gamma_{\text{th}}, \gamma_{sd} + \gamma_{r_*d} < \gamma_{\text{th}}\right] \tag{8.67}$$

$$= \frac{1}{\bar{\gamma}_{sd}} \int_0^{\gamma_{\text{th}}} \Pr\left[\gamma_{sr_*} > \gamma_{\text{th}}, \gamma_{r_*d} < \gamma_{\text{th}} - x\right] e^{-x/\bar{\gamma}_{sd}} dx,$$

where $\Pr\left[\gamma_{sr_*} > \gamma_{\text{th}}, \gamma_{r_*d} < \gamma_{\text{th}} - x\right]$ can be approximately calculated as

$$\Pr\left[\gamma_{sr_*} > \gamma_{\text{th}}, \gamma_{r_*d} < \gamma_{\text{th}} - x\right] \approx \left[1 - F_{\gamma_{sr_*}}(\gamma_{\text{th}})\right] F_{\gamma_{r_*d}}(\gamma_{\text{th}} - x). \tag{8.68}$$

After applying the CDFs of γ_{sr_*} and γ_{r_*d} from the previous section, (8.67) could be rewritten as

$$\Pr\left[\gamma_{sr_*} > \gamma_{\text{th}}, \gamma_{sd} + \gamma_{r_*d} < \gamma_{\text{th}}\right] = \left(1 - F_{\gamma_{sr_*}}(\gamma_{\text{th}})\right)$$
$$\times \left(\sum_{i=1}^{K} \binom{K}{i} \frac{(-1)^{i-1}}{\bar{\gamma}_{sr}} \frac{i\bar{\gamma}}{i\bar{\gamma}_{rd} - \bar{\gamma}} \left[\bar{\gamma}_{rd}\left(1 - e^{-\gamma_{\text{th}}/\bar{\gamma}_{sd}}\right) - \frac{\bar{\gamma}_{rd}^2}{\bar{\gamma}_{rd} - \bar{\gamma}_{sd}} \left(e^{-\gamma_{\text{th}}/\bar{\gamma}_{rd}} - e^{-\gamma_{\text{th}}/\bar{\gamma}_{sd}}\right)\right.\right.$$
$$\left. - \frac{\bar{\gamma}}{i}\left(1 - e^{-\gamma_{\text{th}}/\bar{\gamma}_{sd}}\right) + \frac{\bar{\gamma}^2}{i(\bar{\gamma} - i\bar{\gamma}_{sd})} \left(e^{-i\gamma_{\text{th}}/\bar{\gamma}} - e^{-\gamma_{\text{th}}/\bar{\gamma}_{sd}}\right)\right]$$
$$\left. + \sum_{i=1}^{K} \binom{K}{i} (-1)^{i-1} \frac{\bar{\gamma}}{\bar{\gamma}_{rd}} \left[1 - e^{-\gamma_{\text{th}}/\bar{\gamma}_{sd}} - \frac{\bar{\gamma}}{\bar{\gamma} - i\bar{\gamma}_{sd}} \left(e^{-i\gamma_{\text{th}}/\bar{\gamma}} - e^{-\gamma_{\text{th}}/\bar{\gamma}_{sd}}\right)\right]\right), \tag{8.69}$$

where $\bar{\gamma} = \frac{\bar{\gamma}_{sr}\bar{\gamma}_{rd}}{\bar{\gamma}_{sr}+\bar{\gamma}_{rd}}$. Finally, with (8.65), (8.42), (8.68), and (8.69), the end-to-end outage probability of the cooperative DF opportunistic relaying system can be evaluated.

For the high-SNR regime, the outage probability is approximately given by

$$P_{\text{out}} \approx \left[\frac{1}{(K+1)\sigma_{rd}^2} + \frac{1}{\sigma_{sr}^2}\right]$$
$$\times \left(\frac{1}{\sigma_{sd}^2}\right)\left(\frac{1}{\sigma_{sr}^2} + \frac{1}{\sigma_{rd}^2}\right)^{K-1}\left(\frac{\gamma_{\text{th}}}{E_s/N_0}\right)^{K+1}. \quad (8.70)$$

We can observe from (8.70) that the diversity order of the system is $K+1$, which means a full diversity order is achieved.

8.3.3 Average Error Rate Analysis

In this subsection, we analyze the average error rate performance of the cooperative DF opportunistic relaying system. We consider the BPSK modulation scheme for illustration. Note that the instantaneous error rate of BPSK is given by $Q(\sqrt{2\gamma})$, where γ is the instantaneous received SNR.

With cooperative DF relaying, the destination combines the received signals from the selected relay and the source. As the selected relay may forward an erroneously decoded message, the end-to-end probability of error can be expressed as

$$P_{e,\text{sys}} = P_{sr_*}P_{\text{prop}} + \left(1 - P_{sr_*}\right)P_{\text{mrc}}, \quad (8.71)$$

where P_{sr_*} is the probability of detection error over the hop between the source and the selected relay, P_{prop} is the probability of error propagation, and P_{mrc} is the probability of detection error at the destination, after MRC combining of direct transmission and error-free forwarded signal from the selected relay. P_{sr_*} can be calculated for BPSK modulation as

$$P_{sr_*} = \int_0^\infty \frac{1}{2}\text{erfc}(\sqrt{x})p_{\gamma_{sr_*}}(x)dx, \quad (8.72)$$

where $p_{\gamma_{sr_*}}(.)$ is the PDF of the received SNR at the selected relay, derived as in (8.40) by replacing γ_{r_*d} with γ_{sr_*}, and $\text{erfc}(.)$ is the complementary error function. Hence, performing the integration in (8.72) and using the identity

$$l(\alpha) \triangleq \frac{1}{2}\int_0^\infty \text{erfc}(\sqrt{\beta})e^{-\alpha\beta}d\beta = \frac{1}{2\alpha}\left[1 - \frac{1}{\sqrt{1+\alpha}}\right], \quad (8.73)$$

P_{sr_*} is determined as

$$P_{sr_*} = \sum_{i=1}^{K}\binom{K}{i}\frac{(-1)^{i-1}}{\bar{\gamma}_{rd}}\frac{i\bar{\gamma}}{i\bar{\gamma}_{sr} - \bar{\gamma}}\left[l\left(\frac{1}{\bar{\gamma}_{sr}}\right) - l\left(\frac{i}{\bar{\gamma}}\right)\right] + \sum_{i=1}^{K}\binom{K}{i}\frac{i(-1)^{i-1}}{\bar{\gamma}_{sr}}l\left(\frac{i}{\bar{\gamma}}\right).$$
$$(8.74)$$

The error propagation probability P_{prop} can be approximately calculated for BPSK modulation as

$$P_{\text{prop}} \approx \frac{\bar{\gamma}_{r_*d}}{\bar{\gamma}_{r_*d} + \bar{\gamma}_{sd}}, \quad (8.75)$$

where $\bar{\gamma}_{r_*d}$ denotes the expected value of γ_{r_*d}, which can be calculated as

$$\bar{\gamma}_{r_*d} = \sum_{i=1}^{K} \binom{K}{i} \frac{(-1)^{i-1}}{\bar{\gamma}_{sr}} \frac{i\bar{\gamma}}{i\bar{\gamma}_{rd} - \bar{\gamma}} \left[\bar{\gamma}_{rd}^2 - \left(\frac{\bar{\gamma}}{i}\right)^2\right] + \sum_{i=1}^{K} \binom{K}{i} \frac{(-1)^{i-1}}{i\bar{\gamma}_{rd}} \bar{\gamma}^2. \tag{8.76}$$

Finally, the average error probability after combining direct and opportunistic links in an MRC fashion, denoted by P_{mrc}, is given by

$$P_{mrc} = \int_0^\infty \frac{1}{2} \operatorname{erfc}(\sqrt{\beta}) p_{\gamma_{sd}+\gamma_{r_*d}}(\beta) \mathrm{d}\beta, \tag{8.77}$$

where $p_{\gamma_{sd}+\gamma_{r_*d}}(\cdot)$ is the PDF of the combined SNR after MRC-combining the direct and opportunistic links, which can be obtained by applying a convolution operation on the PDFs of γ_{sd} and γ_{r_*d} as

$$p_{\gamma_{sd}+\gamma_{r_*d}}(\cdot)(\beta) = \sum_{i=1}^{K} \binom{K}{i} \frac{(-1)^{i-1}}{\bar{\gamma}_{sr}} \frac{i\bar{\gamma}}{i\bar{\gamma}_{rd} - \bar{\gamma}}$$
$$\left[\frac{\bar{\gamma}_{rd}}{\bar{\gamma}_{sd} - \bar{\gamma}_{rd}} \left(e^{-\beta/\bar{\gamma}_{sd}} - e^{-\beta/\bar{\gamma}_{rd}}\right) - \frac{\bar{\gamma}}{i\bar{\gamma}_{sd} - \bar{\gamma}} \left(e^{-\beta/\bar{\gamma}_{sd}} - e^{-i\beta/\bar{\gamma}}\right)\right]$$
$$+ \sum_{i=1}^{K} \binom{K}{i} \frac{(-1)^{i-1}}{\bar{\gamma}_{rd}} \frac{i\bar{\gamma}}{i\bar{\gamma}_{sd} - \bar{\gamma}} \left(e^{-\beta/\bar{\gamma}_{sd}} - e^{-i\beta/\bar{\gamma}}\right). \tag{8.78}$$

Substituting (8.96) into (8.77) and carrying out integration, we obtain the closed-form expression of P_{mrc} as

$$P_{mrc} = \sum_{i=1}^{K} \binom{K}{i} \frac{(-1)^{i-1}}{\bar{\gamma}_{sr}} \frac{i\bar{\gamma}}{i\bar{\gamma}_{rd} - \bar{\gamma}}$$
$$\times \left[\frac{\bar{\gamma}_{rd}}{\bar{\gamma}_{sd} - \bar{\gamma}_{rd}} \left(l\left(\frac{1}{\bar{\gamma}_{sd}}\right) - l\left(\frac{1}{\bar{\gamma}_{rd}}\right)\right) - \frac{\bar{\gamma}}{i\bar{\gamma}_{sd} - \bar{\gamma}} \left(l\left(\frac{1}{\bar{\gamma}_{sd}}\right) - l\left(\frac{i}{\bar{\gamma}}\right)\right)\right]$$
$$+ \sum_{i=1}^{K} \binom{K}{i} \frac{(-1)^{i-1}}{\bar{\gamma}_{rd}} \frac{i\bar{\gamma}}{i\bar{\gamma}_{sd} - \bar{\gamma}} \left(l\left(\frac{1}{\bar{\gamma}_{sd}}\right) - l\left(\frac{i}{\bar{\gamma}}\right)\right). \tag{8.79}$$

Finally, with (8.75), (8.76), (8.74), and (8.79), the end-to-end probability of error can be evaluated.[1]

8.3.4 Numerical Results

We again consider a normalized linear network, where the source is located at (0, 0), the destination at (1, 0), and the relays are located at (d, 0). We also assume log-distance path loss model and ignore the shadowing effect. In particular, the mean channel power gains are related to the relay cluster position d as

$$\sigma_{sd}^2 = 1, \ \sigma_{sr_k}^2 = d^{-\nu}, \ \sigma_{r_kd}^2 = (1-d)^{-\nu}, \tag{8.80}$$

where ν is the path loss exponent.

---

[1] P_{sr_*} and P_{mrc} can apply to other modulation schemes. However, P_{prop} must be changed accordingly.

8.3 Cooperative Opportunistic Regenerative Relaying

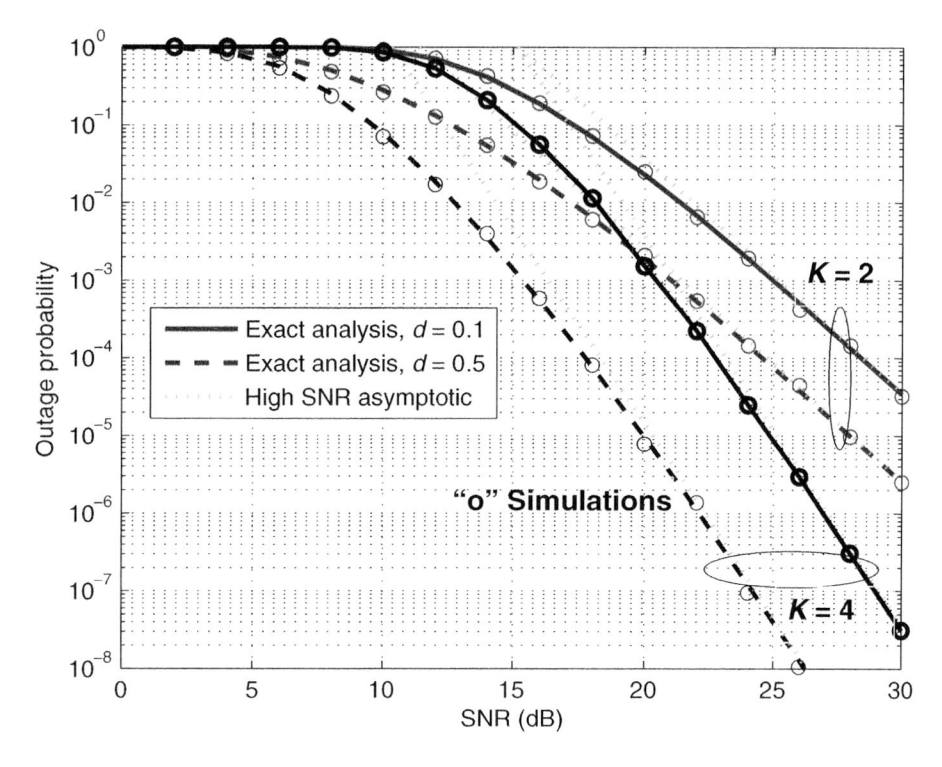

Figure 8.7 End-to-end outage probability of a cooperative DF opportunistic relaying system in a linear network with $K = 2$ and 4. Reprint with permission from [15]. ©2013 Elsevier.

Fig. 8.7 depicts the end-to-end outage performance and the corresponding asymptotic results as a function of SNR in a linear network, where a relay cluster is located at different distances from the source. The targeted transmission rate is $R = 3$ (bits/s/Hz). We compare the result for different numbers of active relays located at different distances from the source. We note that analytical and simulation results coincide, and the diversity order of the system is verified and equal to $K + 1$. We can also observe that the outage performance improves when the relays move closer to the center point between source and destination and when the number of candidate relays increases, as expected.

Fig. 8.8 depicts the end-to-end error-rate performance and the corresponding asymptotic curves as a function of SNR in a linear network, where a relay cluster is located at different distances from the source. Specifically, the average error rate is plotted as a function of the SNR for a varying number of active relays. We can see that more active relays help improve the average BER performance. It could be noted that the diversity order is K when the relay cluster is located at the mid-distance to the destination. This is due to the fact that the error propagation probability is large. The full diversity order is recovered when $d = 0.1$. This is due to the fact that the source to selected relay hop becomes more reliable and the error propagation probability is reduced.

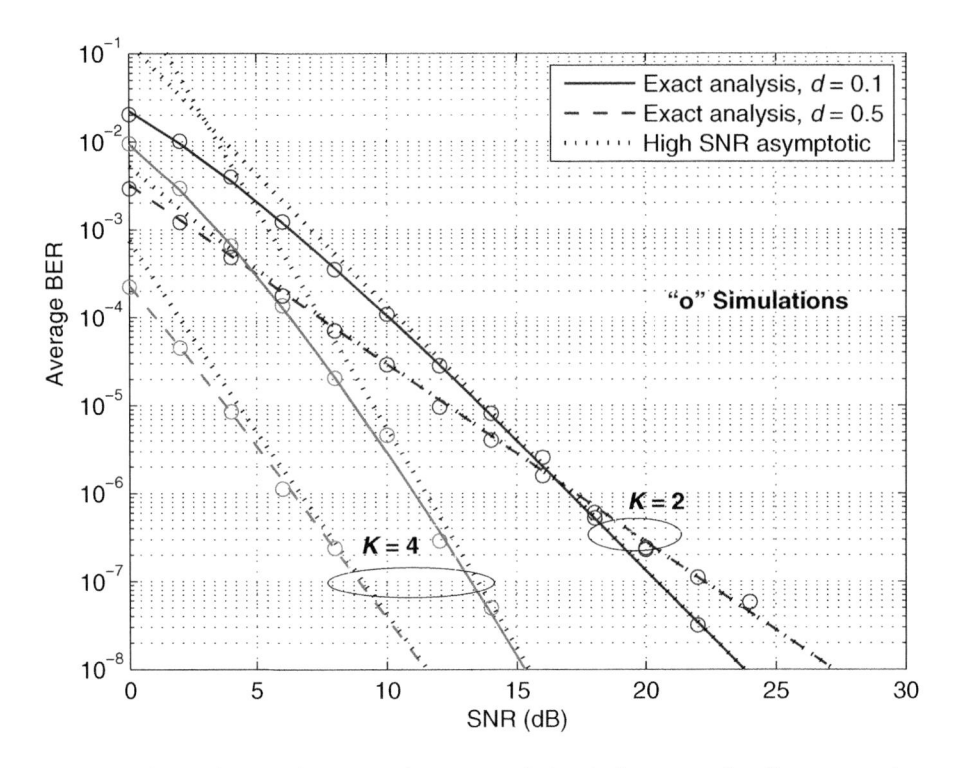

Figure 8.8 End-to-end BER of a cooperative opportunistic relaying system in a linear network. Reprint with permission from [15]. ©2013 Elsevier.

8.4 Incremental Opportunistic Regenerative Relaying

The cooperative diversity gain usually comes at the cost of a certain spectral efficiency loss. Such loss originates from the fact that most relay nodes operate in a half-duplex fashion. As such, the source can only transmit for half of the duration with cooperative relaying. Incremental relaying can reduce such spectral efficiency loss while still enjoying a certain diversity benefit [17]. With incremental relaying, the source node broadcasts the information signal over the first time slot, while both relay and destination nodes receive the transmission. At the end of the first time slot, the destination node will decide if it can decode the source transmission. If yes, the destination will send an acknowledgment to the source and relay nodes. The relay node, upon receiving this acknowledgment, will discard the information it received and the source node will proceed to transmit new data. If the destination cannot decode the source information, it will inform the relay to forward its received signal copy and the source node to hold its transmission of new data. The destination node will perform MRC on the received signal over both links. Assuming the destination node can successfully detect the information over the direct link if and only if the direct link SNR satisfies $\gamma_{sd} \geq \gamma_{th}$, the received SNR at the destination with incremental relaying can be determined as

$$\gamma_c^{\text{IR}} = \begin{cases} \gamma_{sd} + \frac{\gamma_1 \gamma_2}{\gamma_1 + \gamma_2 + 1}, & \gamma_{sd} < \gamma_T; \\ \gamma_{sd}, & \gamma_{sd} \geq \gamma_T. \end{cases} \qquad (8.81)$$

The incremental relaying system achieves a similar diversity benefit as a dual-branch switch-and-stay combining (SSC) scheme.

In this section, we consider opportunistic relaying transmission with regenerative relays that will decode and forward the source information. For DF relaying, the max–min relay selection strategy is optimal in terms of minimizing the end-to-end outage probability and can be implemented in a distributed manner as discussed in the previous section. We assume that incremental relaying is applied to mitigate the spectral efficiency loss due to the half-duplex constraint at the selected relay. We carry out a thorough performance analysis of the resulting relay transmission system.

8.4.1 Mode of Operation

We consider the same system model as the previous section, where one relay out of a cluster of K relays r_k, $k = 1, 2, \ldots, K$, will help the transmission from source s to destination d. All nodes are equipped with a single antenna. The cooperative transmission is triggered only when direct transmission fails. The success/failure of the direct transmission is dependent upon the instantaneous received SNR of the direct source-to-destination link γ_{sd}. When γ_{sd} exceeds a threshold γ_T, we consider that the direct transmission is sufficient and there is no need for relay cooperation. The destination will send binary feedback to the source to proceed with new data transmission. The relays will discard their received data.

Otherwise, cooperation with optimistic relaying is needed. The destination sends binary feedback to the relays requesting their cooperation while informing the source to hold its transmission. With opportunistic DF relaying, only the best relay r_*, which is selected following the rule

$$r_* = \arg \max_k \min \left(\gamma_{sr_k}, \gamma_{r_k d} \right), \qquad (8.82)$$

will forward its decoded information. After the cooperative transmission from the selected relay, the destination eventually combines the received signals over direct link and relay link using an MRC receiver.

The performance of an incremental relaying system depends on the value of threshold γ_T. In general, there are two approaches to setting the threshold value. The first approach is to set γ_T equal to the minimum required SNR value for the direct source to destination link to support the target data rate. In particular, the minimum value from the information theoretical perspective should be $\gamma_T = 2^{R/B} - 1$, where B is the channel bandwidth and R is the target rate. Essentially, if γ_{sd} is smaller than γ_T, then the direct link will enter outage if no relay is available. As such, such threshold value choice is typically adopted in the outage analysis of incremental relaying systems. The other approach is to select the γ_T value to minimize the end-to-end error performance of the transmission systems.

In the following subsections, we will adopt these two approaches to set the γ_T values in the outage and average error rate analysis of the incremental relaying systems.

8.4.2 Outage Analysis

In this subsection, we calculate the outage probability for the incremental DF opportunistic relay transmission system. The exact statistics of γ_{sr_*} and γ_{r_*d} derived in the previous section will be applied in the analysis.

The outage probability of the incremental relay transmission is defined as the probability that the system cannot support a target rate of R. With incremental relaying, the system enters outage if neither the direct transmission nor the cooperative relay transmission can support the target rate. Note that when the relay is activated, the system will use two phases to transmit the information. Each transmission phase should support the rate of $2R$ in order to support the target rate. As such, the outage probability of an incremental DF opportunistic relay transmission system can be calculated as

$$P_{\text{out}} = \Pr\left[\gamma_{sd} < \gamma_T\right]\Pr\left[\gamma_{sr_*} < \gamma_{\text{th}}\right]$$
$$+ \Pr\left[\gamma_{sd} < \gamma_T, \gamma_{sr_*} > \gamma_{\text{th}}, \gamma_{sd} + \gamma_{r_*d} < \gamma_{\text{th}}\right], \qquad (8.83)$$

where the outage threshold $\gamma_{\text{th}} = 2^{2R/B} - 1$ and $\gamma_T = 2^{R/B} - 1$. The first two probabilities in (8.83) can be calculated as

$$\Pr\left[\gamma_{sd} < \gamma_T\right] = 1 - e^{-\gamma_T/\bar{\gamma}_{sd}}, \qquad (8.84)$$

and

$$\Pr\left[\gamma_{sr_*} < \gamma_{\text{th}}\right] \triangleq F_{\gamma_{sr_*}}(\gamma_{\text{th}}), \qquad (8.85)$$

where $F_{\gamma_{sr_*}}(\cdot)$ is given by (8.42). Although γ_{sr_k} and γ_{r_kd} are independent for all relays, γ_{sr_*} and γ_{r_*d} cannot be treated as independent random variables. Conditioning on the SNR of the direct link, the third probability term in (8.83) could be calculated as

$$\Pr\left[\gamma_{sr_*} > \gamma_{\text{th}}, \gamma_{sd} < \gamma_T, \gamma_{sd} + \gamma_{r_*d} < \gamma_{\text{th}}\right] \qquad (8.86)$$
$$= \frac{1}{\bar{\gamma}_{sd}} \int_0^{\gamma_T} \Pr\left[\gamma_{sr_*} > \gamma_{\text{th}}, \gamma_{r_*d} < \gamma_{\text{th}} - x\right] e^{-x/\bar{\gamma}_{sd}} dx,$$

where $\Pr\left[\gamma_{sr_*} > \gamma_{\text{th}}, \gamma_{r_*d} < \gamma_{\text{th}} - x\right]$ is approximately given by

$$\Pr\left[\gamma_{sr_*} > \gamma_{\text{th}}, \gamma_{r_*d} < \gamma_{\text{th}} - x\right] \approx \left[1 - F_{\gamma_{sr_*}}(\gamma_{\text{th}})\right] F_{\gamma_{r_*d}}(\gamma_{\text{th}} - x). \qquad (8.87)$$

Applying the CDFs of γ_{sr_*} and γ_{r_*d}, (8.86) could be rewritten as

$$\Pr\left[\gamma_{sr_*} > \gamma_{\text{th}}, \gamma_{sd} < \gamma_T, \gamma_{sd} + \gamma_{r_*d} < \gamma_{\text{th}}\right] = \left(1 - F_{\gamma_{sr_*}}(\gamma_{\text{th}})\right)\left(\sum_{i=1}^{K} \binom{K}{i} \frac{(-1)^{i-1}}{\bar{\gamma}_{sr}}\right)$$
$$\times \frac{i\bar{\gamma}}{i\bar{\gamma}_{rd} - \bar{\gamma}} \left[\bar{\gamma}_{rd}\left(1 - e^{-\gamma_T/\bar{\gamma}_{sd}}\right) - \frac{\bar{\gamma}_{rd}^2}{\bar{\gamma}_{rd} - \bar{\gamma}_{sd}}\left(e^{-\gamma_{\text{th}}/\bar{\gamma}_{rd}} - e^{-\gamma_T/\bar{\gamma}_{sd}} e^{-2i(\gamma_{\text{th}} - \gamma_T)/\bar{\gamma}}\right)\right]$$

$$-\frac{\bar{\gamma}}{i}(1-e^{-\gamma_T/\bar{\gamma}_{sd}}) + \frac{\bar{\gamma}^2}{i(\bar{\gamma}-i\bar{\gamma}_{sd})}\left(e^{-i\gamma_{th}/\bar{\gamma}} - e^{-\gamma_{th}/\bar{\gamma}_{sd}}e^{-2i(\gamma_{th}-\gamma_T)/\bar{\gamma}}\right)\Bigg]$$

$$+ \sum_{i=1}^{K}\binom{K}{i}(-1)^{i-1}\frac{\bar{\gamma}}{\bar{\gamma}_{rd}}\left[1-e^{-\gamma_T/\bar{\gamma}_{sd}} - \frac{\bar{\gamma}}{\bar{\gamma}-i\bar{\gamma}_{sd}}\left(e^{-i\gamma_{th}/\bar{\gamma}} - e^{-\gamma_T/\bar{\gamma}_{sd}}e^{-2i(\gamma_{th}-\gamma_T)/\bar{\gamma}}\right)\right]\Bigg]). \quad (8.88)$$

Finally, with (8.84), (8.42), (8.87), and (8.88), the end-to-end outage probability of the incremental DF opportunistic relaying system can be evaluated. For the high SNR regime, the outage probability is approximately given by

$$P_{out} \approx \left[\frac{\gamma_{th}^{K+1}}{K+1}\frac{1}{\sigma_{rd}^2} + \frac{\gamma_{th}^{K+1}}{\sigma_{sr}^2}\right] \times \left(\frac{1}{\sigma_{sd}^2}\right)\left(\frac{1}{\sigma_{sr}^2} + \frac{1}{\sigma_{rd}^2}\right)^{K-1}\left(\frac{E_s}{N_0}\right)^{K+1}. \quad (8.89)$$

We can observe from (8.101) that the diversity order of the system is $K+1$, which means a full diversity order is achieved.

8.4.3 Average Error Rate Analysis

In this subsection, we analyze the average error rate performance of an incremental DF opportunistic relaying system. Different from previous subsection, we will determine the optimal value for threshold γ_T to minimize the average error rate. We again consider a BPSK modulation scheme for illustration.

The average end-to-end probability of error with incremental relaying can be expressed as

$$P_{e,sys} = P_{dec}P_e^1 + (1-P_{dec})P_e^2, \quad (8.90)$$

where $P_{dec} = \Pr[\gamma_{sd} \geq \gamma_T]$ is the probability of direct transmission only, which is calculated for the Rayleigh fading as $e^{-\gamma_T/\bar{\gamma}_{sd}}$, P_e^1 is the average probability of error for the direct communication only case, and P_e^2 is the average probability of error with opportunistic cooperative relaying. P_e^1 can be calculated as

$$P_e^1 = \int_{\gamma_T}^{\infty} Q(\sqrt{2\gamma})e^{-\gamma/\bar{\gamma}_{sd}}/\bar{\gamma}_{sd}d\gamma. \quad (8.91)$$

After carrying out integration, we can arrive at the following closed-form result:

$$P_e^1 = Q(\sqrt{2\gamma_T}) - e^{\gamma_T/\bar{\gamma}_{sd}}\sqrt{\frac{1}{1+\frac{1}{\bar{\gamma}_{sd}}}}Q\left(\sqrt{2\gamma_T\left(1+\frac{1}{\bar{\gamma}_{sd}}\right)}\right). \quad (8.92)$$

P_e^2 is the average end-to-end probability of error with cooperative DF relay transmission. With DF relaying, the relay may forward an erroneously decoded message. As such, P_e^2 can be calculated as

$$P_e^2 = P_{prop}P_{sr_*} + (1-P_{sr_*})P_{mrc}, \quad (8.93)$$

where P_{prop} denotes the error propagation probability, P_{sr_*} is the probability of detection error over the source to selected relay link, both of which were calculated in the previous section and given in (8.75) and (8.74), respectively, and P_{mrc} is the error probability of the combined direct and selected relay to destination paths.

Finally, the average error probability P_{mrc} can be calculated as

$$P_{mrc} = \int_0^\infty Q(\sqrt{2\gamma}) p_{\gamma_{sd}+\gamma_{r_*d}}(\gamma) d\gamma, \tag{8.94}$$

where $p_{\gamma_{sd}+\gamma_{r_*d}}(\cdot)$ denotes the PDF of the combined SNR when applying MRC on direct and opportunistic paths. Note that with incremental relaying, the relay is activated only if γ_{sd} is less than γ_T. As such, γ_{sd} has a truncated distribution from above at γ_T. The PDF of γ_{sd} for the Rayleigh fading channel is given by

$$p_{\gamma_{sd}}(\gamma) = \begin{cases} \frac{1}{C}\frac{1}{\bar{\gamma}_{sd}} e^{-\gamma/\bar{\gamma}_{sd}}, & \gamma < \gamma_T \\ 0, & \gamma \geq \gamma_T \end{cases}, \tag{8.95}$$

where $C = 1 - e^{-\gamma_T/\bar{\gamma}_{sd}}$. After convolving with the PDF of γ_{r_*d} given in (8.40), the PDF of γ_c can be obtained as

$$p_{\gamma_c}(\gamma) = \begin{cases} \sum_{i=1}^K \binom{K}{i} \frac{(-1)^{i-1}}{C\bar{\gamma}_{sr}} \frac{i\bar{\gamma}}{i\bar{\gamma}_{rd}-\bar{\gamma}} \left[\frac{\bar{\gamma}_{rd}}{\bar{\gamma}_{sd}-\bar{\gamma}_{rd}} \left(e^{-\gamma/\bar{\gamma}_{sd}} - e^{-\gamma/\bar{\gamma}_{rd}}\right) \right. \\ \left. - \frac{\bar{\gamma}}{i\bar{\gamma}_{sd}-\bar{\gamma}} \left(e^{-\gamma/\bar{\gamma}_{sd}} - e^{-i\gamma/\bar{\gamma}}\right) \right] \\ + \sum_{i=1}^K \binom{K}{i} \frac{(-1)^{i-1}}{C\bar{\gamma}_{rd}} \frac{i\bar{\gamma}}{i\bar{\gamma}_{sd}-\bar{\gamma}} \left(e^{-\gamma/\bar{\gamma}_{sd}} - e^{-i\gamma/\bar{\gamma}}\right), & \gamma < \gamma_T \\ \sum_{i=1}^K \binom{K}{i} \frac{(-1)^{i-1}}{C\bar{\gamma}_{sr}} \frac{i\bar{\gamma}}{i\bar{\gamma}_{rd}-\bar{\gamma}} \left[\frac{\bar{\gamma}_{rd} e^{-\frac{\gamma}{\bar{\gamma}_{rd}}}}{\bar{\gamma}_{sd}-\bar{\gamma}_{rd}} \left(e^{-\gamma_T(\frac{1}{\bar{\gamma}_{sd}}-\frac{1}{\bar{\gamma}_{rd}})} - 1\right) \right. \\ \left. - \frac{\bar{\gamma} e^{-\frac{i\gamma}{\bar{\gamma}}}}{i\bar{\gamma}_{sd}-\bar{\gamma}} \left(e^{-\gamma_T(\frac{1}{\bar{\gamma}_{sd}}-\frac{i}{\bar{\gamma}})} - 1\right) \right] & \gamma \geq \gamma_T \\ + \sum_{i=1}^K \binom{K}{i} \frac{(-1)^{i-1}}{C\bar{\gamma}_{rd}} \frac{i\bar{\gamma} e^{-\frac{i\gamma}{\bar{\gamma}}}}{i\bar{\gamma}_{sd}-\bar{\gamma}} \left(e^{-\gamma_T(\frac{1}{\bar{\gamma}_{sd}}-\frac{i}{\bar{\gamma}})} - 1\right). \end{cases} \tag{8.96}$$

After carrying out integration and some manipulation, we obtain the closed-form expression of P_{mrc} as

$$P_{mrc} = \frac{1}{2C} \sum_{i=1}^K \binom{K}{i} \frac{(-1)^{i-1}}{\bar{\gamma}_{sr}} \frac{i\bar{\gamma}}{i\bar{\gamma}_{rd}-\bar{\gamma}} \left[\frac{\bar{\gamma}_{rd}}{\bar{\gamma}_{sd}-\bar{\gamma}_{rd}} \left(l(\frac{1}{\bar{\gamma}_{sd}},\gamma_T) - l(\frac{1}{\bar{\gamma}_{rd}},\gamma_T)\right) \right.$$
$$\left. - \frac{\bar{\gamma}}{i\bar{\gamma}_{sd}-\bar{\gamma}} \left(l(\frac{1}{\bar{\gamma}_{sd}},\gamma_T) - l(\frac{i}{\bar{\gamma}},\gamma_T)\right) \right]$$
$$+ \frac{1}{2C} \sum_{i=1}^K \binom{K}{i} \frac{(-1)^{i-1}}{\bar{\gamma}_{rd}} \frac{i\bar{\gamma}}{i\bar{\gamma}_{sd}-\bar{\gamma}} \left(l(\frac{1}{\bar{\gamma}_{sd}},\gamma_T) - l(\frac{i}{\bar{\gamma}},\gamma_T)\right)$$
$$+ \frac{1}{2C} \sum_{i=1}^K \binom{K}{i} \frac{(-1)^{i-1}}{\bar{\gamma}_{rd}} \frac{i\bar{\gamma}}{i\bar{\gamma}_{sd}-\bar{\gamma}} \left(e^{-\gamma_T(\frac{1}{\bar{\gamma}_{sd}}-\frac{i}{\bar{\gamma}})} - 1\right) \lambda(\frac{i}{\bar{\gamma}},\gamma_T)$$
$$+ \frac{1}{2C} \sum_{i=1}^K \binom{K}{i} \frac{(-1)^{i-1}}{\bar{\gamma}_{sr}} \frac{i\bar{\gamma}}{i\bar{\gamma}_{rd}-\bar{\gamma}} \frac{\bar{\gamma}_{rd}}{\bar{\gamma}_{sd}-\bar{\gamma}_{rd}} \left(e^{-\gamma_T(\frac{1}{\bar{\gamma}_{sd}}-\frac{1}{\bar{\gamma}_{rd}})} - 1\right) \lambda(\frac{1}{\bar{\gamma}_{rd}},\gamma_T)$$
$$- \frac{1}{2C} \sum_{i=1}^K \binom{K}{i} \frac{(-1)^{i-1}}{\bar{\gamma}_{sr}} \frac{i\bar{\gamma}}{i\bar{\gamma}_{rd}-\bar{\gamma}} \frac{\bar{\gamma}}{i\bar{\gamma}_{sd}-\bar{\gamma}} \left(e^{-\gamma_T(\frac{1}{\bar{\gamma}_{sd}}-\frac{i}{\bar{\gamma}})} - 1\right) \lambda(\frac{i}{\bar{\gamma}},\gamma_T). \tag{8.97}$$

8.4 Incremental Opportunistic Regenerative Relaying

Table 8.1 Optimized SNR threshold γ_T at high SNR (30 dB) when $K = 2, 3$, and 4.

	$d = 0.1$	$d = 0.5$	$d = 0.9$
$K = 2$	13.417	12.301	11.158
$K = 3$	14.484	13.957	12.852
$K = 4$	15.385	15.138	14.02

where

$$l(\alpha, x) = \frac{1}{\alpha}\left[1 - \frac{1}{\sqrt{1+\alpha}} + \frac{2}{\sqrt{1+\alpha}} Q\left(\sqrt{2(1+\alpha)x}\right) - 2e^{-\alpha x} Q\left(\sqrt{2x}\right)\right] \tag{8.98}$$

and

$$\lambda(\alpha, x) = -\frac{2}{\alpha\sqrt{1+\alpha}} Q\left(\sqrt{2(1+\alpha)x}\right) + \frac{2}{\alpha} e^{-\alpha x} Q\left(\sqrt{2x}\right). \tag{8.99}$$

Finally, we can evaluate the end-to-end probability of the incremental opportunistic DF relaying system by combining (8.75), (8.74), (8.93), and (8.97). It is obvious that the performance of the relaying system depends on the threshold γ_T. Intuitively, we would like to choose it optimally to satisfy:

$$\gamma_T^* = \arg\min_{\gamma_T} P_{e,sys}, \tag{8.100}$$

where $\gamma_T \in \mathbb{R}$. Given the complicated form of $P_{e,sys}$, it is evident that this optimization cannot be conducted analytically in a straightforward fashion. Therefore, we can determine the optimum value for γ_T numerically. Table 8.1 summarizes the optimal value of γ_T for different numbers of relays and different positions of relay cluster when the SNR is 30 dB.

For the high SNR regime, the end-to-end error probability of an incremental DF opportunistic relaying scheme is approximately given by

$$P_{e,sys} \approx \left[P_{prop} \frac{\Gamma(K+1/2)}{2\sqrt{\pi}} \frac{\gamma_T}{\sigma_{sr}^2 \sigma_{sd}^2} + \frac{\Lambda(K+1, \gamma_T)}{\sigma_{rd}^2 \sigma_{sd}^2}\right] \\ \times \left(\frac{1}{\sigma_{sr}^2} + \frac{1}{\sigma_{rd}^2}\right)^{K-1} \left(\frac{1}{E_b/N_0}\right)^{K+1}, \tag{8.101}$$

where $\Gamma(.)$ is the gamma function and

$$\Lambda(\alpha, x) = \xi(\alpha, x) + \eta(\alpha, x) - \theta(\alpha, x), \tag{8.102}$$

where

$$\xi(\alpha, x) = \frac{1}{12} \gamma(\alpha, x) + \frac{1}{6}\left(\frac{3}{4}\right)^{K+1} \gamma(\alpha, x), \tag{8.103}$$

$$\eta(\alpha, x) = \frac{1}{12} \Gamma(\alpha, x) + \frac{1}{6}\left(\frac{3}{4}\right)^{K+1} \Gamma(\alpha, x), \tag{8.104}$$

and

$$\theta(\alpha, x) = \frac{1}{12}e^{-x}\Gamma(\alpha) + \frac{1}{6}e^{-4x/3}\left(\frac{3}{4}\right)^{K+1}\Gamma(\alpha). \quad (8.105)$$

Here, $\gamma(\alpha, x)$ and $\Gamma(\alpha, x)$ are the lower and upper incomplete gamma functions, respectively, given by

$$\gamma(\alpha, x) = \int_0^x e^{-t} t^{\alpha-1} dt, \quad (8.106)$$

and

$$\Gamma(\alpha, x) = \int_x^\infty e^{-t} t^{\alpha-1} dt. \quad (8.107)$$

It is straightforward to see that the diversity order of the relay system is $K + 1$, which means a full diversity order is achieved.

8.4.4 Numerical Results

We again consider a normalized linear network, where the source is located at $(0, 0)$, the destination at $(1, 0)$, and the relays at $(d, 0)$. We also assume a log-distance path loss

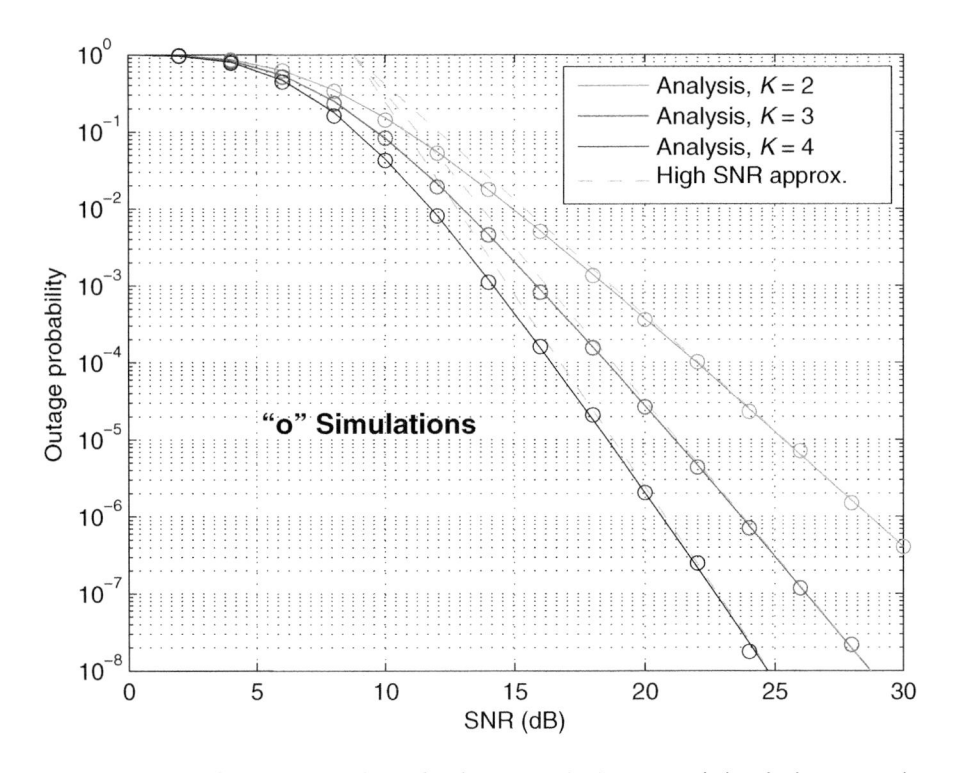

Figure 8.9 End-to-end outage probability of an incremental DF opportunistic relaying system in a linear network with $d = 0.5$, $K = 2, 3$, and 4. Reprint with permission from [17]. ©2011 IEEE.

8.4 Incremental Opportunistic Regenerative Relaying

model and ignore the shadowing effect. In particular, the mean channel power gains are related to the relay cluster position d as

$$\sigma_{sd}^2 = 1, \; \sigma_{sr_k}^2 = d^{-\nu}, \; \sigma_{r_k d}^2 = (1-d)^{-\nu}, \tag{8.108}$$

where ν is the path loss exponent.

Fig. 8.9 depicts the outage probability of an incremental DF opportunistic relaying system in a linear network as a function of SNR $= E_s/N_0$ for a targeted transmission rate $R = 3$ (bits/s/Hz) and different numbers (K) of active relays. We note that analytical and simulations results coincide, and the diversity order of the system is verified and equal to $K + 1$.

We now investigate the end-to-end error rate performance of the relaying system. Fig. 8.10 depicts the end-to-end error rate in a linear network where a relay cluster is located at distance $d = 0.5$ from the source. Specifically, the average error rate is plotted as a function of SNR $= E_b/N_0$ for a preselected threshold γ_T value and different number of active relays K. We note that analytical and simulations results coincide, and the diversity order of the system is verified and equal to $K + 1$.

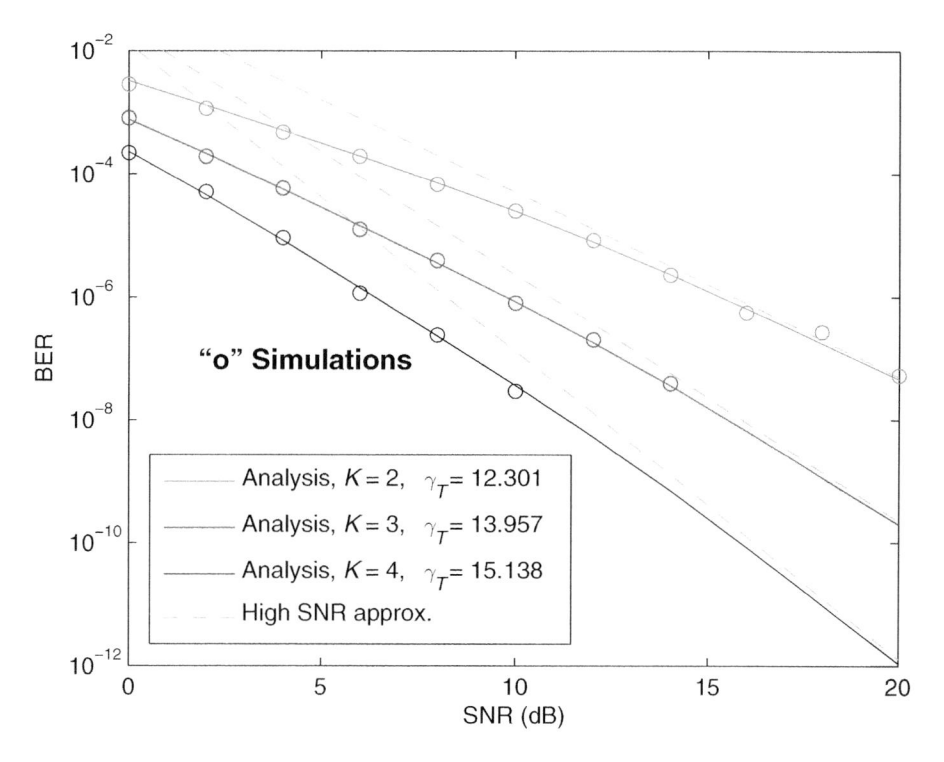

Figure 8.10 End-to-end BER of incremental opportunistic DF relaying system over a linear network with $d = 0.5$. Reprint with permission from [18]. ©2011 IEEE.

8.5 Summary

In this chapter, we presented several relay transmission designs and mathematically characterized their performance in terms of outage probability and average error rate. In particular, after introducing basic relaying strategies for single relay systems, we investigated relay selection, cooperative relaying, and incremental relaying in a multiple-relay setting with either nonregenerative relaying or regenerative relaying strategies. While by no means comprehensive, this chapter aimed to establish a solid foundation for the understanding of advanced relaying transmission strategies.

8.6 Further Reading

For more detailed discussion on conventional relay transmission technologies, the reader can refer to the dedicated textbook by Hong et al. [19]. There is a growing literature on advanced relay transmission. The spectral efficiency loss of conventional half-duplex one-way relaying has motivated a significant amount of interest on two-way relaying [20–22] and full-duplex relaying [5, 23]. Various aspects of relaying transmission are discussed in the edited book of Krikidis and Zheng [24].

References

[1] A. Sendonaris, E. Erkip, and B. Aazhang, "Increasing uplink capacity via user cooperation diversity," in *Proc. 1998 IEEE Int. Sym. Inform. Theor.*, Cambridge, MA, 1998, p. 156.

[2] J. N. Laneman and G. W. Wornell, "Distributed space-time-coded protocols for exploiting cooperative diversity in wireless networks," *IEEE Trans. Inform. Theory*, 49, no. 10, pp. 2415–2425, 2003.

[3] J. N. Laneman, D. N. C. Tse, and G. W. Wornell, "Cooperative diversity in wireless networks: Efficient protocols and outage probability," *IEEE Trans. Inform. Theory*, 50, no. 12, pp. 3062–3080, 2004.

[4] M. O. Hasna and M.-S. Alouini, "Performance analysis of two-hop relayed transmission over Rayleigh fading channels," in *Proc. IEEE Vehicular Technology Conf. (VTC02)*, Vancouver, BC, Canada, September 2002, pp. 1992–1996.

[5] T. Riihonen, S. Werner, and R. Wichman, "Mitigation of loopback self-interference in full-duplex MIMO relays," *IEEE Trans. Signal Processing*, 59, no. 12, pp. 5983–5993, 2011.

[6] M. Abramowitz and I. A. Stegun, *Handbook of Mathematical Functions with Formulas, Graphs, and Mathematical Tables*, 9th ed., Dover, 1970.

[7] M. O. Hasna and M. Alouini, "Harmonic mean and end-to-end performance of transmission systems with relays," *IEEE Trans. Commun.*, 52, no. 1, pp. 130–135, 2004.

[8] A. Bletsas, A. Khisti, D. P. Reed, and A. Lippman, "A simple cooperative diversity method based on network path selection," *IEEE J. Selected Areas in Commun.*, 24, no. 3, pp. 659–672, 2006.

[9] A. Tajer and A. Nosratinia, "Opportunistic cooperation via relay selection with minimal information exchange," in *Proc. IEEE Int. Sym. Inform. Theor. (ISIT'07)*, Nice, France, 2007.

[10] R. Madan, N. B. Mehta, A. F. Molisch, and J. Zhang, "Energy-efficient cooperative relaying over fading channels with simple relay selection," *IEEE Trans. Wireless Commun.*, 7, no. 8, pp. 3013–3025, 2008.

[11] K. Tourki, M.-S. Alouini, K. A. Qaraqe, and H.-C. Yang, "Performance analysis of opportunistic non-regenerative relaying," *Wiley J. Wireless Commun. Mobile Comput.*, doi: 10.1002/wcm. 2347, 2013.

[12] A. P. Prudnikov, Y. A. Brychkov, and O. I. Marichev, *Integrals and Series, vol. 2*, Gordon and Breach Science, 1986.

[13] S. S. Ikki and M. H. Ahmed, "Exact error probability and channel capacity of the best-relay cooperative-diversity networks," *IEEE Signal Processing Lett.*, vol. 16, no. 12, pp. 1051–1054, 2009.

[14] M. Torabi and D. Haccoun, "Performance analysis of cooperative diversity systems with opportunistic relaying and adaptive transmission," *IET Commun.*, 5, no. 3, pp. 264–273, 2011.

[15] K. Tourki, H.-C. Yang, M.-S. Alouini, and K. A. Qaraqe, "New results on performance analysis of opportunistic regenerative relaying," *Phys. Commun.*, doi: 10.1016/j.phycom.2013.03.005, 2013.

[16] I. Krikidis, J. S. Thompson, S. McLaughlin, and N. Goertz, "Max–min relay selection for legacy amplify-and-forward systems with interference," *IEEE Trans. Wireless Commun.*, 8, no. 6, pp. 3016–3027, 2009.

[17] K. Tourki, H.-C. Yang, and M.-S. Alouini, "Accurate outage analysis of incremental decode-and-forward opportunistic relaying," *IEEE Trans. Wireless Commun.*, 10, no. 4, pp. 1021–1025, 2011.

[18] K. Tourki, H.-C. Yang, and M.-S. Alouini, "Error-rate performance analysis of incremental decode-and-forward opportunistic relaying," *IEEE Trans. Commun.*, 59, no. 6, pp. 1519–1524, 2011.

[19] Y.-W. P. Hong, W.-J. Huang, and C.-C. J. Kuo, *Cooperative Communications and Networking: Technologies and System Design*, Springer, 2010.

[20] T. Koike-Akino, P. Popovski, and V. Tarokh, "Optimized constellations for two-way wireless relaying with physical network coding," *IEEE J. Selected Areas Commun.*, 27, no. 5, pp. 773–787, 2009.

[21] Q. F. Zhou, Y. Li, F. C. M. Lau, and B. Vucetic, "Decode-and-forward two-way relaying with network coding and opportunistic relay selection," *IEEE Trans. Commun.*, 58, no. 11, pp. 3070–3076, 2010.

[22] L. Song, "Relay selection for two-way relaying with amplify-and-forward protocols," *IEEE Trans. Vehic. Techn.*, 60, no. 4, pp. 1954–1959, 2011.

[23] H. Q. Ngo, H. A. Suraweera, M. Matthaiou, and E. G. Larsson, "Multipair full-duplex relaying with massive arrays and linear processing," *IEEE J. Selected Areas Commun.*, 32, no. 9, pp. 1721–1737, 2014.

[24] I. Krikidis and G. Zheng, *Advanced Relay Technology in Next Generation Wireless Communications*, IET Press, 2016.

9 Cognitive Transmission

The spectrum resource that is suitable for wireless communications has become increasingly scarce as a result of the rapid deployment in wireless communication systems. In general, there are two approaches to address the spectrum shortage problem. One approach is to explore higher frequency ranges, such as millimeter RF and optical frequencies. An alternative solution is to improve spectrum utilization. Measurements have shown that certain licensed RF spectra are seriously underutilized [1]. The idea of cognitive radio was proposed to address the spectrum underutilization problem. Since its introduction in [2], cognitive radio transmission has received extensive research interest [3–6].

In this chapter, we study the essential analysis and design problems of cognitive radio transmission. In particular, after introducing the basic idea of cognitive radio and key enabling provisions, we investigate two of its most popular implementation strategies, namely opportunistic spectrum access and spectrum sharing. We tackle the fundamental secondary transmission problems through accurate performance analysis in terms of suitable performance metrics. More specifically, we characterize temporal transmission opportunities and transmission delay of secondary users for opportunistic spectrum access implementation. For spectrum sharing implementation, we study the performance-enhancing mechanisms for secondary transmission under primary interference constraints. Wherever feasible, we illustrate the design trade-offs involved in different cognitive transmission strategies.

9.1 Introduction to Cognitive Radio

The basic idea of cognitive radio is to improve radio spectrum utilization by allowing secondary spectrum access [2, 3]. The basic setup for cognitive radio transmission is shown in Fig. 9.1. Specifically, the primary transmitter (PT) is communicating with the primary receiver (PU) over the frequency band that is licensed to the primary system. The secondary users (SU) can access the same frequency band, provided that the secondary transmission (ST) will not generate harmful interference to the primary transmission. There are two major implementation strategies for cognitive radio, *opportunistic spectrum access* and *spectrum sharing*.

With opportunistic spectrum access implementation (also known as interweave), secondary users can access the licensed spectrum only when it is not used by primary users

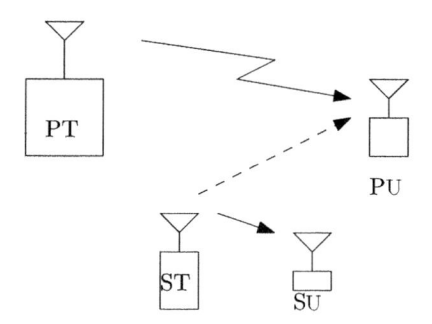

Figure 9.1 Basic structure of a cognitive transmission system.

and must vacate the occupied spectrum when primary users start using the spectrum [7–9]. Spectrum handoff procedures are adapted for returning the channel to PU and then reaccessing it or another channel later to complete secondary transmission. As such, secondary transmission creates no interference to PU. To implement opportunistic spectrum access, the SU need to accurately sense the activity of PU on the target spectrum. Such spectrum sensing problems have been extensively researched in the signal processing community and various solutions are available with different trade-offs between false alarm and miss detection probabilities. Typically, to protect the PU, most sensing strategies target near-zero miss detection probability while accepting some occurrence of false alarms.

With spectrum sharing implementation (or equivalently underlay), PU and SU can simultaneously access the same spectrum, with a stringent constraint on the interference that SU may cause to primary transmission. Typically, the interference generated by the secondary transmission to the primary receiver must be less than a certain threshold level. Since the interference power at the primary receiver depends on the secondary transmission power as well as the channel gain from the secondary transmitter to the primary receiver, the secondary system needs to accurately estimate the forward channel gain to the primary receiver. Such a channel estimation problem can be challenging if no cooperation from the primary receiver is available. Considerable research effort has been directed to develop effective signal processing techniques for forward channel estimation to the primary receiver [10, 11].

Certain cognitive implementations adopt a hybrid interweave/underlay approach to access the licensed spectrum continuously, while respecting the interference constraint when the primary system is using the spectrum. The secondary system will need to perform both spectrum sensing and forward channel estimation. Overlay is the third implementation strategy, where SU help PU's transmission by, for example, acting as relays, while satisfying their own communication needs. Typically, the SU need to detect primary transmission, which entails higher system complexity. In the following sections, we concentrate on interweave and underlay cognitive implementation.

The fundamental provision for cognitive radio is the ability to identify spectrum opportunities. With interweave implementation, accurate sensing of spectrum activity is essential, whereas channel estimation/interference prediction is mandatory for underlay solutions. Once these signal processing problems are effectively solved, the secondary systems can apply most conventional transmission technologies. Meanwhile, secondary cognitive transmission faces some new challenges. In particular, secondary interweave transmission must wait for spectrum opportunities and may be interrupted by primary transmission. Underlay transmission must respect the interference constraint at primary receivers. In what follows, we analyze and design secondary transmission technologies under these unique challenges.

9.2 Temporal Spectrum Opportunity Characterization

In this section, we characterize the temporal spectrum opportunity with interweave cognitive implementation. We introduce two metrics, namely average waiting time and average service time [12], which measure how much time the SU needs to wait before transmission and for how long the SU can transmit on average, respectively. We derive the statistics of these metrics under different PU traffic models with different access schemes.

9.2.1 System Model

We assume that SU try to opportunistically access one of the available channels of primary systems. Specifically, SU have the ability to sense and access only one channel at a time. When the SU is using an idle channel, it will continuously monitor PU activity on that channel and stop accessing it whenever the channel becomes busy. Thus, the secondary transmission causes no interference to the primary system. We assume that the channel sensing is perfect and the channel sensing results at the transmitter and receiver are the same. The occupancy of each channel by the PU evolves independently according to a homogeneous continuous-time Markov chain with idle (OFF) and busy (ON) states. We denote the duration of the ON and OFF period of the primary channel by T_{on}^p and T_{off}^p, respectively.

9.2.2 Single Channel Access

In this subsection, we assume that the SU has access to only one primary channel. When the channel is sensed busy, the SU will periodically sense the channel every T_s period. Once the SU finds that the channel is free, it starts transmission while monitoring the PU activity continuously.

General PU Traffic

We denote by T_{on}^s and T_{off}^s the duration of the ON and OFF periods for the SU, respectively, as illustrated in Fig. 9.2. During the ON period of the PU, the SU performs periodic sensing with period T_s. Thus the OFF duration of the SU can be expressed as $T_{off}^s = N T_s$, where N is a random variable (RV) that represents the number of the sensing periods before the SU switch to the ON state. It is clear that $T_{on}^p + T_{off}^p = T_{on}^s + T_{off}^s$. Thus, the ON duration of the SU can then be expressed as $T_{on}^s = T_{off}^p - \tau$, where

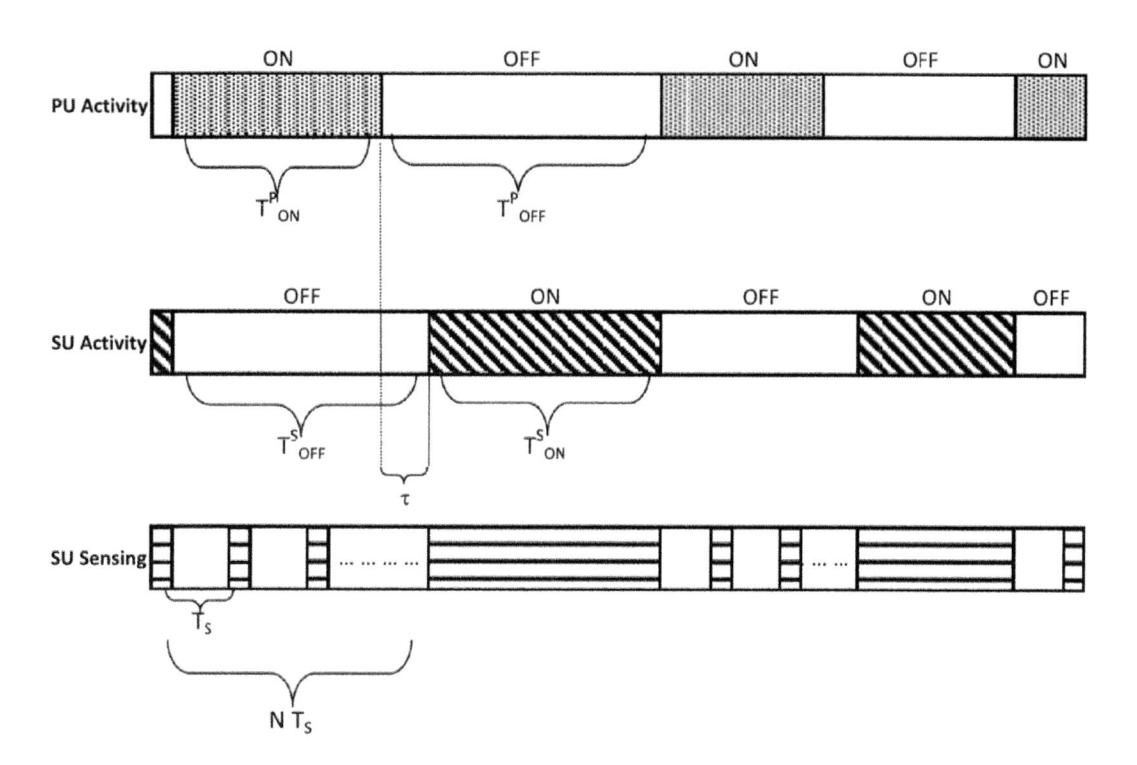

9.2 Temporal Spectrum Opportunity Characterization

Figure 9.2 Illustration of PU and SU activities, and SU sensing. Reprint with permission from [12]. ©2012 IEEE.

$\tau = NT_s - T_{on}^p$ represents the time duration when the PU is OFF and the SU is not transmitting, $\tau \in [0, T_s]$.

Conditioning on N, T_{on}^p is between $(N-1)T_s$ and NT_s. For a general PU activity model, let $F_{T_{on}^p}(.)$ be the cumulative distribution function (CDF) of T_{on}^p; the probability density function (PDF) of T_{on}^p conditioned on N can be expressed as

$$f_{T_{on}^p|N}(t) = \frac{f_{T_{on}^p}(t)}{F_{T_{on}^p}(NT_s) - F_{T_{on}^p}((N-1)T_s)}, \quad NT_s \leq t \leq (N-1)T_s. \tag{9.1}$$

Using this result, we can express the PDF of τ conditioned on N as

$$f_{\tau|N}(t) = \frac{f_{T_{on}^p}(NT_s - t)}{F_{T_{on}^p}(NT_s) - F_{T_{on}^p}((N-1)T_s)}, \quad 0 \leq t \leq T_s. \tag{9.2}$$

The probability mass function (PMF) of N can be expressed as a function of the CDF of T_{on}^p as

$$P_N = F_{T_{on}^p}(NT_s) - F_{T_{on}^p}((N-1)T_s). \tag{9.3}$$

Using the results in (9.2) and (9.3), it follows that

$$f_\tau(t) = \sum_{n=1}^{+\infty} f_{T_{on}^p}(nT_s - t), \quad 0 \leq t \leq T_s. \tag{9.4}$$

The average SU service duration for the general PU traffic model is defined as $\bar{T}^s_{on} = E[T^p_{off}] - E[\tau]$. Assuming that T_s is very small and ignoring the small probability that $T^p_{off} < \tau$, we have

$$\bar{T}^s_{on} = E[T^p_{off}] + E[T^p_{on}] \\ - \sum_{n=1}^{+\infty} n T_s \left(F_{T^p_{on}}(nT_s) - F_{T^p_{on}}((n-1)T_s) \right). \quad (9.5)$$

Using the PMF of N in (9.3), the average SU waiting time can be written as

$$\bar{\delta} = \sum_{n=1}^{+\infty} n T_s \left[F_{T^p_{on}}(nT_s) - F_{T^p_{on}}((n-1)T_s) \right]. \quad (9.6)$$

LT Traffic

Some experimental studies have shown that PU activity in systems such as the 802.11 wireless LAN, and voice-oriented cellular systems can be approximately modeled by light-tailed (LT) traffic [13]. In general, an RV is LT if its distribution decreases exponentially or faster. Some examples include exponential, gamma, and Weibull with shape parameter greater than 1 [14]. One example of LT traffic is the Poisson traffic, in which the ON and OFF periods are exponentially distributed, with PDFs given by

$$f_{T^p_{on}}(t) = \frac{1}{\lambda} e^{-\frac{t}{\lambda}} U(t), \\ f_{T^p_{off}}(t) = \frac{1}{\mu} e^{-\frac{t}{\mu}} U(t), \quad (9.7)$$

respectively, where λ and μ represent the average values of the ON period and the OFF period, respectively, and $U(.)$ is the unit step function. Based on (9.5) for general PU traffic, and using the statistics of the PU activity given in (9.7), the average SU access duration can be obtained as

$$\bar{T}^s_{on} = \mu + \lambda - \frac{T_s}{1 - e^{-\frac{T_s}{\lambda}}}. \quad (9.8)$$

Moreover, for the case under consideration, the PMF of N defined in (9.3) can be expressed as

$$P_N = e^{-(N-1)\frac{T_s}{\lambda}} \left(1 - e^{-\frac{T_s}{\lambda}}\right). \quad (9.9)$$

Therefore, $\bar{\delta}$ is given by

$$\bar{\delta} = \frac{T_s}{1 - e^{-\frac{T_s}{\lambda}}}. \quad (9.10)$$

It can be seen that when $T_s \to 0$, which means that the SU senses the channel continuously, $\bar{T}^s_{on}$ approaches μ and $\bar{\delta}$ approaches λ, as expected intuitively. In this case, the SU will start transmitting exactly when the PU stops transmission, and will stop accessing the channel exactly when the PU becomes active.

HT Traffic

Empirical measurements showed in the case of communication systems for video streaming, and Internet server traffics [15], PU activities should be modeled by heavy-tailed (HT) traffic. Generally speaking, an RV is HT if its distribution decreases slower than exponential. Typical examples include Pareto, log-normal, and Weibull with shape parameter larger than 1 [14]. For HT traffic, the PU busy state can have infinite mean and variance, which can considerably degrade the SU throughput.

As an example of HT PU traffic, we consider the exponential distribution for the OFF duration with mean μ and Pareto distribution for the ON duration with shape parameter $\alpha > 1$ and scale parameter $x_m > 0$, which leads to

$$f_{T_{on}^p}(x) = \frac{\alpha x_m^\alpha}{x^{\alpha+1}} U(x - x_m),$$
$$f_{T_{off}^p}(x) = \frac{1}{\mu} e^{-\frac{x}{\mu}} U(x). \qquad (9.11)$$

Given that $E[T_{on}^p] = \alpha x_m/(\alpha - 1)$, for $\alpha > 1$, and $E[T_{off}^p] = \mu$, the results for $\bar{T}_{on}^s$ and $\bar{\delta}$ can be deduced from (9.5) and (9.6), respectively. When $n_0 T_s < x_m \leq (n_0 + 1)T_s$, for some integer n_0, we have

$$\bar{T}_{on}^s = \mu + \frac{\alpha x_m}{\alpha - 1} - \bar{\delta},$$
$$\bar{\delta} = (n_0 + 1)T_s + \frac{x_m^\alpha}{T_s^{\alpha-1}} \zeta(\alpha, n_0 + 1), \qquad (9.12)$$

where $\zeta(s, q)$ is the Hurwitz zeta function, which is formally defined for real arguments $s > 1$ and $q > 0$ by $\zeta(s, q) = \sum_{n=0}^{+\infty} \frac{1}{(q+n)^s}$.

9.2.3 Multiple Channel Access Based on SEC

In this subsection, we assume that the SU can access the available PU channels in a switch and examine fashion. In particular, if the current-sensing PU channel is found idle, the SU will start using that channel while monitoring the PU activity on it. When PU reappears on that channel, the SU tries to find another idle PU channel for transmission. For simplicity, when SU switches to a new channel, the switching duration is considered as part of the sensing duration T_p.

The service time for the SU, for such SEC-based channel access scheme, is given by $T_{on}^s = T_{off}^p - W - T_p$, where W is a uniform RV defined on the interval $[0, T_{off}^p - T_p]$ with PDF given by

$$f_W(t) = \frac{1}{T_{off}^p - T_p} \left(U(t) - U(t - T_{off}^p + T_p) \right). \qquad (9.13)$$

The average value of T^s_{on} can be determined as

$$\bar{T}^s_{on} = E[T^p_{off}] - \int_0^\infty E[W/T^p_{off}]f_{T^p_{off}}(t)dt - T_p$$
$$= \frac{1}{2}\left(E[T^p_{off}] - T_p\right).$$
(9.14)

The SU waiting time is a multiple of T_p, i.e., $\delta = NT_p$, where N represents the number of PU channels that the SU has to examine before finding a free one. Therefore, the PMF of N becomes

$$P_n = E[T^p_{off}]E[T^p_{on}]^{n-1}\left(E[T^p_{on}] + E[T^p_{off}]\right)^{-n}.$$
(9.15)

Then, the average waiting time can be calculated as

$$\bar{\delta} = \sum_{n=1}^{+\infty} nT_p P_n$$
$$= \frac{T_p}{E[T^p_{off}]}(E[T^p_{on}] + E[T^p_{off}]).$$
(9.16)

For the Poisson PU traffic model considered in previous subsection, the average SU service time is given by $\bar{T}^s_{on} = (1/2)(\mu - T_p)$. Moreover, the average waiting time specializes from (9.16) to $\bar{\delta} = (T_p/\mu)(\lambda + \mu)$. For the sample HT PU traffic model considered earlier, the average SU service time is the same as that for Poisson PU traffic. On the other hand, using (9.16), the SU waiting time can be written as $\bar{\delta} = (T_p/\mu)(\mu + (\alpha x_m/\alpha - 1))$.

9.2.4 Multiple Channel Access Based on SSC

In this subsection, we assume that the SU can sense only one alternative channel in each sensing period in a switch and stay fashion. Specifically, if a channel is found idle, the SU transmits on that channel while monitoring it continuously for PU activity. Once the PU activity is detected, the SU switches to one of the remaining channels. If this switched-to channel is found to be free, the SU can use it directly. Otherwise, the SU stays on that channel and waits until it becomes free. While waiting, the SU senses periodically (every T_s) the PU activity. We note that the SU waiting time for such SSC-based access scheme is exactly the same as that of the single channel access, which is given for general PU traffic in (9.6), and then specialized for LT and HT traffic conditions in (9.10) and (9.12), respectively.

To calculate the average service time for this SSC-based access scheme, two cases need to be considered, which are (1) when the SU switches to another channel and finds it idle; and (2) when the switched-to channel is found to be busy. For the first case, the service duration is denoted by T_1, whereas for the second case by T_2. The service time can then be written as $T^s_{on} = p_1 T_1 + p_2 T_2$, where p_1 and p_2 are the probabilities of finding switched-to channel idle and busy, respectively, and they are given by $p_1 = E[T^p_{off}]/(E[T^p_{on}] + E[T^p_{off}])$ and $p_2 = E[T^p_{on}]/(E[T^p_{on}] + E[T^p_{off}])$.

When the switched-to channel is found to be idle, the SU waits for T_p duration for switching and sensing and then starts transmitting. Thus, the average service time will be identical to that of the SEC scheme, which leads to $\bar{T}_1 = (1/2)(E[T_{off}^p] - T_p)$. When the switched-to channel is found to be busy, the SU will stay in that channel and wait until the channel becomes free. During the waiting time, the SU will sense the channel periodically every T_s. The service time in this case is identical to that of the single channel case, given in (9.5).

For Poisson LT PU traffic, the average SU service duration can be given by

$$\bar{T}_{on}^s = \lambda + \frac{1}{2}\frac{\mu(\mu - T_p)}{\mu + \lambda} - \frac{1}{\mu + \lambda}\left(\frac{T_s}{1 - e^{-\frac{T_s}{\lambda}}}\right). \tag{9.17}$$

On the other hand, for HT PU traffic, the average SU service time can be written as

$$\bar{T}_{on}^s = \frac{\alpha x_m}{\alpha - 1} + \frac{1}{2}\frac{\mu(\alpha - 1)(\mu - T_p)}{\alpha x_m + \mu(\alpha - 1)}$$
$$- \frac{\alpha x_m}{\alpha x_m + \mu(\alpha - 1)} \times \left[(n_0 + 1)T_s + \frac{x_m^\alpha}{T_s^{\alpha-1}}\zeta(\alpha, n_0 + 1)\right]. \tag{9.18}$$

9.2.5 Numerical Results

Fig. 9.3 shows the average SU service time with different access schemes for the LT traffic model as a function of the average OFF duration of the PU μ. It is clear that as μ increases, the service duration of SU increases for all the considered schemes. On the other hand, it can be seen that the single channel access scheme have longer average service duration compared to switching-based schemes.

Fig. 9.4 shows the average waiting time of the SU with different access schemes for the HT traffic model as a function of x_m. It is clear that as x_m increases, the SU waiting duration for all the considered schemes increases, and this is because the ON period of the PU will be longer. On the other hand, the trade-off between the access and waiting durations can be observed. Note that single-channel access scheme achieves the largest access duration, but has the largest waiting duration on average. The advantage of the SEC-based access scheme in terms of reducing the waiting time becomes obvious, at the cost of shorter service duration due to switching overhead.

9.3 Extended Delivery Time Analysis

In this section, we investigate the statistics of the time duration to transmit a given amount of data under interweave cognitive implementation. With interweave implementation, secondary transmission may be interrupted by PU activity. As such, the secondary transmission of a given amount of data may involve multiple spectrum handoffs. The total time required for SU to complete the transmission of a given amount of data will include the waiting periods before accessing the channel. To facilitate the delay analysis of secondary transmission for a fixed amount of data, we study the resulting extended

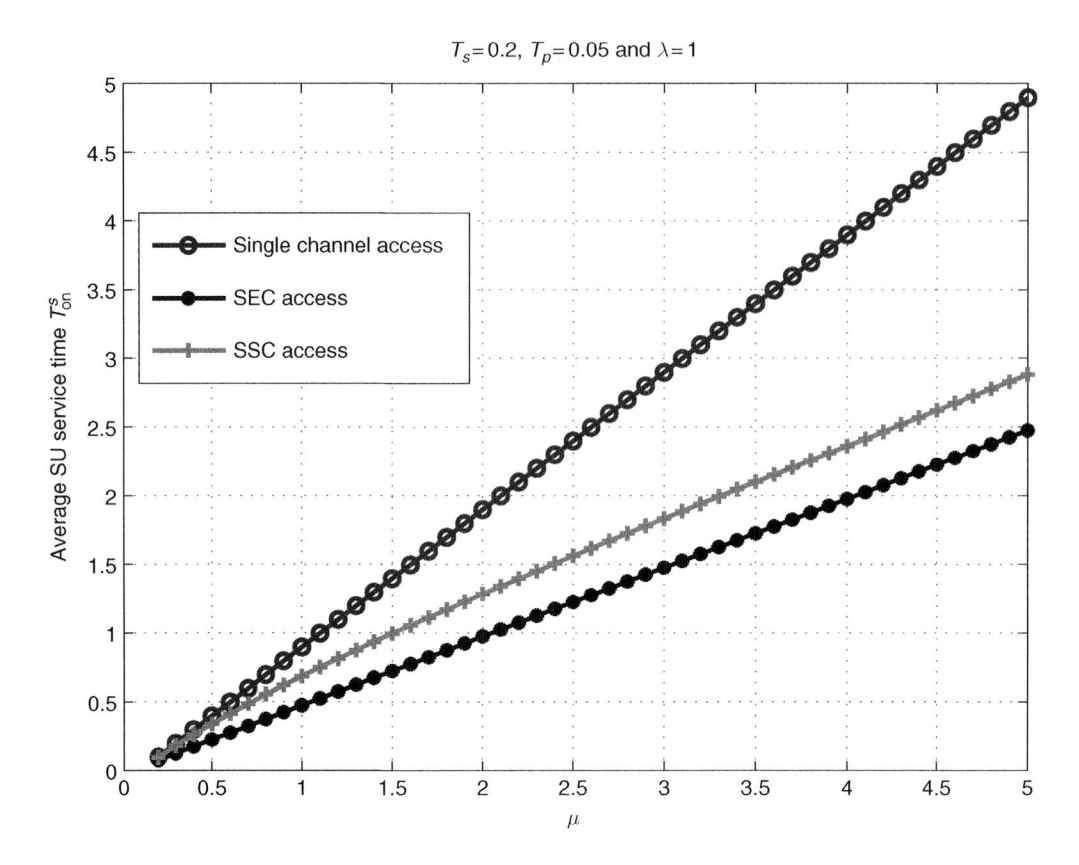

Figure 9.3 Average SU service duration for the different proposed schemes and for the LT traffic model. Reprint with permission from [12]. ©2012 IEEE.

delivery time (EDT), which includes both transmission time and waiting time [16, 17]. In particular, we derive the exact distribution function of EDT for secondary data transmission with work-preserving and non–work-preserving transmission strategies for both the ideal continuous sensing, in which the SU will continuously sense for the channel availability, and the more practical periodic sensing, in which the SU will sense the channel periodically. As an application of these results, we study the secondary queuing performance for Poisson traffic. The effect of sensing imperfection is also examined [18, 19].

9.3.1 System Model and Problem Formulation

We consider a cognitive transmission scenario in which the SU opportunistically accesses a channel of the primary system for data transmission. The occupancy of that channel by the PU evolves according to a homogeneous continuous-time Markov chain with an average busy period of λ and an average idle period of μ. The durations of busy and idle periods are assumed to be exponentially distributed. The SU opportunistically

9.3 Extended Delivery Time Analysis

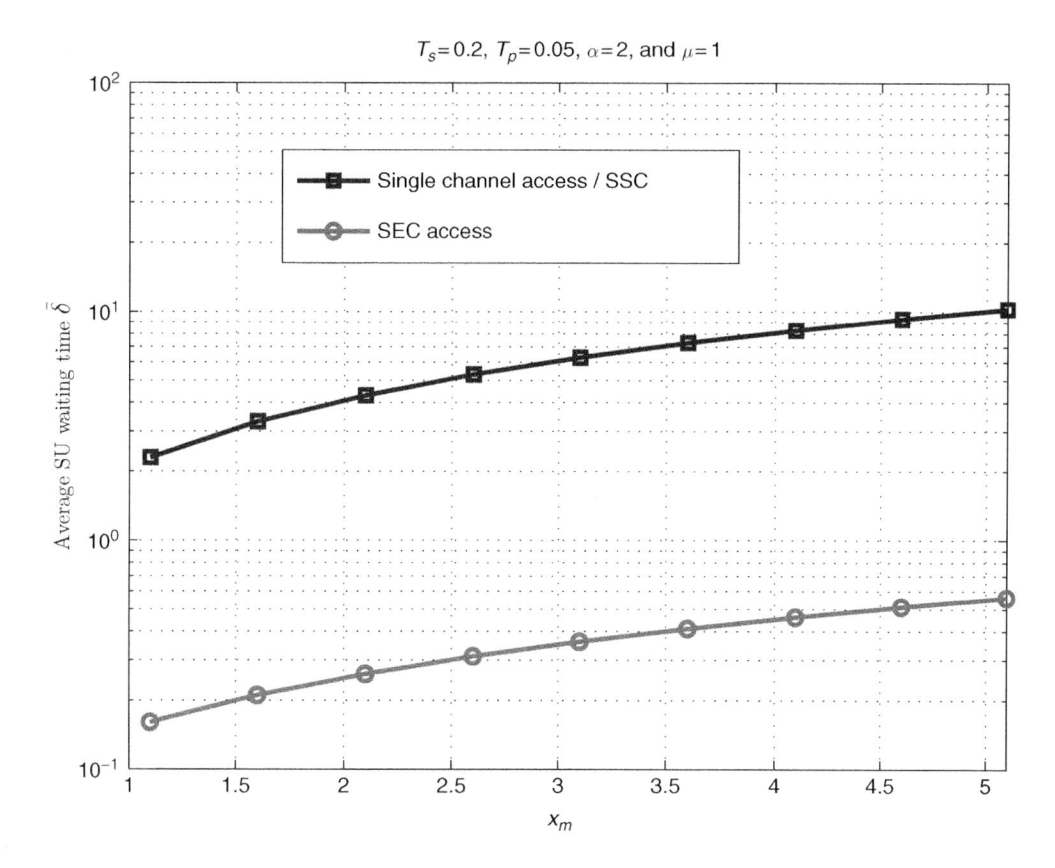

Figure 9.4 Average SU waiting duration for the different proposed schemes and for the HT traffic model. Reprint with permission from [12]. ©2012 IEEE.

accesses the channel in an interweave fashion. Specifically, the SU can use the channel only after PU stops transmission. As soon as the PU restarts transmission, the SU instantaneously stops its transmission, and thus no interference is caused to PU.

The SU monitors PU activity on the target channel through spectrum sensing. With continuous sensing, the SU continuously senses the channel for availability. Thus, the SU starts its transmission as soon as the channel becomes available. We also consider the case in which the SU senses the channel periodically, with a period of T_s. If the channel is sensed to be busy, the SU will wait for T_s time period and re-sense the channel. With periodic sensing, there is a small amount of time when the PU has stopped its transmission but the SU has not yet sensed the channel, as illustrated in Fig. 9.5. During transmission, the SU continuously monitors PU activity. As soon as the PU restarts, the SU stops its transmission. The continuous period of time during which the PU is off and the SU is transmitting is referred to as a transmission slot. Similarly, the continuous period of time during which the PU is transmitting is referred to as a waiting slot. For the periodic sensing case, the waiting slot also includes the time when the PU has stopped transmission, but the SU has not yet sensed the channel.

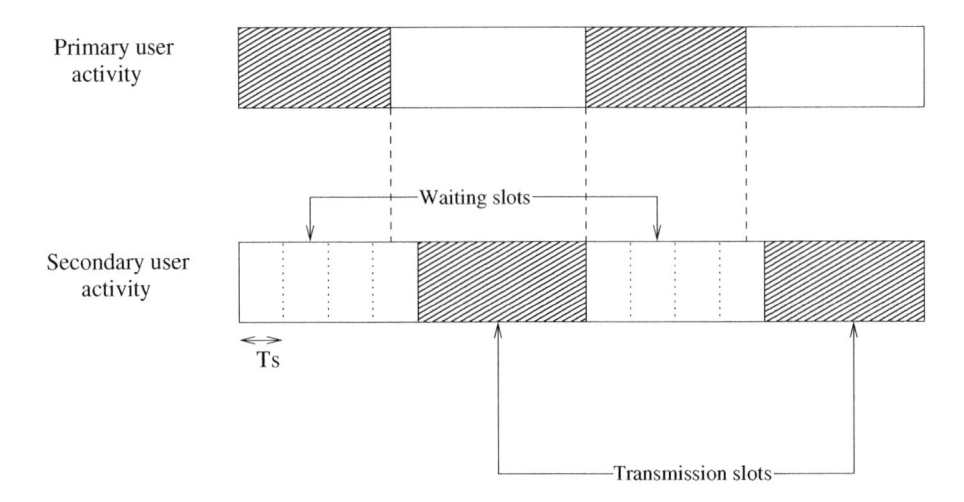

Figure 9.5 Illustration of PU and SU activities and SU sensing for periodic sensing case. Reprint with permission from [18]. ©2015 IEEE.

When the secondary transmission is interrupted by PU activities, the secondary system may adopt either non–work-preserving strategy, where interrupted packets transmission must be repeated [16], or work-preserving strategy, where the secondary transmission can continue from the point at which it was interrupted, without wasting the previous transmission [17]. Work-preserving transmission can be achieved with the application of rateless codes such as Fountain code and Raptor code [20, 21]. The work-preserving strategy also applies to the transmission of small and individually coded mini-packets. The total data delivery time with the work-preserving strategy includes an interleaved sequence of the transmission time and the waiting time. The resulting EDT for the data packet is mathematically given by $T_{ED} = T_w + T_{tr}$, where T_w is the total waiting time and T_{tr} is the packet transmission time. Note that both T_w and T_{tr} are, in general, random variables, with T_w depending on T_{tr}, PU behavior, and sensing strategies, and T_{tr} depending on packet size and secondary channel condition when available. When rateless code is not applied or packet size is not sufficiently small, the non–work-preserving strategy should apply. Analysis with the non–work-preserving strategy is, in general, more challenging as the transmission of a secondary data packet with the non–work-preserving strategy will involve an interleaved sequence of wasted transmission slots and waiting time slots, both of which can have random time durations, followed by the final successful transmission slot.

In the following, we investigate the EDT of a secondary system for a single packet arriving at a random point in time. We concentrate on the case in which T_{tr} can be viewed as a constant. This applies to slow fading channel and very rapidly varying channel when the data transmission will experience different channel realizations. The transmission time for fast fading channel can be estimated by applying the ergodic capacity of wireless channels as [22]

$$T_{tr} \approx \frac{H}{W \int_0^\infty \log_2(1+\gamma) f_\gamma(\gamma) d\gamma}, \qquad (9.19)$$

where H is the entropy of the packet, W is the available bandwidth, and $f_\gamma(\gamma)$ is the PDF of the signal-to-noise ratio (SNR) of the fading channel. For both work-preserving and non–work-preserving strategies with continuous sensing or periodic sensing, we derive the exact distribution of T_{ED}. These analytical results directly characterize the delay of some secondary applications with low traffic intensity. For example, in wireless sensor networks for health care monitoring, forest fire detection, air pollution monitoring, disaster prevention, landslide detection, etc., the transmitter needs to periodically transmit measurement data to the sink with a long duty cycle. The EDT essentially characterizes the delay of measurement data collection. For high-traffic-intensity applications, the EDT characterization will facilitate the queuing analysis, as demonstrated in later subsections.

9.3.2 Work-Preserving Transmission with Continuous Sensing

When the SU adopts the work-preserving strategy, the EDT for secondary packet transmission consists of an interleaved sequence of waiting slots and transmission slots. We assume, without loss of generality, that the packet arrives at $t = 0$. The distribution of T_w depends on whether the PU was on or off at that instant, as illustrated in Figs. 9.6 and 9.7. We denote the PDF of the waiting time of the SU for the case when PU is on at $t = 0$, and for the case when PU is off at $t = 0$, by $f_{T_w, p_{on}}(t)$ and $f_{T_w, p_{off}}(t)$, respectively. The PDF of the waiting time T_w for the SU is then given by

$$f_{T_w}(t) = \frac{\lambda}{\lambda + \mu} f_{T_w, p_{on}}(t) + \frac{\mu}{\lambda + \mu} f_{T_w, p_{off}}(t), \qquad (9.20)$$

where $\frac{\lambda}{\lambda+\mu}$ and $\frac{\mu}{\lambda+\mu}$ are the stationary probabilities that the PU is on or off at $t = 0$, respectively.

When the PU is on at $t = 0$, T_w includes k waiting slots if k transmission slots are needed for packet transmission. Let $\mathcal{P}_k$ represent the probability that the SU completes

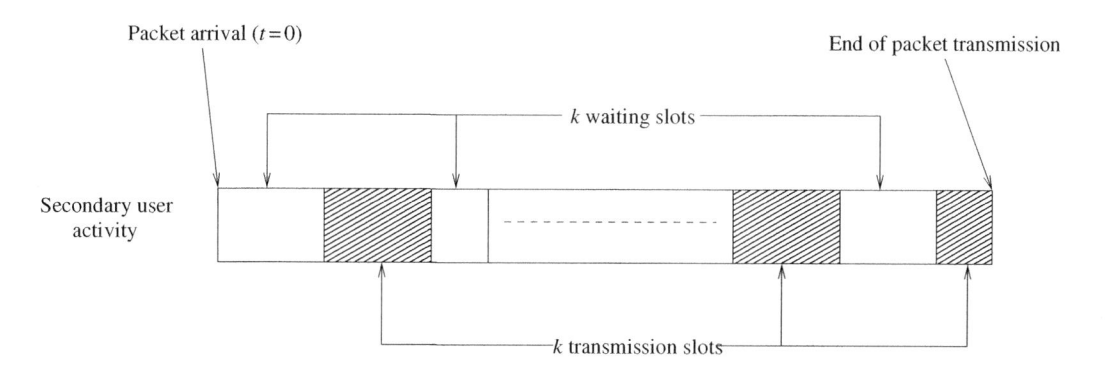

Figure 9.6 Illustration of secondary transmission when the PU is on at $t = 0$. Reprint with permission from [18]. ©2015 IEEE.

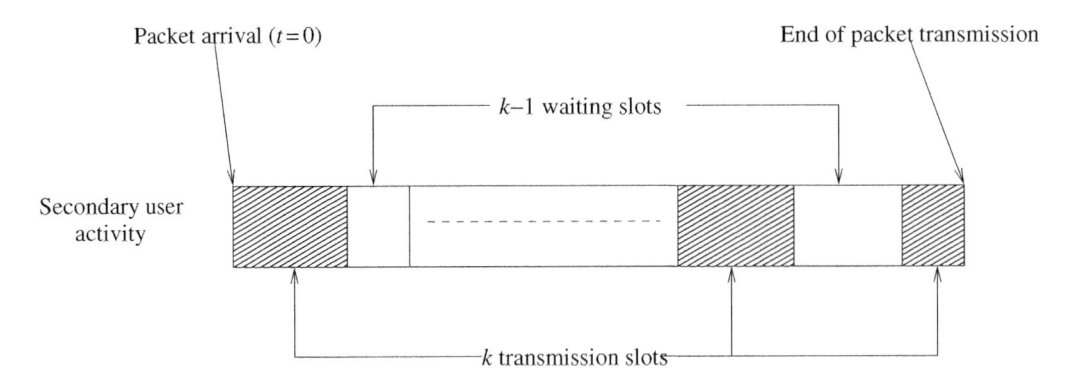

Figure 9.7 Illustration of secondary transmission when the PU is off at $t = 0$. Reprint with permission from [18]. ©2015 IEEE.

packet transmission in k transmission slots, and $f_{T_w,k}(t)$ represent the PDF of the total time duration of k SU waiting slots. Then the PDF of the total SU waiting time, for the case when PU is on at $t = 0$, is given by

$$f_{T_w,p_{on}}(t) = \sum_{k=1}^{\infty} \mathcal{P}_k \times f_{T_w,k}(t). \quad (9.21)$$

Note that $f_{T_w,k}(t)$ is the PDF of the sum of k independent and identically distributed exponential random variables with average λ. Therefore, $f_{T_w,k}(t)$ is given by

$$f_{T_w,k}(t) = \frac{1}{\lambda^k} \frac{t^{k-1}}{(k-1)!} e^{\frac{-t}{\lambda}}. \quad (9.22)$$

$\mathcal{P}_k$ can be calculated as the probability that k SU transmission slots have a total time of more than T_{tr}, whereas $k-1$ transmission slots have a total time of less than T_{tr}. Since the total time for k transmission slots follows the Erlang distribution with PDF

$$f_{T_{tr},k}(t) = \frac{1}{\mu^k} \frac{t^{k-1}}{(k-1)!} e^{\frac{-t}{\mu}}, \quad (9.23)$$

we can show that

$$\mathcal{P}_k = \int_{T_{tr}}^{\infty} \frac{1}{\mu^k} \frac{t^{k-1}}{(k-1)!} e^{\frac{-t}{\mu}} dt - \int_{T_{tr}}^{\infty} \frac{1}{\mu^{k-1}} \frac{t^{k-2}}{(k-2)!} e^{\frac{-t}{\mu}} dt. \quad (9.24)$$

After using integration by parts on the first integral and cancelling the terms, $\mathcal{P}_k$ can be calculated as

$$\mathcal{P}_k = \frac{T_{tr}^{k-1} e^{\frac{-T}{\mu}}}{\mu^{k-1}(k-1)!}. \quad (9.25)$$

After substituting (9.22) and (9.76) into (9.21), we get

$$f_{T_w,p_{on}}(t) = \sum_{k=1}^{\infty} \frac{T_{tr}^{k-1} e^{\frac{-T_{tr}}{\mu}}}{\mu^{k-1}(k-1)!} \times \frac{1}{\lambda^k} \frac{t^{k-1}}{(k-1)!} e^{\frac{-t}{\lambda}}. \quad (9.26)$$

Finally, applying the definition of the Bessel function, we arrive at the following closed-form expression for $f_{T_w, p_{on}}(t)$:

$$f_{T_w, p_{on}}(t) = \frac{1}{\lambda} e^{\frac{-T_{tr}}{\mu}} I_0\left(2\sqrt{\frac{T_{tr} t}{\mu \lambda}}\right) e^{\frac{-t}{\lambda}}, \tag{9.27}$$

where $I_n(.)$ is the modified Bessel function of the first kind of order n.

Similarly, the PDF for T_w when PU is off at $t = 0$ can be obtained as

$$f_{T_w, p_{off}}(t) = \sum_{k=1}^{\infty} \mathcal{P}_k \times f_{T_w, k-1}(t)$$

$$= e^{\frac{-T_{tr}}{\mu}} \delta(t) + \sum_{k=2}^{\infty} \frac{T_{tr}^{k-1} e^{\frac{-T_{tr}}{\mu}}}{\mu^{k-1}(k-1)!} \times \frac{1}{\lambda^{k-1}} \frac{t^{k-2}}{(k-2)!} e^{\frac{-t}{\lambda}}, \tag{9.28}$$

which simplifies to

$$f_{T_w, p_{off}}(t) = e^{\frac{-T_{tr}}{\mu}} \delta(t) + \sqrt{\frac{T_{tr}}{\mu \lambda t}} e^{\frac{-T_{tr}}{\mu}} I_1\left(2\sqrt{\frac{T_{tr} t}{\mu \lambda}}\right) e^{\frac{-t}{\lambda}}, \tag{9.29}$$

where $\delta(t)$ is the delta function. Note that the term $e^{\frac{-T_{tr}}{\mu}} \delta(t)$ corresponds to the case that the number of waiting slots is equal to 0.

After substituting (9.27) and (9.29) into (9.20), and noting $T_{ED} = T_w + T_{tr}$, the PDF for the EDT T_{ED} for the continuous sensing case is given by

$$f_{T_{ED}}(t) = \frac{\mu}{\lambda + \mu} e^{\frac{-T_{tr}}{\mu}} \delta(t - T_{tr}) + \frac{1}{\lambda + \mu} e^{\frac{-t}{\lambda}}$$

$$\times \left[I_0\left(2\sqrt{\frac{T_{tr}(t - T_{tr})}{\mu \lambda}}\right) \right.$$

$$\left. + \sqrt{\frac{T_{tr} \mu}{\lambda(t - T_{tr})}} I_1\left(2\sqrt{\frac{T_{tr}(t - T_{tr})}{\mu \lambda}}\right) \right] u(t - T_{tr}), \tag{9.30}$$

where $u(.)$ is the step function.

9.3.3 Work-Preserving Transmission with Periodic Sensing

In the case of periodic sensing, the waiting time T_w will be an integer multiple of the sensing period T_s. Therefore, T_w will have a discrete distribution. Similar to the continuous sensing case, we can write the probability that the waiting time T_w is nT_s by considering whether the PU is on or off at $t = 0$ separately, as

$$\Pr[T_w = nT_s] = \frac{\lambda}{\lambda + \mu} \Pr[T_w, p_{on} = nT_s] + \frac{\mu}{\lambda + \mu} \Pr[T_w, p_{off} = nT_s], \tag{9.31}$$

where $\Pr[T_w, p_{on} = nT_s]$ is the probability that the total waiting time for the SU is nT_s when the PU is on at $t = 0$, and $\Pr[T_w, p_{off} = nT_s]$ the probability when the PU is off at $t = 0$.

It can be shown based on the illustration in Fig. 9.6 that

$$\Pr[T_w, p_{on} = nT_s] = \sum_{k=1}^{\infty} \mathcal{P}_k \times \Pr[T_{w,k} = nT_s], \qquad (9.32)$$

where $\mathcal{P}_k$ is the probability that the SU completes its transmission in k slots, given in (9.76), and $\Pr[T_{w,k} = nT_s]$ is the probability that the SU waiting time in k slots is nT_s, given by

$$\Pr[T_{w,k} = nT_s] = \binom{n-1}{k-1}(1-\beta)^k(\beta)^{n-k}, \qquad (9.33)$$

where β is the probability that the PU is on at the current sensing instant given that it was on at the previous sensing instant. Due to the memoryless property of exponential distribution, β is a constant. Since the SU only senses every T_s time period, there is a chance that the PU turns off and then back on between two sensing instants. To account for such possibilities, we can model PU activity with a continuous-time Markov process, where the transition rates are given by $\frac{1}{\lambda}$ and $\frac{1}{\mu}$. Therefore, the probability β is given by [23]

$$\beta = \frac{\lambda}{\lambda+\mu} + \frac{\mu}{\lambda+\mu} e^{-(\frac{1}{\lambda}+\frac{1}{\mu})T_s}. \qquad (9.34)$$

If we assume that T_s is very small, and the chance of the PU turning back on again before the SU senses a free channel is negligible, we can approximate β with $e^{\frac{-T_s}{\lambda}}$.

After substituting (9.76) and (9.33) into (9.75), and some manipulation, we can calculate $\Pr[T_w, p_{on} = nT_s]$ as

$$\Pr[T_w, p_{on} = nT_s] = (1-\beta)\beta^{n-1} e^{\frac{-T_{tr}}{\mu}} \times \sum_{0}^{n-1} \left[\frac{T_{tr}(1-\beta)}{\mu\beta}\right]^k \frac{1}{k!}\binom{n-1}{k}, \qquad (9.35)$$

which simplifies to

$$\Pr[T_w, p_{on} = nT_s] = (1-\beta)\beta^{n-1} e^{\frac{-T_{tr}}{\mu}} \times {}_1F_1\left(1-n; 1; \frac{-T_{tr}(1-\beta)}{\mu\beta}\right), \qquad (9.36)$$

where ${}_1F_1(.,.,.)$ is the generalized hypergeometric function. Similarly, $\Pr[T_w, p_{off} = nT_s]$ in (9.31) can be calculated as

$$\Pr[T_w, p_{off} = nT_s] = \sum_{k=1}^{\infty} \mathcal{P}_k \times \Pr[T_{w,k-1} = nT_s]. \qquad (9.37)$$

After substituting (9.76) and (9.33) into (9.37), and some manipulation, we get

$$\Pr[T_w, p_{off} = nT_s] = \left[\frac{T_{tr}(1-\beta)\beta^{n-1}}{\mu}\right] e^{\frac{-T_{tr}}{\mu}}$$
$$\times \sum_{k=0}^{n-1}\left[\frac{T_{tr}(1-\beta)}{\mu\beta}\right]^k \frac{1}{(k+1)!}\binom{n-1}{k} + e^{-\frac{T_{tr}}{\mu}}\delta[n], \qquad (9.38)$$

which eventually simplifies to

9.3 Extended Delivery Time Analysis

$$\Pr[T_w, p_{off} = nT_s] = \left[\frac{T_{tr}(1-\beta)\beta^{n-1}}{\mu}\right] e^{\frac{-T_{tr}}{\mu}}$$
$$\times {}_1F_1\left(1-n; 2; \frac{-T_{tr}(1-\beta)}{\mu\beta}\right) u[n-1] + e^{-\frac{T_{tr}}{\mu}} \delta[n]. \quad (9.39)$$

After substituting (9.36) and (9.39) into (9.31), the PMF of the EDT for the periodic sensing case is given by

$$\Pr[T_{ED} = nT_s + T_{tr}] = \frac{\lambda}{\lambda + \mu}(1-\beta)\beta^{n-1} e^{\frac{-T_{tr}}{\mu}}$$
$$\times {}_1F_1\left(1-n; 1; \frac{-T_{tr}(1-\beta)}{\mu\beta}\right) u[n]$$
$$+ \frac{\mu}{\lambda + \mu}\left[\left(\frac{T_{tr}(1-\beta)\beta^{n-1}}{\mu}\right) e^{\frac{-T_{tr}}{\mu}} {}_1F_1\left(1-n; 2; \frac{-T_{tr}(1-\beta)}{\mu\beta}\right) u[n-1]\right.$$
$$\left. + e^{-\frac{T_{tr}}{\mu}} \delta[n]\right]. \quad (9.40)$$

Fig. 9.8 shows the PMF envelope of the packet delivery time with periodic sensing for various values of sensing period T_s. As can be seen, the performance of periodic sensing improves with reduction in the sensing interval T_s. As T_s approaches 0, the performance of periodic sensing comes close to that of continuous sensing, as expected.

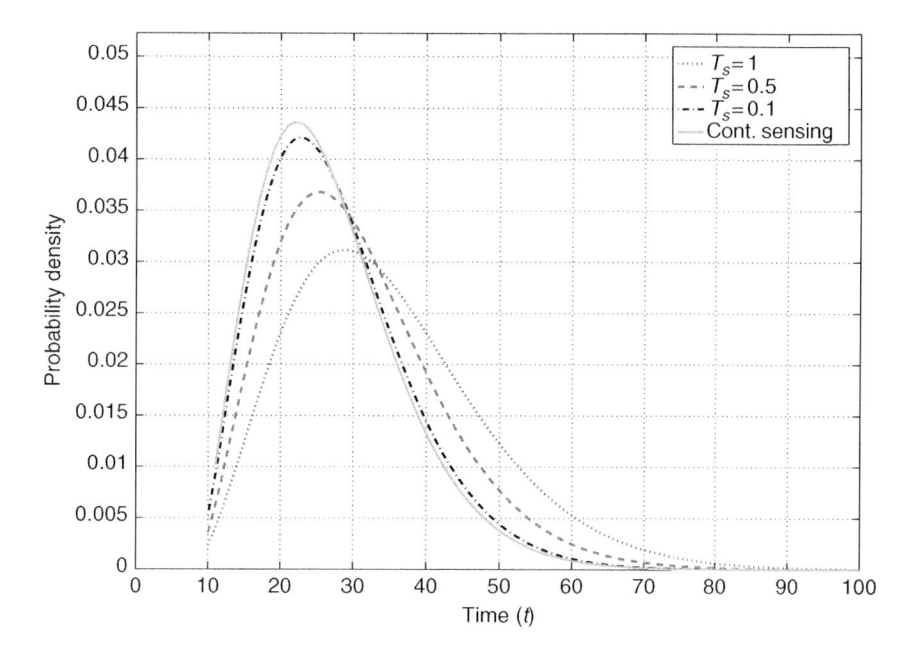

Figure 9.8 Distribution of the EDT with continuous and periodic sensing ($T_{tr} = 10$, $\lambda = 3$, and $\mu = 2$). Reprint with permission from [18]. ©2015 IEEE.

9.3.4 Non–Work-Preserving Transmission with Periodic Sensing

For the non–work-preserving strategy, the EDT for packet transmission by the SU consists of interleaved waiting slots and wasted transmission slots, followed by the final successful transmission slot of duration T_{tr}.

Let $\widehat{\mathcal{P}}_k$ be the probability that the SU was successful in sending the packet in the kth transmission slot with the non–work-preserving strategy. This means that each of the first $(k-1)$ slots had a time duration of less than T_{tr}, while the kth transmission slot had a duration of more than T_{tr}. Thus, $\widehat{\mathcal{P}}_k$ can be calculated, while noting that the duration of secondary transmission slots is exponentially distributed with mean μ, as

$$\widehat{\mathcal{P}}_k = e^{-\frac{T_{tr}}{\mu}} \cdot \left(1 - e^{-\frac{T_{tr}}{\mu}}\right)^{k-1}. \tag{9.41}$$

For the case when PU is off at the instant of packet arrival, if a certain packet is transmitted completely in the kth transmission slot, then the total wait time for that packet includes $(k-1)$ secondary waiting slots and $(k-1)$ wasted transmission slots. Note that the duration of each of these $(k-1)$ waiting slots, denoted by the discrete random variable T_{wait}, which is equal to PU on time, consists of multiple T_s, and follows a geometric distribution with PDF given by $Pr[T_{wait} = nT_s] = (1-\beta)\beta^{n-1}$, where β is given by (9.72), which is a constant due to the memoryless property of exponential distribution, while the duration of each of the previous $(k-1)$ wasted secondary transmission slots, denoted by the random variable T_{waste}, follows a truncated exponential distribution, with PDF given by

$$f_{T_{waste}}(t) = \frac{1}{1 - e^{-\frac{T_{tr}}{\mu}}} \frac{1}{\mu} e^{\frac{-t}{\mu}} \cdot (u(t) - u(t - T_{tr})), \tag{9.42}$$

where $u(t)$ is the unit step function. The moment generating function (MGF) of $T_{w,P_{off}}$ for the perfect periodic sensing case, $\mathcal{M}^{(p)}_{T_w,P_{off}}(s)$, can be calculated as

$$\mathcal{M}^{(p)}_{T_w,P_{off}}(s) = \sum_{k=1}^{\infty} \widehat{\mathcal{P}}_k \cdot \left(\mathcal{M}^{(p)}_{T_{wait}}(s)\right)^{k-1} \cdot \left(\mathcal{M}_{T_{waste}}(s)\right)^{k-1}, \tag{9.43}$$

where $\mathcal{M}^{(p)}_{T_{wait}}(s)$ is the MGF of T_{wait}, given by

$$\mathcal{M}^{(p)}_{T_{wait}}(s) = \sum_{n=1}^{\infty} (1-\beta)\beta^{n-1} e^{nsT_s}, \tag{9.44}$$

and $\mathcal{M}_{T_{waste}}(s)$ is the MGF of T_{waste}, given by

$$\mathcal{M}_{T_{waste}}(s) = \frac{1 - e^{T_{tr}(s - \frac{1}{\mu})}}{(1 - \mu s)(1 - e^{-\frac{T_{tr}}{\mu}})}. \tag{9.45}$$

Substituting (9.41), (9.45), and (9.44) into (9.43), performing some manipulation, and taking the inverse MGF, we obtain the expression for $f^{(p)}_{T_w,P_{off}}(t)$ as

$$f_{T_w,p_{off}}^{(p)}(t) = e^{-\frac{T_{tr}}{\mu}} \delta(t)$$

$$+ \sum_{n=1}^{\infty} \left[\frac{(1-\beta)\beta^{n-1}}{\mu} e^{-\frac{(t-nT_s)}{\mu}} e^{-\frac{T_{tr}}{\mu}} {}_1F_1\left(1-n;1;-\frac{1-\beta}{\beta}\frac{t-nT_s}{\mu}\right)\right.$$

$$+ \sum_{i=1}^{n} \left[(-1)^i e^{-(i+1)\frac{T_{tr}}{\mu}} \binom{n-1}{i-1} \frac{1}{(i-1)!} \frac{(t-iT_{tr}-nT_s)^{i-1}}{\mu^i} (1-\beta)^i \beta^{n-i} e^{\frac{-(t-nT_s-iT_{tr})}{\mu}}\right.$$

$$\left.\left.\times {}_2F_2\left(i+1,i-n;i,i;-\frac{1-\beta}{\beta}\frac{(t-nT_s-iT_{tr})}{\mu}\right)\right]\right], \tag{9.46}$$

where ${}_1F_1(.,.,.)$ is the generalized hypergeometric function [24].

For the case when PU is on at the instant of packet arrival, the PDF of $T_{w,p_{on}}$ for the periodic sensing case can be similarly calculated as

$$f_{T_w,p_{on}}^{(p)}(t) = e^{-\frac{T_{tr}}{\mu}} \sum_{n=2}^{\infty} (n-1) \frac{(1-\beta)^2 \beta^{n-2}}{\mu} e^{-\frac{t-nT_s}{\mu}} {}_1F_1\left(2-n;2;-\frac{1-\beta}{\beta}\cdot\frac{t-nT_s}{\mu}\right)$$

$$+ e^{-\frac{T_{tr}}{\mu}} \sum_{n=1}^{\infty} (1-\beta)\beta^{n-1} \delta(t-nT_s)$$

$$+ \sum_{n=1}^{\infty} \sum_{i=1}^{n-1} \left[(-1)^i e^{-(i+1)\frac{T_{tr}}{\mu}} \binom{n-1}{i} (1-\beta)^{i+1} \beta^{n-i-1}\right.$$

$$\left.\times \frac{t^{i-1} e^{-\frac{t-nT_s-iT_{tr}}{\mu}}}{(i-1)!\mu^i} {}_1F_1\left(i+1-n;i;-\frac{1-\beta}{\beta}\cdot\frac{t-nT_s-iT_{tr}}{\mu}\right)\right]. \tag{9.47}$$

Note that the sequence of impulses corresponds to the case that the packet is transmitted in the first transmission attempt on acquiring the channel after a random number of sensing intervals/attempts. Finally, the PDF of the EDT T_{ED} for the SU is then given by

$$f_{T_{ED}}^{(p)}(t) = \frac{\lambda}{\lambda+\mu} f_{T_w,p_{on}}^{(p)}(t-T_{tr}) + \frac{\mu}{\lambda+\mu} f_{T_w,p_{off}}^{(p)}(t-T_{tr}). \tag{9.48}$$

Fig. 9.9 plots the CDF of the EDT for the non–work-preserving strategy with periodic sensing, $F_{T_{ED}}^{(p)}(t)$, obtained by numerical integration of the analytical PDF expression given by (9.48), for various values of period T_s. As expected, longer T_s leads to slower increase in the CDF values.

9.3.5 Application to Secondary Queuing Analysis

In this subsection, we apply the EDT characterization from previous subsections to study the transmission delay of the secondary system in a queuing setup. In particular, the secondary traffic intensity is high and, as such, a first-in-first-out queue is introduced to hold packets until being transmitted. We assume that packet arrival follows a Poisson process with intensity $\frac{1}{\psi}$, i.e., the average time duration between packet arrivals is ψ. For the

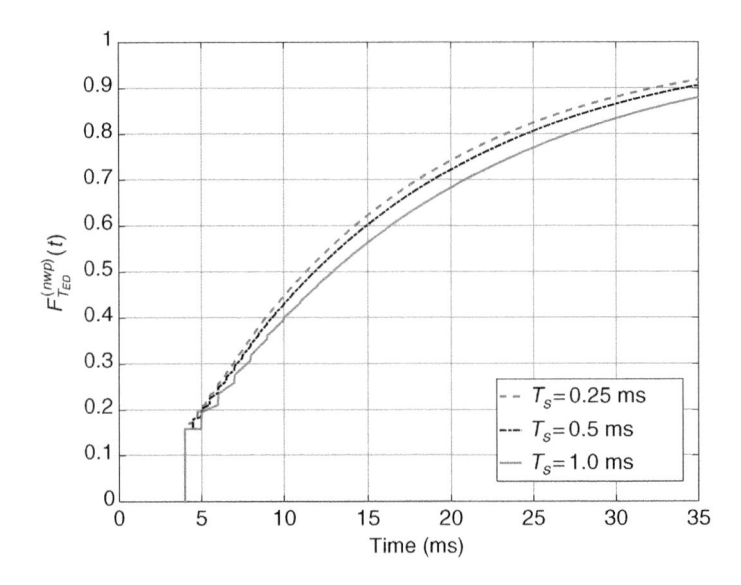

Figure 9.9 Analytical CDF of T_{ED} with the non–work-preserving strategy, perfect periodic sensing ($T_{tr} = 4$ ms, $\lambda = 3$ ms, and $\mu = 2$ ms). Reprint with permission from [19]. ©2017 IEEE.

sake of simplicity, the packets are assumed to be of the same long length, such that their transmission time T_{tr} is a fixed constant. As such, the secondary packet transmission can be modeled as a general M/G/1 queue, where the service time is closely related to the EDT we studied in the previous section. For illustration purposes, we focus on the case of work-preserving transmission with continuous sensing in the following, while noting the analysis for the other scenarios can be similarly solved.

Moments of Packet Service Time

Note also from the EDT analysis in the previous subsection, the waiting time of a packet depends on whether the PU is on or off when the packet is available for transmission. As such, different secondary packets will experience two types of service time characteristics. Specifically, some packets might see that there are one or more packets waiting in the queue or being transmitted upon arrival. Such packets will have to wait in the queue until transmission completion of previous packets. Once all the previous packets are transmitted, the newly arriving packet will find the PU to be off. We term such packets type 1 packets. On the other hand, some packets will arrive when the queue is empty, and will immediately become available for transmission. Such packets might find the PU to be on or off. We call this type of packet a type 2 packet. To facilitate subsequent queuing analysis, we now calculate the first and second moments of the service time for these two types of packet.

The average service time of type 1 packets is equal to the EDT of packets that find PU off at the start of their transmission. Specifically, the first moment of the service time of type 1 packets with continuous sensing can be calculated as

9.3 Extended Delivery Time Analysis

$$E_1[t] = \int_{T_{tr}}^{\infty} t f_{T_w, p_{off}}(t - T_{tr}) dt \triangleq E_{off}^c[t], \tag{9.49}$$

where $E_{off}^c[t]$ denotes the average EDT of a packet that finds PU off at the start of their transmission for continuous sensing. Substituting (9.26) into (9.49) and carrying out integration, we can obtain the closed-form expression of $E_1[t]$ as

$$E_1[t] = E_{off}^c[t] = T_{tr} + \lambda \left(\frac{T_{tr}}{\mu} \right). \tag{9.50}$$

Similarly, the second moment of service time for a type 1 packet with continuous sensing can be calculated as

$$E_1[t^2] = E_{off}^c[t^2] = \lambda^2 \left[\left(\frac{T_{tr}}{\mu} \right)^2 + 2\frac{T_{tr}}{\mu} \right] + 2\frac{\lambda T_{tr}^2}{\mu} + T_{tr}^2. \tag{9.51}$$

Type 2 packets may find the PU on or off at the start of their service upon arrival. Therefore, the service time of type 2 packets is the weighted average of the EDTs of packets that find PU on at the start of their transmission, and those that find PU off. Mathematically speaking, $E_2[t]$ and $E_2[t^2]$ can be calculated as

$$E_2[t] = P_{on,2} \cdot E_{on}[t] + (1 - P_{on,2}) \cdot E_{off}[t], \tag{9.52}$$

and

$$E_2[t^2] = P_{on,2} \cdot E_{on}[t^2] + (1 - P_{on,2}) \cdot E_{off}[t^2], \tag{9.53}$$

where $P_{on,2}$ denotes the probability that a type 2 packet finds the PU on upon arrival, $E_{on}[t]$ and $E_{on}[t^2]$ are the first and second moments of the EDT of a packet that finds the PU on at $t = 0$, respectively, and $E_{off}[t]$ and $E_{off}[t^2]$ are the moments for the PU off case, which can be similarly calculated. We now derive an expression for $P_{on,2}$. Whenever the transmission of the last packet in the queue is completed, due to the memoryless property of exponential distribution, the time taken for the next packet to arrive will follow an exponential distribution with average ψ. At the start of that time interval, it is known that the PU is off. The probability that the PU is on, $P_{p_{on}}(t)$, conditioned on the time elapsed since the completion of last packet transmission, t, is given by [23]

$$P_{p_{on}}(t) = \Pr[\text{PU on at } t_0 + t \mid \text{PU off at } t_0] = \frac{\lambda}{\lambda + \mu} \left[1 - e^{-(\frac{1}{\lambda} + \frac{1}{\mu})t} \right]. \tag{9.54}$$

Removing the conditioning on t, the probability for the PU being on when a type 2 packet arrives, $P_{on,2}$, is obtained as

$$P_{on,2} = E[P_{p_{on}}(t)] = \int_0^{\infty} P_{p_{on}}(t) \cdot \frac{1}{\psi} e^{-\frac{t}{\psi}} dt, \tag{9.55}$$

which simplifies to

$$P_{on,2} = \frac{\lambda \psi}{\lambda \psi + \lambda \mu + \mu \psi}. \tag{9.56}$$

Finally, substituting the moments $E_{on}[t]$, $E_{on}[t^2]$, $E_{off}[t]$, and $E_{off}[t^2]$ for continuous and periodic sensing cases, and $P_{on,2}$ into (9.52) and (9.53), we can obtain the moments of type 2 packet service time.

Queuing Analysis

The average total delay of secondary packet transmission is given by

$$E[D] = E[t] + E[Q], \qquad (9.57)$$

where $E[t]$ is the average service time of an arbitrary packet, and $E[Q]$ is the average wait time in the queue. $E[t]$ is a weighted average of $E_1[t]$ and $E_2[t]$, as defined in (9.50) and (9.52), respectively, given by

$$E[t] = (1 - p_0) \cdot E_1[t] + p_0 \cdot E_2[t], \qquad (9.58)$$

where p_0 is the probability of the queue being empty at any given time instance and $1 - p_0$ is the utilization factor of the queue, which is, in turn, related to $E[t]$ as

$$1 - p_0 = \frac{E[t]}{\psi}. \qquad (9.59)$$

Simultaneously solving (9.58) and (9.59), we can obtain $E[t]$ and p_0 as

$$E[t] = \frac{\psi E_2[t]}{\psi + E_2[t] - E_1[t]}, \qquad (9.60)$$

and

$$p_0 = \frac{\psi - E_1[t]}{\psi + E_2[t] - E_1[t]}, \qquad (9.61)$$

respectively.

The average delay in the queue, $E[D]$, can be calculated using the mean value technique [25] as

$$E[Q] = E[N_Q] \cdot E_1[t] + (1 - p_0) \cdot E[R], \qquad (9.62)$$

where $E[N_Q]$ is the average number of packets waiting in the queue, not including the current packet in service, $E_1[t]$ is the average service time of a packet in the queue (type 1 packet), and $E[R]$ is the mean residual time of the packet currently being served. Specifically, the first addition term corresponds to the average total service time of the packets currently waiting in the queue, if any, and the second term to the waiting time for the currently served packet, if any. Given that a packet is being served at a given instance, the probabilities that the packet is a type 1 packet or type 2 packet are equal to $\frac{(1-p_0)E_1[t]}{(1-p_0)E_1[t]+p_0E_2[t]}$ and $\frac{p_0E_2[t]}{(1-p_0)E_1[t]+p_0E_2[t]}$, respectively. Therefore, mean residual service time, $E[R]$ can be calculated as

$$E[R] = \frac{(1 - p_0)E_1[t]}{(1 - p_0)E_1[t] + p_0E_2[t]} \cdot E[R_1] + \frac{p_0E_2[t]}{(1 - p_0)E_1[t] + p_0E_2[t]} \cdot E[R_2], \qquad (9.63)$$

where $E[R_1]$ and $E[R_2]$ are the mean residual times for type 1 and type 2 packets, respectively, defined by [26]

$$E[R_1] = \frac{E_1[t^2]}{2E_1[t]} \qquad (9.64)$$

and

$$E[R_2] = \frac{E_2[t^2]}{2E_2[t]}. \qquad (9.65)$$

Recalling Little's law stating that

$$E[N_Q] = \frac{E[Q]}{\psi}, \qquad (9.66)$$

$E[Q]$ can be obtained after much simplification as

$$E[Q] = \frac{E[t^2]}{2(\psi - E_1[t])}, \qquad (9.67)$$

where $E[t^2]$ is the second moment of the average service time of all packets, defined by

$$E[t^2] = \frac{(\psi - E_1[t]) \cdot E_2[t^2] + E_2[t] \cdot E_1[t^2]}{\psi + E_2[t] - E_1[t]}. \qquad (9.68)$$

The average number of packets waiting in the queue, not including the packet currently in transmission, is given by

$$E[N_Q] = \frac{E[t^2]}{2\psi(\psi - E_1[t])}. \qquad (9.69)$$

Finally, the average total delay for secondary packets can be simply expressed as

$$E[D] = \frac{\psi E_2[t]}{\psi + E_2[t] - E_1[t]} + \frac{E[t^2]}{2(\psi - E_1[t])}. \qquad (9.70)$$

Fig. 9.10 shows the variation of average total delay against the arrival rate of a data packet with continuous sensing. The graph is based on the assumption that the average delay between packet arrival is greater than the average service time of the SU, as otherwise the queue will become unstable. For comparison purposes, we have included the average queuing delay obtained by modeling the secondary queue by conventional M/G/1 queue with service time moments simply given by (9.58) and (9.68), $E[D] = E[t] + \frac{E[t^2]}{2(\psi - E[t])}$. The simulation results clearly show that our analytical approach is more accurate.

Fig. 9.11 shows the variation of average delay including the queuing delay against the arrival rate of data packet. As the periodic sensing interval becomes small, the periodic sensing curves converge to the continuous sensing curve.

9.3.6 Effect of Imperfect Sensing

In this subsection, we consider the effect of imperfect sensing, taking the probability of false alarm, i.e., sensing a free channel to be busy, into account. We assume that the chance of misdetection, i.e., sensing a busy channel to be free, is negligible, which can be achieved in practice by adjusting the sensing thresholds properly. We illustrate the analysis for work-preserving transmission with the imperfect periodic sensing scenario (Fig. 9.12), while noting other scenarios can be similarly solved.

Following a similar analytical approach as the previous subsections, the PDF of the EDT T_{ED} for the SU can be written as

278 **Cognitive Transmission**

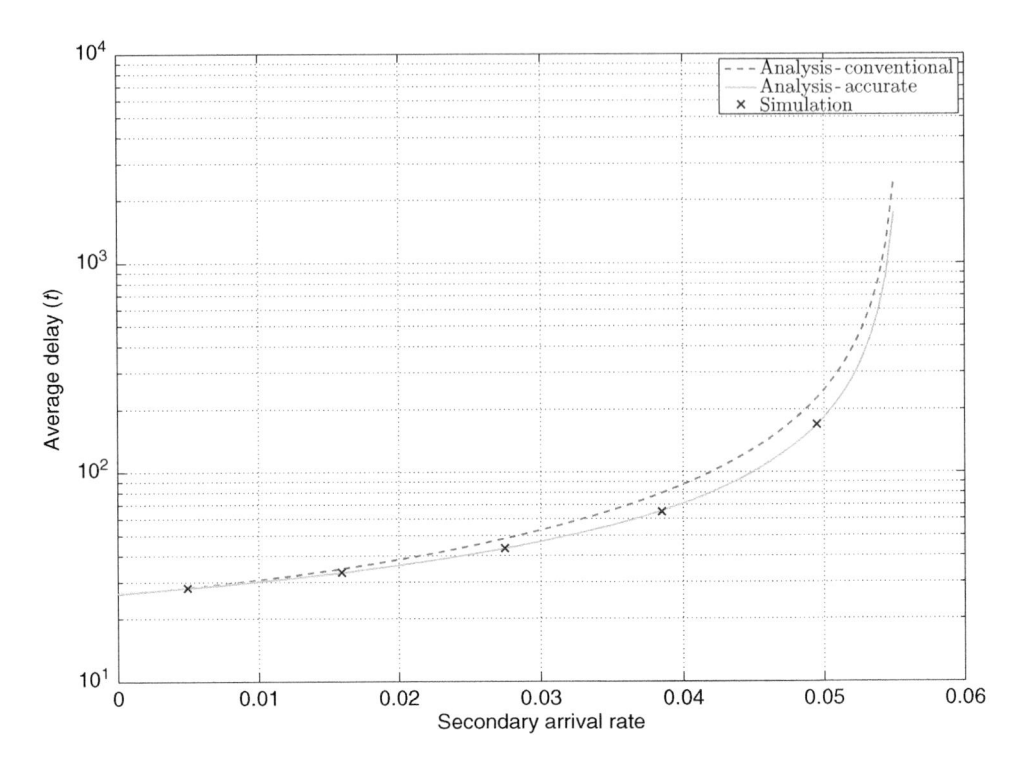

Figure 9.10 Simulation verification for the analytical average queuing delay with continuous sensing ($T_{tr} = 3$, $\lambda = 10$, and $\mu = 2$). Reprint with permission from [18]. ©2015 IEEE.

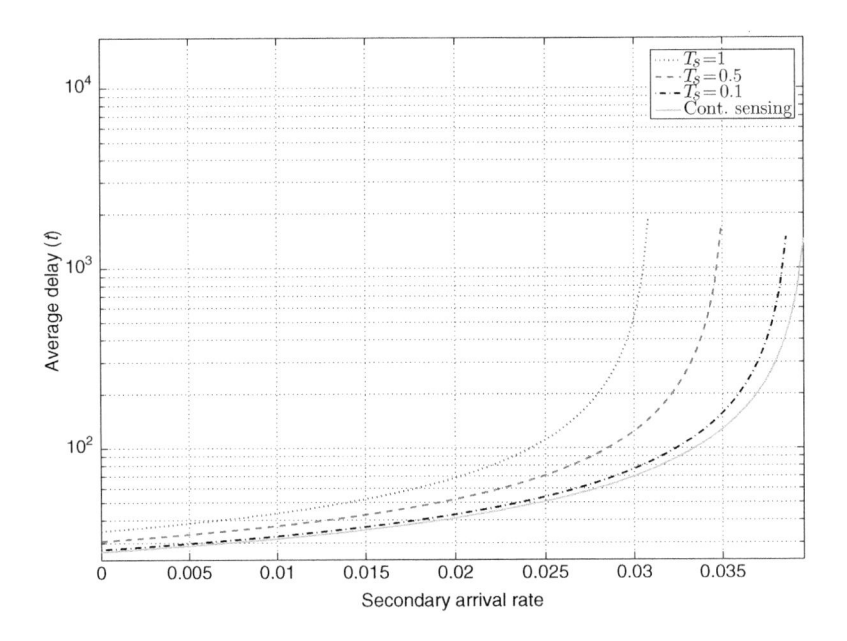

Figure 9.11 Average queuing delay ($T_{tr} = 10$, $\lambda = 3$, and $\mu = 2$). Reprint with permission from [18]. ©2015 IEEE.

9.3 Extended Delivery Time Analysis

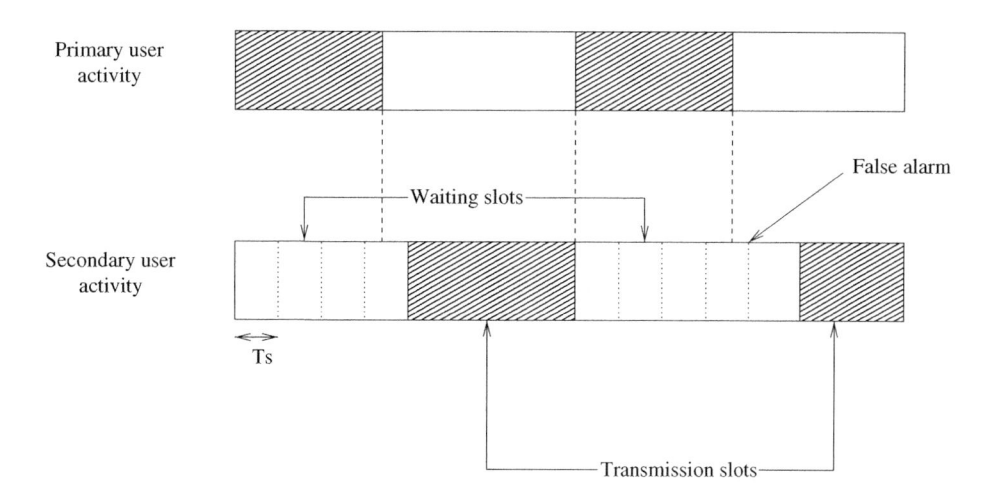

Figure 9.12 Illustration of PU and SU activities and SU sensing: imperfect periodic sensing. Reprint with permission from [19]. ©2017 IEEE.

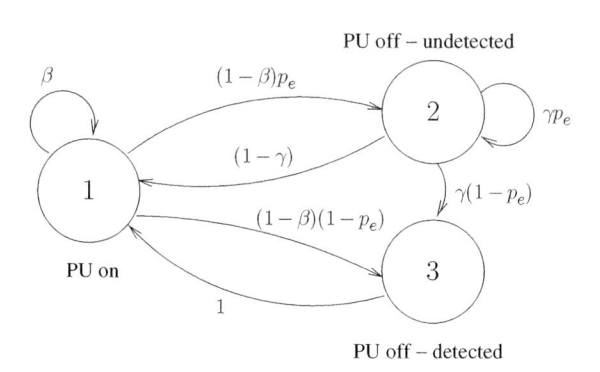

Figure 9.13 State diagram for the cognitive system.

$$\Pr[T_{ED} = nT_s + T_{tr}] = \frac{\lambda}{\lambda + \mu} \Pr[T_{w,p_{on}} = nT_s]$$
$$+ \frac{\mu}{\lambda + \mu} \Pr[T_{w,p_{off}} = nT_s], \quad (9.71)$$

where $\frac{\lambda}{\lambda+\mu}$ and $\frac{\mu}{\lambda+\mu}$ are the stationary probabilities that the PU is on and off at $t = 0$, respectively, $\Pr[T_{w,p_{on}} = nT_s]$ is the probability that the total SU waiting time is nT_s when PU is on at $t = 0$, and $\Pr[T_{w,p_{off}} = nT_s]$ is the probability when PU is off at $t = 0$.

The two probability mass functions (PMFs) $\Pr[T_{w,p_{on}} = nT_s]$ and $\Pr[T_{w,p_{off}} = nT_s]$ above are calculated with the help of a discrete-time Markov chain as depicted in Fig. 9.13. State 1 corresponds to the case that PU is on when the channel is sensed. State 2 corresponds to the case that PU is off when the channel is sensed, but SU has failed to detect the free channel. State 3 represents the case that PU is off and SU has detected the free channel. This Markov chain differs from a conventional discrete-time Markov chain in the sense that it uses the concept of a variable time step, i.e., the period after which a possible transition takes place is not constant and depends on the current state. A time step after state 1 and state 2 is T_s units, which represents the sensing interval. A time step after state 3 is random and exponentially distributed, which represents the SU transmission period. The transition probabilities of the Markov chain are calculated in terms of false alarm probability, denoted by p_e, the probability that the PU is on at a given sensing instant, provided that it was on at the previous sensing instant T_s time units earlier, denoted by β, which is given by

$$\beta = \frac{\lambda}{\lambda + \mu} + \frac{\mu}{\lambda + \mu} e^{-(\frac{1}{\lambda} + \frac{1}{\mu})T_s}, \qquad (9.72)$$

and the probability that PU is off at a given sensing instant, provided that it was off at the previous sensing instant, denoted by γ, which can be calculated as

$$\gamma = \frac{\mu}{\lambda + \mu} + \frac{\lambda}{\lambda + \mu} e^{-(\frac{1}{\lambda} + \frac{1}{\mu})T_s}. \qquad (9.73)$$

Finally, the Markov chain will transit from state 3, after an exponentially distributed random duration, to state 1 with probability 1 when the PU starts transmission.

Based on the above model, the probability that the SU needs to wait n sensing periods before transmitting exactly k times for the case that the PU is on at $t = 0$, denoted by $\Pr[T_{w,p_{on}}^{(k)} = nT_s]$, which is equal to the probability that the Markov chain visits state 3 exactly k times after visiting states 1 and 2 exactly n times, starting from state 1 at $t = 0$, can be calculated as

$$\Pr[T_{w,p_{on}}^{(k)} = nT_s] =$$
$$\sum_{\substack{i+j+2m+z=n-k \\ i,j,m,z \geq 0 \\ i \leq k}} \left[\gamma^{i+j}(1-\gamma)^m \beta^z (1-\beta)^{m+k} p_e^{i+j+m} \right.$$
$$\times (1-p_e)^k \cdot \frac{(m+k-1)!}{m! i! (k-i)!} \cdot k \qquad (9.74)$$
$$\left. \times \binom{i+m+j-1}{j} \cdot \binom{k+m+z-1}{z} \right],$$

where the summation variable i represents the number of transitions from state 2 to state 3, j the number of transitions from state 2 to state 2, m the number of transitions from state 2 to state 1, and z the number of transitions from state 1 to state 1. Note that the number of transitions from state 1 to state 2 is $m + i$, while the number of transitions from state 1 to state 3 is $k - i$. Conditioning on the number of transmitting periods required for completing packet transmission, the PMF of the SU wait time for the PU on at $t = 0$, $\Pr[T_{w,p_{on}} = nT_s]$, can then be calculated as

$$\Pr[T_{w,p_{on}} = nT_s] = \sum_{k=1}^{\infty} \mathcal{P}_k \times \Pr[T_{w,p_{on}}^{(k)} = nT_s], \qquad (9.75)$$

where $\mathcal{P}_k$ is the probability that the SU was successful in sending the packet in the kth transmission slot, given by

$$\mathcal{P}_k = \frac{T_{tr}^{k-1} e^{\frac{-T_{tr}}{\mu}}}{\mu^{k-1}(k-1)!}. \tag{9.76}$$

In a similar manner, we define $\Pr[T_{w,p_{off}}^{(k)} = nT_s]$ as the probability that the time spent by the system in states 1 and 2 is nT_s before it successfully finds the PU to be off exactly k times, for the case when PU is off at $t = 0$. Noting that if PU is off at $t = 0$, there is a probability $(1 - p_e)$ that the SU successfully senses PU to be off at $t = 0$ (which becomes the first of the k successful detections) and the Markov chain starts at state 3, and a probability p_e that the SU fails to sense the free channel, and the Markov chain starts at state 2. $\Pr[T_{w,p_{off}}^{(k)} = nT_s]$ can then be calculated as

$$\Pr[T_{w,p_{off}}^{(k)} = nT_s] = p_e$$

$$\times \left(\sum_{\substack{i+j+z-1=n-k \\ j,z \geq 0, 1 \leq i \leq k}} \left[\gamma^{i+j} \beta^z (1-\beta)^{k-1} p_e^{i+j-1} (1-p_e)^k \right. \right.$$

$$\left. \times \binom{k-1}{i-1} \cdot \binom{i+j-1}{j} \cdot \binom{k+z-2}{z} \right]$$

$$+ \sum_{\substack{i+j+2m+z-1=n-k \\ i,j,z \geq 0, m \geq 1 \\ i \leq k}} \left[\gamma^{i+j}(1-\gamma)^m \beta^z (1-\beta)^{m+k-1} p_e^{i+j+m-1} \right.$$

$$\times (1-p_e)^k \cdot \frac{(m+k-2)!}{m! i! (k-i)!} \cdot (mk+ik-i)$$

$$\left. \left. \times \binom{i+m+j-1}{j} \cdot \binom{k+m+z-2}{z} \right] \right)$$

$$+ (1-p_e) \cdot \Pr[T_{w,k-1,p_{on}} = nT_s]. \tag{9.77}$$

It follows that the PMF of the SU wait time for the PU off at $t = 0$ case, $\Pr[T_{w,p_{off}} = nT_s]$, can be calculated similarly to $\Pr[T_{w,p_{on}} = nT_s]$ in (9.75), but using $\Pr[T_{w,p_{off}}^{(k)} = nT_s]$ in (9.77) in place of $\Pr[T_{w,p_{on}}^{(k)} = nT_s]$.

Fig. 9.14 plots the average queuing delay of secondary packets against the packet arrival rate, for various values of false alarm probability p_e for the work-preserving strategy. As expected, the average delay is always greater for a larger value of the false alarm probability. A similar observation can be made for the non–work-preserving case.

9.4 Spectrum Sharing: Single Antenna Transmitter

We now consider the underlay spectrum sharing cognitive implementation. With spectrum sharing implementation, SU can access the spectrum of PU while ensuring that their communication will not affecting the primary system performance. This can be achieved by limiting the interference generated by the secondary transmission to the

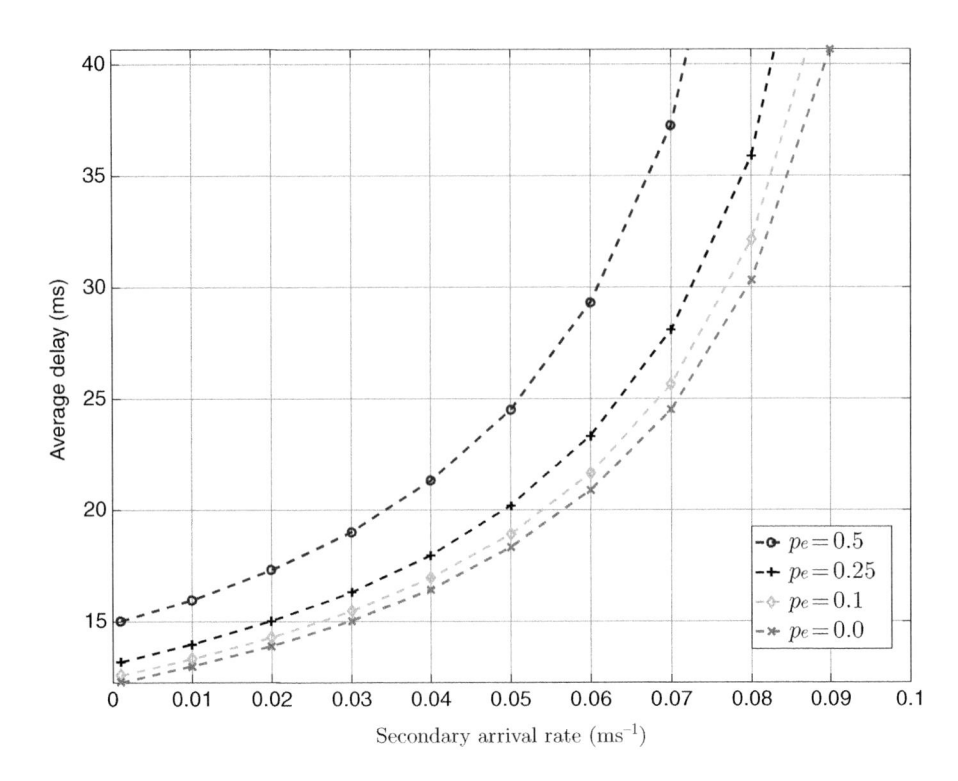

Figure 9.14 Average queuing delay for imperfect periodic sensing with the work-preserving strategy ($T_s = 0.5$ ms, $\lambda = 5$ ms, $\mu = 3$ ms, and $T = 3$ ms). Reprint with permission from [19]. ©2017 IEEE.

primary receiver (PR) below some fixed level determined by, for example, the quality of service (QoS) requirements of the primary system. Researchers have proposed different methods to limit the interference caused by the secondary transmitter (ST) to the PR for an underlay cognitive radio system, with transmit power adaptation being the most intuitive and convenient approach [27], especially when the secondary transmitter is only equipped with a single antenna. Meanwhile, secondary transmission may suffer the interference from the primary transmitter (PT). In this section, we consider the performance of underlay cognitive transmission with single-antenna ST [28].

9.4.1 System Model

Consider an underlay cognitive radio environment in which the ST is equipped with a single antenna while the SR has N antennas. The fading experienced by signal received on different SR antennas are assumed to be independent and identically distributed. The channel gain from the PT to the ith SR antenna is denoted by g_i, $i = 1, 2, \ldots, N$. The channel gains from the ST to the PR and the ith SR antenna are, respectively, denoted by h_0 and h_i, $i = 1, 2, \ldots, N$. The primary network is presumed to be far away from ST and SR. Therefore, the channel gains $|h_0|$ and $|g_i|$ can be presumed to follow Rayleigh fading, implying that the channel power gains, $|h_0|^2$ and $|g_i|^2$ have the exponential PDFs

9.4 Spectrum Sharing: Single Antenna Transmitter

given by $f_0(x) = \eta e^{-\eta x} u(x)$ and $g(x) = \lambda e^{-\lambda x} u(x)$, respectively, where $u(x)$ is the unit step function and the parameters η and λ depend on propagation distance and environment. Meanwhile, the secondary channel gains are assumed to follow Nakagam fading. As such, The channel power gain $|h_i|^2$ are presumed to be i.i.d. gamma RVs with PDF and CDF given by

$$f(x) = \frac{x^{m-1}}{\beta^m \Gamma(m)} e^{-x/\beta} u(x), \tag{9.78}$$

$$F(x) = \left(1 - \frac{\Gamma(m, x/\beta)}{\Gamma(m)}\right) u(x), \tag{9.79}$$

respectively, where the parameters m and β are positive real numbers, $\Gamma(m)$ is the Euler gamma function, and $\Gamma(m, x)$ is the incomplete gamma function [29].

We assume that the ST adaptively changes its transmit power to control the interference of PR. Let P_S be the instantaneous transmit power of ST; with the knowledge of $|h_0|^2$, either through feedback or channel reciprocity, P_S is set to satisfy the constraint $P_S |h_0|^2 \leq T$, where T is the maximum tolerable interference level at the PR. With continuous power adaptation, P_S is set as $P_S = \min\left(P_{\max}, \frac{T}{|h_0|^2}\right)$, where $P_{\max}$ is the maximum transmit power available at the ST. In practice, however, transmit power cannot be changed in a continuous manner. Therefore, we assume that the ST can use $J + 1$ discrete power levels, denoted by $P_j, j = 0, 1, \ldots, J$, where $P_0 = 0 < P_1 < P_2 < \ldots < P_J = P_{\max}$. Based on channel power gain $|h_0|^2$, the transmit power P_S is chosen according to the following rule;

$$P_S = \begin{cases} P_j & \text{for } \frac{T}{P_{j+1}} < |h_0|^2 \leq \frac{T}{P_j}, j = 1, \ldots, J-1 \\ P_J & \text{for } |h_0|^2 \leq \frac{T}{P_J} \\ 0 & \text{for } |h_0|^2 > \frac{T}{P_1} \end{cases}. \tag{9.80}$$

SR applies selection combining to explore diversity benefit. Specifically, the SR will use the signal received the antenna that experiences the highest signal to interference plus noise ratio (SINR) considering the interference power from PT. Mathematically speaking, the index of the chosen antenna is given by

$$i^* = \arg \max_{i=1,2,\ldots,N} \frac{|h_i|^2}{N_0 + P_M |g_i|^2}, \tag{9.81}$$

where N_0 is the common noise variance at each SR's antenna, P_M is the transmit power of the PT, and $P_M |g_i|^2$ is the PT interference power at the ith SR antenna. In what follows, we derive the exact and asymptotic distribution of the received SINR at the chosen SR antenna, which will then be applied to performance analysis of the secondary system.

9.4.2 Statistics of SINR

We first consider the general case in which the noise variance, N_0, cannot be neglected. Let us define a new RV $Y_i = \frac{|h_i|^2}{N_0 + P_M |g_i|^2}$, which is related to the received SINR on the ith

SR antenna, denoted by ξ_i, as $\xi_i = P_S Y_i$. The PDF of Y_i can be found by conditioning on $N_0 + P_M |g_i|^2$ as

$$f_Y(y) = \frac{\lambda e^{\frac{\lambda N_0}{P_M}}}{P_M} \int_{N_0}^{\infty} f(zy) z e^{-\frac{\lambda z}{P_M}} dz. \tag{9.82}$$

Using (9.78), the PDF of Y_i, $f_Y(y)$, can be shown to be given by

$$f_Y(y) = \frac{\lambda N_0^{m+1} e^{\frac{\lambda N_0}{P_M}}}{P_M \Gamma(m) \beta^m} y^{m-1} E_{-m}\left(\frac{N_0 y}{\beta} + \frac{N_0 \lambda}{P_M}\right), y \geq 0, \tag{9.83}$$

where $E_n(z)$ is the exponential integral, defined as $E_n(z) = \int_1^{\infty} t^{-n} e^{-zt} dt$. Using the fact that $\frac{d}{dx} E_n(x) = -E_{n-1}(x)$ and the following recurrence relationship of $E_n(z)$ [30]

$$E_n(z) = \frac{e^{-z}}{n-1} - \frac{z}{n-1} E_{n-1}(z), \tag{9.84}$$

we can show that

$$\frac{E_{-m}(ay+b)}{y^{1-m}} = \frac{\frac{d}{dy}[y^m E_{1-m}(ay+b)] + y^{m-1} e^{-(ay+b)}}{b}. \tag{9.85}$$

Applying (9.85), the CDF of Y_i, $F_Y(y)$, can be found by integrating $f_Y(y)$ as

$$F_Y(y) = \int_0^y f_Y(t) dt = \left(1 - \frac{\Gamma(m, \frac{N_0}{\beta} y)}{\Gamma(m)}\right. \tag{9.86}$$

$$\left. + \left(\frac{N_0 y}{\beta}\right)^m e^{\frac{\lambda N_0}{P_M}} \frac{E_{1-m}\left(\frac{N_0 y}{\beta} + \frac{N_0 \lambda}{P_M}\right)}{\Gamma(m)}\right), y \geq 0. \tag{9.87}$$

Lastly, the CDF of the received SINR at the selected antenna, $\xi_{1:N}$, for $P_S = P_j$ can be shown to be $F_{\xi_{1:N}}(\xi) = \left[F_Y\left(\frac{\xi}{P_j}\right)\right]^N$ [31].

To facilitate the asymptotic analysis, we present the asymptotic distribution of $Y_{(N)}$ in the following proposition.

PROPOSITION 9.1 *The distribution of $Y_{1:N}$ approaches the Gumbel distribution,* $\exp\left[-\exp\left(-\frac{x-b}{a}\right)\right]$, *as the number of ST antennas, N, increases. Here the constants b and a are given as*

$$b = F_Y^{-1}(1 - 1/N), \tag{9.88}$$

$$a = F_Y^{-1}(1 - 1/(Ne)) - b. \tag{9.89}$$

Proof. First of all, we can show that

$$\frac{d}{dy}\left[\frac{1 - F_Y(y)}{f_Y(y)}\right] = -1 - \frac{f_Y'(y)}{f_Y(y)} \cdot \frac{1 - F_Y(y)}{f_Y(y)}. \tag{9.90}$$

Furthermore, it can be shown that

$$\frac{f_Y'(y)}{f_Y(y)} = \frac{m-1}{y} - \frac{N_0}{\beta} \frac{E_{-m-1}\left(\frac{N_0 \lambda}{P_M} + \frac{N_0}{\beta} y\right)}{E_{-m}\left(\frac{N_0 \lambda}{P_M} + \frac{N_0}{\beta} y\right)}. \tag{9.91}$$

By using the recurrence relationship of $E_n(z)$ as given in (9.84) and the asymptotic series expansion of $ze^z E_{n+1}(z)$ [30], we arrive at

$$\lim_{y\to\infty} \frac{f'_Y(y)}{f_Y(y)} = -\frac{N_0}{\beta} \lim_{y\to\infty} \frac{e^{-\left(\frac{N_0\lambda}{P_M}+\frac{N_0}{\beta}y\right)}}{\left(\frac{N_0\lambda}{P_M}+\frac{N_0}{\beta}y\right) E_{-m}\left(\frac{N_0\lambda}{P_M}+\frac{N_0}{\beta}y\right)} \quad (9.92)$$

$$= -\frac{N_0}{\beta} \frac{1}{{}_2F_0(1,-m;0)} = -\frac{N_0}{\beta} \neq 0. \quad (9.93)$$

Here, ${}_2F_0(a,b;x)$ is the generalized hypergeometric function [30]. We can also show that

$$\lim_{y\to\infty} \frac{1-F_Y(y)}{f_Y(y)} = -\lim_{y\to\infty} \frac{f_Y(y)}{f'_Y(y)} = \frac{\beta}{N_0}. \quad (9.94)$$

Substituting (9.93) and (9.94) into (9.90) results in

$$\lim_{y\to\infty} \frac{d}{dy}\left[\frac{1-F_Y(y)}{f_Y(y)}\right] = 0. \quad (9.95)$$

Therefore, the asymptotic distribution of $Y_{1:N}$ is Gumbel with the constants b and a as defined in (9.88) and (9.89), respectively [31, Th. 10.5.2]. □

Interference Limited Scenario

We now discuss the case where the noise at SR is negligible as compared to the interference from the PT. The SINR at the SR can be approximated by the signal-to-interference ratio (SIR), i.e.,

$$Z_i = \frac{|h_i|^2}{P_M|g_i|^2} \approx Y_i. \quad (9.96)$$

The exact statistics of Z_i can easily be computed from (9.83) and (9.86) by setting $N_0 = 0$ and using the fact that $E_\nu(z) = z^{\nu-1}\Gamma(1-\nu,z)$ [30]. For example, the CDF of Z_i, denoted by $F_Z(z)$, is given as

$$F_Z(z) = \left(\frac{P_M z}{\lambda\beta + P_M z}\right)^m u(z). \quad (9.97)$$

We can routinely derive the exact distribution of SIR of the selected antenna, $\xi'_{1:N}$, using the relation $\xi'_{1:N} = P_S Z_{1:N}$, where $Z_{1:N}$ denotes the largest Z_i among N ones.

The asymptotic distribution of $Z_{1:N}$ is derived in the following proposition.

PROPOSITION 9.2 *The asymptotic distribution of the maximum of Z_i, $Z_{(N)}$, is the Fréchet distribution, i.e.,*

$$\lim_{N\to\infty} F_{Z_{1:N}}(z) = \exp\left[\frac{\beta\lambda}{P_M\left(1-\left(1-\frac{1}{N}\right)^{-\frac{1}{m}}\right)z}\right], z \geq 0. \quad (9.98)$$

Proof. It can be shown that

$$\lim_{z\to\infty} \frac{1-F_Z(z)}{1-F_Z(rz)} = r, \quad (9.99)$$

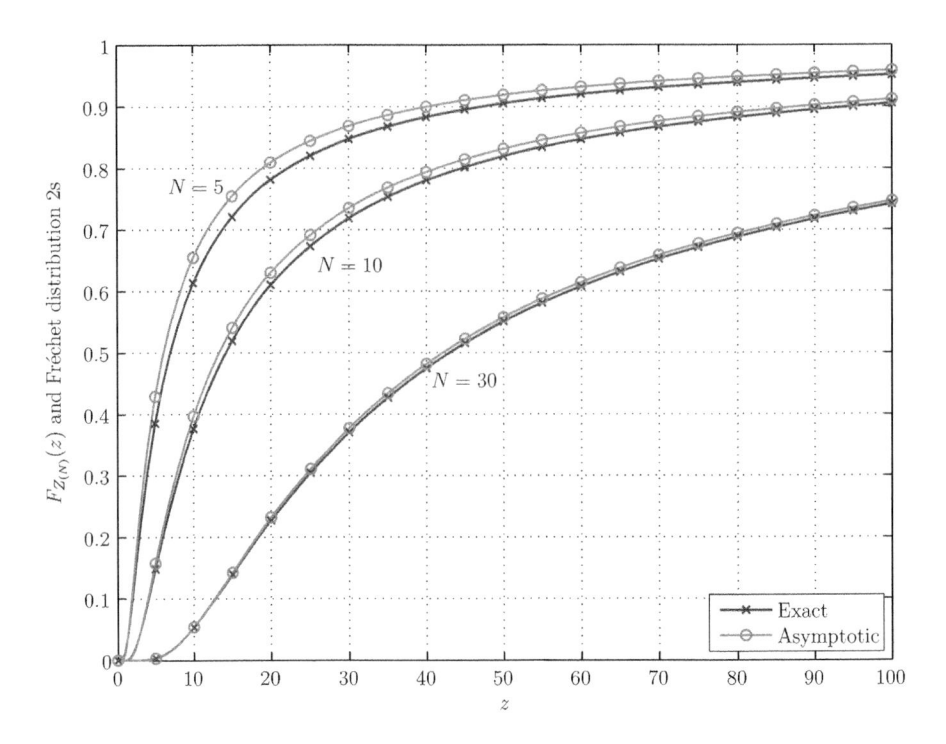

Figure 9.15 Exact and asymptotic distributions of $Z_{(N)}$ for $P_M = 2$, $\lambda = 1$, $\beta = 1$, and $m = 2$. Reprint with permission from [28]. ©2015 IEEE.

which implies that $F_Z(z)$ lies in the domain of maximal attraction of the Fréchet distribution [31, Th. 10.5.2], i.e.,

$$\lim_{N \to \infty} F_{Z_{1:N}}(z) = \exp\left[-\left(\frac{z}{F_Z^{-1}(1 - 1/N)}\right)^{-1}\right], z \geq 0. \quad (9.100)$$

Eq. (9.98) follows directly by observing that $F_Z^{-1}(p) = \frac{\beta\lambda}{P_M(p^{-1/m}-1)}$. □

Fig. 9.15 shows the exact and asymptotic CDFs of $Z_{1:N}$ for different values of N. Observe that the asymptotic distribution in (9.98) is a good approximation even for small values of N and the approximation becomes more accurate by increasing the value of N.

9.4.3 Secondary System Performance Analysis

Outage of secondary system occurs whenever its transmission will generate excess interference to PR or the SINR at SR is below a certain threshold level, ρ. The probability of outage for the secondary system, denoted by P_{out}, can be computed mathematically as

9.4 Spectrum Sharing: Single Antenna Transmitter

$$P_{\text{out}} = p_0 + \sum_{j=1}^{J} p_j \Pr\{P_j Y_{(N)} \leq \rho\} \quad (9.101)$$

$$= p_0 + \sum_{j=1}^{J} p_j \left[F_Y\left(\frac{\rho}{P_j}\right) \right]^N, \quad (9.102)$$

where $F_Y(\cdot)$ is given in (9.86) and $p_j = \Pr\{P_S = P_j\}$ for $j = 0, 1, \ldots, J$. For the Rayleigh fading case, p_j can be computed as

$$p_j = \begin{cases} e^{-\frac{\eta T}{P_1}} & \text{for } j = 0 \\ e^{-\frac{\eta T}{P_{j+1}}} - e^{-\frac{\eta T}{P_j}} & \text{for } j = 1, \ldots, J-1 \\ 1 - e^{-\frac{\eta T}{P_J}} & \text{for } j = J \end{cases} . \quad (9.103)$$

For the interference-limited case, the outage probability can be shown to simplify to

$$P_{\text{out}} \approx p_0 + \sum_{j=1}^{J} p_j \left(\frac{P_M \rho}{\lambda \beta P_j + P_M \rho} \right)^{mN} . \quad (9.104)$$

Fig. 9.16 shows the outage probability of a secondary system against the maximum secondary transmit power, P_S, for different values of the interference threshold, T. The power levels used are $P_j = jP_S/J$, where $J = 5$, and $j = 0, 1, \ldots, J$. The outage threshold level ρ is normalized to 1. We can see that the outage probability first decreases with increasing P_S due to the increase in received SINR at the SR. Also, the outage probability is higher for small values of T, as expected. As P_S further increases, P_{out} starts to rise. This is due to the fact that with five fixed power levels, it becomes increasingly difficult to satisfy the interference constraint at the PR. Monte Carlo simulation for 10^9 trials are also shown, which matches the analytical result very well.

We now consider the error rate performance of secondary transmisson. We consider a general class of modulation schemes whose conditional bit error probability (BEP), $P_b(E|\xi)$, is given by

$$P_b(E|\xi) = C\exp(-\alpha\xi). \quad (9.105)$$

Applying the MGF-based approach, the average BEP can be computed as [9]

$$P_b = CM_{\xi_{1:N}}(-\alpha), \quad (9.106)$$

where $M_{\xi_{1:N}}(s) = E\left[e^{s\xi_{1:N}}\right]$ is the MGF of the SINR at the chosen SR antenna, $\xi_{1:N}$. Using the asymptotic distribution of $Y_{1:N}$, the $M_{Y_{1:N}}(s)$ can be approximately obtained as

$$M_{Y_{1:N}}(s) \approx s \int_0^\infty e^{-e^{-\frac{x-b}{a}} + sx} dx - e^{\frac{b}{a}}. \quad (9.107)$$

Solving the integral in (9.107), we get

$$M_{Y_{1:N}}(s) \approx -ase^{bs}\left[\Gamma(-as) - \Gamma\left(-as, e^{\frac{b}{a}}\right)\right] - e^{\frac{b}{a}}. \quad (9.108)$$

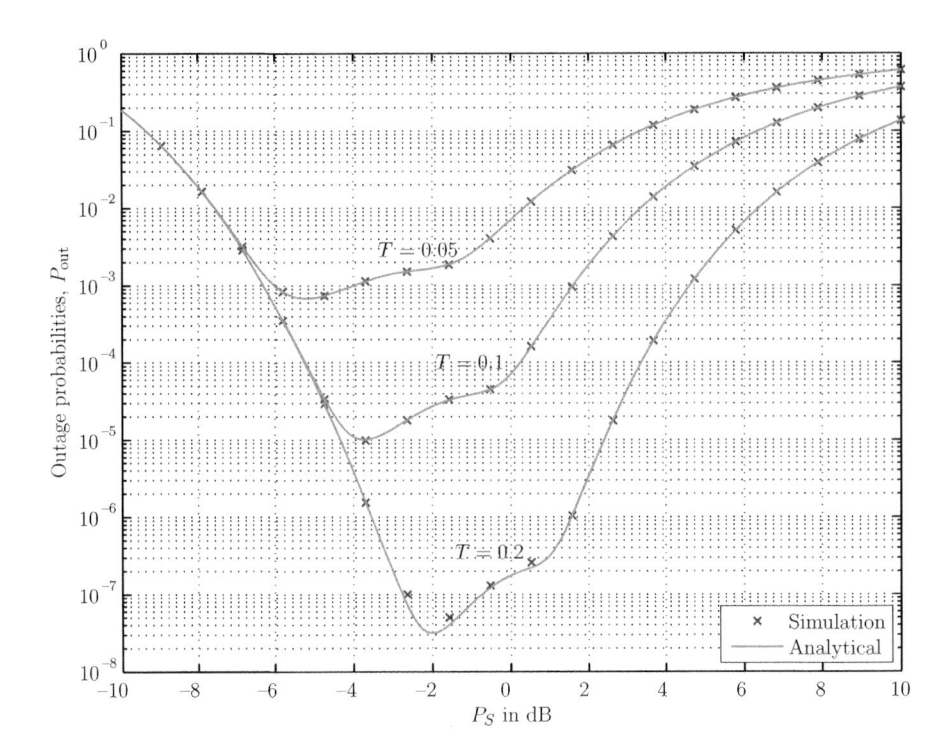

Figure 9.16 Outage probability against maximum secondary transmit power, P_S for $N = 8$, $m = 3.5$, $\beta = 0.1$, $P_M = 1$, $N_0 = 0.025$, and $\lambda = \eta = 20$. Reprint with permission from [28]. ©2015 IEEE.

Finally, noting that $E\left[e^{spY_{(N)}}\right] = E_p\left[E_{Y_{(N)}}\left[e^{spY_{(N)}}\right]\right]$, we can show that

$$P_b \approx C \sum_{j=1}^{J} p_j M_{Y_{1:N}}(-\alpha P_j), \qquad (9.109)$$

where p_j is defined in (9.103).

Fig. 9.17 shows the average BEP of the secondary system against the maximum ST transmit power, P_S, for different values of interference threshold, T. The curves show that the BEP, in general, decreases with increasing P_S, as expected.[1] It is interesting to observe that the BEPs for different values of T are almost same for very small or very large values of P_S. The reason is that when P_S is small, the highest power level is always selected regardless of T as the interference to the PR is negligible. For high enough values of P_S, on the other hand, the smallest power level is selected and hence BEPs turn out to be the same for different values of T. For intermediate values of P_S, lower

---
[1] It should be noted that the curves depict the average BEPs for the secondary transmission when the interference constraint at the PR is satisfied. The outage probability, unlike the average BEP, will increase for significantly large values of P_S due to nonsatisfaction of interference-level constraint at the PR, as in Fig. 9.16.

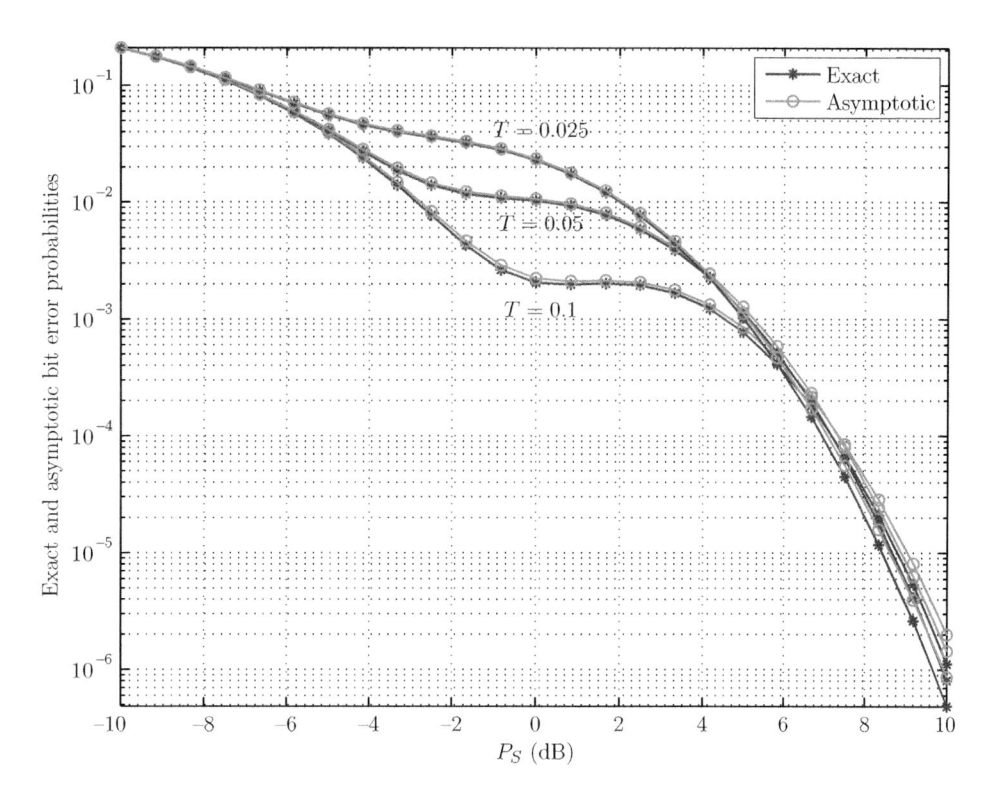

Figure 9.17 Average BEPs of BFSK ($C = \alpha = 0.5$) against P_S for $N = 16$, $m = 3.5$, $\beta = 0.1$, $P_M = 1$, $\lambda = 20$, $\eta = 20$, $N_0 = 0.025$, and $J = 5$. Reprint with permission from [28]. ©2015 IEEE.

BEP is observed for higher T as a higher secondary transmit power level is selected more often.

Fig. 9.18 shows the average BEP of the secondary system against the number of SR antennas, N, for different noise variances, N_0. It is clear from the figure that the approximation of the average BEP improves with increasing N. Also, the approximation is more accurate for larger values of noise variance N_0 as the asymptotic distribution of $Y_{1:N}$ approaches the Fréchet distribution for small values of N_0 (interference-limited scenario) instead of the Gumbel distribution.

9.5 Spectrum Sharing: Transmit Antenna Selection

Another method to effectively control the interference caused by the secondary transmission is to perform beamforming transmission with multiple ST antennas [32–34]. Beamforming transmission, however, entails high hardware cost due to the expensive and power-consuming multiple RF chains. Judicious antenna selection not only reduces the hardware complexity but also achieves diversity benefits. As such, several transmit antenna selection schemes have been proposed for underlay cognitive system [35–40].

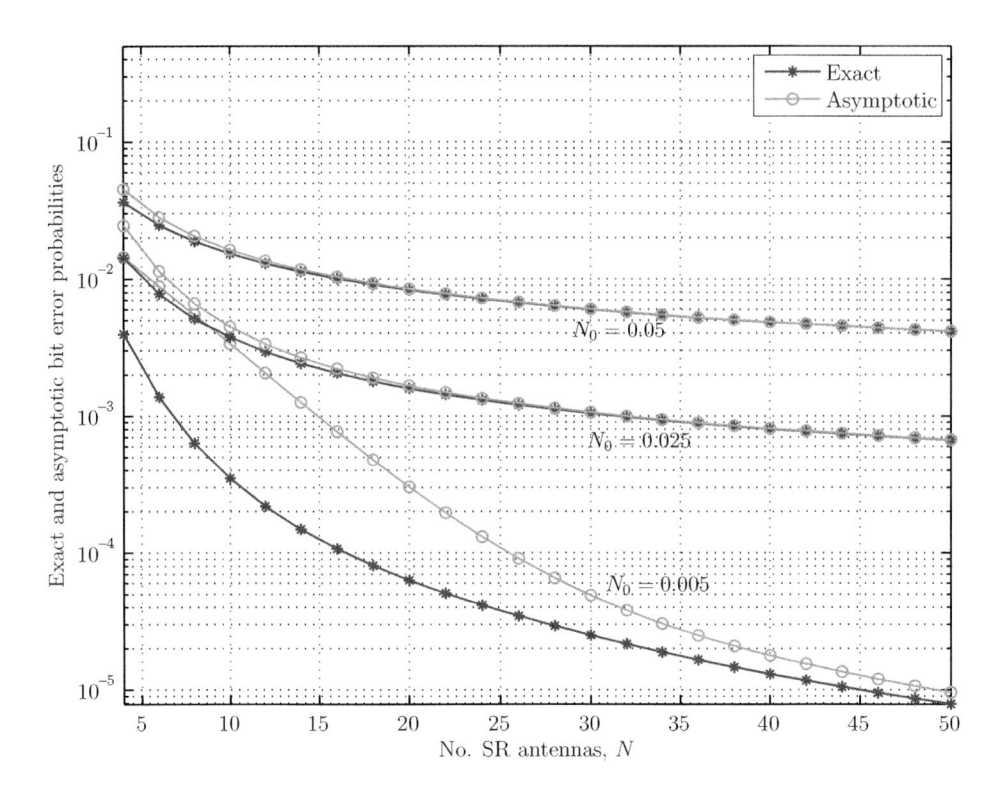

Figure 9.18 Average BEPs of BFSK ($C = \alpha = 0.5$) against number of SR antennas, N, for $m = 3.5, \beta = 0.1, P_M = 1, P_S = 1, \lambda = 20, \eta = 20, J = 5$, and $T = 0.1$. Reprint with permission from [28]. ©2015 IEEE.

In the unconstrained transmit antenna selection (UC) scheme, the antenna leading to the highest instantaneous secondary channel gain is selected to maximize the secondary network throughput. Since the selection does not consider the channel gains from the ST to the PR, the UC scheme results in frequent outages due to the nonsatisfaction of the interference constraint. The minimum interference selection (MI) scheme [37] reduces the probability of outage due to excessive interference to the PR by selecting the antenna that results in minimum interference to the PR. This scheme, however, does not achieve any diversity benefit offered by multiple antennas at the ST. The maximum signal-to-leak interference ratio (MSLIR) scheme overcomes the limitations of the UC and MI schemes by selecting the ST antenna that maximizes the ratio of the ST to the SR and the ST to the PR channel gains [37, 38].

We analyze the performance of transmit antenna selection with both continuous and discrete power adaptation. Specifically, we consider optimal transmit antenna selection that maximizes the received SINR at the SR.

9.5.1 System Model

We consider an underlay CR setup in which the primary network consists of one PT and one PR, whereas the secondary network comprises one ST and one SR, as shown

9.5 Spectrum Sharing: Transmit Antenna Selection

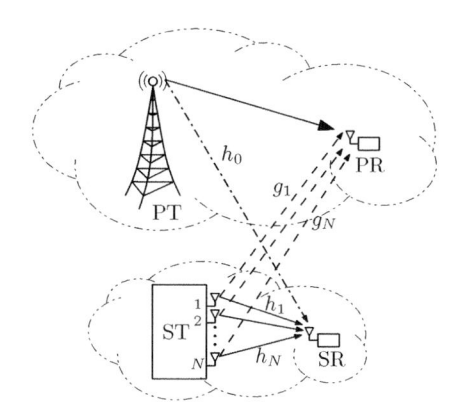

Figure 9.19 An underlay cognitive radio with an ST, equipped with N antennas, serving an SR. Reprint with permission from [41]. ©2017 IEEE.

in Fig. 9.19. Since the primary network is typically a conventional wireless communication system, we presume that both the PT and the PR have only one antenna. On the other hand, we presume that the ST is equipped with N antennas for improving the secondary network performance and reducing the interference to the primary network. Furthermore, the ST communicates with a single-antenna SR by performing judicious transmit antenna selection.

We use h_0 to denote the complex channel gain from the PT to the SR, g_i that from the ith ST antenna to the PR, and h_i that from the ith ST antenna to the SR, where $i = 1, 2, \ldots, N$. Furthermore, we presume that the channel gains are independent Rayleigh random variables. In particular, the channel power gains $|h_0|^2$, $|h_i|^2$, and $|g_i|^2$ are exponential random variables with rates λ_{ps}, λ_{ss}, and λ_{sp}, respectively. These parameters are all positive reals and dependent on the propagation distance and environment. We use the notations $F_{h_0}(x)$, $F_h(x)$, and $F_g(x)$ to denote the CDF of $|h_0|^2$, $|h_i|^2$, and $|g_i|^2$, respectively. The transmit power of the PT is denoted by P_M, while the threshold T denotes the instantaneous interference power limit at the PR, which can be directly related to the interference temperature constraint. Lastly, the noise at the SR is presumed to be a zero mean white Gaussian random variable with variance N_0.

The interference constraint at the PR can be satisfied by appropriately selecting the transmit power level at the ST. Continuous power adaptation with antenna selection at the ST results in the best performance. In particular, with continuous power adaptation at the ST, the transmitting $P_S^{(i)}$ when the ith antenna is selected for transmission is given by

$$P_S^{(i)} = \min\left\{\frac{T}{|g_i|^2}, P_{\max}\right\}, \tag{9.110}$$

where $P_{\max}$ is the maximum ST transmit power. Then the antenna that achieves the highest secondary-link SINR is selected for transmission.

Adapting the transmit power in a continuous fashion, however, is not always possible in practical systems. Therefore, the performance of continuous power adaptation serves only as an upper limit for practical power adaptation schemes. In practice, the transmit power can take on only J discrete values, which we denote by P_j, where $j = 1, 2, \ldots, J$, and $0 < P_1 < \ldots < P_J = P_{\max}$. Depending on the channel conditions and antenna selection criterion, the ST selects one of J power levels for the transmission using the selected antenna. The transmit power level and antenna pair that maximizes the secondary link SINR while satisfying the interference requirement at the PR can be determined as

$$(i^*, j^*) = \arg \max_{i,j} P_j |h_i|^2; \qquad (9.111)$$

$$\text{subject to} : P_j |g_i|^2 \leq T, \qquad (9.112)$$

where $i = 1, 2, \ldots, N$, and $j = 1, 2, \ldots, J$. Finding the optimal pair, however, requires an exhaustive search over all the antennas for all possible power levels, which renders this scheme highly computationally intensive when N and J are large.

We can reduce the complexity by decoupling the antenna selection and power-level adaptation at the ST. Here, we select an antenna first and then adjust the transmit power level to satisfy the interference constraint at the PR. In particular, the antennas are first of all sorted based on the secondary link channel gain. Afterwards, starting from the antenna with the highest secondary link channel gain, we check whether the antenna will result in an acceptable level of interference at the PR should it be selected for transmission with power P_1. The first antenna that satisfies this requirement is the selected antenna. Afterwards, the maximum transmit power level that does not violate the interference constraint at the PR is selected. This sequential antenna and power selection (SAPS) scheme requires knowledge of the highest power level satisfying the interference constraint at the PR for each antenna. This can be accomplished, for example, by estimating PR-to-ST channel with pilot symbol transmitted from PR and employing channel reciprocity to predict ST-to-PR channel. Another option is to rely on the cooperation and feedback from PR.

9.5.2 Statistics of Secondary Received SINR

We now derive the statistics of the secondary received SINR with transmit antenna selection for both continuous and discrete power adaptation schemes.

Adaptation with Unlimited ST Power
We first consider the case where the transmit power of the ST can be adapted without any power limit, i.e., $P_{\max} = \infty$. In this case, the transmit power, P_S, can be written as $P_S^{(i)} = T/|g_i|^2$ when the ith antenna is selected for transmission. After selecting the antenna that achieves the best performance, the received SINR, denoted by γ^{CU}, is given by

$$\gamma^{CU} = \max_i \left\{ \frac{|h_i|^2 \cdot T/|g_i|^2}{N_0 + P_M |h_0|^2} \right\} = \frac{TR}{N_0 + P_M |h_0|^2}, \qquad (9.113)$$

9.5 Spectrum Sharing: Transmit Antenna Selection

where we define $R = \max_i \{|h_i|^2/|g_i|^2\}$. The CDF of R, denoted by $F_R(r)$, can be derived by conditioning on $|g_i|^2$, as

$$F_R(r) = \left(1 - \frac{\lambda_{sp}}{\lambda_{sp} + \lambda_{ss}r}\right)^N. \tag{9.114}$$

Furthermore, the CDF of γ^{CU}, $F^{CU}(x)$, can be computed by conditioning on $|h_0|^2$ as

$$F^{CU}(x) = \lambda_{ps} \int_0^\infty F_R\left(\frac{x(N_0 + P_M y)}{T}\right) e^{-\lambda_{ps} y} dy. \tag{9.115}$$

The closed-from expression of $F^{CU}(x)$ can be found by substituting (9.114) into (9.115), applying the binomial expansion theorem, and using the following relationship:

$$\int_0^\infty \frac{e^{-\alpha y}}{(\beta + \kappa y)^r} dy = \frac{e^{\frac{\alpha\beta}{\kappa}} E_r\left(\frac{\alpha\beta}{\kappa}\right)}{\beta^{r-1}\kappa}. \tag{9.116}$$

Here, $E_k(z) = \int_1^\infty \frac{e^{-zt}}{t^k} dt$ is the generalized exponential integral function [29]. The resulting closed-form expression of $F^{CU}(x)$ is obtained as

$$F^{CU}(x) = \sum_{k=0}^N \binom{N}{k} \frac{T\lambda_{ps}(-\lambda_{sp})^k}{\lambda_{ss}P_M x \left(\lambda_{sp} + N_0\lambda_{ss}x/T\right)^{k-1}}$$
$$\times E_k\left(\frac{T\lambda_{ps}\lambda_{sp} + \lambda_{ps}\lambda_{ss}N_0 x}{\lambda_{ss}P_M x}\right) \exp\left(\frac{T\lambda_{ps}\lambda_{sp} + \lambda_{ps}\lambda_{ss}N_0 x}{\lambda_{ss}P_M x}\right) u(x). \tag{9.117}$$

Adaptation with Limited ST Power

Now we consider a more practical case, where the transmit power is upper bounded. To satisfy both maximum transmit power and interference constraints, the ST transmits with power $P_S^{(i)}$ as given in (9.110) when the ith antenna is selected for transmission. Therefore, the received SINR at SR after best antenna selection, γ^{CL}, is given by

$$\gamma^{CL} = \max_i \left\{\frac{\min\{T/|g_i|^2, P_{\max}\}|h_i|^2}{N_0 + P_M|h_0|^2}\right\}. \tag{9.118}$$

In order to compute the statistics of γ^{CL}, we define i.i.d. random variables Z_i as

$$Z_i = \min\{T/|g_i|^2, P_{\max}\}|h_i|^2, \tag{9.119}$$

where $i = 1, 2, \ldots, N$. The CDF of Z_i, $F_Z(z) = \Pr\{Z_i \leq z\}$, can be computed by using the joint distribution of $|h_i|^2$ and $|g_i|^2$ as

$$F_Z(z) = 1 - \lambda_{ss} \int_{z/P_{\max}}^\infty e^{-\lambda_{ss}y}\left(1 - e^{-\lambda_{sp}Ty/z}\right) dy. \tag{9.120}$$

Upon simplification, (9.120) results in

$$F_Z(z) = \left(1 - e^{-\frac{\lambda_{ss}z}{P_{\max}}} + \frac{\lambda_{ss}z e^{-\frac{\lambda_{sp}T + \lambda_{ss}z}{P_{\max}}}}{\lambda_{sp}T + \lambda_{ss}z}\right) u(z). \tag{9.121}$$

With the closed-form expression of $F_Z(z)$, $F^{CL}(x)$ can be computed by conditioning on $|h_0|^2$ as

$$F^{CL}(x) = \lambda_{ps} \int_0^\infty \left[F_Z(N_0 x + P_M xy)\right]^N e^{-\lambda_{ps} y} dy. \qquad (9.122)$$

Finally, by performing multinomial expansion of $[F_Z(N_0 x + P_M xy)]^N$ and using the integral relationship given in (9.116), a closed-form expression of $F^{CL}(x)$ can be computed as

$$F^{CL}(x) = \sum_{n_1+n_2+n_3=N} \frac{N!}{n_1! n_2! n_3!} \cdot \frac{T\lambda_{ps} \left(e^{-\frac{\lambda_{sp} T}{P_{\max}}} - 1\right)^{n_2} \left(-\lambda_{sp} e^{-\frac{\lambda_{sp} T}{P_{\max}}}\right)^{n_3}}{\lambda_{ss} P_M x (\lambda_{sp} + N_0 \lambda_{ss} x/T)^{n_3-1}}$$

$$\times e^{-\frac{\lambda_{ss}(n_2+n_3)N_0 x}{P_{\max}} + \frac{\left(\lambda_{ps} + \frac{\lambda_{ss} P_M x (n_2+n_3)}{P_{\max}}\right)(\lambda_{sp} T + N_0 \lambda_{ss} x)}{\lambda_{ss} P_M x}}$$

$$\times E_{n_3}\left(\frac{\left(\lambda_{ps} + \frac{\lambda_{ss} P_M x (n_2+n_3)}{P_{\max}}\right)(\lambda_{sp} T + N_0 \lambda_{ss} x)}{\lambda_{ss} P_M x}\right) u(x). \qquad (9.123)$$

Furthermore, it can be verified that $\lim_{P_{\max} \to \infty} F^{CL}(x) = F^{CU}(x)$.

TAS with Discrete Power Adaptation

We let $\gamma_{k,j}^D$ denote the received SINR at the SR when the kth best ST antenna is selected for transmission with the transmit power level of P_j, where $k = 1, 2, \ldots, N$ and $j = 1, 2, \ldots, J$. Mathematically, $\gamma_{k,j}^D$ can be written as

$$\gamma_{k,j}^D = \frac{P_j |h_{k:N}|^2}{N_0 + P_M |h_0|^2}, \qquad (9.124)$$

where $|h_{k:N}|^2$ denotes the kth largest channel power gain with PDF given by [31]

$$f_k(x) = k \binom{N}{k} \lambda_{ss} e^{-\lambda_{ss} k x} \left(1 - e^{-\lambda_{ss} x}\right)^{N-k} u(x). \qquad (9.125)$$

The PDF of $\gamma_{k,j}^D$, denoted by $f_{k,j}^D(x)$, can be computed by conditioning on $|h_0|^2$ as

$$f_{k,j}^D(x) = \frac{P_j}{P_M} \int_{\frac{N_0}{P_j}}^\infty z f_k(xz) f_{h_0}\left(\frac{P_j z - N_0}{P_M}\right) dz. \qquad (9.126)$$

By expanding (9.125) using the binomial theorem and performing integration in (9.126), we obtain a closed-form expression of $f_{k,j}^D(x)$, which is given by

$$f_{k,j}^D(x) = k \binom{N}{k} \lambda_{ss} \lambda_{ps} \sum_{i=0}^{N-k} (-1)^i \binom{N-k}{i}$$

$$\times e^{-\frac{N_0 x \lambda_{ss}(k+i)}{P_j}} \left(\frac{P_j P_M + N_0 P_j \lambda_{ps} + \lambda_{ss}(k+i) N_0 P_M x}{(P_j \lambda_{ps} + \lambda_{ss}(k+i) P_M x)^2}\right) u(x). \qquad (9.127)$$

Finally, the PDF of the received SINR, $f^D(x)$, can be calculated as

$$f^D(x) = p_\phi \delta(x) + \sum_{k=1}^{N} \sum_{j=1}^{J} p_{k,j} f_{k,j}^D(x), \qquad (9.128)$$

where $\delta(x)$ is the Dirac delta function, $p_{k,j}$ is the probability of selecting the kth best ST antenna for data transmission with a transmit power of P_j, and p_ϕ is the probability of the event that none of the ST to PR channel gains satisfies the interference-level requirement, i.e., $P_1|g_i|^2 > T$ for all i. For the SAPS scheme, p_ϕ can be computed for the Rayleigh fading environment as

$$p_\phi = (1 - F_g(T/P_1))^N = e^{-N\lambda_{sp}T/P_1}. \qquad (9.129)$$

The probabilities $p_{k,j}$ can similarly be computed:

$$p_{k,j} = \begin{cases} [1 - F_g(T/P_1)]^{k-1} F_g(T/P_J), & j = J; \\ \left[1 - F_g\left(\frac{T}{P_1}\right)\right]^{k-1} \left(F_g\left(\frac{T}{P_j}\right) - F_g\left(\frac{T}{P_{j+1}}\right)\right), & j \neq J. \end{cases} \qquad (9.130)$$

9.5.3 Performance Analysis

We now apply the statistical results to the performance analysis of the secondary systems.

Outage Probability

Since the interference constraint at PR can be met by adapting the transmit power at the PT, the outage probability, P_{out}, can be calculated using $P_{\text{out}} = F^{CL}(\gamma_T)$ and $P_{\text{out}} = F^{CU}(\gamma_T)$ for transmit antenna selection employing continuous power adaptation with and without peak power limits, respectively.

For the SAPS scheme, the outage probability can be computed as

$$P_{\text{out}} = p_\phi + \sum_{k=1}^{N} \sum_{j=1}^{J} p_{k,j} F_{k,j}^D(\gamma_T), \qquad (9.131)$$

where p_ϕ and $p_{k,j}$ are defined in (9.129) and (9.130), respectively, and $F_{k,j}^D(x)$ is the CDF of $\gamma_{k,j}^D$, which can be computed by conditioning on $|h_0|^2$ and using the following relationship

$$\int_0^\infty e^{-ax}(1 - ce^{-bx})^d dx = \int_0^1 u^{a/b-1}(1 - cu)^d du$$

$$= \frac{{}_2F_1\begin{pmatrix} a/b, & -d \\ a/b+1 \end{pmatrix}; c}{a}. \qquad (9.132)$$

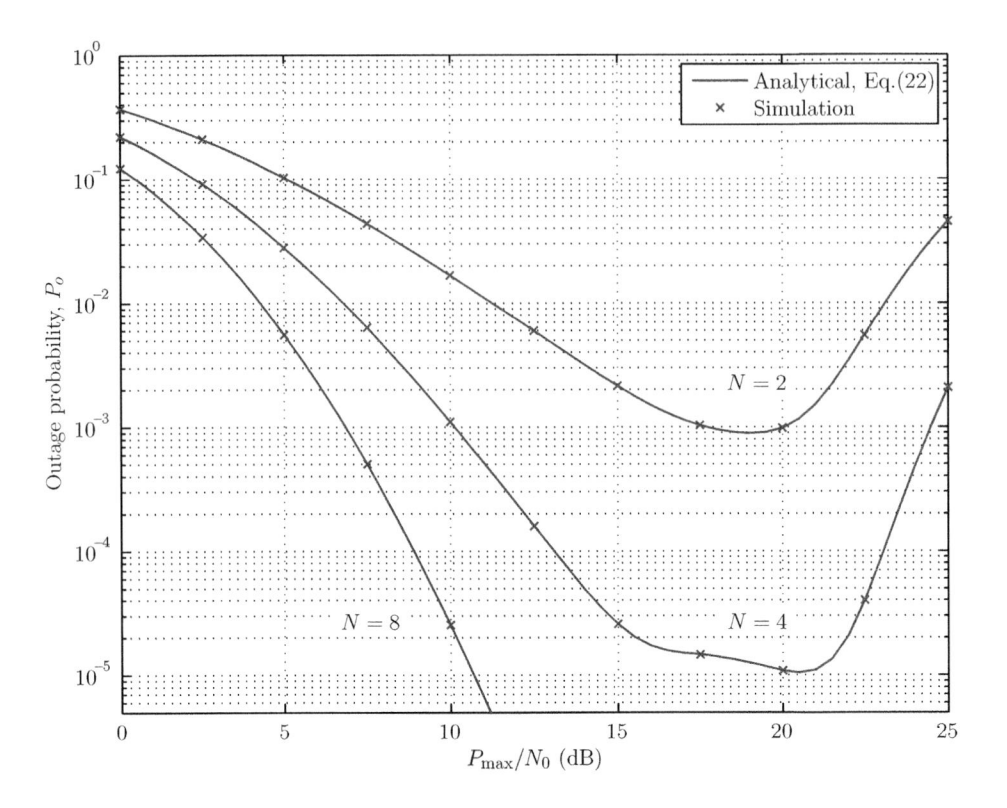

Figure 9.20 Outage probability of the SAPS scheme against $P_{\max}/N_0$ for $P_M = 0$ dBm, $N_0 = -20$ dBm, $\gamma_T = 13$ dBm, $J = 5$, $T = -10$ dBm and $N = 2, 4, 8$ ST antennas. Reprint with permission from [41]. ©2017 IEEE.

Here, $_2F_1\begin{pmatrix} m,n \\ p \end{pmatrix};q)$ is the Gauss hypergeometric function [29]. The closed-form expression of $F_{k,j}^D(x)$ is given by

$$F_{k,j}^D(x) = \lambda_{ps}P_j \sum_{i=0}^{k-1} \binom{N}{i} \exp\left(\frac{-N_0 i \lambda_{ss} x}{P_j}\right)$$
$$\times \frac{1}{\lambda_{ps}P_j + \lambda_{ss}P_M ix} {}_2F_1\begin{pmatrix} i + \lambda_{ps}P_j/(\lambda_{ss}P_M x), i - N \\ i + 1 + \lambda_{ps}P_j/(\lambda_{ss}P_M x) \end{pmatrix}; e^{-\frac{\lambda_{ss}N_0 x}{P_j}}\right) u(x). \quad (9.133)$$

Fig. 9.20 shows the behavior of the outage probability as the function of the normalized peak transmit power, $P_{\max}/N_0$. The fading parameters used for the simulation are $\lambda_{ps} = \lambda_{sp} = 10$, and $\lambda_{ss} = 5$. Also, we used five linearly spaced power levels at the ST, i.e., $P_j = jP_{\max}/J$ for $j = 1, 2, \ldots, J$. Monte Carlo simulations for 10^8 trials are also presented to verify the correctness of the derived analytical result given in (9.131). Since increasing the number of ST antennas increases the chances of meeting the interference constraint at the PR and improves the signal strength at the SR simultaneously, we observe a sharp decrease in the outage probability when N is varied from 2 to 8.

Moreover, increasing the transmit power results in a decrease in the outage probability, Lastly, increasing $P_{\max}$ too much will result in more frequent outages as meeting the interference constraint with high transmit power becomes increasingly difficult. Note that the minimum transmit power level at the ST is $P_{\max}/J$, which increases linearly with $P_{\max}$. We can see that when $P_{\max}$ is sufficiently large, P_{out} starts saturating and eventually increases.

Ergodic Capacity

The ergodic capacity of secondary transmission with continuous power adaptation is difficult to compute in a closed-form expression, but can be evaluated by numerical integration as

$$\overline{C} = \frac{1}{\ln 2} \int_0^\infty \frac{1 - F^C(x)}{1 + x} dx, \qquad (9.134)$$

where $F^C(x) = F^{CL}(x)$, and $F^C(x) = F^{CU}(x)$ for the cases of continuous power adaptation with and without peak power limit, respectively.

For the discrete adaptation case, however, the ergodic capacity can be computed in a closed-form expression as

$$\overline{C} = \sum_{k=1}^{N} \sum_{j=1}^{J} p_{k,j} \overline{C}_{k,j}, \qquad (9.135)$$

where $\overline{C}_{k,j}$ is the ergodic capacity of the secondary link when the kth best antenna is selected for the transmission with the transmit power of P_j, and $p_{k,j}$ is defined in (9.130). In order to derive a closed-form expression of $\overline{C}_{k,j}$, we consider the integral I defined as

$$I = \underbrace{\int_0^\infty \frac{e^{-ax}\ln(1+x)}{(ax+b)^2}dx}_{I_1} + \underbrace{\int_0^\infty \frac{e^{-ax}\ln(1+x)}{ax+b}dx}_{I_2}. \qquad (9.136)$$

By performing the integration by parts, I_1 can be expressed as

$$I_1 = \frac{1}{a} \int_0^\infty \frac{e^{-ax}}{(1+x)(ax+b)} dx - I_2. \qquad (9.137)$$

Lastly, I can be computed by performing partial fraction decomposition and given by

$$I = I_1 + I_2 = \frac{e^b E_1(b) - e^a E_1(a)}{a^2 - ab}. \qquad (9.138)$$

After some manipulations, $f_{k,j}^D(x)$, as given in (9.127), can be written as

$$f_{k,j}^D(x) = A \sum_{i=0}^{N-k} \binom{N-k}{i} (-1)^i \left(\frac{e^{-ax}}{b+ax} + \frac{e^{-ax}}{(b+ax)^2} \right), \qquad (9.139)$$

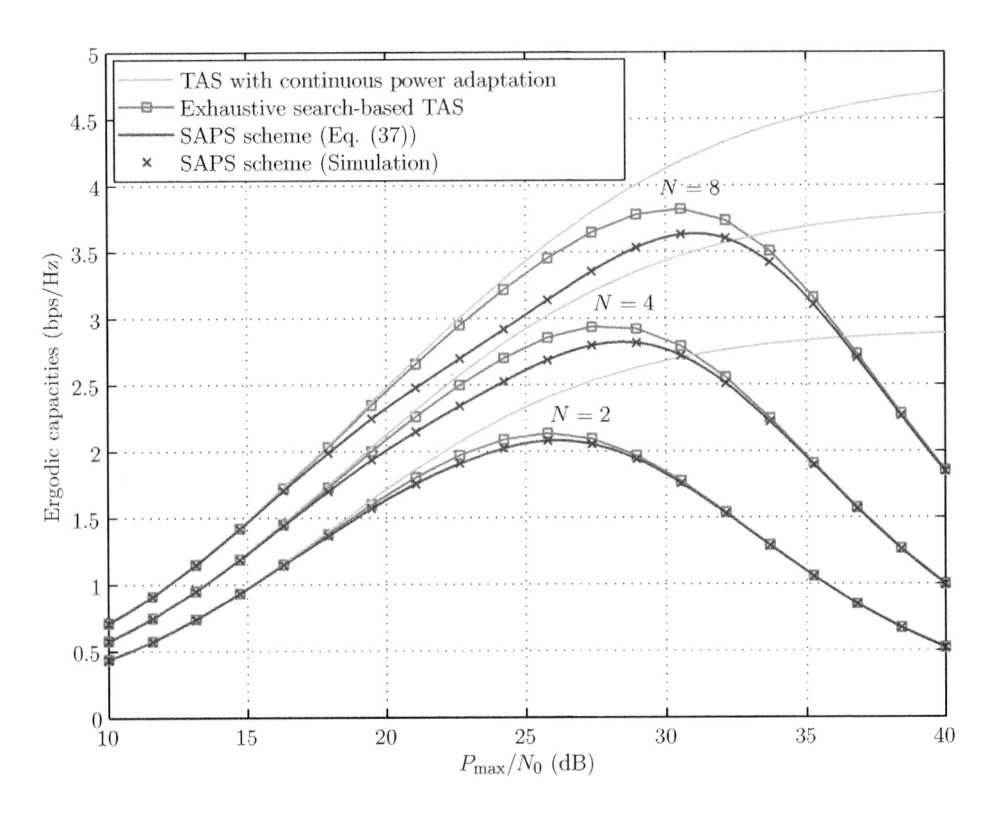

Figure 9.21 Ergodic capacities of different transmit antenna selection with power adaptation schemes against $P_{\max}/N_0$ for $N = 2, 4, 8$ ST antennas, $\lambda_{ps} = \lambda_{sp} = 10$, $\lambda_{ss} = 7$, $T = -10$ dBm, $P_M = 0$ dBm, $N_0 = -20$ dBm, and $J = 5$ power levels with $P_j = jP_{\max}/J$. Reprint with permission from [41]. ©2017 IEEE.

where $A = k\binom{N}{k}\lambda_{ss}\lambda_{ps}\frac{N_0^2}{P_j P_M}$, $a = \frac{\lambda_{ss}(k+i)N_0}{P_j}$, and $b = \frac{N_0\lambda_{ps}}{P_M}$. Using (9.138), we derive a closed-form expression of $\overline{C}_{k,j}$ as given in (9.140):

$$\overline{C}_{k,j} = \int_0^\infty \log_2(1 + x)f_{k,j}^D(x)dx = \frac{k\binom{N}{k}\lambda_{ps}}{\ln 2}\sum_{i=0}^{N-k}(-1)^i\binom{N-k}{i}$$

$$\times \frac{P_j\left[e^{\frac{N_0\lambda_{ps}}{P_M}}E_1\left(\frac{N_0\lambda_{ps}}{P_M}\right) - e^{\frac{N_0\lambda_{ss}(k+i)}{P_j}}E_1\left(\frac{N_0\lambda_{ss}(k+i)}{P_j}\right)\right]}{(k+i)[P_M\lambda_{ss}(k+i) - P_j\lambda_{ps}]}. \quad (9.140)$$

Fig. 9.21 shows the ergodic capacity of SAPS, exhaustive search-based transmit antenna selection, transmit antenna selection with continuous power adaptation schemes as function of normalized peak power for different values of N. The simulation results for the exhaustive search scheme were obtained by averaging the channel capacity over 10^7 different channel realizations. Since increasing N improves the received signal strength at the SR owing to the diversity gain and also reduces the outages caused by the nonsatisfaction of the interference constraint at the PR, we observe an increase in the ergodic capacity for all schemes when N varies from 2 to 8. We also see that transmit antenna selection with continuous power adaptation outperforms the exhaustive search-based transmit antenna selection with discrete power adaptation for all values of $P_{\max}/N_0$. Moreover, we observe a saturation effect for continuous power adaptation for high $P_{\max}$ because the transmit power is limited by the interference constraint and not by $P_{\max}$ when it is very high. For discrete power adaptation, on the other hand, we observe a decrease in the capacity as satisfying the interference constraint becomes increasingly difficult when $P_{\max}$ is very high. Lastly, the SAPS scheme shows a similar behavior as that of the exhaustive search-based scheme with discrete power adaptation, especially for very low and high values of $P_{\max}$.

9.6 Summary

In this chapter, we investigated essential secondary transmission problems assuming the provisions for secondary implementations, including spectrum sensing and channel estimation, have been perfectly achieved. We first characterized temporal transmission opportunities of secondary transmission with interweave implementation. We also analyzed the extended delivery time of secondary interweave transmission in various scenarios and applied their statistical characterization to the queuing performance analysis. For underlay implementation, we studied the performance of secondary transmission with power adaptation and diversity reception. We also analyzed and compared different transmit antenna selection schemes under a primary interference constraint. We showed that the low-complexity discrete power adaptive transmit antenna selection scheme performs nearly as well as exhaustive search-based antenna selection and offers an attractive antenna selection solution.

9.7 Further Reading

Early seminal papers on the concept of cognitive radio include [2, 3, 5]. The dedicated books by Fette [42] and by Hossain et al. [43] provide more comprehensive coverage on various issues of cognitive radio. The performance limits of cognitive transmission are investigated in [6, 44]. [45–47] discuss the challenges and candidate solutions for spectrum sensing. [48] extends the EDT analysis to the scenario in which the transmission time cannot be treated as constant. [49] further investigates the statistics of transmission time over general fading channels.

References

[1] M. Islam, C. Koh, S. W. Oh, et al., "Spectrum survey in Singapore: Occupancy measurements and analyses," in *Proc. 3rd Int. Conf. Cognitive Radio Oriented Wireless Netw. and Commun., 2008. CrownCom 2008*, Singapore, May 2008, pp. 1–7.

[2] J. Mitola III and G. Q. Maguire Jr., "Cognitive radio: Making software radios more personal," *IEEE Pers. Commun.*, 6, pp. 13–18, 1999.

[3] S. Haykin, "Cognitive radio: Brain-empowered wireless communications," *IEEE J. Selected Areas Commun.*, 23, no. 2, pp. 201–220, 2005.

[4] R. Thomas, L. DaSilva, and A. MacKenzie, "Cognitive networks," in *1st IEEE Int. Symp. on New Frontiers in Dynamic Spectrum Access Netw. (DySPAN'2005)*, Baltimore, MO, November 2005, pp. 352–360.

[5] I. F. Akyildiz, W.-Y. Lee, M. C. Vuran, and S. Mohanty, "Next generation/dynamic spectrum access/cognitive radio wireless networks: A survey," *Comput. Netw. J.*, 50, no. 13, pp. 2127–2159, 2006.

[6] A. Goldsmith, S. Jafar, I. Maric, and S. Srinivasa, "Breaking spectrum gridlock with cognitive radios: An information theoretic perspective," *Proc. IEEE*, 97, no. 5, pp. 894–914, 2009.

[7] B. Hamdaoui, "Adaptive spectrum assessment for opportunistic access in cognitive radio networks," *IEEE Trans. Wireless Commun.*, 8, no. 2, pp. 922–930, 2009.

[8] Z. Qianchuan, S. Geirhofer, T. Lang, and B. M. Sadler, "Opportunistic spectrum access via periodic channel sensing," *IEEE Trans. Signal Processing*, 56, no. 2, pp. 785–796, 2008.

[9] Z. Qing, T. Lang, S. Ananthram, and C. Yunxia, "Decentralized cognitive MAC for opportunistic spectrum access in ad hoc networks: A POMDP framework," *IEEE J. Sel. Areas Commun.*, 25, no. 3, pp. 589–600, 2007.

[10] I. Akyildiz, W.-Y. Lee, M. C. Vuran, and S. Mohanty, "A survey on spectrum management in cognitive radio networks," *IEEE Commun. Mag.*, 46, pp. 40–48, 2008.

[11] D. Cabric, S. Mishra, and R. Brodersen, "Implementation issues in spectrum sensing for cognitive radios," in *Conf. Record Thirty-Eighth Asilomar Conf. Signals, Syst. Comput., 2004.*, vol. 1, Pacific Grove, CA November 2004, pp. 772–776.

[12] F. Gaaloul, H.-C. Yang, R. Radaydeh, and M.-S. Alouini, "Switch based opportunistic spectrum access for general primary user traffic model," *IEEE Wireless Commun. Lett.*, 1, pp. 424–427, 2012.

[13] S. Geirhofer, L. Tong, and B. M. Sadler, "Dynamic spectrum access in WLAN channels: Empirical model and its stochastic analysis," *Proc. 1st Int. Workshop Technol. Policy Accessing Spectrum (TAPAS)*, Boston, MA, August 2006.

[14] P. Wang, and I. F. Akyildiz, "On the origins of heavy-tailed delay in dynamic spectrum access networks," *IEEE Trans. Mobile Comput.*, 11, no. 2, pp. 204–217, 2012.

[15] M. Wellens and P. Mahonen, "Lessons learned from an extensive spectrum occupancy measurement campaign and a stochastic duty cycle model," *J. Mobile Netw. Appl.*, 15, no. 3, pp. 461–474, 2010.

[16] F. Borgonovo, M. Cesana, and L. Fratta, "Throughput and delay bounds for cognitive transmissions," in *Advances in Ad Hoc Networking* (P. Cuenca, C. Guerrero, R. Puigjaner, and B. Serra, eds.), Springer, pp. 179–190, 2008.

[17] C.-W. Wang and L.-C. Wang, "Analysis of reactive spectrum handoff in cognitive radio networks," *IEEE J. Sel. Areas Commun.*, 30, pp. 2016–2028, 2012.

[18] M. Usman, H.-C. Yang, and M.-S. Alouini, "Extended delivery time analysis for cognitive packet transmission with application to secondary queuing analysis," *IEEE Trans. Wireless Commun.*, 14, no. 10, pp. 5300–5312, 2015.

[19] M. Usman, H.-C. Yang, and M.-S. Alouini, "Further results on extended delivery time analysis for secondary packet transmission," *IEEE Trans. Wireless Commun.*, 16, no. 10, pp. 6451–6459, 2017.

[20] D. J. C. MacKay, "Fountain codes," *IEE Proc. Commun.*, 152, pp. 1062–1068, 2005.

[21] J. Castura and Y. Mao, "Rateless coding over fading channels," *IEEE Commun. Lett.*, 10, pp. 46–48, 2006.

[22] A. Molisch, N. Mehta, J. S. Yedidia, and J. Zhang, "Performance of fountain codes in collaborative relay networks," *IEEE Trans. Wireless Commun.*, 6, pp. 4108–4119, 2007.

[23] E. Cinlar, *Introduction to Stochastic Processes*, Prentice-Hall, 1975.

[24] M. Abramowitz and I. A. Stegun, *Handbook of Mathematical Functions with Formulas, Graphs, and Mathematical Tables*, 9th ed., Dover, 1970.

[25] R. Boucherie and N. van Dijk, *Queueing Networks: A Fundamental Approach*, Springer, 2010.

[26] I. Adan and J. Resing, *Queueing Theory*, Eindhoven University of Technology, Department of Mathematics and Computing Science, 2001.

[27] X. Kang, Y.-C. Liang, A. Nallanathan, H. Garg, and R. Zhang, "Optimal power allocation for fading channels in cognitive radio networks: Ergodic capacity and outage capacity," *IEEE Trans. Wireless Commun.*, 8, no. 2, pp. 940–950, 2009.

[28] M. Hanif, H.-C. Yang, and M.-S. Alouini, "Receive antenna selection for underlay cognitive radio with instantaneous interference constraint," *IEEE Signal Process. Lett.*, 22, no. 6, pp. 738–742, 2015.

[29] I. S. Gradshteyn and I. M. Ryzhik, *Table of Integrals, Series, and Products*, 7th ed., Elsevier/Academic Press, 2007.

[30] A. Cuyt, F. Backeljauw, and C. Bonan-Hamada, *Handbook of Continued Fractions for Special Functions*, Springer, 2008.

[31] H. A. David and H. N. Nagaraja, *Order Statistics*, Wiley-Interscience, 2003.

[32] R. Zhang and Y.-C. Liang, "Exploiting multi-antennas for opportunistic spectrum sharing in cognitive radio networks," *IEEE J. Sel. Topics Signal Process.*, 2, no. 1, pp. 88–102, 2008.

[33] G. Scutari, D. Palomar, and S. Barbarossa, "Cognitive MIMO radio," *IEEE Signal Process. Mag.*, 25, no. 6, pp. 46–59, 2008.

[34] D. Denkovski, V. Rakovic, V. Atanasovski, L. Gavrilovska, and P. Mähönen, "Generic multiuser coordinated beamforming for underlay spectrum sharing," *IEEE Trans. Commun.*, 64, no. 6, pp. 2285–2298, 2016.

[35] F. A. Khan, K. Tourki, M. S. Alouini, and K. A. Qaraqe, "Performance analysis of a power limited spectrum sharing system with TAS/MRC," *IEEE Trans. Signal Process.*, 62, no. 4, pp. 954–967, 2014.

[36] Y. Deng, L. Wang, M. Elkashlan, K. J. Kim, and T. Q. Duong, "Generalized selection combining for cognitive relay networks over Nakagami-m fading," *IEEE Trans. Signal Process.*, 63, no. 8, pp. 1993–2006, 2015.

[37] J. Zhou, Y. Li, and B. Evans, "Antenna selection for multiple-input and single-output cognitive radio systems," *IET Commun.*, 6, no. 8, pp. 917–930, 2012.

[38] H. Wang, J. Lee, S. Kim, and D. Hong, "Capacity enhancement of secondary links through spatial diversity in spectrum sharing," *IEEE Trans. Wireless Commun.*, 9, no. 2, pp. 494–499, 2010.

[39] R. Sarvendranath and N. Mehta, "Antenna selection in interference-constrained underlay cognitive radios: SEP-optimal rule and performance benchmarking," *IEEE Trans. Commun.*, 61, no. 2, pp. 496–506, 2013.

[40] Y. Wang and J. Coon, "Difference antenna selection and power allocation for wireless cognitive systems," *IEEE Trans. Commun.*, 59, no. 12, pp. 3494–3503, 2011.

[41] M. Hanif, H.-C. Yang, and M.-S. Alouini, "Transmit antenna selection for power adaptive underlay cognitive radio with instantaneous interference constraint," *IEEE Trans. Commun.*, 65, no. 6, pp. 2357–2367, 2017.

[42] B. A. Fette, *Cognitive Radio Technology*, 2nd ed., Academic Press, 2009.

[43] E. Hossain, D. Niyato, and Z. Han, *Dynamic Spectrum Access and Management in Cognitive Radio Networks*, Cambridge University Press, 2009.

[44] N. Devroye, P. Mitran, and V. Tarokh, "Achievable rates in cognitive radio channels," *IEEE Trans. Inform. Theor.*, 52, no. 5, pp. 1813–1827, 2006.

[45] A. Ghasemi and E. S. Sousa, "Spectrum sensing in cognitive radio networks: Requirements, challenges and design trade-offs," *IEEE Commun. Mag.*, 46, no. 4, pp. 32–39, 2008.

[46] K. B. Letaief and W. Zhang, "Cooperative communications for cognitive radio networks," *Proc. IEEE*, 97, no. 5, pp. 878–893, 2009.

[47] T. Yucek and H. Arslan, "A survey of spectrum sensing algorithms for cognitive radio applications," *IEEE Commun. Surv. Tut.*, 11, no. 1, pp. 116–130, 2009.

[48] W.-J. Wang, M. Usman, H.-C. Yang, and M.-S. Alouini, "Extended delivery time analysis for secondary packet transmission with adaptive modulation under interleave cognitive implementation," *IEEE Trans. Cognitive Commun. Netw.*, 3, no. 2, pp. 180–189, 2017.

[49] W.-J. Wang, H.-C. Yang, and M.-S. Alouini, "Wireless transmission of big data: A transmission time analysis over fading channel," *IEEE Trans. Wireless Commun.*, 17, no. 7, pp. 4315–4325, 2018.

10 Application: Hybrid FSO/RF Transmission

Free space optical (FSO) communications and millimeter wave (mmWave) radio frequency (RF) communications have emerged as effective solutions for high-data-rate wireless transmission over short distances. These systems operate over unlicensed (free) RF and optical frequencies and can work up to a few kilometers [1, 2]. They can help to address the continuous demand for higher-data-rate transmission in the presence of the growing scarcity of RF spectrum. Meanwhile, FSO channel and mmWave RF channel exhibit complementary characteristics to atmospheric and weather effects. In particular, FSO link performance degrades significantly due to fog, but is not sensitive to rain [3, 4]. Contrarily, mmWave RF is very sensitive to rain but is quite indifferent to fog. Thus, FSO and RF transmission systems are good candidates for joint deployment to provide reliable high-data-rate wireless transmission solution.

In this chapter, we apply selected transmission technologies in previous chapter to design several practical transmission schemes for hybrid FSO/RF systems. We first present a switching-based hybrid FSO/RF transmission scheme. Then we design a hybrid FSO/RF transmission scheme with the application of adaptive combining. Finally, the joint design of adaptive transmission and adaptive combining is introduced to hybrid FSO/RF systems. We carry out a thorough performance analysis of the resulting designs and discuss their achieved performance versus complexity trade-offs. Our goal is to demonstrate the key challenges of hybrid FSO/RF transmission and their potential solutions.

10.1 Switching-Based Hybrid FSO/RF Transmission

Most previous work on hybrid FSO/RF systems focus on the joint design of coding schemes over FSO and RF links, which achieves soft-switching between two links [5–8]. On another front, diversity combining can be introduced over parallel FSO and RF channels, assuming both links transmit identical information simultaneously. Such hard-switching schemes typically lead to some rate loss compared to soft-switching schemes. Meanwhile, both classes of transmission schemes above require the FSO and RF links to be active continuously, even when experiencing poor quality, which will lead to wasted transmission power and generate unnecessary interference to other systems.

Here, we consider a low-complexity hard-switching scheme for hybrid FSO/RF transmission, which will transmit data using either the FSO link or the mmWave RF link [9].

The FSO link will be used as long as its link quality is above a certain threshold. When the FSO link becomes unacceptable, the system will resort to the RF link. Using only a single link at a time will result in lower power consumption at the transmitter, and will not require the use of combining or multiplexing operations at the receiver. In fact, such implementations are widely adopted in commercially available hybrid FSO/RF products, [10].

We analyze the performance of such a low-complexity transmission scheme for hybrid FSO/RF systems. We derive closed-form analytical expressions for the outage probability, the average bit error rate (BER), and the ergodic capacity for the hybrid system under practical fading channel models. In addition to the single FSO threshold implementation, we consider the dual FSO threshold implementation, which can avoid the frequent on–off transitions of the FSO link. Specifically, exact expressions are obtained for the outage probabilities and BERs of both FSO and RF links, and capacity for the RF link. An approximate expression has been obtained for the capacity of the FSO link. These expressions are then applied to analyze the overall system.

10.1.1 System and Channel Model

We consider a hybrid FSO/RF system in which the FSO link works in parallel with an RF link, as shown in Fig. 10.1. To keep the receiver implementation simple, the transmission occurs only on one of the links at a time. Due to its generally higher data rate FSO will be given a higher priority and will be used for transmission whenever its link quality is acceptable.

FSO Link Modeling
The FSO link adopts intensity modulation and direct detection (IM/DD), together with a quadrature modulation scheme [11, 12]. Specifically, the information is first modulated

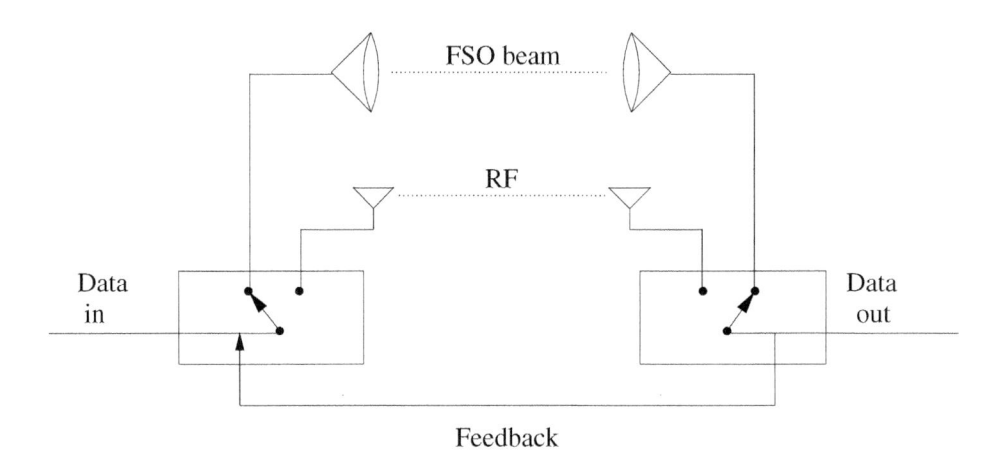

Figure 10.1 System block diagram of a hybrid FSO/RF system. Reprint with permission from [9]. ©2014 IEEE.

10.1 Switching-Based Hybrid FSO/RF Transmission

using a quadrature modulation scheme. The modulated electrical signal is then directly modulated on the transmitter's laser intensity. To avoid any clipping while modulating onto the laser intensity, a DC bias must be added to the modulated electrical signal to ensure that the value of the signal is nonnegative. Hence, the intensity of the transmitted optical signal can be written as

$$I(t) = P_{FSO}(1 + \mu x(t)), \tag{10.1}$$

where P_{FSO} is the average transmitted optical power, μ is the modulation index ($0 < \mu < 1$) which is introduced to eliminate over-modulation-induced clipping, and $x(t)$ is the quadrature modulated electrical signal. At the receiver end of the FSO subsystem, the incident optical power on the photodetector is converted into an electrical signal through direct detection. After the DC bias is filtered out, the electrical signal is demodulated to obtain the discrete-time equivalent baseband signal, given by

$$r[k] = \mu \eta P_{FSO} \cdot h_{FSO} \cdot G_{FSO} \cdot x[k] + n[k], \tag{10.2}$$

where η is the receiver's optical-to-electrical conversion efficiency, G_{FSO} represents the average gain of the FSO link [7], h_{FSO} is the turbulence-induced random fading channel gain with $E[h] = 1$, $x[k]$ represents the complex baseband information symbol over the kth symbol period with average electrical symbol energy E_s, and $n[k]$ is the zero mean circularly symmetric complex Gaussian noise component with variance σ_{FSO}^2. The instantaneous electrical signal-to-noise ratio (SNR) at the output of the FSO receiver, denoted by γ_{FSO}, can be shown to be equal to

$$\gamma_{FSO} = \frac{\mu^2 \eta^2 P_{FSO}^2 G_{FSO}^2 h_{FSO}^2 E_s}{\sigma_{FSO}^2}. \tag{10.3}$$

For analytical tractability, we adopt log-normal fading model for turbulence-induced fading [13, 14], which is valid particularly for light atmospheric turbulence conditions, while assuming that there are no pointing errors associated with the laser beam. Specifically, the fading coefficient h_{FSO} has the probability density function (PDF)

$$f_{h_{FSO}}(h) = \frac{1}{\sqrt{8\pi} h \sigma_x} e^{-\frac{[\ln(h) - 2\mu_x]^2}{8\sigma_x^2}}, \tag{10.4}$$

where μ_x and σ_x represent the mean and variance of the log-amplitude fading, respectively. It has been shown that the condition $E[h_{FSO}] = 1$ implies $\mu_x = -\sigma_x^2$ [15]. Based on (10.3) and (10.4), the instantaneous SNR of the FSO link can be shown to have the following PDF

$$f_{\gamma_{FSO}}(\gamma) = \frac{1}{\sqrt{32\pi} \gamma \sigma_x} e^{-\frac{\left[\ln\left(\frac{\gamma}{\bar{\gamma}_{FSO}}\right) + 8\sigma_x^2\right]^2}{32\sigma_x^2}}, \tag{10.5}$$

where $\bar{\gamma}_{FSO}$ is the average electrical SNR given by

$$\bar{\gamma}_{FSO} = E[\gamma_{FSO}] = \frac{\mu^2 \eta^2 P_{FSO}^2 G_{FSO} E_s}{\sigma_{FSO}^2} e^{4\sigma_x^2}. \tag{10.6}$$

RF link modeling

The electrical modulated signal is up-converted to a mmWave RF carrier frequency and transmitted over the RF link. The received discrete-time signal, after the demodulation process, can be written as

$$r_{RF}[k] = \sqrt{G_{RF}P_{RF}} h_{RF} x[k] + n_{RF}[k], \quad (10.7)$$

where G_{RF} is the average power gain of the RF link, P_{RF} is the RF transmit power, and h_{RF} is the fading gain over the RF channel, with $E[h_{RF}^2]$ normalized to unity. $n_{RF}[k]$ is the zero mean circularly symmetric AWGN component with variance σ_{RF}^2. The average power gain G_{RF} is determined as

$$G_{RF}[dB] = G_T + G_R - 20\log_{10}\left(\frac{4\pi z}{\lambda_{RF}}\right) - \alpha_{oxy}z - \alpha_{rain}z, \quad (10.8)$$

where G_T and G_R denote the transmit and receive antenna gains, respectively and λ_{RF} is the wavelength of the RF subsystem. α_{oxy} and α_{rain} are the attenuation caused by oxygen absorption and rain, respectively.[1]

The instantaneous received SNR of the RF link γ_{RF} is given by $\gamma_{RF} = \bar{\gamma}_{RF} h_{RF}^2$, where $\bar{\gamma}_{RF}$ is the average SNR of the RF link defined as $\bar{\gamma}_{RF} = E_s P_{RF} G_{RF}/\sigma_{RF}^2$. The fading gain h_{RF} is modeled by a Nakagami-m distribution, which represents a wide variety of realistic line-of-sight (LOS), and non-LOS (NLOS) fading channels encountered in practice [27]. Accordingly, the received SNR γ_{RF} will have the following PDF [28]:

$$f_{\gamma_{RF}}(\gamma_{RF}) = \frac{\left(\frac{m}{\bar{\gamma}_{RF}}\right)^m \gamma_{RF}^{m-1}}{\Gamma(m)} \exp\left(\frac{-m\gamma_{RF}}{\bar{\gamma}_{RF}}\right), \gamma_{RF} \geq 0. \quad (10.9)$$

After applying [22, Eq. (3.351.1)], and some simple algebraic manipulations, the CDF of γ_{RF} can be expressed as

$$F_{\gamma_{RF}}(\gamma_{RF}) = \frac{1}{\Gamma(m)} \gamma\left(m, \frac{m\gamma_{RF}}{\bar{\gamma}_{RF}}\right), \gamma_{RF} \geq 0, \quad (10.10)$$

where $\gamma(\cdot,\cdot)$ is the lower incomplete gamma function defined in [22, Eq. (8.350.1)].

10.1.2 Single Switching Threshold Design

In this scenario, the FSO link will be used if the instantaneous SNR of the FSO channel is above a threshold γ_{th}^{FSO}. If the SNR of the FSO link is below γ_{th}^{FSO}, then the system will check the RF link; if the RF link SNR is above a threshold γ_{th}^{RF}, the RF link will be used for transmission. If the SNRs of both links are below their respective thresholds, an outage will be declared. In the following, we calculate the outage probability, the average BER during the non-outage period of time, and the ergodic capacity for this implementation scenario.

---

[1] Oxygen absorption at 60 GHz attenuates the signal at all times, regardless of the weather [26].

Outage Probability

The outage probability can be calculated based on the above mode of operation as

$$P_{\text{out}} = F_{\gamma_{FSO}}(\gamma_{\text{th}}^{FSO}) \times F_{\gamma_{RF}}(\gamma_{\text{th}}^{RF}), \qquad (10.11)$$

where $F_{\gamma_{FSO}}(.)$ is the cumulative distribution function (CDF) of the FSO link SNR, given by

$$F_{\gamma_{FSO}}(\gamma_{th}) = \int_0^{\gamma_{th}} f_{\gamma_{FSO}}(\gamma)d\gamma = 1 - \frac{1}{2}\text{erfc}\left(\frac{\ln\left(\frac{\gamma_{th}}{\gamma_{FSO}}\right) + 8\sigma_x^2}{\sqrt{32}\sigma_x}\right), \qquad (10.12)$$

$F_{\gamma_{RF}}(.)$ is the CDF of the RF link SNR, given by

$$F_{\gamma_{RF}}(\gamma_{th}) = \int_0^{\gamma_{th}} f_{\gamma_{RF}}(\gamma)d\gamma = \frac{\gamma\left[m, \frac{\gamma_{th}}{\gamma_{RF}}m\right]}{\Gamma[m]}, \qquad (10.13)$$

erfc(.) denotes the complementary error function, and $\gamma[.,.]$ is the lower incomplete gamma function.

Fig. 10.2 shows the variation of the outage probability with the average SNR of the FSO link, for a fixed value of γ_{th}^{FSO} and γ_{th}^{RF}. The varying average SNR of the FSO link corresponds to a varying weather condition. The three curves correspond to FSO only, hybrid FSO/RF where the RF link has an average SNR of 5 dB, and hybrid FSO/RF where the RF link has an average SNR of 10 dB cases. As can be seen, using the hybrid system improves the outage performance of the system, particularly when a high-quality RF link is used. Even with a low-quality RF link, some improvement can be observed. The simulation results included validate our analytical results.

Average Bit Error Rate

For the purpose of the average BER calculation, we will assume that the data are modulated using M-phase-shift-keying (PSK), and then transmitted through the FSO link or the RF link. We will assume that both the FSO and RF links operate at the same data rates. The generalization to the different rate case and other modulation schemes is straightforward. The BER for M-PSK with Gray code, as a function of the instantaneous SNR, is given by [16, Eq. (5.2.61)]

$$p(e|\gamma) = \frac{A}{2}\text{erfc}(\sqrt{\gamma}B), \qquad (10.14)$$

where $A = 1$ and $B = 1$ when $M = 2$ (binary PSK), and $A = \frac{2}{\log_2 M}$ and $B = \sin\frac{\pi}{M}$ when $M > 2$.

The average BER during the nonoutage period can be calculated in terms of the average BER when the FSO link is active and that when the RF link is active, as

$$\overline{BER} = \frac{BER_{FSO}(\gamma_{th}^{FSO}) + F_{\gamma_{FSO}}(\gamma_{th}^{FSO}) \cdot BER_{RF}(\gamma_{th}^{RF})}{1 - P_{out}}, \qquad (10.15)$$

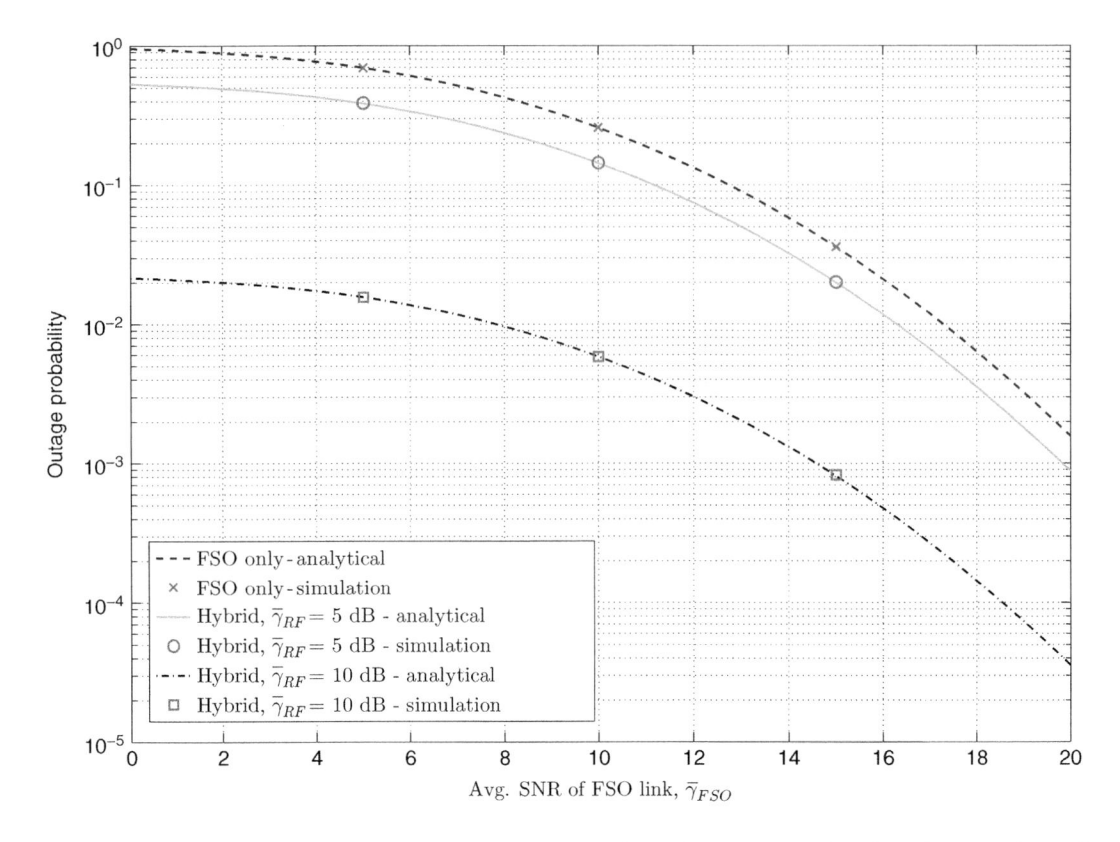

Figure 10.2 Outage probability for the single FSO threshold case as a function of average SNR of the FSO link: $\gamma_{th}^{FSO} = \gamma_{th}^{RF} = 5$ dB, $m = 5$, $\sigma_x = 0.25$. Reprint with permission from [9]. ©2014 IEEE.

where P_{out} was given in (10.11), $F_{\gamma_{FSO}}$ was given in (10.12), and $BER_{FSO}(.)$ and $BER_{RF}(.)$ are defined as

$$BER_{FSO}(\gamma_{th}) = \int_{\gamma_{th}}^{\infty} p(e|\gamma) f_{\gamma_{FSO}}(\gamma) d\gamma \qquad (10.16)$$

and

$$BER_{RF}(\gamma_{th}) = \int_{\gamma_{th}}^{\infty} p(e|\gamma) f_{\gamma_{RF}}(\gamma) d\gamma, \qquad (10.17)$$

respectively.

Substituting (10.14) into (10.16), and applying the series expansion of the complementary error function, we get

$$BER_{FSO}(\gamma_{th}) = \int_{\gamma_{th}}^{\infty} \frac{A}{2} \text{erfc}\left(B\sqrt{\gamma}\right) f_{\gamma_{FSO}}(\gamma) d\gamma$$

$$= \int_{\gamma_{th}}^{\infty} \frac{A}{2} \left[1 - \frac{2}{\sqrt{\pi}} \sum_{j=0}^{\infty} \frac{(-1)^j (B\sqrt{\gamma})^{2j+1}}{j!(2j+1)} \right] \times f_{\gamma_{FSO}}(\gamma) d\gamma.$$

After some manipulation, and using the change of variable $y = \frac{\ln(\frac{\gamma}{\bar{\gamma}_{FSO}})+8\sigma_x^2}{\sqrt{32}\sigma_x}$, we arrive at

$$BER_{FSO}(\gamma_{th}) = \frac{A}{4}\text{erfc}\,(\psi(\gamma_{th})) - \frac{A}{\sqrt{\pi}}\sum_{j=0}^{\infty}\left[\frac{(-1)^j B^{2j+1}}{j!(2j+1)}\right.$$
$$\left.\times \int_{\psi(\gamma_{th})}^{\infty} \frac{1}{\sqrt{\pi}}e^{-y^2}e^{(j+0.5)(\ln(\bar{\gamma}_{FSO})-8\sigma_x^2+\sqrt{32}\sigma_x y)}dy\right], \quad (10.18)$$

where

$$\psi(\gamma_{th}) = \frac{\ln(\frac{\gamma_{th}}{\bar{\gamma}_{FSO}}) + 8\sigma_x^2}{\sqrt{32}\sigma_x}. \quad (10.19)$$

It can be shown, with the application of the definition of the erfc(.) function, that the above expression finally simplifies to

$$BER_{FSO}(\gamma_{th}) = \frac{A}{4}\text{erfc}\,(\psi(\gamma_{th}))$$
$$- \frac{A}{2\sqrt{\pi}}\sum_{j=0}^{\infty}\left[\frac{(-1)^j B^{2j+1}}{j!(2j+1)}\bar{\gamma}_{FSO}^{(j+0.5)}e^{(8j^2-2)\sigma_x^2}\right.$$
$$\left.\times \text{erfc}\left(\psi(\gamma_{th}) - (j+0.5)\sqrt{8}\sigma_x\right)\right]. \quad (10.20)$$

For the RF link, substituting (10.9) and (10.14) into (10.17), we get

$$BER_{RF}(\gamma_{th}) = \int_{\gamma_{th}}^{\infty} \frac{A}{2}\text{erfc}(\sqrt{\gamma}B)\left(\frac{m}{\bar{\gamma}_{RF}}\right)^m \frac{\gamma^{m-1}}{\Gamma[m]} e^{-\frac{\gamma}{\bar{\gamma}_{RF}}m}d\gamma. \quad (10.21)$$

For integer values of m, the above integral becomes [17]

$$BER_{RF}(\gamma_{th}) = \left\{-\frac{A}{2}\frac{\text{erfc}(\sqrt{\gamma}B)}{\Gamma[m]}\Gamma\left[m,\frac{\gamma}{\bar{\gamma}_{RF}}m\right]\right\}_{\gamma_{th}}^{\infty}$$
$$+ \left\{\frac{AB}{2\sqrt{\pi}}\sum_{n=0}^{m-1}\frac{\Gamma\left[n+\frac{1}{2},\gamma\cdot(B^2+\frac{m}{\bar{\gamma}_{RF}})\right]}{\Gamma(n+1)(B^2+\frac{m}{\bar{\gamma}_{RF}})^{n+\frac{1}{2}}}\left(\frac{m}{\bar{\gamma}_{RF}}\right)^n\right\}_{\gamma_{th}}^{\infty}, \quad (10.22)$$

which, after the substitution of the limits, simplifies to

$$BER_{RF}(\gamma_{th}) = \frac{A}{2}\frac{\text{erfc}(\sqrt{\gamma_{th}}B)}{\Gamma[m]}\Gamma\left[m,\frac{\gamma_{th}}{\bar{\gamma}_{RF}}m\right]$$
$$- \frac{AB}{2\sqrt{\pi}}\sum_{n=0}^{m-1}\frac{\Gamma\left[n+\frac{1}{2},\gamma_{th}\cdot(B^2+\frac{m}{\bar{\gamma}_{RF}})\right]}{\Gamma(n+1)(B^2+\frac{m}{\bar{\gamma}_{RF}})^{n+\frac{1}{2}}}\left(\frac{m}{\bar{\gamma}_{RF}}\right)^n.$$

Fig. 10.3 shows the variation of the average BER against the average SNR of the FSO link, for a fixed value of γ_{th}^{FSO} and γ_{th}^{RF}. As seen in the figure, when using a low-quality RF link, the BER performance deteriorates slightly. This is expected because

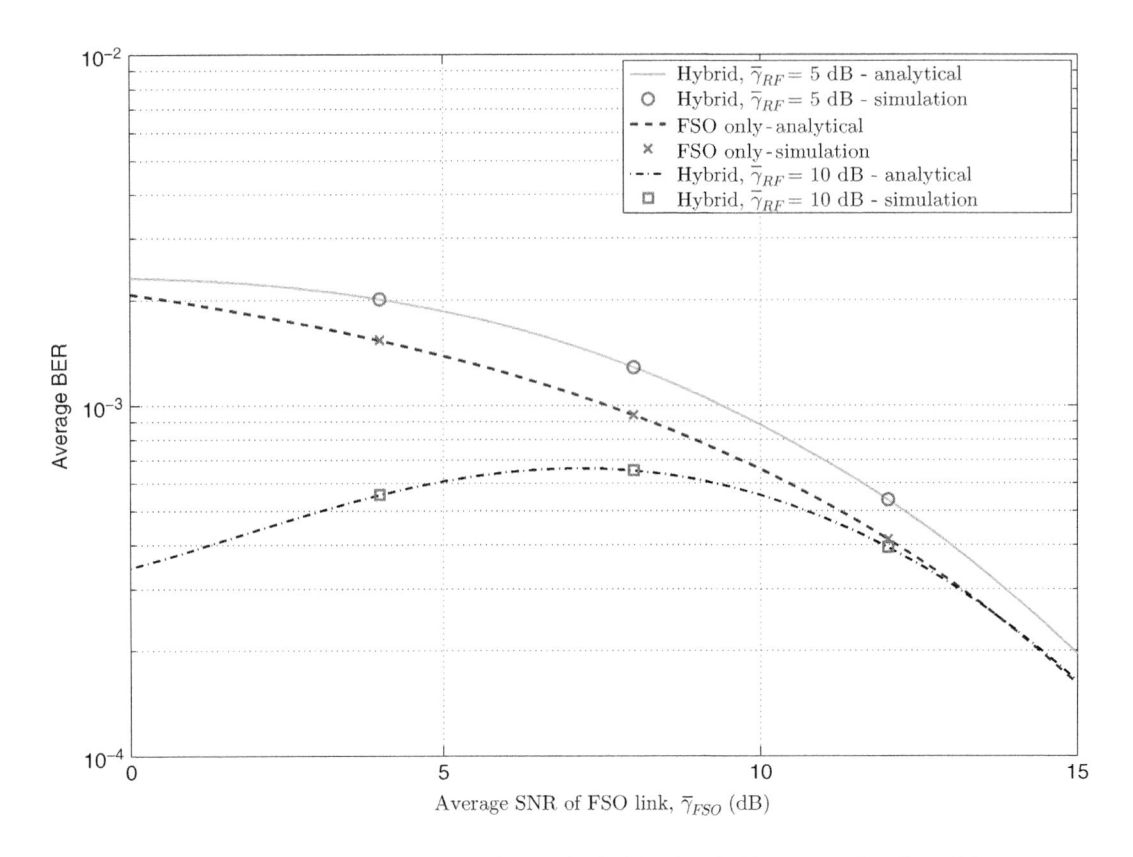

Figure 10.3 Average bit error rate for the single FSO threshold case as a function of the average SNR of the FSO link: $\gamma_{th}^{FSO} = \gamma_{th}^{RF} = 5$ dB, $m = 5$, $\sigma_x = 0.25$. Reprint with permission from [9]. ©2014 IEEE.

now, instead of turning the transmission off, a weak channel is being used for transmission, which affects the average BER. With a high-quality RF link, considerable improvement is seen, especially over the low FSO SNR region. In fact, as the average SNR of the FSO link improves, the BER performance of the overall system slightly deteriorates at first. This is because at a very low value of $\bar{\gamma}_{FSO}$, the high-quality RF link is being used frequently. As $\bar{\gamma}_{FSO}$ becomes larger, the weak FSO link is used more often, which increases the average BER. When $\bar{\gamma}_{FSO}$ increases further, the FSO link becomes better and the BER performance of the system again improves. The simulation results have also been included, which are found to conform to the analytical results.

Ergodic Capacity

The ergodic capacity of the hybrid FSO/RF system for the single FSO threshold case under consideration can be computed as

$$C^{(1)} = C_{FSO}(\gamma_{th}^{FSO}) + F_{\gamma_{FSO}}(\gamma_{th}^{FSO}) \cdot C_{RF}(\gamma_{th}^{RF}), \tag{10.23}$$

10.1 Switching-Based Hybrid FSO/RF Transmission

where $C_{FSO}(\gamma_{th}^{FSO})$ and $C_{RF}(\gamma_{th}^{RF})$ are the capacities of the FSO and RF link respectively, when they are active, and are given by

$$C_{FSO}(\gamma_{th}) = \int_{\gamma_{th}}^{\infty} W_{FSO} \cdot \log_2(1 + \gamma) f_{\gamma_{FSO}}(\gamma) d\gamma \tag{10.24}$$

and

$$C_{RF}(\gamma_{th}) = \int_{\gamma_{th}}^{\infty} W_{RF} \cdot \log_2(1 + \gamma) f_{\gamma_{RF}}(\gamma) d\gamma, \tag{10.25}$$

respectively, W_{FSO} is the bandwidth of the FSO link, and W_{RF} is the bandwidth of the RF link.

Substituting (10.5) into (10.24), we have

$$C_{FSO}(\gamma_{th}) = \int_{\gamma_{th}}^{\infty} \frac{W_{FSO} \log_2(1+\gamma)}{\sqrt{32\pi}\gamma\sigma_x} e^{-\frac{\left[\ln\left(\frac{\gamma}{\bar{\gamma}_{FSO}}\right)+8\sigma_x^2\right]^2}{32\sigma_x^2}} d\gamma. \tag{10.26}$$

After applying the high SNR approximation $\ln(\gamma) \approx \ln(1+\gamma)$, and then using the replacement $y = \frac{\ln\left(\frac{\gamma}{\bar{\gamma}_{FSO}}\right)+8\sigma_x^2}{\sqrt{32}\sigma_x}$, the above expression becomes

$$C_{FSO}(\gamma_{th}) \approx \int_{\psi(\gamma_{th})}^{\infty} \frac{W_{FSO}}{\ln(2)} \frac{\sqrt{32}\sigma_x y + \ln(\bar{\gamma}_{FSO}) - 8\sigma_x^2}{\sqrt{\pi}} e^{-y^2} dy, \tag{10.27}$$

where $\psi(\gamma_{th})$ is defined in (10.19). Carrying out integration, it can be shown that the above expression simplifies to

$$C_{FSO}(\gamma_{th}) \approx \frac{W_{FSO}}{\ln(2)} \frac{\sqrt{8}\sigma_x}{\sqrt{\pi}} e^{-[\psi(\gamma_{th})]^2} + \frac{W_{FSO}}{\ln(2)} \frac{\ln(\bar{\gamma}_{FSO}) - 8\sigma_x^2}{2} \operatorname{erfc}(\psi(\gamma_{th})). \tag{10.28}$$

For the RF link, substituting (10.9) into (10.25), we have

$$C_{RF}(\gamma_{th}) = \int_{\gamma_{th}}^{\infty} W_{RF} \cdot \log_2(1+\gamma) \left(\frac{m}{\bar{\gamma}_{RF}}\right)^m \frac{\gamma^{m-1}}{\Gamma[m]} e^{-\frac{\gamma}{\bar{\gamma}_{RF}}m} d\gamma. \tag{10.29}$$

After using the replacement $y = 1 + \gamma$, and performing binomial expansion and some manipulation, we arrive at

$$C_{RF}(\gamma_{th}) = \left(\frac{m}{\bar{\gamma}_{RF}}\right)^m \frac{W_{RF} \cdot e^{\frac{m}{\bar{\gamma}_{RF}}}}{\ln(2)} \sum_{p=0}^{m-1} \left[\frac{(-1)^{m-p-1}}{\Gamma(p+1)\Gamma(m-p)}\right.$$
$$\left. \times \int_{1+\gamma_{th}}^{\infty} y^p \ln(y) e^{-\frac{y}{\bar{\gamma}_{RF}}m} dy \right].$$

Finally, we can obtain the closed-form expression for $C_{RF}(.)$ as

$$C_{RF}(\gamma_{th}) = \left(\frac{m}{\bar{\gamma}_{RF}}\right)^m \frac{W_{RF} \cdot e^{\frac{m}{\bar{\gamma}_{RF}}}}{\ln(2)} \sum_{p=0}^{m-1} \left[\frac{(-1)^{m-p-1}}{\Gamma(p+1)\Gamma(m-p)}\right.$$
$$\left. \times M\left(p+1, \frac{\bar{\gamma}_{RF}}{m}, 1 + \gamma_{th}\right)\right], \tag{10.30}$$

where

$$M(\alpha, \theta, \beta) = \int_{\beta}^{\infty} y^{\alpha-1} \ln(y) e^{-\frac{y}{\theta}} dy$$

$$= \sum_{n=1}^{\alpha-1} \left[\frac{\Gamma(\alpha)}{\Gamma(n+1)} \left[\theta^{\alpha-n} e^{-\frac{\beta}{\theta}} \beta^n \ln(\beta) + \theta^{\alpha} \Gamma(n, \frac{\beta}{\theta}) \right] \right]$$

$$+ \theta^{\alpha} \Gamma(\alpha) \left[e^{-\frac{\beta}{\theta}} \ln(\beta) - \text{Ei}(-\frac{\beta}{\theta}) \right], \quad (10.31)$$

Ei(.) is the exponential integral, and $\Gamma(.,.)$ is the upper incomplete gamma function.

Fig. 10.4 shows the variation of the ergodic capacity against the average SNR of the FSO link, for a fixed value of γ_{th}^{FSO} and γ_{th}^{RF}. It is evident that using the hybrid scheme improves the capacity of the overall system, particularly at lower SNRs. In fact, when

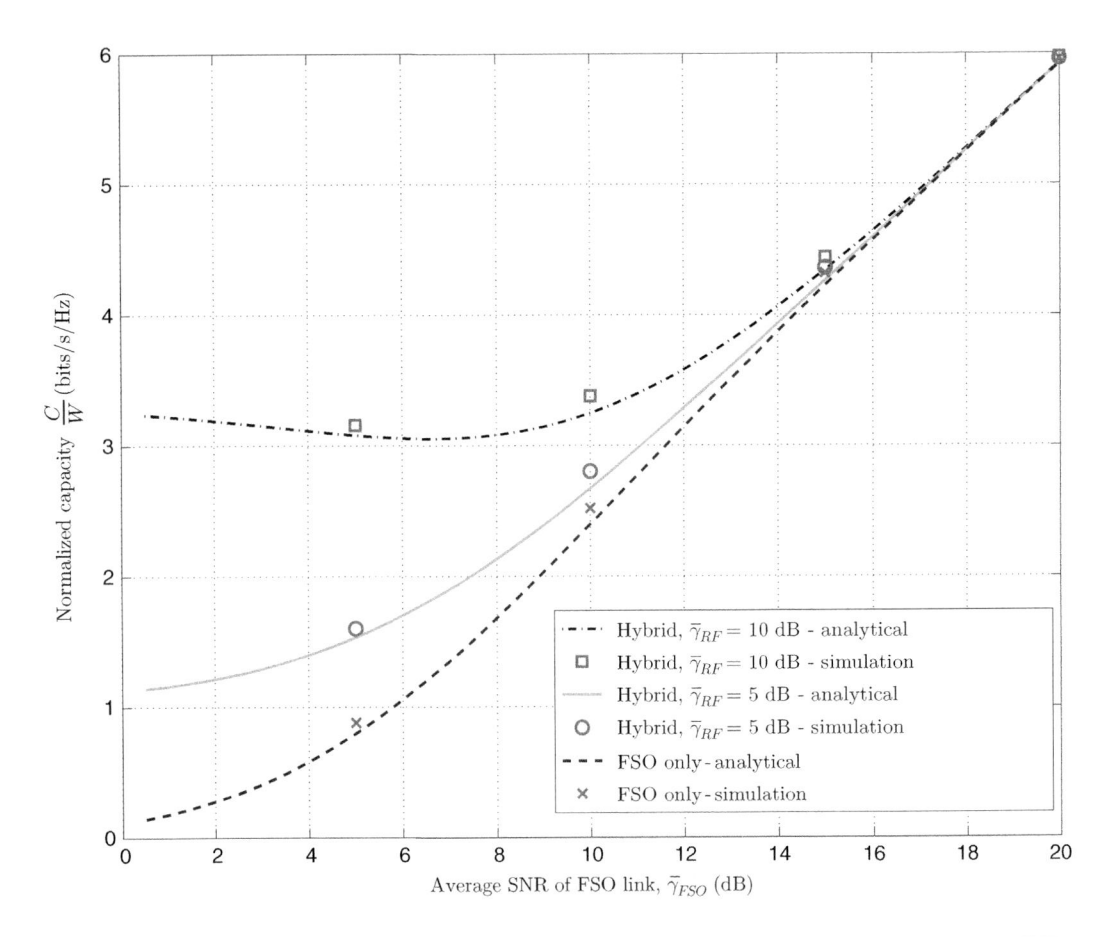

Figure 10.4 Ergodic capacity for the single FSO threshold case as a function of the average SNR of the FSO link: $\gamma_{th}^{FSO} = \gamma_{th}^{RF} = 5$ dB, $m = 5$, $\sigma_x = 0.25$. Reprint with permission from [9]. ©2014 IEEE.

the RF link quality is high, the capacity over the low $\bar{\gamma}_{FSO}$ range is slightly higher than the median $\bar{\gamma}_{FSO}$ range, as the system benefits from the frequent switching to the RF link over the low $\bar{\gamma}_{FSO}$ range. In order to show the variation of normalized capacity, we have assumed $W_{FSO} = W_{RF}$. Since the calculation of the capacity for the FSO link involved an approximation, the analytical results provide a lower bound on the capacity, as validated by simulation results. At high SNR values, the approximation becomes more accurate, as is evident from the graph.

10.1.3 Dual Switching Threshold Design

A practically useful variation on the scheme considered in the previous subsection is the use of two thresholds for the FSO link to reduce frequent on–off transitions of the FSO link. This is especially important for extending the lifetime of the FSO communication link. In this scenario, the FSO link continues operation until the SNR of the FSO link falls below $\gamma_{th,l}^{FSO}$, in which case it will turn off. The FSO link will only turn back on when the SNR reaches above $\gamma_{th,u}^{FSO}$, where $\gamma_{th,u}^{FSO} > \gamma_{th,l}^{FSO}$. As in the single threshold scenario, the RF link will be used for data transmission only if its SNR is above a threshold γ_{th}^{RF} and the FSO link is off. If both links are off, then an outage will be declared. A sample illustration of the above mode of operation is depicted in Fig. 10.5. Note that when the

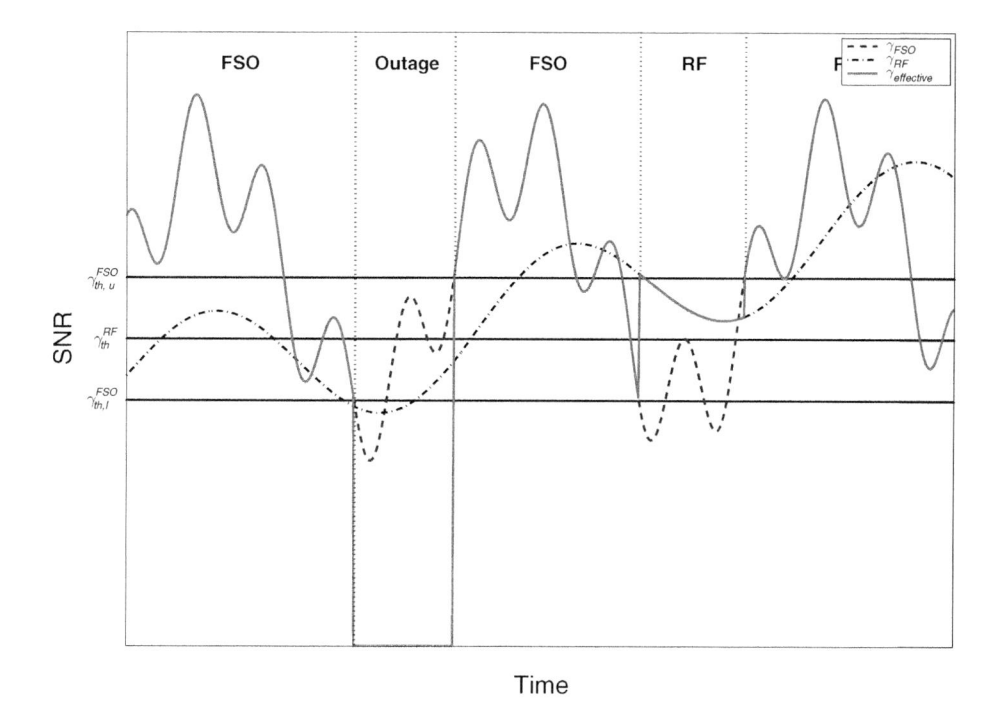

Figure 10.5 Operation with a dual FSO threshold. Reprint with permission from [9]. ©2014 IEEE.

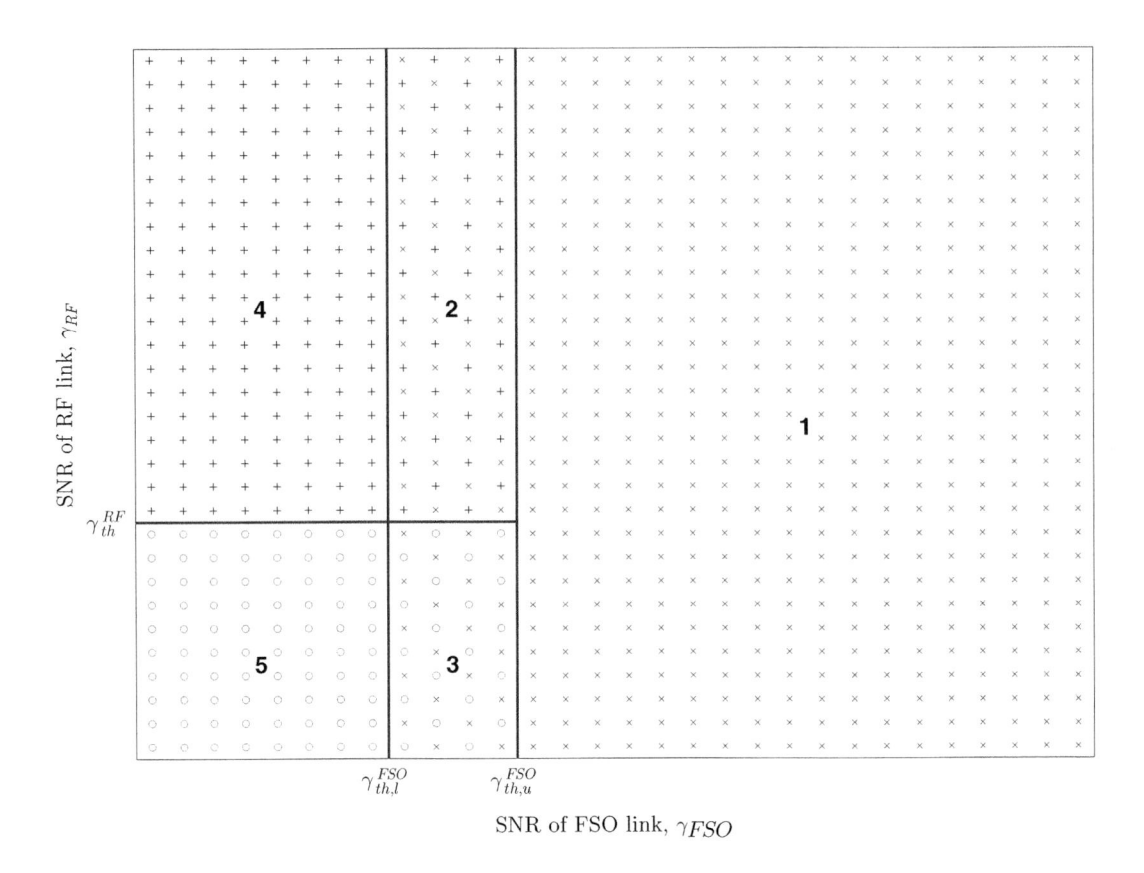

Figure 10.6 Operation regions of the dual FSO threshold case. Reprint with permission from [9]. ©2014 IEEE.

SNR of the FSO link is between $\gamma_{th,l}^{FSO}$ and $\gamma_{th,u}^{FSO}$, either the RF link or the FSO link may be used, depending on the previous status of the FSO link. The gap between the lower and upper thresholds will affect the frequency of on–off transitions. The larger the gap, the lower the transition frequency.

To facilitate the understanding of the following analysis, we further elucidate the operation modes of the hybrid FSO/RF system with the dual FSO threshold using Fig. 10.6. The SNR thresholds $\gamma_{th,u}^{FSO}$, $\gamma_{th,l}^{FSO}$, and γ_{th}^{RF} divide the possible SNR values of the FSO and RF links into five regions. In region 1, since the FSO link SNR is above the upper threshold, FSO will be used for transmission. In region 2, the on–off status of the FSO link depends on the last state of the FSO SNR, i.e., whether it was above $\gamma_{th,u}^{FSO}$ or below $\gamma_{th,l}^{FSO}$. Both the FSO link and the RF link may be used for transmission. In region 3, the chances of the FSO link being on or off are the same as those in region 2, but the RF link is off in this region. Thus, either the FSO link is being used for transmission or an outage is declared. In region 4, the FSO link is certainly off, while the RF link is on, and hence the RF link will be used for transmission. In region 5, both the links are off and an outage is declared.

Outage Probability

Based on the above mode of operation, outage occurs when both links are off. The FSO link is off if $\gamma_{FSO} < \gamma_{th,l}^{FSO}$ or $\gamma_{th,l}^{FSO} < \gamma_{FSO} < \gamma_{th,u}^{FSO}$ but γ_{FSO} was previously less than $\gamma_{th,l}^{FSO}$. The probability that the FSO link is off, denoted by $P_{out}^{FSO}(\gamma_{th,l}^{FSO}, \gamma_{th,u}^{FSO})$, can be calculated as

$$P_{out}^{FSO}(\gamma_{th,l}^{FSO}, \gamma_{th,u}^{FSO}) = P_{low}^{FSO} + P_{med}^{FSO} \frac{P_{low}^{FSO}}{P_{hi}^{FSO} + P_{low}^{FSO}}, \quad (10.32)$$

where P_{low}^{FSO} is the probability that the FSO link SNR is below $\gamma_{th,l}^{FSO}$, given by

$$P_{low}^{FSO} = F_{\gamma_{FSO}}(\gamma_{th,l}^{FSO}), \quad (10.33)$$

P_{med}^{FSO} is the probability that the FSO link SNR is between $\gamma_{th,l}^{FSO}$ and $\gamma_{th,u}^{FSO}$, given by

$$P_{med}^{FSO} = F_{\gamma_{FSO}}(\gamma_{th,u}^{FSO}) - F_{\gamma_{FSO}}(\gamma_{th,l}^{FSO}), \quad (10.34)$$

P_{hi}^{FSO} is the probability that the FSO link SNR is above $\gamma_{th,u}^{FSO}$, given by

$$P_{hi}^{FSO} = 1 - F_{\gamma_{FSO}}(\gamma_{th,u}^{FSO}), \quad (10.35)$$

and $F_{\gamma_{FSO}}(.)$ was given in (10.12). Specifically, $\frac{P_{low}^{FSO}}{P_{hi}^{FSO} + P_{low}^{FSO}}$ represents the probability that the FSO link SNR was previously below $\gamma_{th,l}^{FSO}$, given the FSO link SNR was not between $\gamma_{th,l}^{FSO}$ and $\gamma_{th,u}^{FSO}$.

The outage probability of the overall hybrid system is hence given by

$$P_{out} = P_{out}^{FSO}(\gamma_{th,l}^{FSO}, \gamma_{th,u}^{FSO}) \times F_{\gamma_{RF}}(\gamma_{th}^{RF}), \quad (10.36)$$

where $F_{\gamma_{RF}}(.)$ is given by (10.13).

Fig. 10.7 shows the variation of the outage probability with the average SNR of the FSO link, for fixed values of γ_{th}^{RF} and $\bar{\gamma}_{RF}$. The three curves correspond to different gaps between the upper and lower FSO thresholds, i.e., $\gamma_{th,u}^{FSO} = \gamma_{th,l}^{FSO} = 5$ dB; $\gamma_{th,u}^{FSO} = 5.5$ dB and $\gamma_{th,l}^{FSO} = 4.5$ dB; and $\gamma_{th,u}^{FSO} = 6$ dB and $\gamma_{th,l}^{FSO} = 4$ dB cases, respectively. The first curve essentially corresponds to the single threshold case. We see that the performance difference between the single FSO threshold case and the dual FSO threshold case is minimal. In fact, at high values of average FSO link SNR, $\bar{\gamma}_{FSO}$, minor improvement is seen for the dual FSO threshold case, which may be attributable to the fact that with high $\bar{\gamma}_{FSO}$, most of the operation in region 2 of Fig. 10.6 is preceded by operation in region 1, which means that the FSO link will be transmitting for a greater amount of time.

Average Bit Error Rate

Based on the mode of operation of the hybrid system with the dual FSO threshold, the average BER of the system can be calculated as

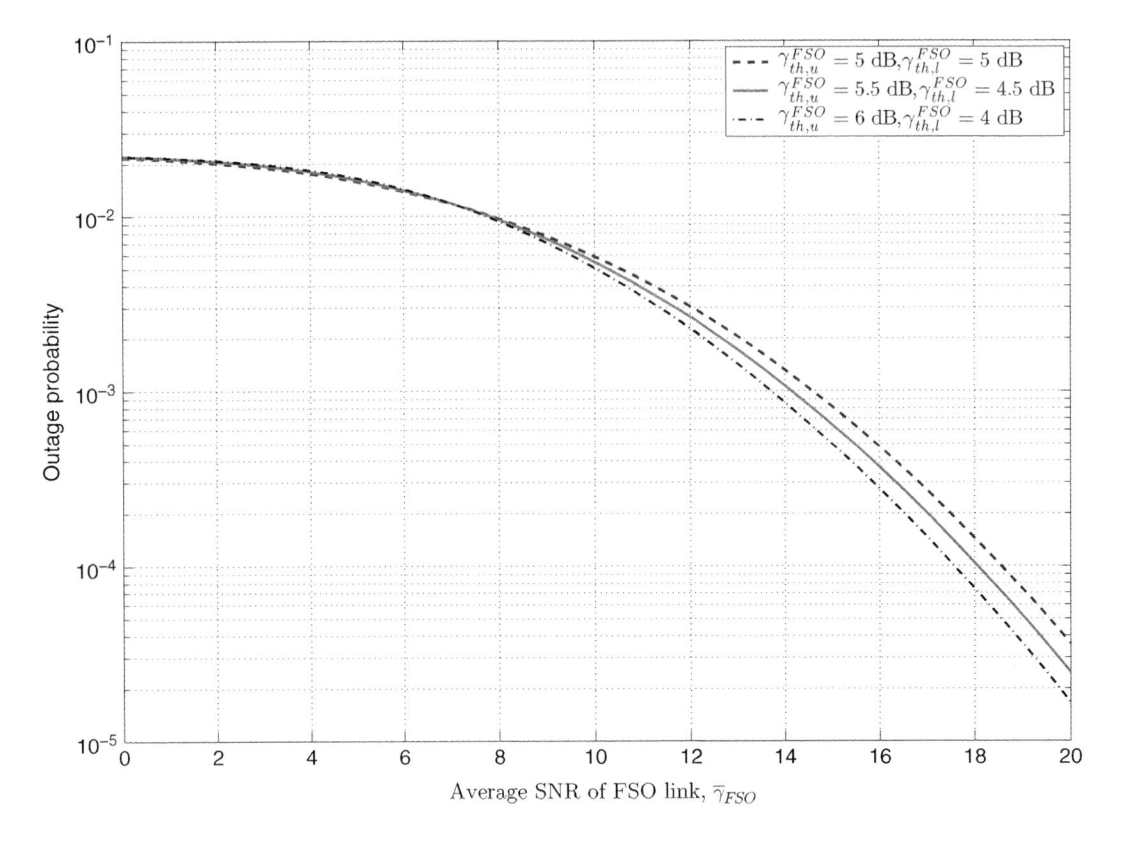

Figure 10.7 Outage probability for the dual FSO threshold case as a function of the average SNR of the FSO link: $\bar{\gamma}_{RF} = 8$ dB, $\gamma_{th}^{RF} = 5$ dB, $m = 5$, $\sigma_x = 0.25$. Reprint with permission from [9]. ©2014 IEEE.

$$\overline{BER} = \frac{1}{1-P_{out}} \times \left[BER_{FSO}(\gamma_{th,u}^{FSO}) + P_{out}^{FSO}(\gamma_{th,l}^{FSO}, \gamma_{th,u}^{FSO}) \cdot BER_{RF}(\gamma_{th}^{RF}) \right.$$

$$\left. + \left[BER_{FSO}(\gamma_{th,l}^{FSO}) - BER_{FSO}(\gamma_{th,u}^{FSO}) \right] \cdot \frac{P_{hi}^{FSO}}{P_{hi}^{FSO} + P_{low}^{FSO}} \right], \quad (10.37)$$

where $B_{FSO}(.)$ and $B_{RF}(.)$ are defined in (10.16) and (10.17) respectively. The first addition term in the bracket corresponds to the operation in region 1 of Fig. 10.6, the second term in regions 2 and 4 when the FSO link is off, and the third term in regions 2 and 3 if the FSO link is on.

Fig. 10.8 shows the variation of the average BER against the average SNR of the FSO link for fixed values of γ_{th}^{RF} and $\bar{\gamma}_{RF}$. Again, there is very little difference between the single FSO threshold and the dual FSO threshold cases in terms of BER performance. Some minor increase in the average BER is seen at high values of $\bar{\gamma}_{FSO}$, which may be understood with a similar argument as follows. Specifically, with the dual FSO threshold and high $\bar{\gamma}_{FSO}$, the FSO link is on for some extra time when the SNR is between the

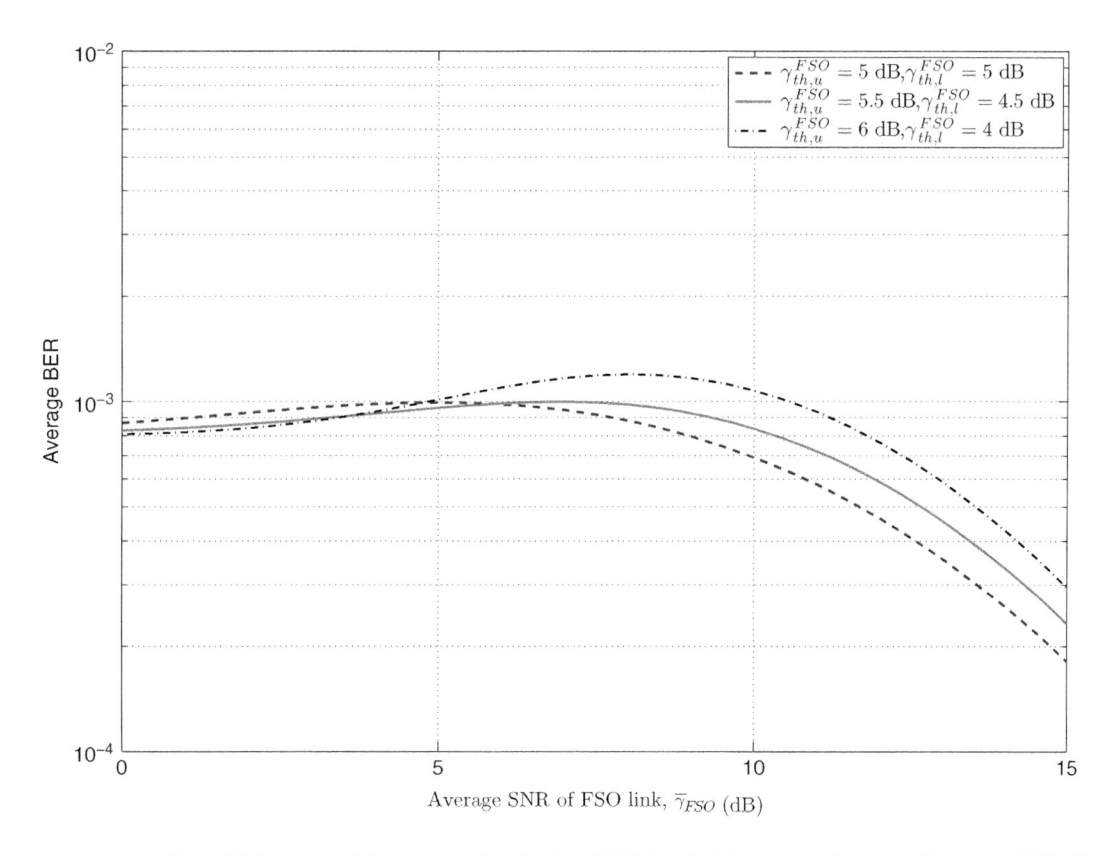

Figure 10.8 Average bit error rate for the dual FSO threshold case as a function of average SNR of the FSO link - $\bar{\gamma}_{RF} = 8$ dB, $\gamma_{th}^{RF} = 5$ dB, $m = 5$, $\sigma_x = 0.25$. Reprint with permission from [9]. ©2014 IEEE.

lower and upper FSO thresholds, instead of it using a better-quality RF link. This results in some increase in the average BER. Slightly improved performance is also seen at very low values of $\bar{\gamma}_{FSO}$, which can be explained with the following argument. Now most of the operations in regions 2 and 3 of Fig. 10.6 are preceded by operations in regions 4 and 5, respectively. As such, the FSO link is mostly off in these regions, resulting in either shifting to a better-quality RF link (region 2), or an outage being declared (region 3), both of which lead to an improved average BER.

Ergodic Capacity

Based on similar reasoning for BER analysis, the capacity of the hybrid system for the dual FSO threshold case is given by

$$C = C_{FSO}(\gamma_{th,u}^{FSO}) + P_{out}^{FSO}(\gamma_{th,l}^{FSO}, \gamma_{th,u}^{FSO}) \cdot C_{RF}(\gamma_{th}^{RF}) \\ + \left[C_{FSO}(\gamma_{th,l}^{FSO}) - C_{FSO}(\gamma_{th,u}^{FSO}) \right] \cdot \frac{P_{hi}^{FSO}}{P_{hi}^{FSO} + P_{low}^{FSO}}, \qquad (10.38)$$

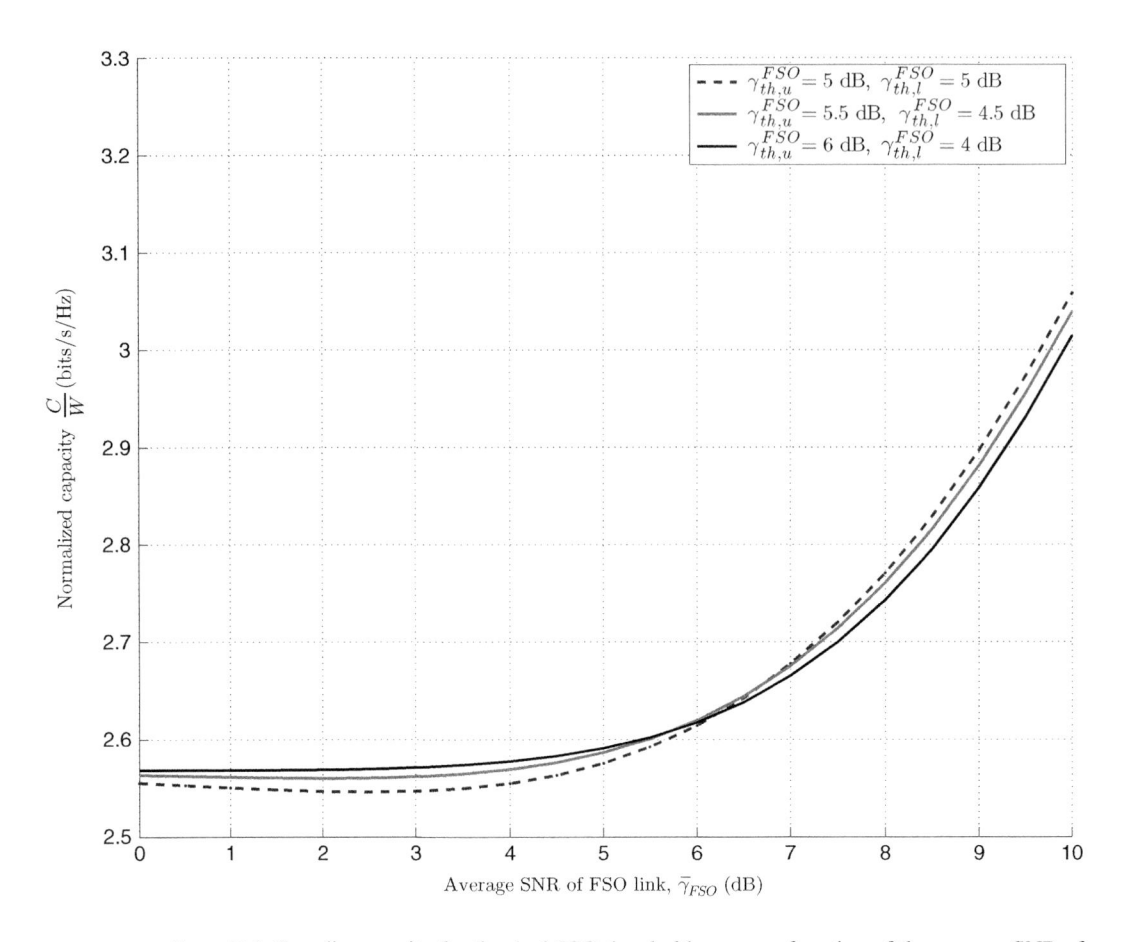

Figure 10.9 Ergodic capacity for the dual FSO threshold case as a function of the average SNR of the FSO link: $\bar{\gamma}_{RF} = 8$ dB, $\gamma_{th}^{RF} = 5$ dB, $m = 5$, $\sigma_x = 0.25$. Reprint with permission from [9]. ©2014 IEEE.

where $C_{FSO}(.)$ and $C_{RF}(.)$ are defined in (10.24) and (10.25) respectively. Specifically, the first addition term in the above expression corresponds to the operation in region 1 of Fig. 10.6, the second term in regions 2 and 4 when the FSO link is off, and the third term in regions 2 and 3 if the FSO link is on.

Fig. 10.9 shows the variation of the ergodic capacity against the average SNR of the FSO link for fixed values of γ_{th}^{RF} and $\bar{\gamma}_{RF}$. The capacity performance of the dual FSO threshold case is essentially the same as that of the single FSO threshold case. Minor deterioration at high values of $\bar{\gamma}_{FSO}$ and minor improvement at low values of $\bar{\gamma}_{FSO}$ are observed, similar to the average BER performance. Specifically, with high $\bar{\gamma}_{FSO}$, the FSO link is on more often when the system operates in region 2 of Fig. 10.6 instead of shifting to a better-quality RF link, which results in certain capacity reduction. Similarly, with low $\bar{\gamma}_{FSO}$, the FSO link is mostly off, and the system switches to a better-quality RF link when operating in region 2, which increases the capacity.

10.2 Hybrid FSO/RF Transmission with Adaptive Combining

In this section, we present a scheme for hybrid FSO/RF transmission systems, named hybrid FSO/RF transmission system with adaptive combining [18]. In this scheme, the FSO link is used alone as long as its quality is acceptable. When the FSO link's quality becomes unacceptable, the system activates the RF link and applies the MRC (maximal ratio combining) scheme on signals received from both FSO and RF links. When the quality of the FSO link alone becomes acceptable again, the RF link is deactivated to save power and spectrum utilization. We drive the CDF of the receiver SNR, which is then used to study the outage performance of the resulting transmission system.

The operation of the hybrid FSO/RF system is summarized as follows. Only the FSO link is active as long as its instantaneous SNR at the optical receiver, denoted by γ_{FSO}, is above a certain threshold, defined by γ_T. When γ_{FSO} falls below the predetermined threshold γ_T, the receiver sends a 1-bit feedback signal to activate the RF link along with the FSO link for simultaneous transmission of the same data. In this case, the data received from both links are combined using an MRC scheme. The receiver SNR, denoted by γ_c, will equal γ_{FSO} when $\gamma_{FSO} \geq \gamma_T$. On the other hand, when $\gamma_{FSO} < \gamma_T$, γ_c will equal the sum of γ_{FSO} and γ_{RF}, where γ_{RF} is the receiver instantaneous SNR of the RF link.

10.2.1 Channel Model

We assume the same IM/DD transmission as previous subsection for FSO link.

Assuming perfect alignment between the FSO transmitter and receiver apertures,[2] and considering gamma–gamma turbulence-induced fading, the fading channel gain for FSO link h_{FSO} will have the following PDF [21]:

$$f_{h_{FSO}}(h_{FSO}) = \frac{2(\alpha\beta)^{\frac{\alpha+\beta}{2}}}{\Gamma(\alpha)\Gamma(\beta)} h_{FSO}^{\frac{\alpha+\beta}{2}-1} K_{\alpha-\beta}(2\sqrt{\alpha\beta h_{FSO}}), \; h_{FSO} \geq 0, \quad (10.39)$$

where α and β are parameters related to the atmospheric turbulences and $K_\nu(.)$ is the νth-order modified Bessel function of the second kind defined as [22, Eq. (8.407)]. Typically, α and β are the effective number of small-scale and large-scale eddies of the turbulent environment, respectively. According to the values of α and β, the atmospheric turbulence can be modeled from weak to strong turbulence regimes because these parameters are directly related to the atmospheric turbulence conditions. Expressions for calculating the parameters α and β for different propagation conditions can be found in [23]. Assuming spherical wave propagation, expressions for calculating α and β in (10.39) are given by [23]

$$\alpha = \left[\exp\left(\frac{0.49\chi^2}{(1+0.18d^2+0.56\chi^{\frac{12}{5}})^{\frac{7}{6}}}\right) - 1\right]^{-1}, \quad (10.40)$$

---
[2] Perfect alignment between the FSO transmitter and receiver apertures can be achieved by using pointing, acquisition, and tracking (PAT) systems. However, these PAT systems add extra hardware complexity to FSO systems.

$$\beta = \left[\exp\left(\frac{0.51\chi^2(1+0.69\chi^{\frac{12}{5}})^{-\frac{5}{6}}}{(1+0.9d^2+0.62d^2\chi^{\frac{12}{5}})^{\frac{5}{6}}}\right)-1\right]^{-1}, \quad (10.41)$$

where $\chi^2 = 0.5 C_n^2 k^{7/6} L^{11/6}$ is the Rytov variance and $d = (kD^2/4L)^{1/2}$ with $k = 2\pi/\lambda_{FSO}$ is the optical wave number. Here, C_n^2, D, and λ_{FSO} are respectively the refractive index structure parameter, the diameter of the optical receiver aperture, and the optical wavelength.

The instantaneous received electrical SNR of the FSO link is related to h_{FSO} as $\gamma_{FSO} = \bar{\gamma}_{FSO} h_{FSO}^2$, where $\bar{\gamma}_{FSO}$ is the average electrical SNR which is defined as $\bar{\gamma}_{FSO} = E_s \mu^2 \eta^2 P_{FSO}^2 G_{FSO}^2 / \sigma_{FSO}^2$. Using power transformation of random variables, it is easy to show that the PDF of γ_{FSO} is given by:

$$f_{\gamma_{FSO}}(\gamma_{FSO}) = \frac{(\alpha\beta/\sqrt{\bar{\gamma}_{FSO}})^{\frac{\alpha+\beta}{2}}}{\Gamma(\alpha)\Gamma(\beta)} \gamma_{FSO}^{\frac{\alpha+\beta}{4}-1} K_{\alpha-\beta}\left(2\sqrt{\frac{\alpha\beta}{\sqrt{\bar{\gamma}_{FSO}}}\gamma_{FSO}^{\frac{1}{2}}}\right), \gamma_{FSO} \geq 0. \quad (10.42)$$

By using [24, Eq. (14)] to express $K_{\alpha-\beta}(.)$ in terms of the Meijer G-function $G_{p,q}^{m,n}\left(z \Big| {a_1,...,a_p \atop b_1,...,b_q}\right)$ defined as [22, Eq. (9.301)] and [22, Eq. (9.31.5)], (10.42) can be expressed as:

$$f_{\gamma_{FSO}}(\gamma_{FSO}) = \frac{\gamma_{FSO}^{-1}}{2\Gamma(\alpha)\Gamma(\beta)} G_{0,2}^{2,0}\left(\frac{\alpha\beta}{\sqrt{\bar{\gamma}_{FSO}}} \gamma_{FSO}^{\frac{1}{2}} \Big| {- \atop \alpha, \beta}\right). \quad (10.43)$$

By using [25, Eq. (07.34.21.0084.01)], and some simple algebraic manipulations, the CDF of γ_{FSO} can be expressed as:

$$F_{\gamma_{FSO}}(\gamma_{FSO}) = \frac{2^{\alpha+\beta-2}}{\pi\Gamma(\alpha)\Gamma(\beta)} G_{1,5}^{4,1}\left(\frac{(\alpha\beta)^2}{16\bar{\gamma}_{FSO}}\gamma_{FSO} \Big| {1 \atop \frac{\alpha}{2},\frac{\alpha+1}{2},\frac{\beta}{2},\frac{\beta+1}{2},0}\right). \quad (10.44)$$

10.2.2 Outage Analysis

When the instantaneous output SNR γ_c falls below a given threshold γ_{out}, the communication system enters a state called outage, in which the received SNR cannot support the target BER of the system. The probability that the SNR γ_c falls below the outage threshold γ_{out} can be simply calculated by evaluating the CDF of γ_c at γ_{out} as $P_{out} = F_{\gamma_c}(\gamma_{out})$.

Based on the modes of operation of the proposed hybrid FSO/RF system, with adaptive combining the CDF of γ_c, is given by

$$F_{\gamma_c}(x) = \Pr[\gamma_{FSO} \geq \gamma_T, \gamma_{FSO} < x] + \Pr[\gamma_{FSO} < \gamma_T, \gamma_{FSO} + \gamma_{RF} < x]$$
$$= \begin{cases} F_1(x) & \text{if } x \leq \gamma_T \\ F_{\gamma_{FSO}}(x) - F_{\gamma_{FSO}}(\gamma_T) + F_2(x) & \text{if } x > \gamma_T, \end{cases} \quad (10.45)$$

where $F_1(x)$ is defined as

$$F_1(x) = \int_0^x f_{\gamma_{FSO}+\gamma_{RF}}(y) dy \qquad (10.46)$$

and $F_2(x)$ is defined as:

$$F_2(x) = \int_0^{\gamma_T} f_{\gamma_{FSO}}(\gamma_{FSO}) F_{\gamma_{RF}}(x - \gamma_{FSO}) d\gamma_{FSO}, \qquad (10.47)$$

with $F_{\gamma_{FSO}}(.)$, and $F_{\gamma_{RF}}(.)$ given by (10.44) and (10.10), respectively.

Noting that the FSO and RF links are statistically independent, $f_{\gamma_{FSO}+\gamma_{RF}}(y)$ in (10.46) can be evaluated as:

$$f_{\gamma_{FSO}+\gamma_{RF}}(y) = \int_0^y f_{\gamma_{FSO}}(\gamma_{FSO}) f_{\gamma_{RF}}(y - \gamma_{FSO}) d\gamma_{FSO}. \qquad (10.48)$$

After substituting (10.43) and (10.9) into (10.48) and applying the binomial expansion defined in [22, Eq. (1.111)], and the series expansion of the exponential defined in [22, Eq. (1.211.1)], along with [25, Eq. (07.34.21.0084.01)], $f_{\gamma_{FSO}+\gamma_{RF}}(y)$ can be evaluated as:

$$f_{\gamma_{FSO}+\gamma_{RF}}(y) = \frac{2^{\alpha+\beta-2} e^{\frac{-my}{\bar{\gamma}_{RF}}} (\frac{m}{\bar{\gamma}_{RF}})^m y^{m-1}}{\pi \Gamma(\alpha)\Gamma(\beta)\Gamma(m)} \left\{ \sum_{n=0}^{\infty} \frac{(\frac{my}{\bar{\gamma}_{RF}})^n}{n!} \sum_{i=0}^{m-1} \binom{m-1}{i} \right.$$

$$\left. \times (-1)^i G_{1,5}^{4,1}\left(\frac{(\alpha\beta)^2 y}{16\bar{\gamma}_{FSO}} \mid \begin{matrix} K_1 \\ K_2 \end{matrix} \right) \right\}, \qquad (10.49)$$

where $K_1 = 1 - n - i$ and $K_2 = \frac{\alpha}{2}, \frac{\alpha+1}{2}, \frac{\beta}{2}, \frac{\beta+1}{2}, -n - i$.

Substituting (10.49) in (10.46), using the series expansion of the exponential, and applying [24, Eq. (26)], $F_1(x)$ can be expressed as

$$F_1(x) = \frac{2^{\alpha+\beta-2}(\frac{mx}{\bar{\gamma}_{RF}})^m}{\pi \Gamma(\alpha)\Gamma(\beta)\Gamma(m)} \left\{ \sum_{n=0}^{\infty} \frac{(\frac{mx}{\bar{\gamma}_{RF}})^n}{n!} \sum_{i=0}^{m-1} \binom{m-1}{i} (-1)^i \sum_{k=0}^{\infty} \frac{(\frac{-mx}{\bar{\gamma}_{RF}})^k}{k!} \right.$$

$$\left. \times G_{2,6}^{4,2}\left(\frac{(\alpha\beta)^2 x}{16\bar{\gamma}_{FSO}} \mid \begin{matrix} K_3 \\ K_4 \end{matrix} \right) \right\}, \qquad (10.50)$$

where $K_3 = 1 - n - i, 1 - k - n - m$, and $K_4 = \frac{\alpha}{2}, \frac{\alpha+1}{2}, \frac{\beta}{2}, \frac{\beta+1}{2}, -k - n - m, -n - i$.

Substituting (10.43) and (10.10) in (10.47), $F_2(x)$ can be expressed in the integral form as

$$F_2(x) = \int_0^{\gamma_T} \frac{\gamma(m, \frac{m(x-\gamma_{FSO})}{\bar{\gamma}_{RF}}) \gamma_{FSO}^{-1}}{2\Gamma(\alpha)\Gamma(\beta)\Gamma(m)} G_{0,2}^{2,0}\left(\frac{\alpha\beta \gamma_{FSO}^{1/2}}{\sqrt{\bar{\gamma}_{FSO}}} \mid \begin{matrix} - \\ \alpha, \beta \end{matrix} \right) d\gamma_{FSO}. \qquad (10.51)$$

By using the series representation of $\gamma(\cdot,\cdot)$, defined in [22, Eq. (8.352.1)], and then applying the binomial expansion rule, along with the series expansion of the exponential, the term $\gamma\left(m, \frac{m(x-\gamma_{FSO})}{\bar{\gamma}_{RF}}\right)$ in (10.51) can be represented by:

$$\gamma\left(m, \frac{m(x-\gamma_{FSO})}{\bar{\gamma}_{RF}}\right) = (m-1)!\left[1 - e^{\frac{-mx}{\bar{\gamma}_{RF}}} \sum_{n=0}^{\infty} \frac{(\frac{m\gamma_{FSO}}{\bar{\gamma}_{RF}})^n}{n!} \sum_{k=0}^{m-1} \frac{(\frac{m}{\bar{\gamma}_{RF}})^k}{k!} \right.$$
$$\left. \times \sum_{j=0}^{k} \binom{k}{j} x^{k-j}(-\gamma_{FSO})^j \right]. \quad (10.52)$$

By plugging (10.52) into (10.51) and applying [25, Eq. (07.34.21.0084.01)], $F_2(x)$ can be evaluated as:

$$F_2(x) = \frac{(m-1)!}{\Gamma(m)} F_{\gamma_{FSO}}(\gamma_T) - \left\{ \frac{2^{\alpha+\beta-2} e^{\frac{-mx}{\bar{\gamma}_{RF}}} (m-1)!}{\pi \Gamma(\alpha)\Gamma(\beta)\Gamma(m)} \sum_{n=0}^{\infty} \frac{(m\gamma_T/\bar{\gamma}_{RF})^n}{n!} \right.$$
$$\left. \times \sum_{k=0}^{m-1} \frac{(mx/\bar{\gamma}_{RF})^k}{k!} \sum_{j=0}^{k} \binom{k}{j} \left(\frac{-\gamma_T}{x}\right)^j G_{1,5}^{4,1}\left(\frac{(\alpha\beta)^2 \gamma_T}{16\bar{\gamma}_{FSO}} \bigg| \begin{array}{c} K_5 \\ K_6 \end{array} \right) \right\},$$
$$(10.53)$$

where $K_5 = 1 - n - j$ and $K_6 = \frac{\alpha}{2}, \frac{\alpha+1}{2}, \frac{\beta}{2}, \frac{\beta+1}{2}, -n-j$.

Finally, the CDF of γ_c is obtained after substituting (10.50) and (10.53) into (10.45).

10.2.3 Numerical Results

In Figs. 10.10 and 10.11, we consider clear weather conditions, which is the hybrid FSO/RF system's operational condition most of the time. We use typical values of α and β for strong atmospheric turbulence ($\alpha = 2.064$, and $\beta = 1.342$ [29]), which has the dominant effect on a hybrid system's performance in this case. The fading severity over the RF link m is set to 5.

It can be seen from Fig. 10.10 that the hybrid FSO/RF system gives much better outage performance than using FSO-only or RF-only systems in clear weather conditions. Also, as expected, it can be observed that the performance of the hybrid system is improved with the increase of $\bar{\gamma}_{RF}$. The numerical results shown in Fig. 10.10 are obtained using $n = 30$ and $k = 30$ in (10.50) and $n = 30$ in (10.53). As can be observed from Fig. 10.10, evaluating the outage probability using the truncated values of (10.50) and (10.53) gives accurate results that coincide with the values of the outage probability obtained by evaluating the integrals in (10.46) and (10.47) using numerical methods.

It can be seen from Fig. 10.11 that, when γ_{out} is less than γ_T, the outage performance of the hybrid FSO/RF system is improved, because the system activates the RF link before the FSO link goes into outage. In this case, and as expected, outage probability decreases as $\bar{\gamma}_{RF}$ increases. On the other hand, when γ_{out} is greater than γ_T, the outage performance of the hybrid system does not decrease, as the RF link quality improves because the system goes into outage before it activates the RF link.

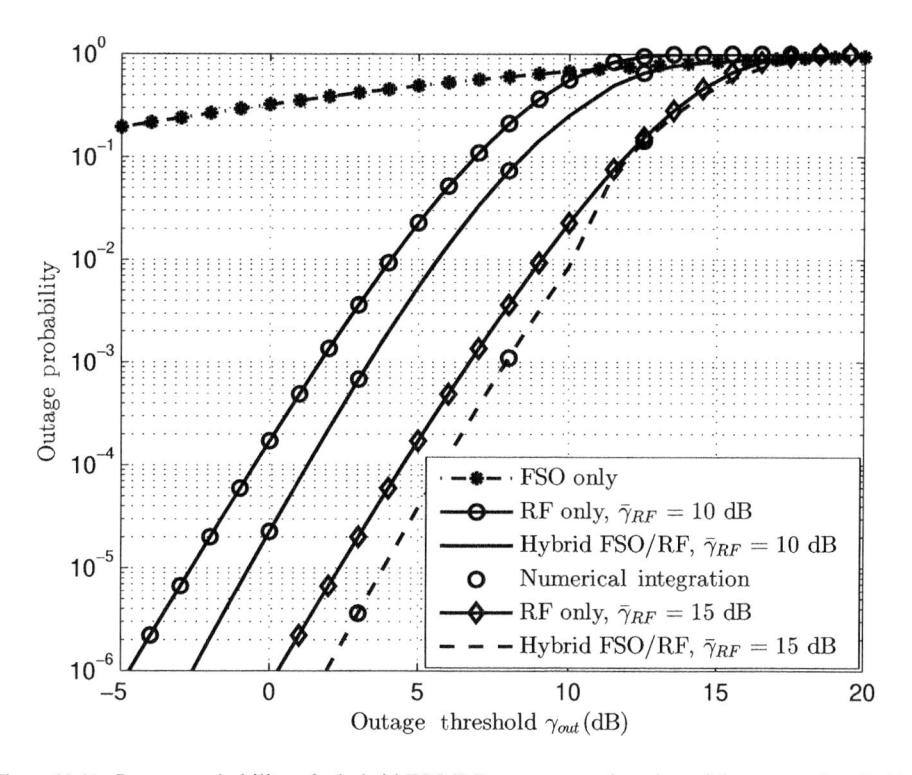

Figure 10.10 Outage probability of a hybrid FSO/RF system as a function of the outage threshold with $\bar{\gamma}_{FSO} = 10$ dB.

It can be seen from Fig. 10.12 that the FSO link's quality degrades due to weak atmospheric turbulence in rain conditions. As observed, considering the "five nines" reliability criterion, which implies an outage performance of 10^{-6}, there is an improvement by using the hybrid system of about 4 dB in transmit power over the FSO-only system.

10.3 Joint Adaptive Modulation and Combining for Hybrid FSO/RF Transmission

In this section, we present and analyze a transmission scheme for a hybrid FSO/RF communication system based on joint adaptive modulation and adaptive combining [30]. Specifically, the data rate on the FSO link is adjusted in a discrete manner according to the FSO link's instantaneous received SNR. If the FSO link's quality is too poor to maintain the target BER, the system activates the RF link along with the FSO link. When the RF link is activated, simultaneous transmission of the same modulated data takes place on both links, where the received signals from both links are combined using the MRC scheme. In this case, the data rate of the system is adjusted according to the instantaneous combined SNRs. We study the spectral efficiency and outage performance of the resulting hybrid FSO/RF system.

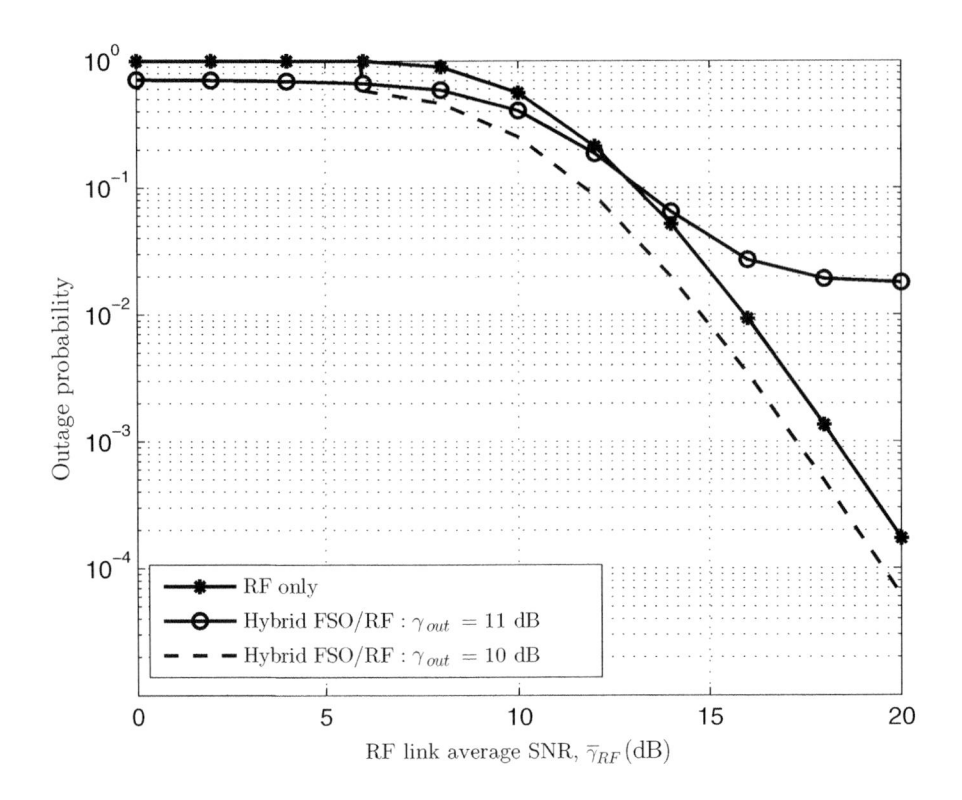

Figure 10.11 Outage probability of a hybrid FSO/RF system as a function of the average SNR of the RF link with $\bar{\gamma}_{FSO} = 10$ dB compared to the RF-only system. Reprint with permission from [18]. ©2015 IEEE.

10.3.1 System and Channel Model

We consider a hybrid FSO/RF system, which is composed of coherent/heterodyne FSO and RF communication subsystems. A coded digital baseband signal, created by the signal source, is converted to an analog electrical signal through an electrical modulator, which can adaptively use one of N different M-square quadrature amplitude modulation (QAM) schemes. M-QAM is widely used in high-rate data transmissions over FSO links [31] and RF links [32] because of its high spectral efficiency and ease of the signal modulation/demodulation process. At the FSO transmitter, the QAM electrical signal is mixed with an optical carrier, produced by an optical frequency local oscillator (LO) to produce the optical signal.

At the FSO receiver, the received optical signal undergoes the heterodyne detection process. The instantaneous SNR per symbol of the FSO receiver is given by [34, 35]

$$\gamma_{FSO} = \bar{\gamma}_{FSO} h_{FSO}, \qquad (10.54)$$

where $\bar{\gamma}_{FSO}$ and, h_{FSO} are, respectively, the average SNR and the fading gain over the FSO link, with $E[h_{FSO}]$ normalized to unity, where $E[.]$ is the expectation operator. We assume using a phase-locked loop (PLL) to compensate for phase noise in

10.3 Joint Adaptive Modulation and Combining

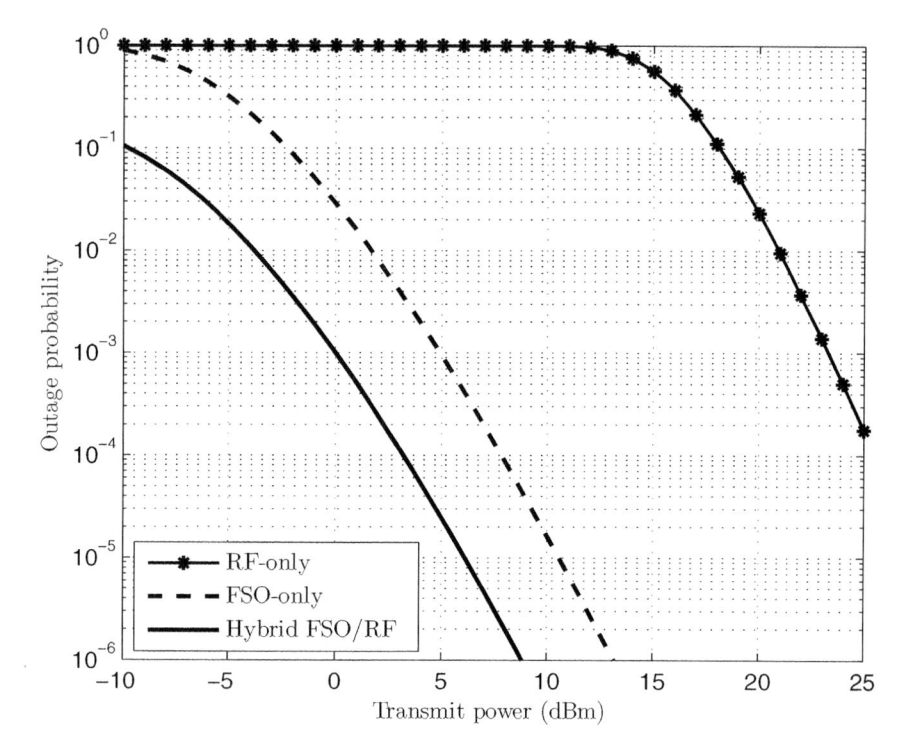

Figure 10.12 Outage probability of a hybrid FSO/RF system as a function of transmit power in moderate rain conditions, with $\gamma_{out} = 10$ dB, and link range $z = 4000$ m. Reprint with permission from [18]. ©2015 IEEE.

the received optical signal and using large enough LO power, such that thermal and background noises can be neglected. In this case, the average SNR $\bar{\gamma}_{FSO}$ is given by $\bar{\gamma}_{FSO} = 2E_{avg}\eta^2 P_{LO}P_{FSO}G_{FSO}/\sigma_{FSO}^2$, where E_{avg}, η, P_{LO}, P_{FSO}, G_{FSO}, and σ_{FSO}^2 are the average QAM symbol energy, photodetector responsivity, LO power, average transmitted optical power, optical power attenuation, and variance of shot noise, which is modeled as AWGN, respectively. The optical power attenuation G_{FSO} is given by the Beers–Lambert law as $G_{FSO} = \alpha_{FSO}z$ [7], with α_{FSO} being the weather-dependent attenuation coefficient (in dB/km) and z being the link range from the transmitter to the receiver. The attenuation G_{FSO} is considered as a fixed scaling factor, and no randomness exists in its behavior [20].

We assume the FSO link experience gamma-gamma fading and pointing error. The fading channel gain of FSO link is defined as $h_{FSO} = h_a h_p$, where h_a is gamma–gamma atmospheric turbulence-induced fading gain factor [23] and h_p is the Gaussian pointing errors-induced fading gain factor [20]. Following the same procedure used in [36], we can show that the PDF of γ_{FSO} is given by:

$$f_{\gamma_{FSO}}(\gamma_{FSO}) = \frac{\xi^2 \gamma_{FSO}^{-1}}{\Gamma(\alpha)\Gamma(\beta)} G_{1,3}^{3,0}\left[\frac{\xi^2 \alpha\beta\gamma_{FSO}}{(\xi^2+1)\bar{\gamma}_{FSO}} \Big|_{\xi^2,\,\alpha,\,\beta}^{\xi^2+1}\right], \quad (10.55)$$

where ξ is the ratio between the equivalent beam radius ω_{eq} and the pointing error (jitter) standard deviation σ_s given by $\xi = \omega_{eq}/2\sigma_s$. Here, $\omega_{eq}^2 = \omega_z^2 \sqrt{\pi} \mathrm{erf}(v)/2v \exp(-v^2)$, where erf(.) is the error function and ω_z is the beam radius calculated at distance z from the transmitter aperture and $v = \sqrt{\pi}D/2\sqrt{2}\omega_z$ where D is the photodetector diameter. ω_z is given by $\omega_z = \theta_0 z$, where θ_0 is the transmit divergence at $1/e^2$. $\Gamma(.)$ in (10.55) is the standard gamma function where α and β are the scintillation parameters that are related to the refractive index structure parameter C_n^2 [23]. $G[.]$ is the Meijer G-function as defined in [22, Eq. (9.301)]. By using [25, Eq. (07.34.21.0084.01)] and some simple algebraic manipulations, the CDF of γ_{FSO} can be expressed as:

$$F_{\gamma_{FSO}}(\gamma_{FSO}) = \frac{\xi^2}{\Gamma(\alpha)\Gamma(\beta)} G_{2,4}^{3,1}\left[\frac{\xi^2 \alpha\beta \gamma_{FSO}}{(\xi^2+1)\bar{\gamma}_{FSO}} \Big|_{\xi^2, \alpha, \beta, 0}^{1, \xi^2+1}\right]. \tag{10.56}$$

We adopt the same Nakagami-m fading model used in the previous section for the RF link. Specifically, the PDF and CDF of the instantaneous received SNR over the RF link, denoted by γ_{RF}, are respectively given by [28]

$$f_{\gamma_{RF}}(\gamma_{RF}) = \left(\frac{m}{\bar{\gamma}_{RF}}\right)^m \frac{\gamma_{RF}^{m-1}}{\Gamma(m)} \exp\left(\frac{-m\gamma_{RF}}{\bar{\gamma}_{RF}}\right) \tag{10.57}$$

and

$$F_{\gamma_{RF}}(\gamma_{RF}) = \frac{1}{\Gamma(m)}\gamma\left(m, \frac{m\gamma_{RF}}{\bar{\gamma}_{RF}}\right), \tag{10.58}$$

where $\gamma(\cdot,\cdot)$ is the lower incomplete gamma function defined in [22, Eq. (8.350.1)].

With QAM based adaptive modulation, a particular constellation size M is chosen to achieve the highest possible spectral efficiency, while maintaining the instantaneous BER below the target value of BER$_0$. Let $\gamma_{T_1}, \gamma_{T_2}, \ldots, \gamma_{T_N}$ be the N different thresholds corresponding to constellation sizes of $M = 4, 16, \ldots, 2^{2N}$, respectively such that $\gamma_{T_1} < \gamma_{T_2} < \ldots < \gamma_{T_N}$. Note that $\gamma_{T_{N+1}} = \infty$. To meet a BER requirement of BER$_0$, the thresholds are set using [33]

$$\gamma_{T_n} = (2^{2n} - 1)[-\frac{2}{3}\ln(5\,\mathrm{BER}_0)], \quad n \geq 1. \tag{10.59}$$

To achieve the maximum spectral efficiency, the FSO link uses the modulation scheme 2^{2N}-QAM as long as γ_{FSO} is greater than or equal to γ_{T_N}. If γ_{FSO} falls below γ_{T_N}, the receiver finds the largest γ_{T_n} that satisfies $\gamma_{FSO} \geq \gamma_{T_n}$. In this case, the receiver sends a feedback signal to the transmitter indicating the modulation scheme 2^{2n}-QAM should be used without activating the RF subsystem. If γ_{FSO} is less than γ_{T_1}, the receiver sends a feedback signal to activate the RF link for simultaneous transmission of the same data along with the FSO link. Therefore, the feedback required is $\lceil \log_2(N+1) \rceil$ bits for the first stage. At the receiver, the received signal along both links will be combined using an MRC combiner. In this case, the receiver SNR, denoted by γ_c, will be equal to the sum of γ_{FSO} and γ_{RF}. Note that γ_c is equal to γ_{FSO} as long as $\gamma_{FSO} \geq \gamma_{T_1}$. To this end, the receiver checks whether γ_c is greater than or equal to γ_{T_N}. If so, the receiver sends a feedback signal to the transmitter to selecting the modulation scheme 2^{2N}-QAM on both the FSO link and the RF link. If not, the receiver checks another threshold γ_{T_n} in descending order until one threshold satisfies $\gamma_c \geq \gamma_{T_n}$. In this case, the receiver sends a feedback signal to the transmitter to select the modulation scheme 2^{2n}-QAM. If the receiver thresholds-checking process reaches γ_{T_1} and $\gamma_c < \gamma_{T_1}$, the receiver sends a signal to suspend data transmission over both FSO and RF links.[3] In this second stage, if necessary, the feedback load is again $\lceil \log_2(N+1) \rceil$ bits. The average feedback of the proposed adaptive hybrid system is $\lceil \log_2(N+1) \rceil \left(1 + \Pr[\gamma_{FSO} < \gamma_{T_1}]\right)$ bits.

10.3.2 Performance Analysis

To study the performance of this hybrid transmission scheme, we need to deduce the CDF of γ_c. Based on the modes of operation of this joint adaptive scheme, the CDF of γ_c is given by

$$F_{\gamma_c}(x) = \Pr[\gamma_{FSO} \geq \gamma_{T_1}, \gamma_{FSO} < x] + \Pr[\gamma_{FSO} < \gamma_{T_1}, \gamma_{FSO} + \gamma_{RF} < x]$$
$$= \begin{cases} F_1(x), & \text{if } x \leq \gamma_{T_1} \\ F_{\gamma_{FSO}}(x) - F_{\gamma_{FSO}}(\gamma_{T_1}) + F_2(x), & \text{if } x > \gamma_{T_1}, \end{cases} \quad (10.60)$$

where $F_1(x)$ is defined as:

$$F_1(x) = \int_0^x f_{\gamma_{FSO}+\gamma_{RF}}(y) dy \quad (10.61)$$

and $F_2(x)$ is defined as

$$F_2(x) = \int_0^{\gamma_{T_1}} f_{\gamma_{FSO}}(\gamma_{FSO}) F_{\gamma_{RF}}(x - \gamma_{FSO}) d\gamma_{FSO}. \quad (10.62)$$

Noting that the FSO and RF links are statistically independent, $f_{\gamma_{FSO}+\gamma_{RF}}(y)$ in (10.61) can be evaluated as

$$f_{\gamma_{FSO}+\gamma_{RF}}(y) = \int_0^y f_{\gamma_{FSO}}(\gamma_{FSO}) f_{\gamma_{RF}}(y - \gamma_{FSO}) d\gamma_{FSO}. \quad (10.63)$$

After substituting (10.55) and (10.57) into (10.63) and applying the binomial expansion, and the series expansion of the exponential function, along with [25, Eq. (07.34.21.0084.01)], $f_{\gamma_{FSO}+\gamma_{RF}}(y)$ can be evaluated as:

$$f_{\gamma_{FSO}+\gamma_{RF}}(y) = \frac{\xi^2 e^{\frac{-my}{\bar{\gamma}_{RF}}} \left(\frac{m}{\bar{\gamma}_{RF}}\right)^m y^{m-1}}{\Gamma(\alpha)\Gamma(\beta)\Gamma(m)} \left\{ \sum_{n=0}^{\infty} \frac{\left(\frac{my}{\bar{\gamma}_{RF}}\right)^n}{n!} \sum_{i=0}^{m-1} \binom{m-1}{i}(-1)^i \right. \\ \left. \times G_{2,4}^{3,1}\left[\frac{\xi^2 \alpha \beta y}{(\xi^2+1)\bar{\gamma}_{FSO}} \Big|_{\xi^2,\alpha,\beta,-n-i}^{1-n-i,\xi^2+1}\right] \right\}. \quad (10.64)$$

---
[3] When data transmission is suspended, the pilot signal is assumed to be continuously transmitted over the FSO link to check its status.

Substituting (10.64) in (10.61), using the series expansion of the exponential function, and applying [25, Eq. (07.34.21.0084.01)], $F_1(x)$ can be expressed as

$$F_1(x) = \frac{\xi^2 (\frac{mx}{\bar{\gamma}_{RF}})^m}{\Gamma(\alpha)\Gamma(\beta)\Gamma(m)} \left\{ \sum_{n=0}^{\infty} \frac{(\frac{mx}{\bar{\gamma}_{RF}})^n}{n!} \sum_{i=0}^{m-1} \binom{m-1}{i} (-1)^i \sum_{k=0}^{\infty} \frac{(\frac{-mx}{\bar{\gamma}_{RF}})^k}{k!} \right.$$
$$\left. \times G_{3,5}^{3,2} \left[\frac{\xi^2 \alpha \beta x}{(\xi^2 + 1)\bar{\gamma}_{FSO}} \Big|_{\xi^2, \alpha, \beta, -k-n-m, -n-i}^{1-n-i, 1-k-n-m, \xi^2+1} \right] \right\}. \tag{10.65}$$

Substituting (10.55) and (10.58) in (10.62), $F_2(x)$ can be expressed in the integral form as:

$$F_2(x) = \int_0^{\gamma_{T_1}} \frac{\gamma(m, \frac{m(x-\gamma_{FSO})}{\bar{\gamma}_{RF}}) \xi^2 \gamma_{FSO}^{-1}}{\Gamma(\alpha)\Gamma(\beta)\Gamma(m)} \times G_{1,3}^{3,0} \left[\frac{\xi^2 \alpha \beta}{(\xi^2+1)} \frac{\gamma_{FSO}}{\bar{\gamma}_{FSO}} \Big|_{\xi^2, \alpha, \beta}^{\xi^2+1} \right] d\gamma_{FSO}. \tag{10.66}$$

By using the series representation of $\gamma(a,x)$ defined in [22, Eq. (8.352.1)], and the binomial expansion rule along with the series expansion of the exponential function, the term $\gamma\left(m, \frac{m(x-\gamma_{FSO})}{\bar{\gamma}_{RF}}\right)$ can be represented by:

$$\gamma\left(m, \frac{m(x-\gamma_{FSO})}{\bar{\gamma}_{RF}}\right) = (m-1)! \left\{ 1 - e^{\frac{-mx}{\bar{\gamma}_{RF}}} \sum_{n=0}^{\infty} \frac{(\frac{m\gamma_{FSO}}{\bar{\gamma}_{RF}})^n}{n!} \sum_{k=0}^{m-1} \frac{(\frac{m}{\bar{\gamma}_{RF}})^k}{k!} \right.$$
$$\left. \sum_{j=0}^{k} \binom{k}{j} x^{k-j} (-\gamma_{FSO})^j \right\}. \tag{10.67}$$

After plugging (10.67) in (10.66) and applying [25, Eq. (07.34.21.0084.01)], $F_2(x)$ can be evaluated as

$$F_2(x) = \frac{(m-1)!}{\Gamma(m)} F_{\gamma_{FSO}}(\gamma_{T_1}) - \left\{ \frac{\xi^2 e^{\frac{-mx}{\bar{\gamma}_{RF}}} (m-1)!}{\Gamma(\alpha)\Gamma(\beta)\Gamma(m)} \sum_{n=0}^{\infty} \frac{(m\gamma_{T_2}/\bar{\gamma}_{RF})^n}{n!} \sum_{k=0}^{m-1} \right.$$
$$\left. \frac{(mx/\bar{\gamma}_{RF})^k}{k!} \sum_{j=0}^{k} \binom{k}{j} \times \left(\frac{-\gamma_{T_1}}{x}\right)^j G_{2,4}^{3,1} \left[\frac{\xi^2 \alpha \beta \gamma_{T_1}}{(\xi^2+1)\bar{\gamma}_{FSO}} \Big|_{\xi^2, \alpha, \beta, -n-j}^{1-n-j, \xi^2+1} \right] \right\}. \tag{10.68}$$

Finally, the CDF of γ_c is obtained after substituting (10.65) and (10.68) into (10.60).

The average spectral efficiency of the proposed adaptive hybrid FSO/RF system, defined as the average number of bits transmitted over each symbol period [33], is given by

$$\eta^{Adapt} = \sum_{n=1}^{N} 2n[F_{\gamma_c}(\gamma_{T_{n+1}}) - F_{\gamma_c}(\gamma_{T_n})]. \tag{10.69}$$

When $\gamma_{FSO} < \gamma_{T_1}$, and γ_c also falls below γ_{T_1}, the communication system goes into the outage state, in which the received SNR cannot support the target BER of the system and data transmission is suspended over both links. The outage probability of the adaptive hybrid FSO/RF system can be simply calculated by evaluating the CDF of γ_c at γ_{T_1} as:

$$P_{out}^{Adapt} = F_{\gamma_c}(\gamma_{T_1}) = F_1(\gamma_{T_1}). \tag{10.70}$$

10.3.3 Switch-over Hybrid FSO/RF Transmission with Adaptive Modulation

For comparison purposes, we analyze the performance of a switch-over hybrid FSO/RF system [37], where the system will rely only on the RF link when the FSO link cannot support the lowest-order modulation scheme. To have fair comparison, we assume the same adaptive modulation schemes. The average spectral efficiency of the switch-over hybrid FSO/RF system in this case is given by

$$\eta^{Switch} = \sum_{n=1}^{N} 2n \Big\{ \big[F_{\gamma_{FSO}}(\gamma_{T_{n+1}}) - F_{\gamma_{FSO}}(\gamma_{T_n})\big] + F_{\gamma_{FSO}}(\gamma_{T_1}) \\ \times \big[F_{\gamma_{RF}}(\gamma_{T_{n+1}}) - F_{\gamma_{RF}}(\gamma_{T_n})\big] \Big\}. \qquad (10.71)$$

When $\gamma_{FSO} < \gamma_{T_1}$, and γ_{RF} also falls below γ_{T_1}, the communication system goes into outage. In this case, the outage probability of the hybrid system can be evaluated as

$$P_{out}^{Switch} = F_{\gamma_{FSO}}(\gamma_{T_1}) F_{\gamma_{RF}}(\gamma_{T_1}). \qquad (10.72)$$

10.3.4 Numerical Results

In this section we present several numerical examples to illustrate the math formulation and to study the proposed system's performance. We will consider three different 2^{2n}-QAM modulation schemes with $n = 1, 2$, and 3, where the corresponding thresholds γ_{T_n}, assuming $BER_0 = 10^{-6}$, can be calculated using (10.59). The relevant parameters of the FSO and RF subsystems considered for the numerical results are given in Table 10.1. The strongest atmospheric turbulence commonly occurs when weather is clear and becomes weaker as the weather condition gets worse by means of either fog or rain. We consider in our numerical examples the scenario of clear weather conditions and strong atmospheric turbulence ($C_n^2 = 2 \times 10^{-13}$), with weather-dependent attenuation coefficient $\alpha_{FSO} = 0.44$ dB/km and RF rain attenuation $\alpha_{rain} = 0$ dB/km, along with pointing error effects [20]. We assume RF channel fading severity of $m = 5$ in all figures.

In Fig. 10.13, we plot the average spectral efficiency of the proposed adaptive hybrid FSO/RF system as a function of the transmitted power of the FSO link P_{FSO}. It can be seen from Fig. 10.13 that average spectral efficiency increases as the FSO link's quality improves by increasing P_{FSO}. Also, we can observe that with sufficiently high P_{FSO}, the average spectral efficiency of the adaptive hybrid system is the same as that of the FSO-only system and switch-over hybrid system, which in this case relies on the FSO link. This is because the FSO link's quality is good enough to support the target BER alone without the need to activate the RF link. On the other hand, as the FSO link's quality degrades due to strong atmospheric turbulence in clear weather conditions, the RF link is activated along with the FSO link, which leads to a significant improvement in the average spectral efficiency of the adaptive hybrid system over both switch-over hybrid FSO/RF and FSO-only systems, as can be seen from Fig. 10.13. The numerical results shown in Fig. 10.13 for the average spectral efficiency of the proposed adaptive hybrid FSO/RF system are obtained using $n = 30$ and $k = 30$ in (10.65) and $n = 30$

Table 10.1 Parameters of FSO and RF subsystems. Reprint with permission from [18]. ©2015 IEEE.

Parameter	Symbol	Value
FSO subsystem		
Wavelength	λ_{FSO}	1550 nm
Shot noise variance	σ_{FSO}^2	2×10^{-14}
Data rate	$1/T_s$	1 Gbit/s
Responsivity	η	0.5 A/W
Photodetector diameter	D	20 cm
Transmit divergence at $1/e^2$	θ_0	2.5 mrad
Jitter standard deviation	σ_s	30 cm
Link distance	z	1000 m
RF subsystem		
Carrier frequency	f_{RF}	60 GHz
Bandwidth	W	250 MHz
Transmit antenna gain	G_T	43 dBi
Receive antenna gain	G_R	43 dBi
Noise power spectral density	N_0	−114 dBm/MHz
Receiver noise figure	N_F	5 dB
Oxygen attenuation	α_{oxy}	15.1 dB/Km

in (10.68), which are then plugged into (10.60) to be used in evaluating (10.69). As can be observed, evaluating the average spectral efficiency using the truncated values of (10.65) and (10.68) gives accurate results that coincide with the values that are obtained by evaluating the integrals in (10.61) and (10.62) using numerical methods before being plugged into (10.60) to evaluate (10.69).

In Fig. 10.14 we plot the outage probability of the proposed adaptive hybrid system as a function of P_{FSO}. It can be seen that the proposed adaptive hybrid system gives an outage performance much better than that of the switch-over hybrid system or FSO-only system with the same adaptive modulation scheme.

It is worth mentioning that similar superior performance of the proposed adaptive hybrid FSO/RF system over both FSO-only and switch-over hybrid FSO/RF systems is expected in other weather conditions, mainly in fog and rain. In rain conditions, the FSO link is not affected and thus the communication system can use it without needing to activate the RF link. On the other hand, the FSO link is affected by fog, which will degrade its performance, thus activating the RF link, which is not affected by fog and will support the FSO link to maintain a reliable communication channel.

10.4 Summary

In this chapter, we studied the design and analysis of practical transmission schemes for hybrid FSO/RF systems. We presented several low-complexity hybrid FSO/RF

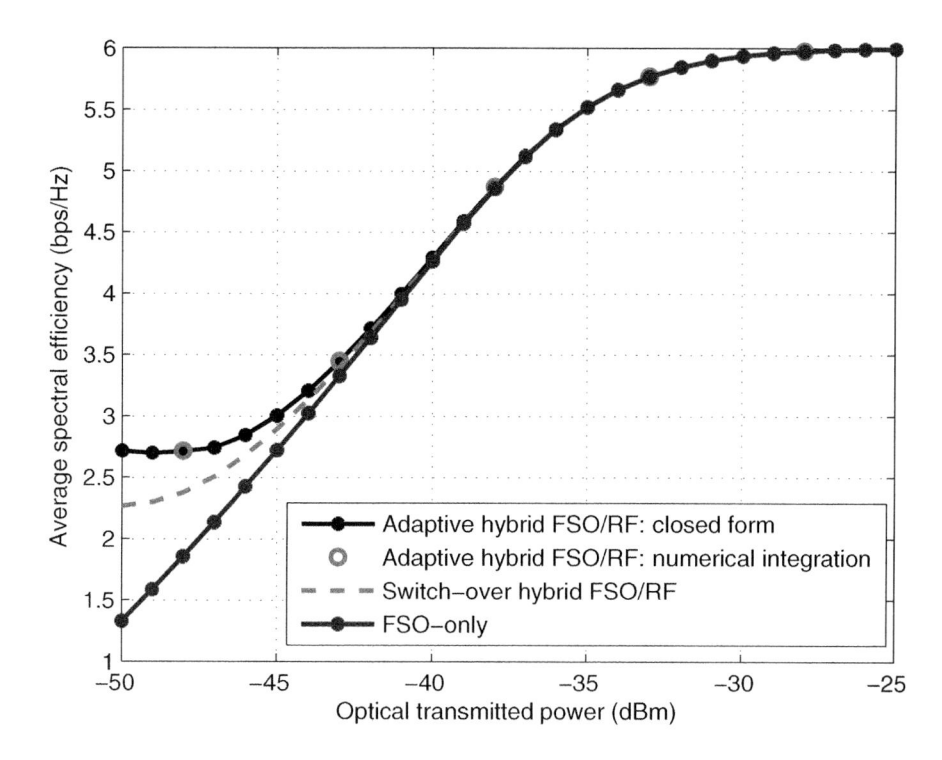

Figure 10.13 Average spectral efficiency of the proposed hybrid FSO/RF system as a function of the transmitted power of the FSO link. Reprint with permission from [30]. ©2015 IEEE.

transmission schemes, which incorporate advanced transmission technologies discussed in previous chapters. For the switching-based hybrid transmission scheme, we considered both single and dual threshold designs. To further improve transmission performance, we design novel hybrid FSO/RF transmission schemes with the application of adaptive transmission and combining. The performance of each design is accurately quantified through mathematical analysis assuming popular statistical channel models.

10.5 Further Reading

The reader can refer to [38, 39] for further details on the general subject of optical wireless communications and to [40] for free-space optical communications in particular. [41, 42] present additional analytical results on the performance of FSO transmissions with pointing errors. Some capacity results on free-space optical intensity channels can be found in [43, 44]. Further to the hybrid FSO/RF schemes presented in this chapter, [45] integrates power adaptation with adaptive combining in the design. The analysis of hybrid FSO/RF schemes is extended into a point-to-multi-point scenario in [46].

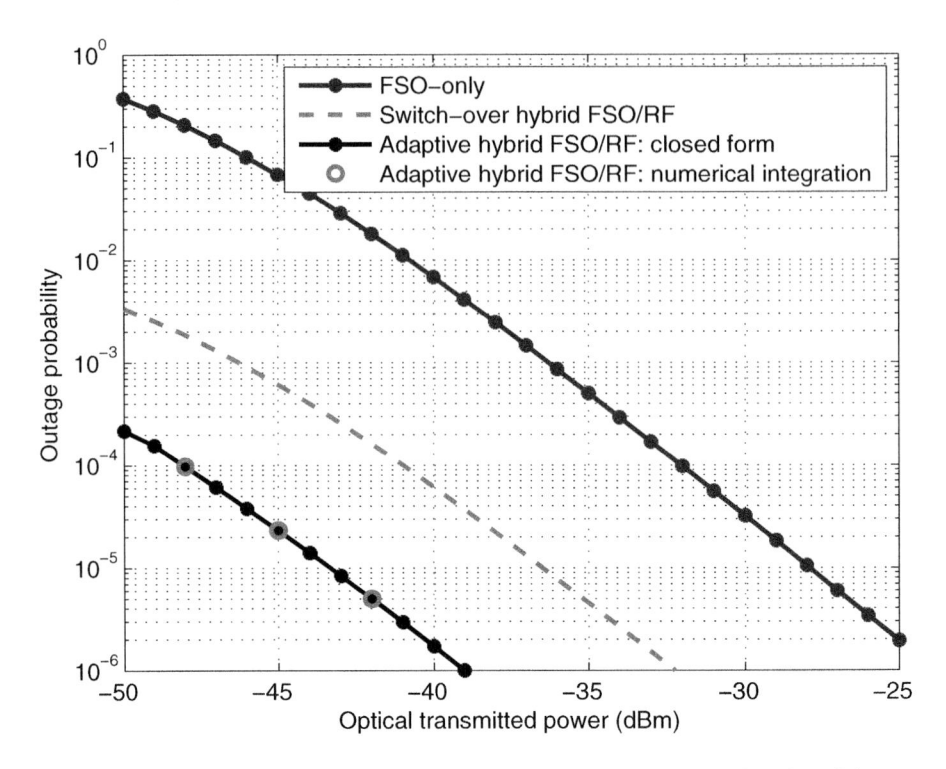

Figure 10.14 Outage probability of the proposed hybrid FSO/RF system as a function of the transmitted power of the FSO link. Reprint with permission from [30]. ©2015 IEEE.

References

[1] L. C. Andrews, R. L. Phillips, and C. Y. Hopen, *Laser Beam Scintillation with Applications*, SPIE Publications, 2001.
[2] F. Giannetti, M. Luise, and R. Reggiannini, "Mobile and personal communications in the 60 GHz band: A survey," *Wireless Pers. Commun.*, 10, pp. 207–243, 1999.
[3] F. Nadeem, V. Kvicera, M. Awan, et al., "Weather effects on hybrid FSO/RF communication link," *IEEE J. Sel. Areas Commun.*, 27, pp. 1687–1697, 2009.
[4] H. Wu and M. Kavehrad, "Availability evaluation of ground-to-air hybrid FSO/RF links," *Int. J. Wireless Inf. Netw.*, 14, pp. 33–45, 2007.
[5] S. Vangala and H. Pishro-Nik, "A highly reliable FSO/RF communication system using efficient codes," in *Proc. IEEE Global Telecommun. Conf., 2007. GLOBECOM '07*, Washington, DC, November 2007, pp. 2232–2236.
[6] A. Abdulhussein, A. Oka, T. T. Nguyen, and L. Lampe, "Rateless coding for hybrid free-space optical and radio-frequency communication," *IEEE Trans. Wireless Commun.*, 9, pp. 907–913, 2010.
[7] B. He and R. Schober, "Bit-interleaved coded modulation for hybrid RF/FSO systems," *IEEE Trans. Commun.*, 57, pp. 3753–3763, 2009.
[8] W. Zhang, S. Hranilovic, and C. Shi, "Soft-switching hybrid FSO/RF links using short-length raptor codes: Design and implementation," *IEEE J. Sel. Areas Commun.*, 27, pp. 1698–1708, 2009.

References

[9] M. Usman, H.-C. Yang, and M.-S. Alouini, "Practical switching based hybrid FSO/RF transmission and its performance analysis," *IEEE Photonics J.*, 6, no. 5, 2902713, 2014.

[10] M. Khalighi and M. Uysal, "Survey on free space optical communication: A communication theory perspective," *IEEE Commun. Surv. Tut.*, no. 99, p. 1, 2014.

[11] N. Chatzidiamantis, G. Karagiannidis, E. Kriezis, and M. Matthaiou, "Diversity combining in hybrid RF/FSO systems with PSK modulation," in *Proc. IEEE Int. Conf. Commun. (ICC), 2011*, Kyoto, June 2011, pp. 1–6.

[12] J. Li, J. Q. Liu, and D. Taylor, "Optical communication using subcarrier PSK intensity modulation through atmospheric turbulence channels," *IEEE Trans. Commun.*, 55, pp. 1598–1606, 2007.

[13] S. Navidpour, M. Uysal, and M. Kavehrad, "BER performance of free-space optical transmission with spatial diversity," *IEEE Trans. Wireless Commun.*, 6, pp. 2813–2819, 2007.

[14] S. Wilson, M. Brandt-Pearce, Q. Cao, and M. Baedke, "Optical repetition MIMO transmission with multipulse PPM," *IEEE J. Sel. Areas Commun.*, 23, pp. 1901–1910, 2005.

[15] S. Haas, *Capacity of and Coding for Multiple-Aperture Wireless Optical Communications*, PhD dissertation, Massachusetts Institute of Technology, 2003.

[16] J. Proakis, *Digital Communications*, 4th ed., McGraw Hill, 2000.

[17] J. Hossain, P. Vitthaladevuni, M.-S. Alouini, V. Bhargava, and A. Goldsmith, "Adaptive hierarchical modulation for simultaneous voice and multiclass data transmission over fading channels," *IEEE Trans. Veh. Technol.*, 55, pp. 1181–1194, 2006.

[18] T. Rakia, H.-C. Yang, F. Gebali, and M.-S. Alouini, "Outage analysis of practical FSO/RF hybrid system with adaptive combining," *IEEE Commun. Lett.*, 19, no. 8, pp. 1366–1269, 2015.

[19] N. D. Chatzidiamantis, G. K. Karagiannidis, E. E. Kriezis, and M. Matthaiou, "Diversity combining in hybrid RF/FSO systems with PSK modulation," in *2011 IEEE Int. Conf. Commun. (ICC)*, Kyoto, June 2011.

[20] A. A. Farid, and S. Hranilovic, "Outage capacity optimization for free-space optical links with pointing errors," *J. Lightwave Technol.*, 25, no. 7, pp. 1702–1710, 2007.

[21] A. Al-Habash, L. C. Andrews, and R. L. Philips, "Mathematical model for the irradiance probability density function of a laser beam propagating through turbulent media," *Opt. Eng.*, 40, no. 8, pp. 1554–1562, 2001.

[22] I. S. Gradshteyn and I. M. Ryzhikhev, *Table of Integrals, Series, and Products*, 6th ed., Academic, 2000.

[23] L. C. Andrews and R. L. Phillips, *Laser Beam Propagation Through Random Media*, 2nd ed., SPIE Optical Engineering Press, 2005.

[24] V. S. Adamchik and O. I. Marichev, "The algorithm for calculating integrals of hypergeometric type functions and its realization in REDUCE system," in *Proc. Int. Symp. Symbolic Algebr. Comput.*, Tokyo, 1990, pp. 212–224.

[25] Mathematica v. 8.0, Wolfram Research, Inc., 2010.

[26] S. Bloom and D. J. T. Heatley, "The last mile solution: Hybrid FSO radio," White Paper, AirFiber Inc., 802-0008-000 M-A1, 2002.

[27] H. A. Suraweera, P. J. Smith, and J. Armstrong, "Outage probability of cooperative relay networks in Nakagami-*m* fading channels," *IEEE Commun. Lett.*, 10, no. 12, pp. 834–836, 2006.

[28] M. K. Simon and M.-S. Alouini, *Digital Communication over Fading Channels*, Wiley, 2005.

[29] X. Tang, Z. Ghassemlooy, S. Rajbhandari, W. O. Popoola, and C. G. Lee, "Coherent polarization shift keying modulated free space optical links over a gamma–gamma turbulence channel," *Am. J. Eng. Appl. Sci.*, 4, no. 4, pp. 520–530, 2012.

[30] T. Rakia, H.-C. Yang, F. Gebali, and M.-S. Alouini, "Joint adaptive modulation and combining for hybrid FSO/RF systems," *Proc. IEEE Int. Conf. Ubiquitous Wireless Broadband (ICUWB'2015)*, Montreal, October 2015, pp. 1–5.

[31] B. T. Vu, N. T. Dang, T. G. Thang, and A. T. Pham, "Bit error rate analysis of rectangular QAM/FSO systems using APD receiver over atmospheric turbulence channels," *J. Opt. Commun. Netw.*, 5, no. 5, pp. 437–446, 2013.

[32] X. Lei, P. Fan, and L. Hao, "Exact symbol error probability of general order rectangular QAM with MRC diversity reception over Nakagami-m fading channels," *IEEE Commun. Lett.*, 11, no. 12, pp. 958–960, 2007.

[33] A. Goldsmith, *Wireless Communication*, 1st ed., Cambridge University Press, 2005.

[34] M. Niu, J. Cheng, and J. F. Holzman, "Error rate analysis of M-ary coherent free-space optical communication systems with K-distributed turbulence," *IEEE Trans. Commun.*, 59, no. 3, pp. 664–668, 2011.

[35] M. Niu, J. Cheng, and J. F. Holzman, "Error rate performance comparison of coherent and subcarrier intensity modulated optical wireless communications," *J. Opt. Commun. Netw.*, 5, no. 6, pp. 554–564, 2013.

[36] H. G. Sandalidis, T. A. Tsiftsis, G. K. Karagiannidis, and M. Uysal, "BER performance of FSO links over strong atmospheric turbulence channels with pointing errors," *IEEE Commun. Lett.*, 12, no. 1, pp. 44–46, 2008.

[37] V. V. Mai and A. T. Pham, "Adaptive multi-rate designs for hybrid FSO/RF systems over fading channels," in *2014 IEEE Globecom Workshops*, Austin, TX, 2014.

[38] Z. Ghassemlooy, W. Popoola, and S. Rajbhandari, *Optical Wireless Communications: System and Channel Modelling with MATLAB*, CRC Press, 2013.

[39] M. Uysal, C. Capsoni, S. Ghassemlooy, A. Boucouvalas, and E. Udvary, (eds.), *Optical Wireless Communications: An Emerging Technology*, Springer, 2016.

[40] H. Kaushal, V. K. Jain, and S. Kar, *Free Space Optical Communication*, Springer, 2017.

[41] I. S. Ansari, F. Yilmaz, and M.-S. Alouini, "Performance analysis of free-space optical links over Malaga-(M) turbulence channels with pointing errors," *IEEE Trans. Wireless Commun.*, 15, no. 1, pp. 91–102, 2016.

[42] H. AlQuwaiee, H.-C. Yang, and M.-S. Alouini, "On the asymptotic capacity of dual-aperture FSO systems with generalized pointing error model," *IEEE Trans. Wireless Commun.*, 15, no. 9, pp. 6502–6512, 2016.

[43] S. Hranilovic and F. R. Kschischang, "Capacity bounds for power- and band-limited optical intensity channels corrupted by Gaussian noise," *IEEE Trans. Inform. Theor.*, 50, no. 5, pp. 784–795, 2004.

[44] A. Chaaban, Z. Rezki, and M.-S. Alouini, "On the capacity of the intensity-modulation direct-detection optical broadcast channel," *IEEE Trans. Wireless Commun.*, 15, no. 5, pp. 3114–3130, 2016.

[45] T. Rakia, H.-C. Yang, M.-S. Alouini, and F. Gebali, "Power adaptation based on truncated channel inversion for hybrid FSO/RF transmission with adaptive combining," *IEEE Photonics J.*, 7, no. 4, 7903012, 2015.

[46] T. Rakia, F. Gebali, H.-C. Yang, and M.-S. Alouini, "Cross layer analysis of point-to-multipoint hybrid FSO/RF networks," *J. Opt. Commun. Netw.*, 9, no. 3, pp. 234–243, 2017.

11 Application: Sensor Transmission with RF Energy Harvesting

Wireless sensors are used in a wide range of applications, such as environment monitoring, surveillance, health care, and intelligent building [1]. The sensor nodes are usually powered by batteries with finite lifetimes, which is an important limiting factor to the functionality of wireless sensor networks (WSNs). Replacing or charging the batteries may either incur high costs or be impractical for certain application scenarios (e.g., applications that require sensors to be embedded into structures). Powering sensor nodes through ambient energy harvesting has therefore received a lot of attention in both academia and industrial communities [2, 3]. Various techniques have been developed to harvest energy from conventional ambient energy sources, including solar power, wind power, thermoelectricity, and vibrational excitation [4–7]. RF (radio frequency) energy is another candidate ambient energy source for powering sensor nodes. There has been a continuing interest in RF energy harvesting due to the intensive deployment of cellular/WiFi wireless systems in addition to traditional radio/TV broadcasting systems [8]. It has been experimentally proved that RF energy harvesting is feasible from the hardware implementation viewpoint [9–11].

In this chapter, we study the performance of sensor transmission with harvested RF energy over a wireless environment. We first present an analytical framework to derive the statistics of harvested RF energy over a certain time period. With this statistical characterization, we analyze the performance of secondary sensor transmission for both delay-sensitive and delay-insensitive traffics. As an application of the advanced wireless transmission technologies in previous chapters, we investigate the energy harvesting performance improvement with cooperative beam selection at the primary transmitters. Both single and multiple energy harvesting sensor scenarios are considered, for which secondary energy harvesting performance versus primary system sum-rate performance trade-offs are quantitatively characterized.

11.1 Cognitive Transmission with Harvested RF Energy

In this section, we consider an overlaid sensor transmission scenario in which a sensor-to-sink communication link operates in the coverage of an existing wireless system over the same frequency. Unlike conventional WSN implementations, where both the sensor and the sink are equipped with reliable power supplies, we assume that only the sink has a constant power source and that the sensor needs to harvest

RF energy from the transmitted signals of existing wireless systems. Specifically, the sensor node can only harvest RF energy when its received signal power is larger than a certain sensitivity level [12]. As such, the existing system, being either cellular, WiFi, or TV broadcasting systems, serves as the ambient source for the sensor energy harvesting and acts as an interference source during sensor transmission. Such an overlaid implementation strategy of RF-energy powered WSN has the potential to offer attractive green solutions to a wide range of sensing applications, particularly in view of the increasingly severe spectrum scarcity. While scavenging the radiated RF energy from an existing system, the RF energy-powered sensor transmission will introduce very limited interference to existing systems due to its low transmission power and short transmission duration.

We investigate the packet transmission performance of the sensor-to-sink link over Rayleigh fading wireless channels. Considering the sensor's limited energy harvesting capability, we assume that the link is used to support low-rate sensing applications with low traffic intensity. Specifically, we first consider a delay-sensitive scenario in which the sensor needs to periodically transmit a new packet to the sink with a hard delay constraint. We evaluate the packet loss probability assuming no retransmission is allowed. For delay-insensitive traffic, where the sensing data must be delivered to the sink without error at the expense of a certain delay, we calculate the average delay of packet transmission over the wireless link with harvested energy. Whenever feasible, we derive the exact analytical expression for performance metrics of interest in simple closed form, which facilitates fast evaluation and convenient applications to parameter optimization. These analytical results will help determine what type of sensing applications the overlaid sensor implementation strategy with RF energy harvesting can effectively support.

11.1.1 System and Channel Model

We consider a point-to-point packet transmission from a single-antenna wireless sensor to its sink over a flat Rayleigh fading channel. The sink and the sensor are deployed in the coverage area of an existing wireless system. We assume that the sensor can harvest RF energy from the transmitted signal of the existing system and use it as its sole energy source for transmission, as illustrated in Fig. 11.1.

We assume that the sensor works in two stages within one duty cycle, i.e., an energy harvesting stage and a packet transmission stage. The energy harvesting stage and the packet transmission stage may correspond to a sleep stage and a wake-up stage, respectively, for traditional wireless sensors. In the energy harvesting stage, the sensor harvests RF energy from the radio transmission of existing wireless systems over multiple channel coherence time.

During the packet transmission stage, the sensor will transmit its collected information to the sink using harvested energy. We assume that the energy consumed for information collection is negligible compared with the energy used for transmission [14]. Then the energy that can be used for transmission is approximately equal to

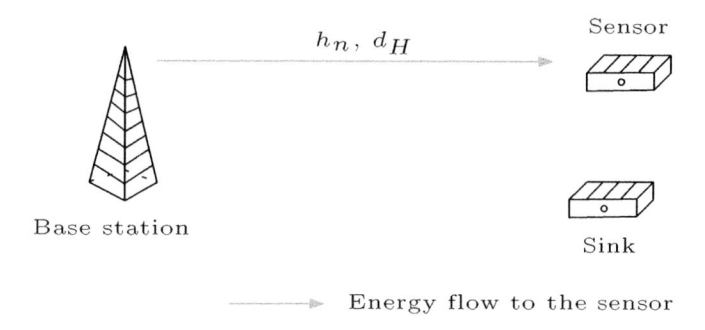

11.1 Cognitive Transmission with Harvested RF Energy

(a) Energy harvesting stage

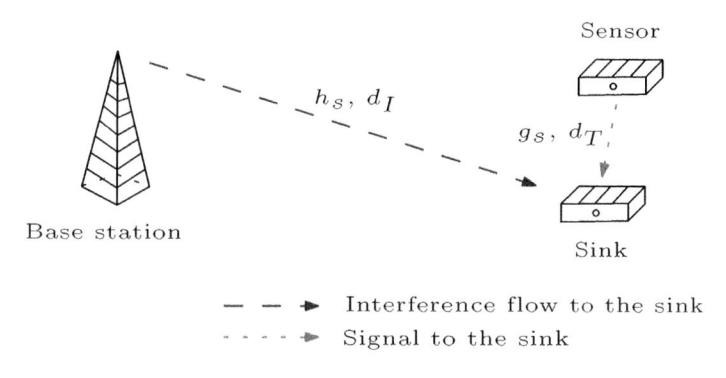

(b) Packet transmission stage

Figure 11.1 System model for two-stage sensor transmission with RF energy harvesting. Reprint with permission from [13]. ©2015 IEEE.

the harvested energy. Also note that the sensor transmission will suffer interference from the existing system in this stage, the effect of which will be further discussed in the following sections. Due to the low transmission power and short transmission duration, we ignore the interference that the sensor transmission may generate to the existing system. We also ignore the delay for access control and/or carrier sensing

We adopt a log-distance path loss plus Rayleigh block fading channel model for the operating environment [15] while ignoring the shadowing effect for the sake of presentation clarity. In particular, the channel gain between the base station and the sensor remains constant over one channel coherence time, denoted by T_c, and changes to an independent value afterwards. Let h_n denote the fading channel gain over the nth coherence time, where $h_n \in \mathcal{CN}(0,1)$. For notational conciseness, we use α_n to denote its amplitude square, i.e., $\alpha_n = ||h_n||^2$, whose PDF for the Rayleigh fading channel under consideration is given by

$$f_{\alpha_n}(x) = e^{-x}. \tag{11.1}$$

Then the instantaneous received signal power at the sensor over the nth coherence time is given by $P_n = \overline{P}\alpha_n$, where $\overline{P}$ is the average received power at the sensor due to path loss, given by

$$\overline{P} = \frac{P_T}{\Gamma d_H^\lambda}, \tag{11.2}$$

where P_T is the constant transmission power of the BS, d_H is the distance from the BS to the sensor, λ is the path loss exponent of the environment, ranging from 2 to 5, and Γ is a constant parameter of the log-distance path loss model. Specifically, $\Gamma = \frac{PL(d_0)}{d_0^\lambda}$, where d_0 is a reference distance of the antenna far field, and $PL(d_0)$ is linear path loss at distance d_0, depending on the propagation environment.

Typically, the sensor can harvest RF energy only when the received signal power is strong enough [10, 11]. As such, we assume that the sensor can only harvest energy when the instantaneous received signal power P_n is greater than the sensitivity level P_{th} which is greater than the receiver sensitivity for information reception. The harvested energy is proportional to $P_n - P_{th}$. Consequently, the amount of energy that the sensor can harvest during the nth coherence time can be represented as [12]

$$E_n = \begin{cases} \eta T_c (P_n - P_{th}), & P_n \geq P_{th}, \\ 0, & P_n < P_{th}, \end{cases} \tag{11.3}$$

where $0 \leq \eta \leq 1$ is RF energy harvesting efficiency. It follows that the amount of energy harvested by the sensor over N consecutive coherence times can be given by

$$E_h^{(N)} = \min\left(\sum_{n=1}^N E_n, E_c\right), \tag{11.4}$$

where E_c is the energy storage capacity of the sensor.[1]

The transmission power of the sensor when it uses the harvested energy over N coherence time is equal to $\frac{E_h^{(N)}}{T_s}$, where T_s denotes the transmission time duration. We assume, with the notion of low-rate sensing applications, that T_s is much smaller than the channel coherence time T_c. Let h_s and g_s denote the fading channel gains from BS to the sink and from the sensor to the sink, respectively, where $h_s \in \mathcal{CN}(0, 1)$ and $g_s \in \mathcal{CN}(0, 1)$. The received SINR at the sink can be calculated as

$$\gamma_s = \frac{\frac{E_h^{(N)}}{T_s d_T^\lambda}\|g_s\|^2}{\frac{P_T}{d_I^\lambda}\|h_s\|^2 + \Gamma\sigma^2}, \tag{11.5}$$

where d_T is the distance from the sensor to the sink, d_I is the distance from the BS to the sink, and σ^2 is the variance of the additive noise at the sink. In general, the sensor and the sink are very close to each other, i.e., $d_T \ll d_H \approx d_I$.

---

[1] E_c can also be viewed as the energy threshold, above which the sensor can carry out packet transmission. We will examine the optimization of this important design parameter in the following sections.

In the following, we study the performance of such overlaid sensor transmission when it is used to support low-rate data traffics. These analytical results can help design and optimize various system parameters, such as the number of channel coherence times for energy harvesting, N, and sensor transmission power. In this chapter, we focus on properly designing energy storage capacity E_c, while assuming system parameters, such as BS transmit power P_T, sensor sensitivity P_{th}, sensor energy harvesting efficiency η, transmission time duration T_s, and channel model parameters to be fixed.

11.1.2 Performance Analysis for Delay-Sensitive Traffic

For certain sensing applications, such as smart metering and environment monitoring, the sensor node needs to periodically send their collected information (e.g., energy usage, temperature, humidity information) to the sink. Any delay in the delivery of this information may render them useless. Therefore, the goal is to successfully transmit these information packets within a fixed time duration. As such, an important performance metric for such applications is the packet loss probability, i.e., the percentage of packets that could not be delivered to the sink in time. We analyze the packet loss probability of the overlaid sensing implementation with RF energy harvesting.

We assume that the sensor must collect and transmit one packet to the sink over a fixed time duration T_F. The number of coherence times in T_F, denoted by N, is approximately equal to $\lfloor \frac{T_F}{T_c} \rfloor$. The sensor will first harvest RF energy for N channel coherence times and then transmit the packet to the sink using the harvested energy. We focus on low-rate sensing applications and ignore the potential packet collision with other sensors. We also assume that, with adoption of a certain error correction coding scheme, the packet can be successfully received by the sink if the received signal to interference plus noise ratio (SINR) at the sink during packet transmission is above γ_T. As such, packet loss will occur if and only if the received SINR at the sink during packet transmission is below the threshold γ_T. This may be due to insufficient harvested energy, poor sensor to sink channel quality, as well as strong interference from the BS. Mathematically, the packet loss probability of the sensor transmission is given by

$$P_{PL} = \Pr[\gamma_s < \gamma_T] = \Pr\left[\frac{\frac{E_h^{(N)}}{T_s d_T^\lambda}||g_s||^2}{\frac{P_T}{d_I^\lambda}||h_s||^2 + \Gamma\sigma^2} < \gamma_T\right]. \quad (11.6)$$

Conditioning on $E_h^{(N)}$, the packet loss probability can be rewritten in terms of the PDFs of $E_h^{(N)}$, $||g_s||^2$, and $||h_s||^2$, denoted by $f_{E_h^{(N)}}(\cdot), f_{||g_s||^2}(\cdot)$, and $f_{||h_s||^2}(\cdot)$, respectively, as

$$P_{PL} = \int_0^{E_c} \int_0^\infty F_{||g_s||^2}\left(\frac{T_s \gamma_T d_T^\lambda (\frac{P_T y}{d_I^\lambda} + \Gamma\sigma^2)}{z}\right) f_{||h_s||^2}(y) f_{E_h^{(N)}}(z) dy dz. \quad (11.7)$$

The PDF of $||h_s||^2$ and $||g_s||^2$ for the Rayleigh fading channel model under consideration are commonly given by

$$f_{||h_s||^2}(x) = f_{||g_s||^2}(x) = e^{-x}. \qquad (11.8)$$

After proper substitution and some manipulations, we can rewrite P_{PL} as

$$P_{PL} = \int_0^{E_c} \left(1 - \frac{ze^{-\frac{T_s \gamma_T \Gamma d_T^\lambda \sigma^2}{z}}}{z + \frac{P_T}{d_I^\lambda} T_s \gamma_T d_T^\lambda}\right) f_{E_h^{(N)}}(z) dz. \qquad (11.9)$$

To proceed further, we need the PDF of the harvested energy over N coherence times, $f_{E_h^{(N)}}(z)$. We first consider the one coherence time case, i.e., $N = 1$. The CDF of the harvested energy can be simply represented as

$$F_{E_h^{(1)}}(x) = \Pr[E_h^{(1)} < x] = \Pr[E_1 < x], x \leq E_c. \qquad (11.10)$$

After substituting (11.3) into (11.10) and some manipulation, we have

$$F_{E_h^{(1)}}(x) = 1 - e^{-\frac{x}{\eta T_c \overline{P}_R} - \frac{P_{th}}{\overline{P}_R}}, x \leq E_c. \qquad (11.11)$$

For the multiple channel coherence time case, i.e., $N > 1$, we denote the number of channel coherence times in which the sensor can harvest energy by N_a. According to the total probability theorem, the CDF of the harvested energy is shown as

$$F_{E_h^{(N)}}(x) = \Pr[E_h^{(N)} < x] \qquad (11.12)$$

$$= \sum_{i=0}^N \Pr[\sum_{n=1}^N E_n < x, N_a = i].$$

When the ith largest received power is larger than P_{th} and the $(i+1)$th largest one is lower than P_{th}, the number of coherence times that the sensor can harvest energy is $N_a = i$. We denote the ordered version of N i.i.d. random variables α_n as $\alpha_{1:N} \geq \alpha_{2:N} \geq \ldots \geq \alpha_{N:N}$, and the sum of the $i-1$ largest variables as $\beta_i = \sum_{j=1}^{i-1} \alpha_{j:N}$. We can show that $N_a = i$ if and only if $\alpha_{i:N} \geq \frac{\Gamma d_H^\lambda P_{th}}{P_T}$ and $\alpha_{i+1:N} < \frac{\Gamma d_H^\lambda P_{th}}{P_T}$. Therefore, $F_{E_h}(x)$ can be calculated as

$$F_{E_h^{(N)}}(x) = \sum_{i=2}^{N-1} \Pr[\beta_i + \alpha_{i:N} < \frac{x}{\eta T_c \overline{P}_R} + \frac{i P_{th}}{\overline{P}_R}, \alpha_{i:N} \geq \frac{P_{th}}{\overline{P}_R}, \alpha_{i+1:N} < \frac{P_{th}}{\overline{P}_R}] \qquad (11.13)$$

$$+ \Pr[\alpha_{1:N} < \frac{P_{th}}{\overline{P}_R}] + \Pr[\frac{P_{th}}{\overline{P}_R} \leq \alpha_{1:N} < \frac{x}{\eta T_c \overline{P}_R} + \frac{P_{th}}{\overline{P}_R}, \alpha_{2:N} < \frac{P_{th}}{\overline{P}_R}]$$

$$+ \Pr[\beta_N + \alpha_{N:N} < \frac{x}{\eta T_c \overline{P}_R} + \frac{N P_{th}}{\overline{P}_R}, \alpha_{N:N} \geq \frac{P_{th}}{\overline{P}_R}]$$

11.1 Cognitive Transmission with Harvested RF Energy

$$= \sum_{i=2}^{N-1} \int_{\frac{P_{th}}{\bar{P}_R}}^{\frac{x}{i\eta T_c \bar{P}_R}+\frac{P_{th}}{\bar{P}_R}} \int_{(i-1)y}^{\frac{x}{\eta T_c \bar{P}_R}+\frac{iP_{th}}{\bar{P}_R}-y} \int_0^{\frac{P_{th}}{\bar{P}_R}} f_{\beta_i,\alpha_{i:N},\alpha_{i+1:N}}(t,y,z) dt dy dz$$

$$+ \int_0^{\frac{P_{th}}{\bar{P}_R}} f_{\alpha_{1:N}}(t) dt + \int_0^{\frac{P_{th}}{\bar{P}_R}} \int_{\frac{P_{th}}{\bar{P}_R}}^{\frac{x}{\eta T_c \bar{P}_R}+\frac{P_{th}}{\bar{P}_R}} f_{\alpha_{1:N},\alpha_{2:N}}(t,y) dt dy$$

$$+ \int_{\frac{P_{th}}{\bar{P}_R}}^{\frac{x}{N\eta T_c \bar{P}_R}+\frac{P_{th}}{\bar{P}_R}} \int_{(N-1)y}^{\frac{x}{\eta T_c \bar{P}_R}+\frac{NP_{th}}{\bar{P}_R}-y} f_{\beta_N,\alpha_{N:N}}(t,y) dt dy,$$

where $f_{\alpha_{1:N}}(x,y)$, $f_{\alpha_{1:N},\alpha_{2:N}}(x,y)$, $f_{\beta_N,\alpha_{N:N}}(x,y)$, and $f_{\beta_i,\alpha_{i:N},\alpha_{i+1:N}}(x,y,z)$ are the marginal and joint PDFs of $\alpha_{i:N}$ and β_i. In particular, their closed-form expression can be obtained as [16]

$$f_{\beta_N,\alpha_{N:N}}(x,y) = \frac{N}{(N-2)!}[x-(N-1)y]^{N-2}e^{-x-y}, x \geq (N-1)y, \quad (11.14)$$

$$f_{\beta_i,\alpha_{i:N},\alpha_{i+1:N}}(x,y,z)$$
$$= \frac{N! e^{-x-y-z}(1-e^{-z})^{N-i-1}[x-(i-1)y]^{i-2}}{(i-1)!(i-2)!(N-i-1)!}, \frac{x}{i-1} > y > z. \quad (11.15)$$

$$f_{\alpha_{1:N}}(x) = N e^{-x}(1-e^{-x})^{N-1}, \quad (11.16)$$

$$f_{\alpha_{1:N},\alpha_{2:N}}(x,y) = \frac{N!}{(N-2)!} e^{-x-y}(1-e^{-y})^{N-2}. \quad (11.17)$$

By properly substituting (11.14), (11.1.2), (11.16), and (11.17) into (11.13) and carrying out integration, the closed-form expression of the CDF of harvested energy is obtained as

$$F_{E_h^{(N)}}(x) = \begin{cases} \sum_{i=2}^{N} \frac{N!(1-e^{-\frac{P_{th}}{\bar{P}_R}})^{N-i}e^{-\frac{iP_{th}}{\bar{P}_R}}}{(i-1)!(i-2)!(N-i)!} \sum_{m=0}^{i-2}(1-i)^{i-2-m} C_{i-2}^m \sum_{j=0}^{m} \frac{m!}{(m-j)!} \\ \left\{ (i-1)^{m-j} \sum_{k=0}^{i-2-j} \frac{(i-2-j)!}{(i-2-j-k)! i^{k+1}} \left[\left(\frac{P_{th}}{\bar{P}_R}\right)^{i-2-j-k} \right. \right. \\ \left. - e^{-\frac{x}{\eta T_c \bar{P}_R}} \left(\frac{x}{i\eta T_c \bar{P}_R}+\frac{P_{th}}{\bar{P}_R}\right)^{i-2-j-k} \right] - e^{-\frac{x}{\eta T_c \bar{P}_R}} \sum_{s=0}^{m-j}(-1)^{m-j-s} C_{m-j}^s \\ \left(\frac{x}{\eta T_c \bar{P}_R}+\frac{iP_{th}}{\bar{P}_R}\right)^s \frac{\left(\frac{x}{i\eta T_c \bar{P}_R}+\frac{P_{th}}{\bar{P}_R}\right)^{i-1-j-s}-\left(\frac{P_{th}}{\bar{P}_R}\right)^{i-1-j-s}}{i-1-j-s} \right\} + (1-e^{-\frac{P_{th}}{\bar{P}_R}})^N \\ + N(1-e^{-\frac{P_{th}}{\bar{P}_R}})^{N-1}(e^{-\frac{P_{th}}{\bar{P}_R}}-e^{-\frac{x}{\eta T_c \bar{P}_R}-\frac{P_{th}}{\bar{P}_R}}), \quad x \leq E_c; \\ 1. \quad x > E_c. \end{cases} \quad (11.18)$$

After taking the derivative with respect to x, the PDF of $F_{E_h}(x)$ is derived as

$$f_{E_h^{(N)}}(x) = \begin{cases} \sum_{i=2}^{N} \frac{N!(1-e^{-\frac{P_{th}}{\bar{P}}})^{N-i} e^{-\frac{x}{\eta T_c \bar{P}} - \frac{iP_{th}}{\bar{P}}}}{(i-1)!(i-2)!(N-i)!} \sum_{m=0}^{i-2}(1-i)^{i-2-m} C_{i-2}^m \\ \sum_{j=0}^{m} \frac{m!}{(m-j)!} \left\{ (i-1)^{m-j} \sum_{k=0}^{i-2-j} \frac{(i-2-j)!}{(i-2-j-k)! i^{k+1}} (\frac{x}{i\eta T_c \bar{P}} + \frac{P_{th}}{\bar{P}})^{i-2-j-k-1} \right. \\ (\frac{x}{i\eta^2 T_c^2 \bar{P}^2} + \frac{P_{th}}{\eta T_c \bar{P}^2} - \frac{i-2-j-k}{i\eta T_c \bar{P}}) + \sum_{s=0}^{m-j}(-1)^{m-j-s} C_{m-j}^s \frac{i^s}{i-1-j-s} \\ \{(\frac{x}{i\eta T_c \bar{P}} + \frac{P_{th}}{\bar{P}})^{i-2-j} (\frac{x}{i\eta^2 T_c^2 \bar{P}^2} + \frac{P_{th}}{\eta T_c \bar{P}^2} - \frac{i-1-j}{i\eta T_c \bar{P}}) \\ \left. + (\frac{P_{th}}{\bar{P}})^{i-1-j-s} (\frac{x}{i\eta T_c \bar{P}} + \frac{P_{th}}{\bar{P}})^{s-1} (-\frac{x}{i\eta^2 T_c^2 \bar{P}^2} - \frac{P_{th}}{\eta T_c \bar{P}^2} + \frac{s}{i\eta T_c \bar{P}}) \right\} \right\} \\ + \frac{N}{\eta T_c \bar{P}} (1 - e^{-\frac{P_{th}}{\bar{P}}})^{N-1} e^{-\frac{x}{\eta T_c \bar{P}} - \frac{P_{th}}{\bar{P}}} + (1 - e^{-\frac{P_{th}}{\bar{P}}})^N \delta(x) \\ + [1 - F_{E_h}(E_c)]\delta(x - E_c), \qquad N > 1; \\ \frac{1}{\eta T_c \bar{P}} e^{-\frac{x}{\bar{P}\eta T_c} - \frac{P_{th}}{\bar{P}}} + (1 - e^{-\frac{P_{th}}{\bar{P}}})^N \delta(x) + [1 - F_{E_h}(E_c)]\delta(x - E_c), \quad N = 1, \end{cases}$$

(11.19)

where $\delta(\cdot)$ denotes the impulse function. Note that the PDF involves two impulse functions at 0 and E_c due to the capacity constraints.

Finally, the packet loss probability for delay-sensitive traffic can be calculated by substituting (11.19) into (11.9) and carrying out numerical integration. Note that only finite integration of some basic functions is involved in the calculation.

We assume the same parameters for RF energy harvesting system as in [9]. In particular, the transmission power of the BS is $P_T = 10$ kW. The distance from the BS to the sensor, the BS to the sink, and the sensor to the sink are set as $d_H = 100$ meters, $d_I = 100$ meters, and $d_T = 1$ meter, respectively. The path loss exponent λ is assumed to be 3, the channel coherence time T_c to be 100 ms, and the transmission time of the sensor T_s to be 1 ms. The sensitivity of the sensor is assumed to be $P_{th} = -10$ dBm $= 0.1$ mW [10]. For simplicity, we assume harvesting efficiency $\eta = 1$ and path loss constant $\Gamma = 1$.

In Fig. 11.2, we plot the packet loss probability at the sink as a function of the SINR threshold for different energy capacity E_c with $N = 3$. We can see that when γ_T is small, the packet loss probability shows approximately linear degradation. We also observe that larger energy capacity E_c leads to smaller packet loss probability. However, the benefit of packet loss probability shrinks with the increase of the energy capacity E_c. This is because when E_c gets larger, the sensor has smaller probability to be fully charged, such that the effect of the energy capacity on the packet loss probability gradually reduces.

In Fig. 11.3, we plot the packet loss probability at the sink as a function of the number of the channel coherence time before each packet transmission. We can see that the packet loss probability at the sink gradually reduces as N increases, and converges to a constant value when N is very large. This is due to the existence of energy storage capacity E_c, which limits the total harvested energy and in turn the transmission power. Moreover, we notice that higher SINR threshold leads to higher packet loss probability, as expected by intuition. Note that our analysis results accurately reflect *how* the packet

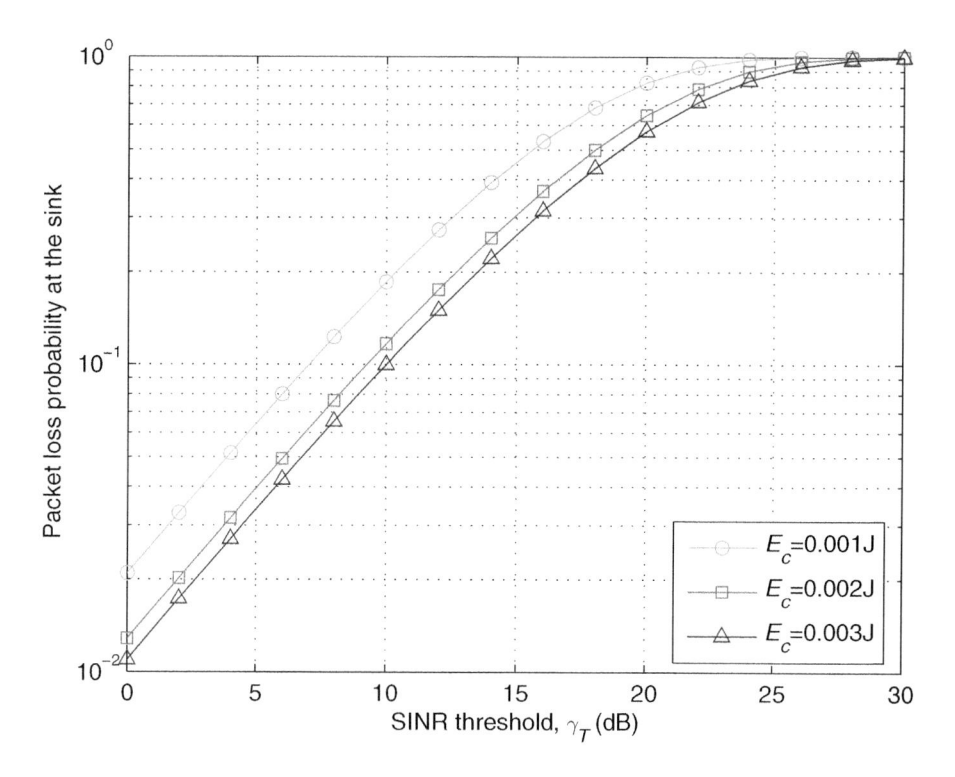

Figure 11.2 Packet loss probability at the sink for different energy storage capacity ($N = 3$). Reprint with permission from [13]. ©2015 IEEE.

loss probability decreases with the increase of energy harvesting time NT_c, which can be used to predict whether the packet loss probability requirement can be satisfied with a given maximum delay requirement NT_c.

11.1.3 Channel-Blind Transmission for Delay-Insensitive Traffic

We now consider the transmission of delay-insensitive traffic with the overlaid sensing implementation with RF energy harvesting. We focus on low-intensity traffic, whose packets need to be delivered to the sink reliably while suffering certain delay. The delay-insensitive traffic applies to the sensing scenarios that require high-level data integrity but are less time-critical. The performance metric of primary interest becomes the average delay for packet delivery, which for the sensor transmission with energy harvesting includes the energy harvesting time as well as the transmission delay.

In this subsection, we adopt a channel-blind transmission strategy in which the sensor will transmit the packet immediately after being fully charged, i.e., harvesting at least E_c amount of energy. If the transmission is not successful, the sensor will restart the energy harvesting and packet transmission process until the packet is successfully received. We analyze the average delay performance for packet delivery with the channel-blind strategy in the following. We first derive the distribution of the time required to fully

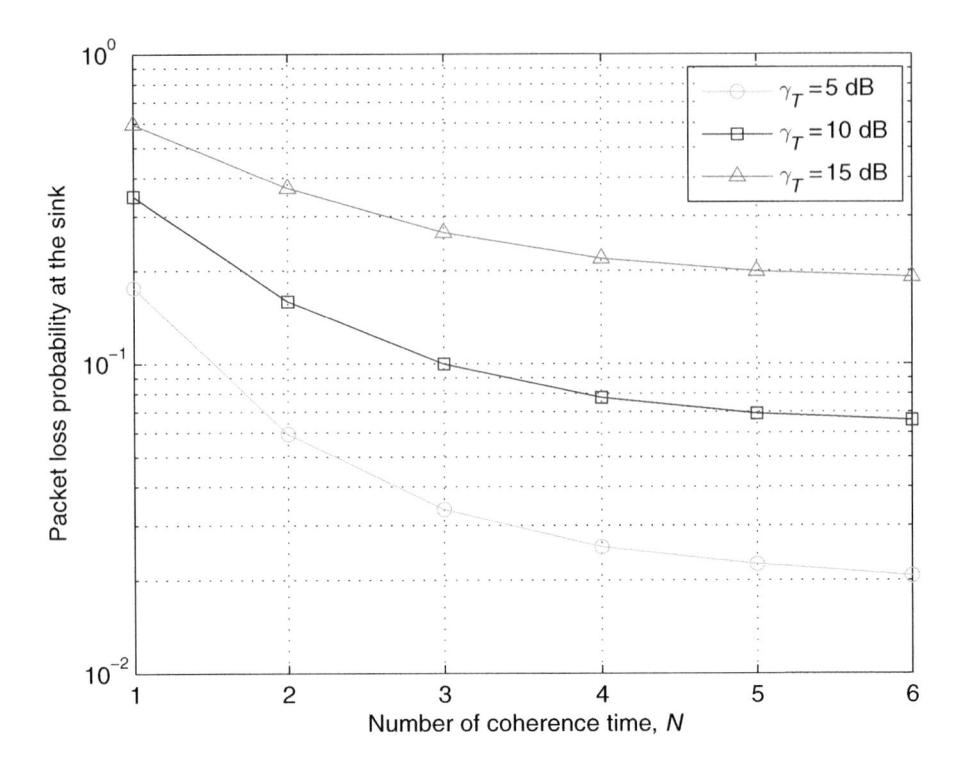

Figure 11.3 Packet loss probability at the sink over N coherence times. Reprint with permission from [13]. ©2015 IEEE.

charge the sensor, based on which we obtain the exact closed-form expression on the average packet transmission delay with retransmission. We also investigate the effect of energy storage capacity E_c on the average packet delivery delay and its optimization. We ignore the queuing delay in the analysis in this work since our focus is on sensing applications with low traffic intensity. The results here can, however, be applied to the queuing delay analysis of high-intensity traffic by adopting certain queuing models.

We first derive the statistics of the charging time, i.e., the number of T_c that it takes to fully charge the sensor. Let $\Pr[T_d = KT_c]$ denote the probability that the sensor can be fully charged in K channel coherence time, which is equal to the probability that the harvested energy during the first $K-1$ coherence time is less than E_c, and that during the first K coherence time larger than E_c.[2] Therefore, we can calculate $\Pr[T_d = KT_c]$ as

$$\Pr[T_d = KT_c] = \Pr[\sum_{j=1}^{K-1} E_j + E_K \geq E_c, \sum_{j=1}^{K-1} E_j < E_c]$$

$$= \int_0^{E_c} \int_{E_c-z}^{E_c} f_{E_h^{(1)}}(z) f_{E_h^{(K-1)}}(x) \, dx dz, \qquad (11.20)$$

---

[2] Random walk theory or Brownian motion cannot be used here, as the distribution of harvested energy over one coherence time is mixed with impulse at 0 and E_c.

11.1 Cognitive Transmission with Harvested RF Energy

where $f_{E_h^{(1)}}(x)$ denotes the PDF of the harvested energy in one coherence time, and $f_{E_h^{(K-1)}}(x)$ denotes the PDF of the harvested RF energy distribution over the first $K-1$ coherence time, both of which can be obtained from (11.19). After carrying out integration and some manipulation, the closed-form expression of $\Pr[T_d = KT_c]$ is given by

$$\Pr[T_d = KT_c] =$$

$$\begin{cases} e^{-\frac{E_c}{P\eta T_c}-\frac{P_{th}}{P}} \sum_{i=2}^{K-1} \frac{(K-1)!(1-e^{-\frac{P_{th}}{P}})^{K-1-i}e^{-\frac{iP_{th}}{P}}}{(i-1)!(i-2)!(K-1-i)!} \sum_{m=0}^{i-2}(1-i)^{i-2-m} \\ C_{i-2}^m \sum_{j=0}^m \frac{m!}{(m-j)!}\left\{(i-1)^{m-j}\sum_{k=0}^{i-2-j}\frac{(i-2-j)!}{(i-2-j-k)!i^{k+1}}\left\{\frac{i}{i-1-j-k}\right.\right. \\ \left[(\frac{E_c}{i\eta T_c \bar{P}} + \frac{P_{th}}{\bar{P}})^{i-1-j-k} - (\frac{P_{th}}{\bar{P}})^{i-1-j-k}\right] - \left[(\frac{E_c}{i\eta T_c \bar{P}} + \frac{P_{th}}{\bar{P}})^{i-2-j-k}\right. \\ \left.\left.-(\frac{P_{th}}{\bar{P}})^{i-2-j-k}\right]\right\} + \sum_{s=0}^{m-j}(-1)^{m-j-s}C_{m-j}^s \frac{i^s}{i-1-j-s}\left\{\frac{i}{i-j}\right. \\ \left[(\frac{E_c}{i\eta T_c \bar{P}} + \frac{P_{th}}{\bar{P}})^{i-j} - (\frac{P_{th}}{\bar{P}})^{i-j}\right] - \left[(\frac{E_c}{i\eta T_c \bar{P}} + \frac{P_{th}}{\bar{P}})^{i-1-j}\right. \\ \left.-(\frac{P_{th}}{\bar{P}})^{i-1-j}\right] - (\frac{P_{th}}{\bar{P}})^{i-1-j-s}\frac{i}{s+1}\left[(\frac{E_c}{i\eta T_c \bar{P}} + \frac{P_{th}}{\bar{P}})^{s+1} - (\frac{P_{th}}{\bar{P}})^{s+1}\right] \\ \left.+(\frac{P_{th}}{\bar{P}})^{i-1-j-s}\left[(\frac{E_c}{i\eta T_c \bar{P}} + \frac{P_{th}}{\bar{P}})^s - (\frac{P_{th}}{\bar{P}})^s\right]\right\}\right\} \\ +\frac{(K-1)E_c}{\eta T_c \bar{P}}(1-e^{-\frac{P_{th}}{\bar{P}}})^{K-2}e^{-\frac{P_{th}}{\bar{P}}} + e^{-\frac{E_c}{P\eta T_c}-\frac{P_{th}}{\bar{P}}}(1-e^{-\frac{P_{th}}{\bar{P}}})^{K-1}, \quad K>2; \\ e^{-\frac{E_c}{P\eta T_c}-\frac{P_{th}}{\bar{P}}}(1-e^{-\frac{P_{th}}{\bar{P}}} + \frac{E_c}{P\eta T_c}e^{-\frac{P_{th}}{\bar{P}}}), \quad K=2. \end{cases}$$
(11.21)

It is easy to show that the probability that the sensor can be fully charged in one coherence time ($K=1$) can be calculated as

$$\Pr[T_d = T_c] = \Pr[E_h^{(1)} \geq E_c] = e^{-\frac{E_c}{\eta T_c \bar{P}} - \frac{P_{th}}{\bar{P}}}. \tag{11.22}$$

Finally, the average number of T_c for fully charging the sensor is calculated as

$$\overline{K} = \sum_{K=1}^{\infty} K\Pr[T_d = KT_c] = \frac{e^{\frac{P_{th}}{\bar{P}}}}{\eta T_c \bar{P}} E_c + e^{\frac{P_{th}}{\bar{P}}}, \tag{11.23}$$

which is a linear function of the capacity of the sensor, as expected by intuition.

The sensor node will start packet transmission after being fully charged. We again assume that for the sink to successfully decode a packet, the received SINR should be above a threshold γ_T. The packet error probability at the sink is mathematically given by

$$P_E = \Pr[\gamma_s < \gamma_T] = \Pr\left[\frac{\frac{E_c}{T_s d_T^\lambda}\|g_s\|^2}{\frac{P_T}{d_I^\lambda}\|h_s\|^2 + \Gamma\sigma^2} < \gamma_T\right], \tag{11.24}$$

where $\frac{E_c}{T_s}$ is the sensor transmission power. The closed-form expression of P_E can be obtained using the PDFs of $||h_s||^2$ and $||g_s||^2$, given in (11.8), as

$$P_E = \int_0^\infty \int_0^{T_s\gamma_T d_T^\lambda(\frac{P_T y}{d_I^\lambda}+\Gamma\sigma^2)/E_c} f_{||h_s||^2}(x) f_{||g_s||^2}(y) \, dx \, dy$$

$$= 1 - \frac{E_c e^{-\frac{\Gamma\sigma^2 T_s \gamma_T d_T^\lambda}{E_c}}}{E_c + \frac{P_T}{d_I^\lambda} T_s \gamma_T d_T^\lambda}. \quad (11.25)$$

If the received SINR at the sink is below γ_T, the sensor will repeat the energy harvesting and packet transmission until the packet is decoded correctly. Consequently, the average delay for successfully transmitting a packet to the sink can be calculated as

$$\overline{T}_{CB} = (\overline{K}T_c + T_s) \sum_{i=1}^\infty i P_E^{i-1}(1 - P_E) = \frac{\overline{K}T_c + T_s}{1 - P_E}, \quad (11.26)$$

where the term $\overline{K}T_c + T_s$ represents the average time duration for one charging and packet transmission cycle and $P_E^{i-1}(1 - P_E)$ denotes the probability that the packet can be successfully delivered after the ith transmission. Then, by substituting (11.5) and (11.23) into (11.26), the closed-form expression of the overall average delay for successful packet transmission can be given by

$$\overline{T}_{CB} = \left[\left(\frac{E_c}{\eta\overline{P}} + T_c\right) e^{\frac{P_{th}}{\overline{P}}} + T_s\right]\left(1 + \frac{P_T T_s \gamma_T d_T^\lambda}{d_I^\lambda E_c}\right) e^{\frac{\Gamma\sigma^2 \gamma_T T_s d_T^\lambda}{E_c}}. \quad (11.27)$$

Intuitively, a larger energy storage capacity of the sensor will lead to longer time required for fully charging the sensor, but smaller probability for transmission failure, which will help reduce the average number of retransmissions for packet delivery. Thus, we are interested in finding an optimal energy capacity of the sensor, denoted by E_{CB}^*, to minimize $\overline{T}_{CB}$. By calculating the derivative of $\overline{T}_{CB}$ with respect to E_c and setting it to zero, we have

$$E_c^3 - \Gamma\sigma^2 T_s \gamma_T d_T^\lambda E_c^2 - T_s \gamma_T d_T^\lambda \left[\eta\overline{P}(\frac{P_T}{d_I^\lambda} + \Gamma\sigma^2)(T_c + T_s e^{-\frac{P_{th}}{\overline{P}}})\right.$$

$$\left. + \frac{P_T \Gamma\sigma^2 T_s \gamma_T d_T^\lambda}{d_I^\lambda}\right] E_c - \eta\overline{P}P_T \Gamma\sigma^2 T_s^2 \gamma_T^2 d_T^{2\lambda}(T_c + T_s e^{-\frac{P_{th}}{\overline{P}}}) \frac{P_T}{d_I^\lambda} = 0, \quad (11.28)$$

which is the homogeneous equation of E_c with order 3. The optimal energy capacity E_{CB}^* is the solution of (11.28) that satisfies $\frac{dT_p(E_c^-)}{dE_c} \leq 0$, $\frac{dT_p(E_c^+)}{dE_c} \geq 0$, and $E_c > 0$. We found that for most practical values of system parameters, there exists only one unique solution. One can easily find the ultimate minimum if more than one solution exists, given the simplicity of the objective function.

Since in general the packet transmission time T_s is much smaller than the channel coherence time T_c for the low-rate sensing application targeted here, (11.28) can be further simplified by omitting the terms involving $\mathcal{O}(T_s^2)$ as

11.1 Cognitive Transmission with Harvested RF Energy

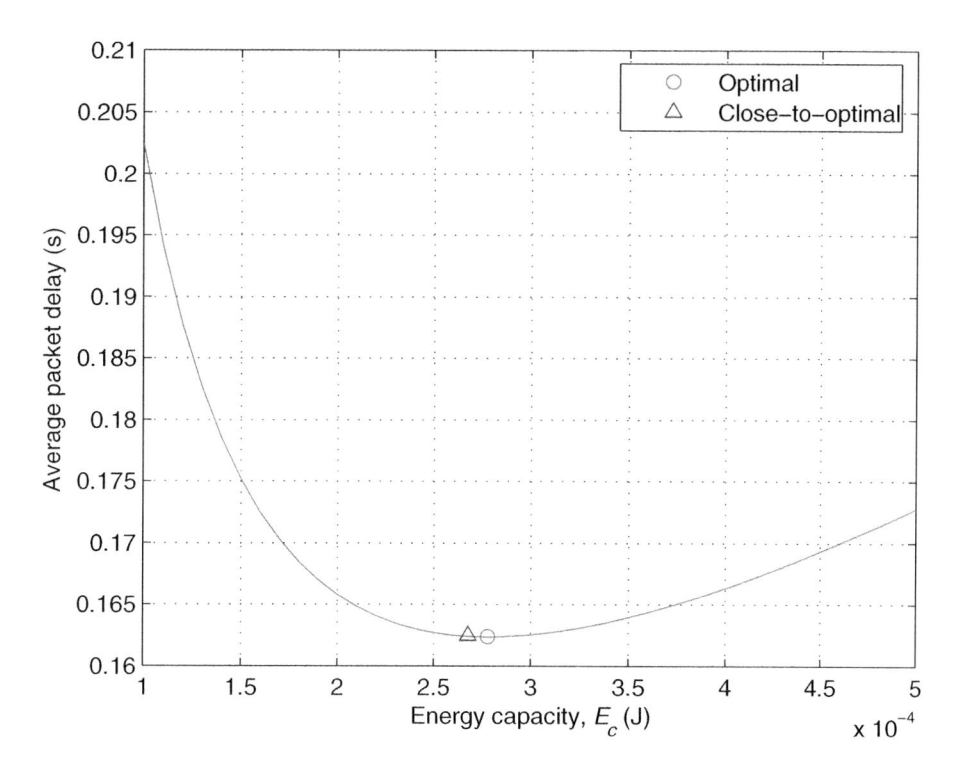

Figure 11.4 Average packet delay versus energy capacity E_c for the channel-blind strategy for delay-insensitive traffic. Reprint with permission from [13]. ©2015 IEEE.

$$E_c^2 - \Gamma\sigma^2 T_s \gamma_T d_T^\lambda E_c - \eta \overline{P} T_c T_s \gamma_T d_T^\lambda \left(\frac{P_T}{d_I^\lambda} + \Gamma\sigma^2 \right) = 0. \tag{11.29}$$

Then the close-to-optimal energy capacity to minimize the average transmission delay can be calculated as

$$E_{CB}^* \approx \frac{\Gamma\sigma^2 T_s \gamma_T d_T^\lambda + \sqrt{(\Gamma\sigma^2 T_s \gamma_T d_T^\lambda)^2 + 4\eta \overline{P} T_c T_s \gamma_T d_T^\lambda (\frac{P_T}{d_I^\lambda} + \Gamma\sigma^2)}}{2}. \tag{11.30}$$

In Fig. 11.4, we plot the average packet transmission delay for the channel-blind strategy as a function of the sensor energy storage E_c with $\gamma_T = 20$ dB. It is shown that with the increase of E_c, the packet delay first quickly reduces, and then gradually increases. The reason is that, when E_c is small, the sensor can be fully charged quickly, but the sink cannot decode the packet due to low received SINR. The sensor has to retransmit the packet multiple times, leading to large average delay. When E_c increases, the delay for packet retransmission reduces as the transmission typically enjoys high SINR, but the sensor now needs more time to harvest enough energy. We also mark the approximate and exact values of the optimal energy capacity E_{CB}^*. We observe that the near-to-optimal solution achieves almost the same average delay performance as the exact optimal energy capacity value.

11.1.4 Channel-Aware Transmission for Delay-Insensitive Traffic

In this section, we consider an alternative transmission strategy for delay-insensitive traffic and evaluate its delay performance. Note that with the channel-blind strategy, the sensor will transmit its packet immediately after harvesting enough energy, which may not be successful due to poor signal channel or strong interference in the coming channel coherence time, and thus leads to the waste of harvested RF energy. To better utilize the harvested energy, we consider a channel-aware transmission strategy in this section. Specifically, we again assume that the sensor will harvest RF energy from the BS until it is fully charged. After that, instead of transmitting its packet immediately, the sensor will decide whether to transmit or not according to the instantaneous received SINR estimation at the sink. A packet will be transmitted only if the experienced SINR at the sink is larger than γ_T, which will lead to successful delivery. If the SINR at the sink is below γ_T, then the sensor will hold its transmission for future channel coherence time until the SINR is above γ_T. We assume the sink can precisely estimate the signal and interference channel, and feed back the SINR to the sensor at the beginning of each channel coherence time.

We again focus on the average packet delay performance for packet delivery and would like to quantify the benefit, if any, that the channel information can bring in terms of delay reduction. Note that total packet transmission delay becomes the sum of the sensor charging time, the waiting time for packet transmission, and the packet transmission time T_s.

We first derive the statistics of the waiting time for packet transmission, i.e., the number of channel coherence times T_c that the sensor needs to wait before the SINR at the sink is larger than γ_T and packet transmission can start. Let $\Pr[T_w = MT_c]$ denote the probability that the sensor needs to wait M channel coherence times before packet transmission, which is equal to the probability that the SINR at the sink during each of the first M coherence times is less than γ_T, and that during the $(M+1)$th coherence time larger than γ_T. Therefore, we can mathematically calculate $\Pr[T_w = MT_c]$ as

$$\Pr[T_w = MT_c] = P_E^M (1 - P_E)$$

$$= \left(1 - \frac{E_c e^{-\frac{\Gamma \sigma^2 T_s \gamma_T d_T^\lambda}{E_c}}}{E_c + \frac{P_T}{d_I^\lambda} T_s \gamma_T d_T^\lambda}\right)^M \left(\frac{E_c e^{-\frac{\Gamma \sigma^2 T_s \gamma_T d_T^\lambda}{E_c}}}{E_c + \frac{P_T}{d_I^\lambda} T_s \gamma_T d_T^\lambda}\right), \quad (11.31)$$

where P_E denotes the probability that the instantaneous received SINR at the sink is below γ_T, which has been calculated in (11.25).

The distribution of the total delay for packet transmission can be given by

$$\Pr[T_{CA} = LT_c + T_s] = \sum_{K=1}^{L} \Pr[T_d = KT_c] \Pr[T_w = (L-K)T_c], \quad (11.32)$$

where $\Pr[T_d = KT_c]$ denotes the probability that K channel coherence time is needed to fully charge the sensor, which has been derived in (11.21) and (11.22). After substituting (11.21), (11.22), and (11.31) into (11.32) and some manipulations, the closed-form

11.1 Cognitive Transmission with Harvested RF Energy

expression for the probability mass function (PMF) of the total packet transmission delay with channel-aware transmission strategy can be calculated as

$$\Pr[T_{CA} = LT_c + T_s] =$$

$$\begin{cases}
e^{-\frac{E_c}{\bar{P}\eta T_c} - \frac{P_{th}}{\bar{P}}} \left(\frac{E_c e^{-\frac{\Gamma \sigma^2 T_s \gamma_T d_T^\lambda}{E_c}}}{E_c + \frac{P_T}{d_I^\lambda} T_s \gamma_T d_T^\lambda} \right), & L = 1; \\[1em]
e^{-\frac{E_c}{\bar{P}\eta T_c} - \frac{P_{th}}{\bar{P}}} (1 - e^{-\frac{P_{th}}{\bar{P}}} + \frac{E_c}{\bar{P}\eta T_c} e^{-\frac{P_{th}}{\bar{P}}}) \left(\frac{E_c e^{-\frac{\Gamma \sigma^2 T_s \gamma_T d_T^\lambda}{E_c}}}{E_c + \frac{P_T}{d_I^\lambda} T_s \gamma_T d_T^\lambda} \right) \\[1em]
+ e^{-\frac{E_c}{\bar{P}\eta T_c} - \frac{P_{th}}{\bar{P}}} (1 - \frac{E_c e^{-\frac{\Gamma \sigma^2 T_s \gamma_T d_T^\lambda}{E_c}}}{E_c + \frac{P_T}{d_I^\lambda} T_s \gamma_T d_T^\lambda}) \left(\frac{E_c e^{-\frac{\Gamma \sigma^2 T_s \gamma_T d_T^\lambda}{E_c}}}{E_c + \frac{P_T}{d_I^\lambda} T_s \gamma_T d_T^\lambda} \right), & L = 2; \\[1em]
\sum_{K=3}^{L} \left\{ e^{-\frac{E_c}{\bar{P}\eta T_c} - \frac{P_{th}}{\bar{P}}} \sum_{i=2}^{K-1} \frac{(K-1)!(1 - e^{-\frac{P_{th}}{\bar{P}}})^{K-1-i} e^{-\frac{i P_{th}}{\bar{P}}}}{(i-1)!(i-2)!(K-1-i)!} \right. \\[1em]
\sum_{m=0}^{i-2} (1-i)^{i-2-m} C_{i-2}^m \sum_{j=0}^{m} \frac{m!}{(m-j)!} \left\{ (i-1)^{m-j} \sum_{k=0}^{i-2-j} \right. \\[1em]
\frac{(i-2-j)!}{(i-2-j-k)! i^{k+1}} \left\{ \frac{i}{i-1-j-k} \left[(\frac{E_c}{i \eta T_c \bar{P}} + \frac{P_{th}}{\bar{P}})^{i-1-j-k} - (\frac{P_{th}}{\bar{P}})^{i-1-j-k} \right] \right. \\[1em]
\left. - \left[(\frac{E_c}{i \eta T_c \bar{P}} + \frac{P_{th}}{\bar{P}})^{i-2-j-k} - (\frac{P_{th}}{\bar{P}})^{i-2-j-k} \right] \right\} \\[1em]
+ \sum_{s=0}^{m-j} (-1)^{m-j-s} C_{m-j}^s \frac{i^s}{i-1-j-s} \left\{ \frac{i}{i-j} \left[(\frac{E_c}{i \eta T_c \bar{P}} + \frac{P_{th}}{\bar{P}})^{i-j} - (\frac{P_{th}}{\bar{P}})^{i-j} \right] \right. \\[1em]
\left. - \left[(\frac{E_c}{i \eta T_c \bar{P}} + \frac{P_{th}}{\bar{P}})^{i-1-j} - (\frac{P_{th}}{\bar{P}})^{i-1-j} \right] \right. \\[1em]
- (\frac{P_{th}}{\bar{P}})^{i-1-j-s} \frac{i}{s+1} \left[(\frac{E_c}{i \eta T_c \bar{P}} + \frac{P_{th}}{\bar{P}})^{s+1} - (\frac{P_{th}}{\bar{P}})^{s+1} \right] \\[1em]
\left. + (\frac{P_{th}}{\bar{P}})^{i-1-j-s} \left[(\frac{E_c}{i \eta T_c \bar{P}} + \frac{P_{th}}{\bar{P}})^s - (\frac{P_{th}}{\bar{P}})^s \right] \right\} \right\} \\[1em]
+ \frac{(K-1) E_c}{\eta T_c \bar{P}} (1 - e^{-\frac{P_{th}}{\bar{P}}})^{K-2} e^{-\frac{P_{th}}{\bar{P}}} + (1 - e^{-\frac{P_{th}}{\bar{P}}})^{K-1} \right\} \\[1em]
(1 - \frac{E_c e^{-\frac{\Gamma \sigma^2 T_s \gamma_T d_T^\lambda}{E_c}}}{E_c + \frac{P_T}{d_I^\lambda} T_s \gamma_T d_T^\lambda})^{L-K} (\frac{E_c e^{-\frac{\Gamma \sigma^2 T_s \gamma_T d_T^\lambda}{E_c}}}{E_c + \frac{P_T}{d_I^\lambda} T_s \gamma_T d_T^\lambda}) \\[1em]
+ e^{-\frac{E_c}{\bar{P}\eta T_c} - \frac{P_{th}}{\bar{P}}} (1 - e^{-\frac{P_{th}}{\bar{P}}} + \frac{E_c}{\bar{P}\eta T_c} e^{-\frac{P_{th}}{\bar{P}}}) \\[1em]
(1 - \frac{E_c e^{-\frac{\Gamma \sigma^2 T_s \gamma_T d_T^\lambda}{E_c}}}{E_c + \frac{P_T}{d_I^\lambda} T_s \gamma_T d_T^\lambda})^{L-2} (\frac{E_c e^{-\frac{\Gamma \sigma^2 T_s \gamma_T d_T^\lambda}{E_c}}}{E_c + \frac{P_T}{d_I^\lambda} T_s \gamma_T d_T^\lambda}) \\[1em]
+ e^{-\frac{E_c}{\bar{P}\eta T_c} - \frac{P_{th}}{\bar{P}}} (1 - \frac{E_c e^{-\frac{\Gamma \sigma^2 T_s \gamma_T d_T^\lambda}{E_c}}}{E_c + \frac{P_T}{d_I^\lambda} T_s \gamma_T d_T^\lambda})^{L-1} (\frac{E_c e^{-\frac{\Gamma \sigma^2 T_s \gamma_T d_T^\lambda}{E_c}}}{E_c + \frac{P_T}{d_I^\lambda} T_s \gamma_T d_T^\lambda}), & L > 2.
\end{cases}$$

(11.33)

With the PMF of T_{CA}, we can readily calculate the average packet transmission delay, $\overline{T}_{CA}$. Alternatively, we can first calculate the average number of T_c that the sensor needs to wait for packet transmission using (11.31) as

$$\overline{M} = \sum_{M=0}^{\infty} M \Pr[T_w = MT_c] = \frac{P_E}{1-P_E}$$
$$= \left(1 + \frac{P_T T_s \gamma_T d_T^\lambda}{E_c d_I^\lambda}\right) e^{\frac{\sigma^2 \gamma_T T_s \Gamma d_T^\lambda}{E_c}} - 1. \quad (11.34)$$

Finally, the average delay for successfully transmitting a packet to the sink can be calculated by combining the results in (11.23) and (11.34) as

$$\overline{T}_{CA} = (\overline{K} + \overline{M})T_c + T_s$$
$$= \left[\frac{e^{\frac{P_{th}}{P}}}{\overline{P}\eta T_c}E_c + e^{\frac{P_{th}}{P}}\right.$$
$$\left. + \left(1 + \frac{P_T T_s \gamma_T d_T^\lambda}{E_c d_I^\lambda}\right) e^{\frac{\Gamma\sigma^2 T_s \gamma_T d_T^\lambda}{E_c}} - 1\right]T_c + T_s. \quad (11.35)$$

Intuitively, larger energy storage capacity at the sensor E_c will lead to longer charging time, but smaller waiting time until the received SINR at the sink becomes larger than γ_T. Thus, we expect there is also an optimal value of sensor energy capacity to minimize $\overline{T}_{CA}$. By calculating the derivative of $\overline{T}_{CA}$ with respect to E_c and setting it to zero, we have the equation

$$E_c^3 - \eta \overline{P} T_c \left[T_s \gamma_T d_T^\lambda \left(\frac{P_T}{d_I^\lambda} + \Gamma\sigma^2\right)E_c \right.$$
$$\left. + \frac{\Gamma\sigma^2 P_T (T_s \gamma_T d_T^\lambda)^2}{d_I^\lambda}\right] e^{\frac{\sigma^2 \gamma_T T_s \Gamma d_T^\lambda}{E_c} - \frac{P_{th}}{P}} = 0. \quad (11.36)$$

To proceed further, we apply a Taylor series expansion to the exponential term around T_s, which is typically very small, as

$$e^{\frac{\Gamma\sigma^2 \gamma_T d_T^\lambda}{E_c}T_s} = 1 + \frac{\Gamma\sigma^2 \gamma_T d_T^\lambda}{E_c}T_s + \mathcal{O}(T_s^2). \quad (11.37)$$

By substituting (11.37) into (11.36) and omitting the terms involving $\mathcal{O}(T_s^2)$, we have

$$E_c^2 - \eta \overline{P} T_c T_s \gamma_T d_T^\lambda \left(\frac{P_T}{d_I^\lambda} + \Gamma\sigma^2\right) e^{-\frac{P_{th}}{P}} = 0. \quad (11.38)$$

Then the close-to-optimal energy capacity to minimize the average transmission delay, denoted by E_{CA}^*, can be calculated as

$$E_{CA}^* \approx \sqrt{\eta \overline{P} T_c T_s \gamma_T d_T^\lambda \left(\frac{P_T}{d_I^\lambda} + \Gamma\sigma^2\right) e^{-\frac{P_{th}}{P}}}. \quad (11.39)$$

In Fig. 11.5, we plot the average packet delay as a function of the energy storage capacity of the sensor for both channel-aware and channel-blind transmission strategies. It is observed that with the increase of E_c, the packet delay for the channel-aware strategy first quickly reduces, and then gradually increases in a close-to-linear fashion. The reason is that, when E_c is small, the sensor can be fully charged quickly, but has to wait

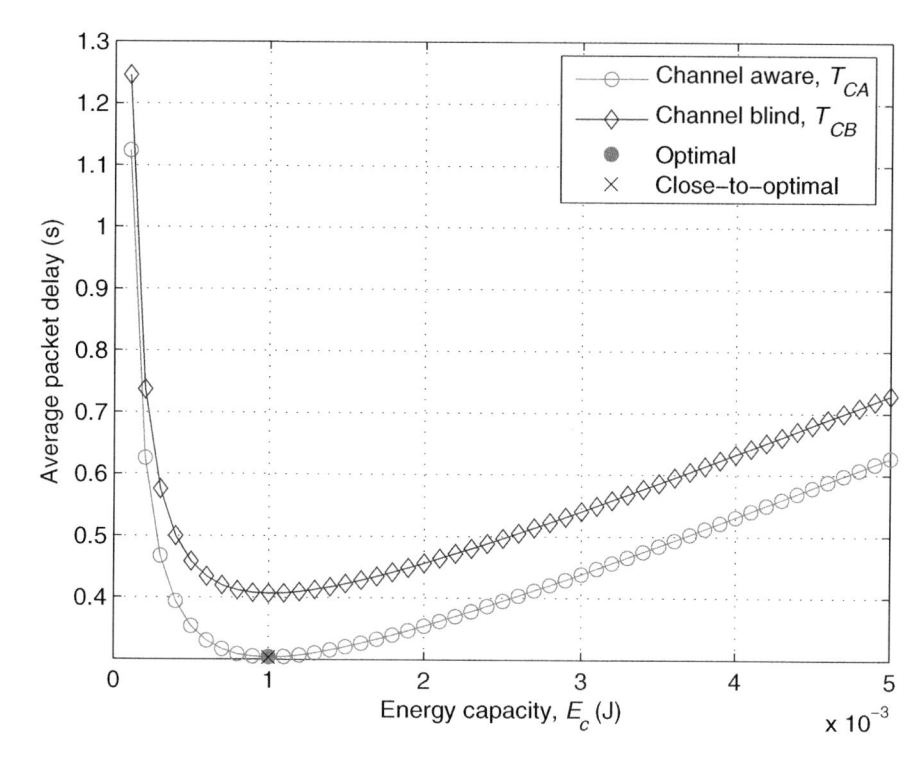

Figure 11.5 Average packet delay for two transmission strategies for delay-insensitive traffic. Reprint with permission from [13]. ©2015 IEEE.

for a long time until the received SINR at the sink is larger than γ_T, leading to large average total delay. When E_c increases, the delay of waiting for packet transmission reduces, but the sensor needs more time to become fully charged. We mark the approximated values of the optimal energy capacity E_{CA}^*. The near-to-optimal solution achieves almost the same average delay performance as the exact value of optimal energy capacity. Fig. 11.5 also shows that the average packet delay of the channel-aware strategy is much smaller than that of the channel-blind strategy. The instantaneous SINR obtained at the beginning of each coherence time not only helps avoid wasted packet transmission and the associated recharging, but also reduces the average total packet transmission delay. This observation is further confirmed in Fig. 11.6, where we plot the average packet delay as a function of the SINR threshold γ_T for both channel-aware and channel-blind strategies. It is observed that with the increase of γ_T, the packet transmission delay for both strategies increase quickly, and the packet delay for the channel-blind strategy grows faster than that for the channel-aware strategy.

11.2 Cooperative Beam Selection for RF Energy Harvesting

A wireless transmitter can serve as both a data source for its own users and an energy source for RF-energy-powered sensor nodes [12]. Inspired by [12], we consider a

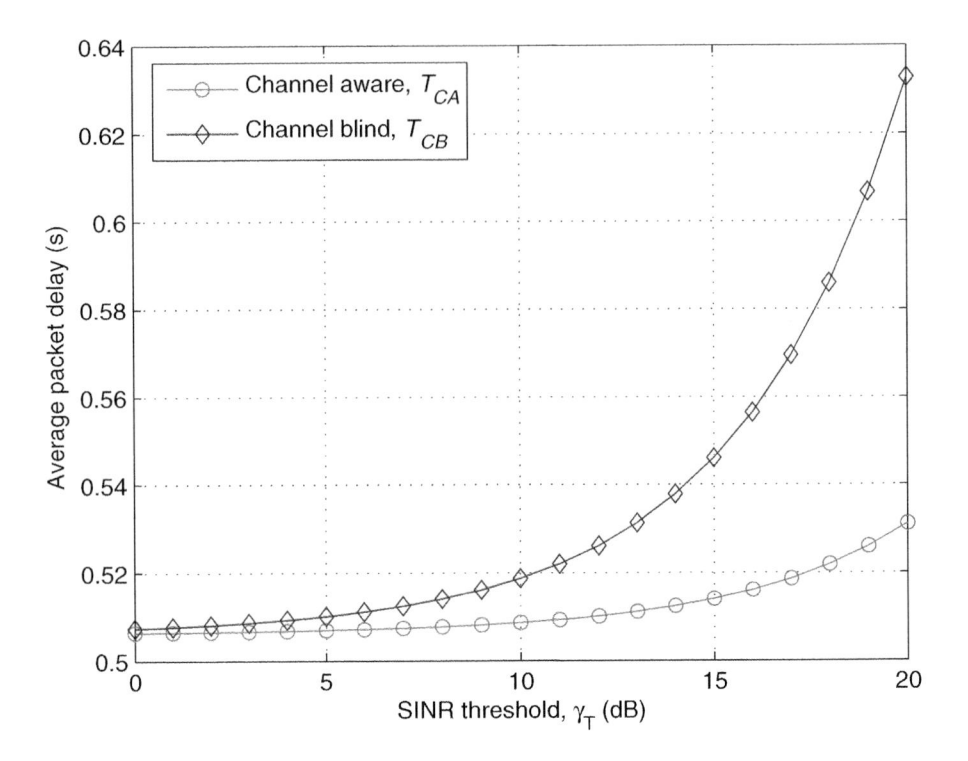

Figure 11.6 Average packet delay versus packet loss threshold γ_T for two transmission strategies for delay-insensitive traffic. Reprint with permission from [13]. ©2015 IEEE.

practical cooperative charging scenario in which an existing multiuser MIMO system helps the energy harvesting of an RF-energy-powered sensor node, while simultaneously serving its own users. We adopt random unitary beamforming (RUB) as the transmission scheme for multiuser MIMO systems, which requires very low feedback load and has been incorporated in several wireless standards [17, 18]. We study a cooperative beam selection scheme in which the BS of the existing system selects the best beams for transmission, while trying to satisfy the energy harvesting requirements of the sensor, i.e., the harvested energy over each coherence time is above a predefined threshold. Specifically, for the single sensor case, the BS of the MISO system selects the best beam for transmission. The number of usable beams that the BS can select from to serve its user is reduced. We derive the statistical distribution of the amount of energy that can be harvested at the sensor, and throughput of the existing MISO system. For the multiple sensors case, the BS of the multiuser MIMO system selects a maximal number of active beams for transmission, while trying to satisfy the energy harvesting requirements. With a constant total transmission power, the BS can enhance energy harvesting at the sensor by concentrating the transmission power on selected beams. Meanwhile, the number of users that the BS can serve simultaneously is reduced. To evaluate the trade-off between the average harvested energy at the sensor and the throughput performance of existing multiuser MIMO system, we derive the statistical distribution of the amount of energy that can be harvested, as well as the sum rate of the existing MIMO system with the proposed cooperative RF energy harvesting scheme. These analytical results will help determine the optimal energy threshold value that can satisfy the requirements of certain sensing applications, while considering the negative effect on the multiuser MIMO system.

11.2.1 System and Channel Model

We consider a single-antenna wireless sensor node deployed in the coverage area of an existing RUB-based multiuser MIMO system. The sensor, which can also be a special user of a multiuser MIMO system, can harvest RF energy from the transmitted signal of the multiuser MIMO system, and use it as its sole energy source. The multiuser MIMO system consists of a single BS with M antennas and K single-antenna users. The BS can serve up to M selected users simultaneously using random orthonormal beams generated from an isotropic distribution. Let $\mathcal{W} = [\mathbf{w}_1, \mathbf{w}_2, \ldots, \mathbf{w}_M]^T$ denote the set of beam vectors, assumed to be known to both the BS and its users. The transmitted signal vector from M antennas over one symbol period can be written, assuming m beams are active, as $\mathbf{x} = \sum_{j=1}^{m} \sqrt{\frac{P_T}{m}} \mathbf{w}_j s_j$, where s_j denotes the information symbol for the jth selected user. Here, we assume that the transmission power P_T is constant and equally allocated to different active beams.

We adopt a log-distance path loss plus Rayleigh block slow fading channel model for the operating environment while ignoring the shadowing effect [15]. In particular, the channel gain between the BS and the sensor remains constant over one channel coherence time, denoted by T_c, and changes to an independent value afterward. Let $\mathbf{h}_e = [h_{e_1}, h_{e_2}, \ldots, h_{e_M}]^T$ denote the fading channel gain vector from the BS to the sensor, where $h_{e_m} \in \mathcal{CN}(0, 1)$. Then the harvested energy at the sensor from the ith beam, when m beams are active, can be given by

$$E_i = \left(\frac{\eta T_c}{\Gamma d_H^\lambda}\right)\left(\frac{P_T}{m}\right) |\mathbf{h}_e^T \mathbf{w}_i|^2, \ i = 1, 2, \ldots, m, \quad (11.40)$$

where d_H is the distance from the BS to the sensor, η is the energy harvesting efficiency, λ is the path loss exponent, ranging from 2 to 5, and Γ is a constant parameter of the log-distance model. Specifically, $\Gamma = \frac{PL(d_0)}{d_0^\lambda}$, where d_0 is a reference distance in the antenna far field, and $PL(d_0)$ is the linear path loss at distance d_0, depending on the propagation environment. For notational conciseness, we use α_m to denote the amplitude square of the projection of $\mathbf{h}_e$ onto $\mathbf{w}_m$, i.e., $\alpha_m = |\mathbf{h}_e^T \mathbf{w}_m|^2$, whose PDF for the Rayleigh fading channel under consideration is given by

$$f_{\alpha_m}(x) = e^{-x}. \quad (11.41)$$

11.2.2 Cooperative Energy Harvesting with Single Beam Selection

With the proposed cooperative energy harvesting scheme, the BS will select the best beam to serve the user, while ensuring that the harvested energy at the sensor node during each coherence time is above a predefined energy threshold E_{th}.

At the beginning of each channel coherence time, the BS first estimates the channel vector from the BS to the sensor. The BS then calculates and ranks the projection amplitude square α_m for each beam, the order version of which is denoted by $\alpha_{m:M}$, where $\alpha_{1:M} \geq \alpha_{2:M} \geq \ldots \geq \alpha_{M:M}$. After that, the BS calculates the amount of RF energy that the sensor can harvest when the BS uses each beam, corresponding to $\alpha_{1:M}$ to $\alpha_{M:M}$. Specifically, the harvested energy denoted by $E_{i:M}$, when the ith best beam is used for transmission, is given by

$$E_{i:M} = \left(\frac{\eta P_T T_c}{\Gamma d_H^\lambda}\right)\alpha_{i:M}, \; i = 1, 2, \ldots, M. \quad (11.42)$$

If the harvested energy from the ith best beam is larger than the predefined energy threshold E_{th}, whereas the harvested energy from the $(i + 1)$th best beam is less than E_{th}, i.e., $E_{i:M} \geq E_{th}$, and $E_{i+1:M} < E_{th}$, then the BS selects one beam from the best i beams, corresponding to $\alpha_{1:M}$ to $\alpha_{i:M}$, to serve its user. It is worth noting that the amount of harvested energy at the sensor may be smaller than E_{th} even when the BS allocates the best beam j^* corresponding to $\alpha_{1:M}$, i.e., $j^* = \arg\max_j(|\mathbf{h}_e^T \mathbf{w}_j|^2)$. In this case, the BS will still use beam j^* with transmission power P_T to charge the sensor as well as serve its user.

Distribution of the Number of Usable Beams

In the following, we derive the PMF of the number of usable beams M_a ($1 \leq M_a \leq M$) that the BS can use, which will be applied to the throughput analysis for the MISO system.

According to our proposed cooperative beam selection scheme, the number of usable beams M_a is equal to m ($1 < m < M$) if and only if $E_{m:M} \geq E_{th}$, and $E_{m+1:M} < E_{th}$. Furthermore, the number of usable beams M_a is equal to 1 if the energy threshold cannot be satisfied with all transmission power P_T allocated to the best beam, i.e., $E_{1:M} < E_{th}$, or if only the best beam can lead to harvest energy larger than E_{th}, i.e., $E_{1:M} \geq E_{th}$, and $E_{2:M} < E_{th}$. The number of usable beams M_a is equal to M if the harvested energy is larger than E_{th} with P_T allocated to the worst beam, i.e., $E_{M:M} \geq E_{th}$. Therefore, the probability that M_a beams are usable can be given by

$$\Pr[M_a = i] = \begin{cases} \Pr[E_{1:M} < E_{th}] + \Pr[E_{1:M} \geq E_{th}, E_{2:M} < E_{th}], & i = 1, \\ \Pr[E_{i:M} \geq E_{th}, E_{i+1:M} < E_{th}], & 1 < i < M, \\ \Pr[E_{M:M} \geq E_{th}], & i = M. \end{cases} \quad (11.43)$$

After substituting (11.42) into (11.43) and some manipulations, (11.43) can be rewritten as

$$\Pr[M_a = i] = \begin{cases} \int_0^{\frac{E_{th}}{\Lambda}} f_{\alpha_{1:M}}(x)dx + \int_0^{\frac{E_{th}}{\Lambda}} \int_{\frac{E_{th}}{\Lambda}}^{\infty} f_{\alpha_{1:M},\alpha_{2:M}}(x,y)dxdy, & i = 1, \\ \int_0^{\frac{E_{th}}{\Lambda}} \int_{\frac{E_{th}}{\Lambda}}^{\infty} f_{\alpha_{i:M},\alpha_{i+1:M}}(x,y)dxdy, & 1 < i < M, \\ \int_{\frac{E_{th}}{\Lambda}}^{\infty} f_{\alpha_{M:M}}(x)dx, & i = M, \end{cases} \quad (11.44)$$

where Λ is a constant parameter equal to $\frac{\eta P_T T_c}{\Gamma d_H^\lambda}$, the PDF of $\alpha_{1:M}$, and $\alpha_{M:M}$, and the joint PDF of $\alpha_{i:M}$ and $\alpha_{i+1:M}$, can be given by [16]

$$f_{\alpha_{1:M}}(x) = M(1-e^{-x})^{M-1}e^{-x}, \tag{11.45}$$

$$f_{\alpha_{i:M},\alpha_{i+1:M}}(x,y) = \frac{M!e^{-ix-y}(1-e^{-y})^{M-i-1}}{(i-1)!(M-i-1)!}, \quad x > y, \tag{11.46}$$

and

$$f_{\alpha_{M:M}}(x) = Me^{-Mx}, \tag{11.47}$$

respectively. By substituting (11.45), (11.46), and (11.47) into (11.44) and carrying out integration, the closed-form expression of $\Pr[M_a = i]$ is calculated as

$$\Pr[M_a = i] = \begin{cases} M\sum_{j=0}^{M-1}\binom{M-1}{j}\frac{(-1)^j}{j+1}(1-e^{-(j+1)\frac{E_{th}}{\Lambda}}) \\ +\frac{M!e^{-\frac{E_{th}}{\Lambda}}}{(M-2)!}\sum_{j=0}^{M-2}\binom{M-2}{j}\frac{(-1)^j}{j+1}(1-e^{-(j+1)\frac{E_{th}}{\Lambda}}), & i=1, \\ \frac{M!e^{-i\frac{E_{th}}{\Lambda}}}{i!(M-i-1)!}\sum_{j=0}^{M-i-1}(-1)^j\binom{M-i-1}{j}\frac{1}{j+1}(1-e^{-(j+1)\frac{E_{th}}{\Lambda}}), & 1<i<M, \\ e^{-M\frac{E_{th}}{\Lambda}} & i=M, \end{cases} \tag{11.48}$$

Throughput Performance Analysis

We are interested in the average throughput of the MISO system, which can be calculated as

$$R = \sum_{i=1}^{M} \Pr[M_a = i] R_i, \tag{11.49}$$

where $\Pr[M_a = i]$ denotes the probability that i beams are usable, given in (11.48), R_i is the average throughput when i beams are usable, which can be calculated using the distribution of the largest SNR among all usable beams, as

$$R_i = \int_0^\infty \log_2(1+x) f_{\gamma_{1:M_a}}(x) dx, \quad i=1,2,\ldots,M, \tag{11.50}$$

where $f_{\gamma_{1:M_a}}(x)$ is the PDF of the largest received SNR $\gamma_{1:M_a}$ at the user, given by [16]

$$f_{\gamma_{1:M_a}}(x) = \frac{M_a}{\bar{\gamma}} e^{-\frac{x}{\bar{\gamma}}} (1 - e^{-\frac{x}{\bar{\gamma}}})^{M_a-1}, \tag{11.51}$$

where $\bar{\gamma}$ denotes the common average received SNR for each beam. By substituting (11.51) into (11.49) and some manipulation, the closed-form expression of the throughput of the MISO system can be calculated as

$$R = \sum_{i=1}^{M} \Pr[M_a = i] \left\{ \frac{M_a}{\ln 2} \sum_{n=0}^{M_a-1} (-1)^{n+1} C_{M_a-1}^n \frac{e^{\frac{n+1}{\bar{\gamma}s}}}{n+1} \mathrm{Ei}\left(-\frac{n+1}{\bar{\gamma}}\right) \right\}, \tag{11.52}$$

where $\mathrm{Ei}(\cdot)$ is the exponential integral function.

Energy Harvesting Performance Analysis

To evaluate the energy harvesting performance, we derive the exact statistical distribution of the harvested energy over one coherence time T_c at the sensor, which can be used for calculating the average harvested energy, as well as packet transmission performance of the sensor [13]. Conditioning on the number of usable beams for transmission, the CDF of E_H can be represented as

$$F_{E_H}(x) = \sum_{m=1}^{M} \Pr[E_H < x, M_a = m]. \tag{11.53}$$

According to our proposed cooperative beam selection scheme, the BS selects the best beam from all M_a usable beams to achieve the largest throughput, whereas the probability that each of M_a usable beams is selected to charge the sensor is equal to $\frac{1}{M_a}$. Therefore, we can rewrite (11.53) as

$$F_{E_H}(x) = \sum_{m=1}^{M-1} \frac{1}{m} \sum_{i=1}^{m} \Pr\left[E_{i:M} < x, E_{m:M} \geq E_{th}, E_{m+1:M} < E_{th}\right]$$

$$+ \Pr[E_{1:M} < x, E_{1:M} < E_{th}] + \frac{1}{M} \sum_{i=1}^{M} \Pr\left[E_{i:M} < x, E_{M:M} \geq E_{th}\right]. \tag{11.54}$$

For the case of $x \leq E_{th}$, (11.54) can be simply calculated as

$$F_{E_H}(x) = \Pr[E_{1:M} < x] = \int_0^{\frac{x}{\Lambda}} f_{\alpha_{1:M}}(y) dy, \; x \leq E_{th}. \tag{11.55}$$

By substituting (11.45) into (11.55), $F_{E_H}(x)$ can be calculated as

$$F_{E_H}(x) = M \sum_{j=0}^{M-1} \binom{M-1}{j} \frac{(-1)^j}{j+1} (1 - e^{-(j+1)\frac{x}{\Lambda}}), \; x \leq E_{th}. \tag{11.56}$$

For the case of $x > E_{th}$, (11.65) can be rewritten as

$$F_{E_H}(x) = \sum_{m=1}^{M-1} \frac{1}{m} \Bigg\{ \int_{\frac{E_{th}}{\Lambda}}^{\frac{x}{\Lambda}} \int_0^{\frac{E_{th}}{\Lambda}} f_{\alpha_{m:M}, \alpha_{m+1:M}}(y,z) dy dz$$

$$+ \sum_{i=1}^{m-1} \int_{\frac{E_{th}}{\Lambda}}^{\frac{x}{\Lambda}} \int_{\frac{E_{th}}{\Lambda}}^{w} \int_0^{\frac{E_{th}}{\Lambda}} f_{\alpha_{i:M}, \alpha_{m:M}, \alpha_{m+1:M}}(w, y, z) dw dy dz \Bigg\}$$

$$+ \frac{1}{M} \Bigg\{ \int_{\frac{E_{th}}{\Lambda}}^{\frac{x}{\Lambda}} f_{\alpha_{M:M}}(y) dy + \sum_{i=1}^{M-1} \int_{\frac{E_{th}}{\Lambda}}^{\frac{x}{\Lambda}} \int_{\frac{E_{th}}{\Lambda}}^{y} f_{\alpha_{i:M}, \alpha_{M:M}}(y,z) dy dz \Bigg\}$$

$$+ \int_0^{\frac{E_{th}}{\Lambda}} f_{\alpha_{1:M}}(y) dy, \; x > E_{th}, \tag{11.57}$$

where the joint PDF of $\alpha_{i:M}$, $\alpha_{m:M}$, and $\alpha_{m+1:M}$, and the joint PDF of $\alpha_{i:M}$ and $\alpha_{M:M}$, can be given by

11.2 Cooperative Beam Selection for RF Energy Harvesting

$$f_{\alpha_{i:M},\alpha_{m:M},\alpha_{m+1:M}}(x,y,z)$$
$$= \frac{M!e^{-ix-y-z}(e^{-y}-e^{-x})^{m-i-1}(1-e^{-z})^{M-m-1}}{(i-1)!(m-i-1)!(M-m-1)!}, x > y > z, \quad (11.58)$$

$$f_{\alpha_{i:M},\alpha_{M:M}}(y,z)$$
$$= \frac{M!}{(i-1)!(M-i-1)!}e^{-iy-z}(e^{-z}-e^{-y})^{M-i-1}, y > z, \quad (11.59)$$

respectively [16]. By substituting (11.45), (11.46), (11.58), and (11.59) into (11.57) and carrying out integrations, we can obtain the closed-form expression of $F_{E_H}(x)$ for $x > E_{th}$ as

$$F_{E_H}(x) = \sum_{m=1}^{M-1} \frac{1}{m} \left\{ \sum_{j=0}^{M-m-1} \binom{M-m-1}{j} \frac{(-1)^j}{j+1}(1-e^{-(j+1)\frac{E_{th}}{\Lambda}}) \right\}$$
$$\times \left\{ \frac{M!(e^{-m\frac{E_{th}}{\Lambda}} - e^{-m\frac{x}{\Lambda}})}{m!(M-m-1)!} \right.$$
$$+ \sum_{i=1}^{m-1} \frac{M!}{(i-1)!(m-i-1)!(M-m-1)!} \sum_{k=0}^{m-i-1} \binom{m-i-1}{k} \frac{(-1)^{m-i-k-1}}{k+1}$$
$$\times \left[e^{-(k+1)\frac{E_{th}}{\Lambda}} \frac{e^{-(m-k-1)\frac{E_{th}}{\Lambda}} - e^{-(m-k-1)\frac{x}{\Lambda}}}{m-k-1} - \frac{e^{-m\frac{E_{th}}{\Lambda}} - e^{-m\frac{x}{\Lambda}}}{m} \right] \right\} + \frac{1}{M} \left\{ e^{-M\frac{E_{th}}{\Lambda}} \right.$$
$$- e^{-M\frac{x}{\Lambda}} + \sum_{i=1}^{M-1} \frac{M!}{(i-1)!(M-i-1)!} \sum_{j=0}^{M-i-1} \binom{M-i-1}{j} \frac{(-1)^{M-i-j-1}}{j+1}$$
$$\left[\frac{e^{-(j+1)\frac{E_{th}}{\Lambda}}}{M-j-1}(e^{-(M-j-1)\frac{E_{th}}{\Lambda}} - e^{-(M-j-1)\frac{x}{\Lambda}}) - \frac{1}{M}(e^{-M\frac{E_{th}}{\Lambda}} - e^{-M\frac{x}{\Lambda}}) \right] \right\}$$
$$+ M \sum_{j=0}^{M-1} \binom{M-1}{j} \frac{(-1)^j}{j+1}(1-e^{-(j+1)\frac{E_{th}}{\Lambda}}), \quad x \geq E_{th}. \quad (11.60)$$

After taking the derivative of (11.68) and (11.60), the closed-form expression of the PDF of the harvested energy over one coherence time can be calculated as

$$f_{E_H}(x) =$$
$$\begin{cases} \frac{M}{\Lambda} \sum_{j=0}^{M-1} (-1)^j \binom{M-1}{j} e^{-(j+1)\frac{x}{\Lambda}}, & x < E_{th}, \\ \sum_{m=1}^{M-1} \frac{e^{-m\frac{x}{\Lambda}}}{m\Lambda} \left\{ \sum_{j=0}^{M-m-1} \binom{M-m-1}{j} \frac{(-1)^j}{j+1}(1-e^{-(j+1)\frac{E_{th}}{\Lambda}}) \right\} \\ \times \left\{ \frac{M!}{(m-1)!(M-m-1)!} + \sum_{i=1}^{m-1} \frac{M!}{(i-1)!(m-i-1)!(M-m-1)!} \right. \\ \sum_{k=0}^{m-i-1} \binom{m-i-1}{k} \frac{(-1)^{m-i-k-1}}{k+1} \left[e^{-\frac{(k+1)(E_{th}-x)}{\Lambda}} - 1 \right] \right\} \\ + \frac{e^{-M\frac{x}{\Lambda}}}{M\Lambda} \left\{ M + \sum_{i=1}^{M-1} \frac{M!}{(i-1)!(M-i-1)!} \sum_{k=0}^{M-i-1} \binom{M-i-1}{k} \frac{(-1)^{M-i-k-1}}{k+1} \right. \\ \left[e^{-\frac{(k+1)(E_{th}-x)}{\Lambda}} - 1 \right] \right\}, & x \geq E_{th}, \end{cases} \quad (11.61)$$

which can be used to calculate the average harvested energy $\overline{E}_H$ in closed form as

$$\begin{aligned}\overline{E}_H &= \int_0^\infty x f_{E_H}(x)dx \\ &= M \sum_{j=0}^{M-1} (-1)^j \binom{M-1}{j} \left[\frac{-E_{th}}{j+1} e^{-(j+1)\frac{E_{th}}{\Lambda}} + \frac{\Lambda}{(j+1)^2}(1 - e^{-(j+1)\frac{E_{th}}{\Lambda}}) \right] \\ &+ \sum_{m=1}^{M-1} \frac{e^{-m\frac{E_{th}}{\Lambda}}}{m} \Bigg\{ \sum_{j=0}^{M-m-1} \binom{M-m-1}{j} \frac{(-1)^j}{j+1}(1 - e^{-(j+1)\frac{E_{th}}{\Lambda}}) \\ &\quad \times \Bigg\{ \frac{M!(E_{th} + \frac{\Lambda}{m})}{m!(M-m-1)!} + \sum_{i=1}^{m-1} \frac{M!}{(i-1)!(m-i-1)!(M-m-1)!} \\ &\quad \sum_{k=0}^{m-i-1} \binom{m-i-1}{k} \frac{(-1)^{m-i-k-1}}{k+1} \left[\frac{E_{th}}{m-k-1} + \frac{\Lambda}{(m-k-1)^2} - \frac{E_{th}}{m} - \frac{\Lambda}{m^2} \right] \Bigg\} \\ &+ \frac{e^{-M\frac{E_{th}}{\Lambda}}}{M} \Bigg\{ E_{th} + \frac{\Lambda}{M} + \sum_{i=1}^{M-1} \frac{M!}{(i-1)!(M-i-1)!} \sum_{k=0}^{M-i-1} \binom{M-i-1}{k} \frac{(-1)^{M-i-k-1}}{k+1} \\ &\quad \left[\frac{E_{th}}{M-k-1} + \frac{\Lambda}{(M-k-1)^2} - \frac{E_{th}}{M} - \frac{\Lambda}{M^2} \right] \Bigg\}. \end{aligned} \quad (11.62)$$

In Fig. 11.7, we plot the average harvested energy $\overline{E}_H$ as a function of the energy threshold E_{th} for different antenna number M. We can observe that more antennas lead to larger average harvested energy, as expected. We can also see that the average harvested energy at the sensor quickly increases as E_{th} increased, and gradually converges to a constant value when E_{th} is large. This is because when E_{th} is large enough, the BS will only use the best beam to charge the sensor.

In Fig. 11.8, we plot the average throughput of the MISO system as a function of the energy threshold E_{th} for different antenna number M. We can observe that a larger antenna number leads to larger throughput, due to the beam selection benefit. We also observe that the throughput reduces gradually to a constant value with the increase of E_{th}. This is because when E_{th} is large, the BS will only use the best beam, from the energy harvesting perspective, to serve its selected user. Combined with Fig. 11.7, we can see there exists a trade-off of average harvested energy at the sensor versus throughput of the MISO system. In particular, larger E_{th} leads to larger average harvested energy, but smaller throughput. We can achieve the desired energy harvesting performance by properly adjusting E_{th} at the expense of certain throughput degradation in the MISO system.

11.2.3 Cooperative Energy Harvesting with Multiple Beam Selection

In this case, the BS will select a maximal number of active beams to serve its users, while ensuring that the harvested energy at the sensor node during each coherence time is above E_{th}.

At the beginning of each channel coherence time, the BS first estimates the channel vector from the BS to the sensor. The BS then calculates and ranks the projection

11.2 Cooperative Beam Selection for RF Energy Harvesting

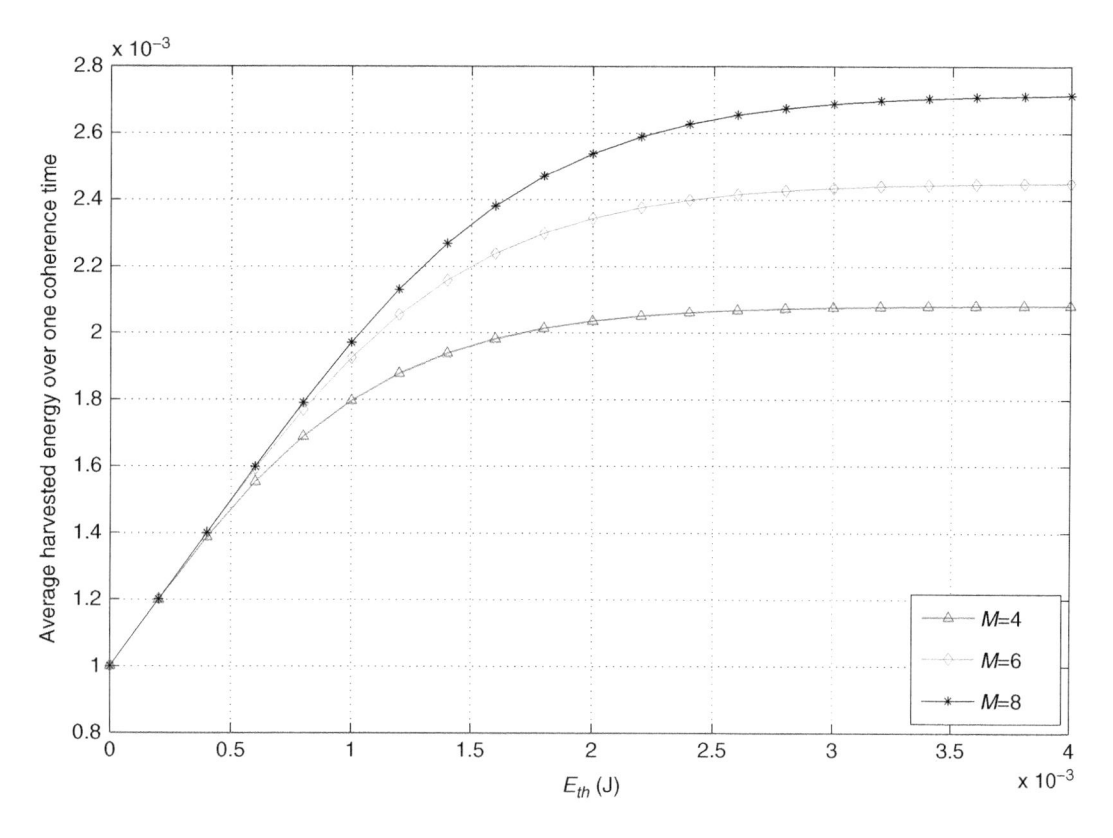

Figure 11.7 Average harvested energy at the sensor. Reprint with permission from [19]. ©2015 IEEE.

amplitude square α_m for each beam, the order version of which is denoted by $\alpha_{m:M}$, where $\alpha_{1:M} \geq \alpha_{2:M} \geq \ldots \geq \alpha_{M:M}$. After that, the BS calculates the total amount of RF energy that the sensor can harvest when the BS uses m best beams, corresponding to $\alpha_{1:M}$ to $\alpha_{m:M}$. The total harvested energy, denoted by E_H, can be given by $E_H = \sum_{i=1}^{m} E_{i,m}$, where $E_{i,m}$ denotes the harvested energy from the ith best beam with projection amplitude square $\alpha_{i:M}$, when m best beams are used for transmission, given by

$$E_{i,m} = \left(\frac{\eta T_c}{\Gamma d_H^\lambda}\right)\left(\frac{P_T}{m}\right)\alpha_{i:M}, i = 1, 2, \ldots, m. \quad (11.63)$$

If the harvested energy with m best beams is larger than the predefined energy threshold E_{th}, whereas the harvested energy with $m+1$ best beams is less than E_{th}, i.e., $\sum_{i=1}^{m} E_{i,m} \geq E_{th}$, and $\sum_{i=1}^{m+1} E_{i,m+1} < E_{th}$, then the BS uses the m best beams to serve its users. Note that with constant total transmission power used at the BS and uniform power allocation, the sensor can harvest more energy from less active beams, because the transmission power concentrates on the better beams, i.e., with larger projection power.

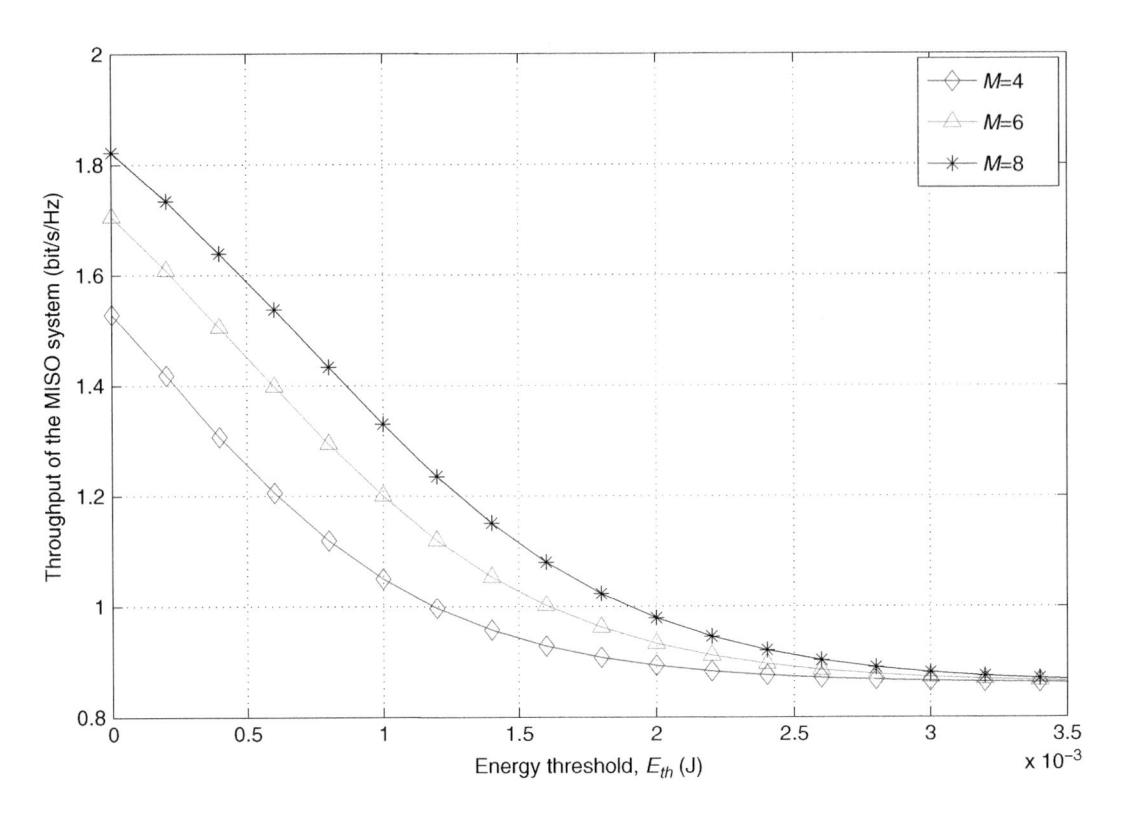

Figure 11.8 Throughput of the MISO system. Reprint with permission from [19]. ©2015 IEEE.

Energy Harvesting Performance Analysis

Conditioning on the number of active beams for transmission, denoted by M_a, the CDF of E_H can be represented as

$$F_{E_H}(x) = \sum_{m=1}^{M} \Pr[E_H < x, M_a = m]. \tag{11.64}$$

According to our proposed cooperative beam selection scheme, the number of active beams M_a is equal to m ($1 < m < M$) if and only if $\sum_{i=1}^{m} E_{i,m} \geq E_{th}$, and $\sum_{i=1}^{m+1} E_{i,m+1} < E_{th}$. Furthermore, the number of active beams M_a is equal to 1 if the energy threshold E_{th} cannot be satisfied with all transmission power P_T allocated to the best beam, i.e., $E_{1,1} < E_{th}$, and equal to M if the harvested energy is larger than E_{th} with P_T allocated to all M beams, i.e., $\sum_{i=1}^{M} E_{i,M} \geq E_{th}$. Therefore, we can rewrite (11.64) as

$$\begin{aligned}F_{E_H}(x) = &\sum_{m=1}^{M-1} \Pr\left[\sum_{i=1}^{m} E_{i,m} < x, \sum_{i=1}^{m} E_{i,m} \geq E_{th}, \sum_{i=1}^{m+1} E_{i,m+1} < E_{th}\right] \\ &+ \Pr[E_{1,1} < x, E_{1,1} < E_{th}] \\ &+ \Pr\left[\sum_{i=1}^{M} E_{i,M} < x, \sum_{i=1}^{M} E_{i,M} \geq E_{th}\right]. \end{aligned} \tag{11.65}$$

11.2 Cooperative Beam Selection for RF Energy Harvesting

For notation conciseness, we denote the sum of the m largest projection amplitude square from $\alpha_{1:M}$ to $\alpha_{m:M}$ as z_m, i.e., $z_m = \sum_{i=1}^{m} \alpha_{i:M}$. It follows that $\sum_{i=1}^{m} E_{i,m} = \left(\frac{\eta T_c}{\Gamma d_H^\lambda}\right)\left(\frac{P_T}{m}\right) z_m$. For the case of $x \leq E_{th}$, (11.65) can be simply calculated as

$$F_{E_H}(x) = \Pr[E_{1,1} < x] = \int_0^{\frac{\mu x}{E_{th}}} f_{z_1}(y) dy, \quad x \leq E_{th}, \tag{11.66}$$

where $f_{z_1}(x)$ denotes the PDF of z_1, given by [14, eq. (3.1)]

$$f_{z_1}(x) = Me^{-x}(1 - e^{-x})^{M-1}, \tag{11.67}$$

μ is a constant value equal to $\frac{E_{th}\Gamma d_H^\lambda}{\eta T_c P_T}$ for notational conciseness. By substituting (11.67) into (11.66), $F_{E_H}(x)$ can be calculated as

$$F_{E_H}(x) = \frac{\mu M}{E_{th}} e^{-\frac{\mu x}{E_{th}}} (1 - e^{-\frac{\mu x}{E_{th}}})^{M-1}, \quad x \leq E_{th}. \tag{11.68}$$

For the case of $x > E_{th}$, (11.65) can be mathematically calculated, while noting $\Pr[E_{1,1} < x, E_{1,1} < E_{th}]$ is equal to 0, as

$$F_{E_H}(x) = \sum_{m=1}^{M-1} \left\{ \int_0^{\mu} \int_{m\mu}^{(m+1)\mu - y} f_{\alpha_{m+1:M}, z_m}(y, z) dy dz \right.$$
$$\left. - \int_0^{(m+1)\mu - \frac{m\mu x}{E_{th}}} \int_{\frac{m\mu x}{E_{th}}}^{(m+1)\mu - y} f_{\alpha_{m+1:M}, z_m}(y, z) dy dz \right\} \tag{11.69}$$
$$+ \int_{M\mu}^{\frac{M\mu x}{E_{th}}} f_{z_M}(y) dy, \quad x > E_{th},$$

where $f_{z_M}(x)$ and $f_{\alpha_{m+1}, z_m}(x)$ denote the PDF of z_M, and the joint PDF of α_{m+1} and z_m, respectively, the closed-form expression of which can be obtained as [14, Eqs. (3.19) and (3.31)]

$$f_{z_M}(x) = \frac{x^{M-1} e^{-x}}{(M-1)!} \tag{11.70}$$

and

$$f_{\alpha_{m+1:M}, z_m}(x, y) = \sum_{i=0}^{M-m-1} \frac{(-1)^i M!(y - mx)^{m-1} e^{-y-(i+1)x}}{(M-m-1-i)! m!(m-1)! i!}, \quad y \geq mx, \tag{11.71}$$

respectively. By substituting (11.70) and (11.71) into (11.69) and carrying out integrations, we can obtain the closed-form expression of $F_{E_H}(x)$ for $x > E_{th}$. After taking the derivative of (11.68) and (11.69), the closed-form expression of the PDF of the harvested energy over one coherence time can be calculated as

$$f_{E_H}(x) =$$

$$\begin{cases}
\left\{\sum_{m=1}^{M-1}\left\{\sum_{i=1}^{M-m-1}\frac{(-1)^{i+1}M!}{(M-m-1-i)!m!(m-1)!i!}\sum_{j=0}^{m-1}\binom{m-1}{j}(-m)^{m-1-j}\sum_{t=0}^{j}\frac{j!}{(j-t)!}\right.\right. \\
\left\{\frac{(m-1-j)!}{(i+1)^{m-j}}\left(\frac{m\mu}{E_{th}}\right)^{j-t}e^{-\frac{m\mu x}{E_{th}}}x^{j-t-1}\left(j-t-\frac{m\mu x}{E_{th}}\right)-\left(\frac{m\mu}{E_{th}}\right)^{j-t}e^{-(i+1)(m+1)\mu}\right. \\
\sum_{r=0}^{m-j-1}\frac{(m-1-j)!}{(m-1-j-r)!(i+1)^{r+1}}\sum_{u=0}^{m-1-j-r}(-1)^u\binom{m-1-j-r}{u}\left(\frac{m\mu}{E_{th}}\right)^u[(m+1)\mu]^{m-1-j-r-u} \\
e^{im\mu\frac{x}{E_{th}}}x^{j-t+u-1}\left(j-t+u+\frac{im\mu x}{E_{th}}\right)+e^{-(i+1)(m+1)\mu}\sum_{s=0}^{j-t}\binom{j-t}{s}(-1)^s[(m+1)\mu]^{j-t-s} \\
\sum_{r=0}^{m-1-j+s}\frac{(m-1-j+s)!}{(m-1-j+s-r)!i^{r+1}}\sum_{u=0}^{m-1-j+s-r}(-1)^u\binom{m-1-j+s-r}{u} \\
\left.\left(\frac{m\mu}{E_{th}}\right)^u[(m+1)\mu]^{m-1-j+s-r-u}x^{u-1}e^{im\mu\frac{x}{E_{th}}}\left(u+\frac{im\mu}{E_{th}}x\right)\right\}-\frac{M!}{(M-m-1)!m!(m-1)!} \\
\sum_{j=0}^{m-1}\binom{m-1}{j}(-m)^{m-1-j}\sum_{t=0}^{j}\frac{j!}{(j-t)!}\left\{e^{-(m+1)\mu}\sum_{s=0}^{j-t}\binom{j-t}{s}(-1)^s\frac{m\mu}{E_{th}}[(m+1)\mu]^{j-t-s}\right. \\
\left[(m+1)\mu-\frac{m\mu x}{E_{th}}\right]^{m-1-j+s}+(m-1-j)!\left(\frac{m\mu}{E_{th}}\right)^{j-t}e^{-m\mu\frac{x}{E_{th}}}x^{j-t-1}\left(j-t-\frac{m\mu}{E_{th}}x\right) \\
-\left(\frac{m\mu}{E_{th}}\right)^{j-t}e^{-(m+1)\mu}\sum_{r=0}^{m-j-1}\frac{(m-1-j)!}{(m-1-j-r)!}\sum_{u=0}^{m-1-j-r}(-1)^u\binom{m-1-j-r}{u} \\
\left.\left.\left.\left(\frac{m\mu}{E_{th}}\right)^u[(m+1)\mu]^{m-1-j-r-u}(j-t+u)x^{j-t+u-1}\right\}\right\}\mathcal{U}\left(\left(1+\frac{1}{m}\right)E_{th}-x\right) \\
+\sum_{s=0}^{M-1}\left\{-(M-1-s)+\frac{M\mu x}{E_{th}}\right\}\frac{(M\mu)^{M-1-s}x^{M-2-s}e^{-\frac{M\mu x}{E_{th}}}}{(M-1-s)!E_{th}^{M-1-s}}\mathcal{U}(x-E_{th}) \\
+\frac{\mu M}{E_{th}}e^{-\frac{\mu x}{E_{th}}}\left(1-e^{-\frac{\mu x}{E_{th}}}\right)^{M-1}\mathcal{U}(E_{th}-x),
\end{cases}$$ (11.72)

which can be used to calculate the average harvested energy $\overline{E}_H$ in closed form.

In Fig. 11.9, we plot the average harvested energy $\overline{E}_H$ as a function of the energy threshold E_{th} for different antenna number M. We can see that the average harvested energy at the sensor quickly increases as E_{th} increases, and gradually converges to a constant value when E_{th} is large. This is because when E_{th} is large enough, the BS will only use the best beam to charge the sensor. We also observe that more antennas lead to smaller average harvested energy when E_{th} is small, and larger average harvested energy when E_{th} is large. This is because when E_{th} is small, more antennas lead to more potential active beams, which leads to wider distribution of the BS transmit power. When E_{th} is large, the sensor can enjoy more benefits from best beam selection. When E_{th} is 0, the MIMO system serves its users with all beams, and the amount of energy that the sensor can harvest is the same as [13] without considering energy sensitivity and storage capacity.

In Fig. 11.10, we plot the average sum rate of the multiuser MIMO system as a function of the energy threshold E_{th} for different numbers of users K with $M=4$ antennas, assuming the user selection scheme proposed in [20] to maximize the sum rate of the multiuser MIMO system. We can observe that larger user numbers lead to larger sum rate due to user selection. We also observe that the sum rate reduces gradually to a constant value with the increase of E_{th}. This is because when E_{th} is large, the BS will only use the best beam, from the energy harvesting perspective, to serve its selected user. Especially, while the sum rate for the case with $K = 100$ users remains larger than the sum rate for the case with $K = 50$ users for any

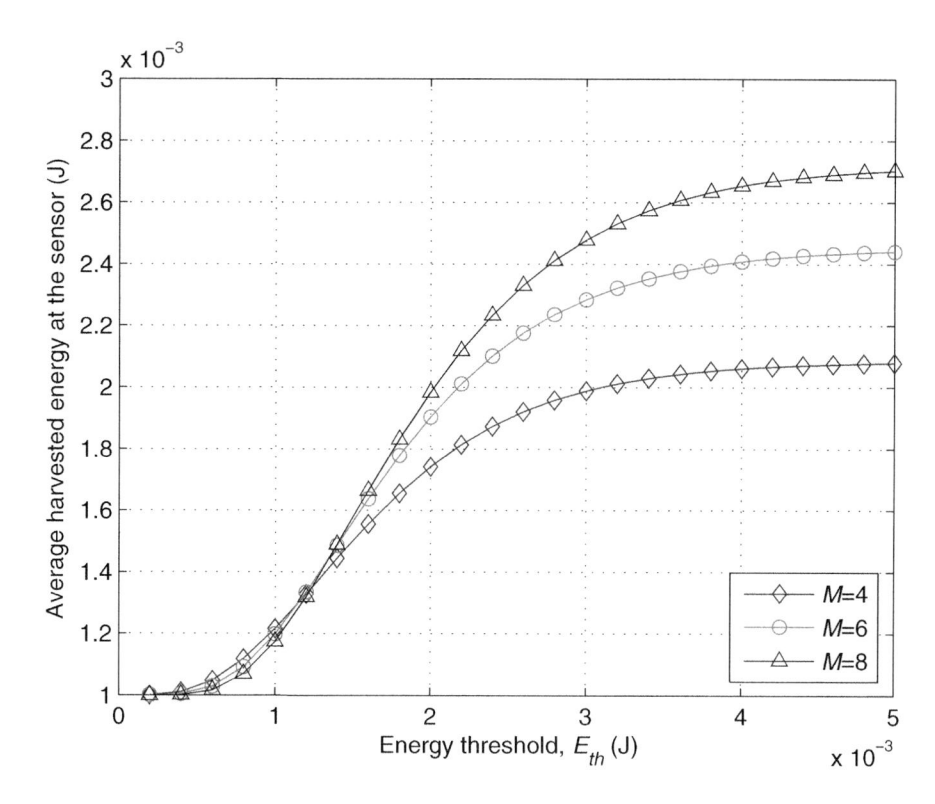

Figure 11.9 Average harvested energy at the sensor. Reprint with permission from [21]. ©2014 IEEE.

value of E_{th}, the sum-rate difference gradually converges to a constant value when E_{th} goes to infinity. Combined with Fig. 11.9, we can see there exists a trade-off of average harvested energy at the sensor versus sum rate of the multiuser MIMO system. In particular, larger E_{th} leads to larger average harvested energy, but smaller sum rate. We can achieve the desired energy harvesting performance by properly adjusting E_{th} at the expense of certain sum-rate degradation in the multiuser MIMO system.

11.3 Summary

In this chapter, we investigated the packet transmission performance of wireless sensor nodes powered by harvesting RF energy from existing wireless systems. We considered the effect of fading channel variation on both the energy harvesting stage and the information transmission stage. For the energy harvesting stage, we obtained the closed-form expression of the distribution of harvested energy over multiple channel coherence times. For the packet transmission stage, we considered the effects of interference from the existing network. We also studied an RUB-based cooperative beam selection

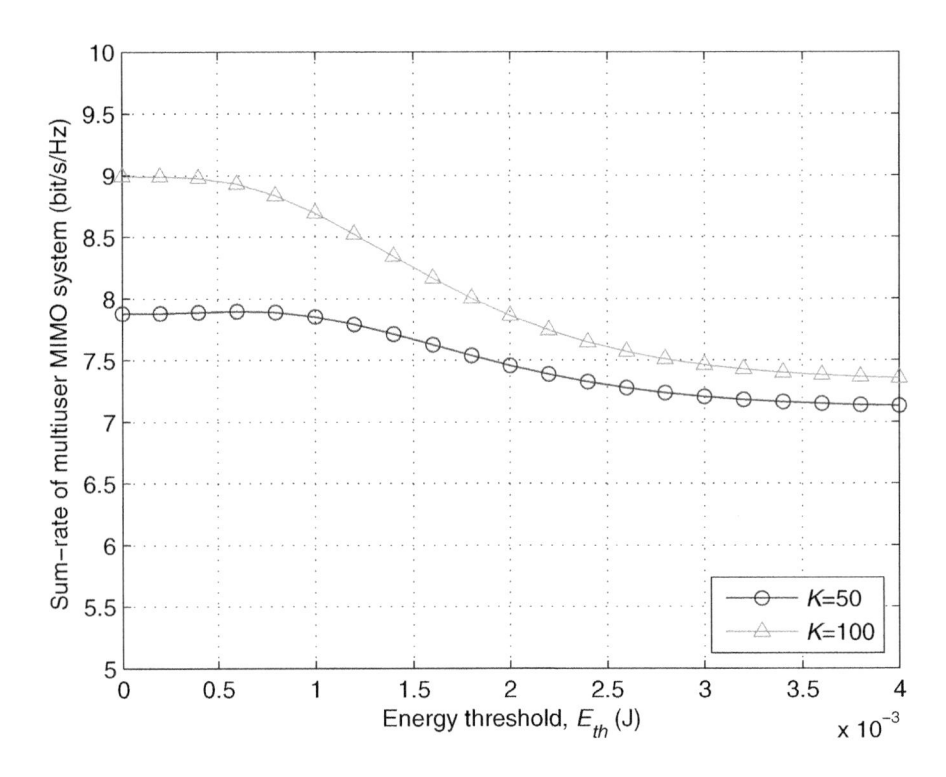

Figure 11.10 Sum rate of multiuser MIMO system for $M = 4$ antennas. Reprint with permission from [21]. ©2014 IEEE.

scheme, where an existing multiuser MIMO system can help increase the amount of harvested energy of wireless sensor nodes. We obtained the closed-form expression for the distribution functions of harvested energy, based on which we investigated the trade-off of the average harvested energy versus the throughput of the multiuser MIMO systems.

11.4 Further Reading

The reader can refer to [22–24] for the general challenges, performance limits, and candidate solutions in wireless communication systems with energy harvesting. The optimal packet scheduling policies assuming predictable energy arrival were investigated in [25, 26]. In [27–29], throughput maximization and packet delay minimization problems with energy harvesting constraints are studied for different channel environments. The fundamental performance limits of simultaneous wireless information and energy transfer systems over point-to-point links are studied in [30, 31]. [32] investigates mode switching between information decoding and energy harvesting, based on the instantaneous channel and interference condition over a point-to-point link. Simultaneous wireless information and power transfer over a cooperative relaying network is studied in [33].

References

[1] I. Akyildiz, W. Su, Y. Sankarasubramaniam, and E. Cayirci, "A survey on sensor networks," *IEEE Commun. Mag.*, 40, no. 8, pp. 102–114, 2002.

[2] W. Seah, Z. A. Eu, and H.-P. Tan, "Wireless sensor networks powered by ambient energy harvesting (WSN-HEAP) Survey and challenges," in *Proc. 1st Int. Conf. Wireless Commun., Vehic. Technol., Inform. Theor. Aerosp. Electron. Syst. Technol. (Wireless VITAE'2009)*, Aalborg May 2009, pp. 1–5.

[3] S. Sudevalayam and P. Kulkarni, "Energy harvesting sensor nodes: survey and implications," *IEEE Commun. Surv. Tut.*, 13, no. 3, pp. 443–461, 2011.

[4] C. Alippi and C. Galperti, "An adaptive system for optimal solar energy harvesting in wireless sensor network nodes," *IEEE Trans. Circuits Syst.*, 55, no. 6, pp. 1742–1750, 2008.

[5] M. Weimer, T. Paing, and R. Zane, "Remote area wind energy harvesting for low-power autonomous sensors," in *37th IEEE Power Electronics Specialists Conf.*, June 2006, pp. 1–5.

[6] L. Mateu, C. Codrea, N. Lucas, M. Pollak, and P. Spies, "Energy harvesting for wireless communication systems using thermogenerators," in *Proc. XXI Conf. on Design Circuits Integ. Syst. (DCIS)*, Barcelona, Spain, November 2006.

[7] Y. K. Tan, K. Y. Hoe, and S. K. Panda, "Energy harvesting using piezoelectric igniter for self-powered radio frequency (RF) wireless sensors," in *Proc. IEEE Intl. Conf. Industrial Technol. (ICIT)*, December 2006, pp. 1711–1716.

[8] T. Le, K. Mayaram, and T. Fiez, "Efficient far-field radio frequency energy harvesting for passively powered sensor networks," *IEEE J. Solid-State Circuits*, 43, no. 5, pp. 1287–1302, 2008.

[9] V. Liu, A. Parks, V. Talla, et al., "Ambient backscatter: Wireless communication out of thin air," in *Proc. ACM SIGCOMM*, Hong Kong, August 2013, pp. 1–13.

[10] U. Baroudi, A. Qureshi, V. Talla, et al., "Radio frequency energy harvesting characterization: An experimental study," in *Proc. IEEE TSPCC*, February 2012, pp. 1976–1981.

[11] Powercast Corporation, "TX91501 users manual and P2110s datasheet," www.powercastco.com/resources.

[12] R. Zhang and C. K. Ho, "MIMO broadcasting for simultaneous wireless information and power transfer," *IEEE Trans. Wireless Commun.*, 12, no. 5, pp. 1989–2001, 2013.

[13] T. Wu and H. Yang, "On the performance of overlaid wireless sensor transmission with RF energy harvesting," *IEEE J. Sel. Area. Commun.*, 33, no. 8, pp. 1693–1705, 2015.

[14] V. Raghunathan, S. Ganeriwal, and M. Srivastava, "Emerging techniques for long lived wireless sensor networks," *IEEE Commun. Mag.*, 44, no. 4, pp. 108–114, 2006.

[15] A. Goldsmith, *Wireless Communications*, Cambridge University Press, 2005.

[16] H.-C. Yang and M.-S. Alouini, *Order Statistics in Wireless Communications*, Cambridge University Press, 2011.

[17] K. K. J. Chung, C.-S. Hwang, and Y. K. Kim, "A random beamforming technique in MIMO systems exploiting multiuser diversity," *IEEE J. Sel. Areas Commun.*, 21, no. 5, pp. 848–855, 2003.

[18] M. Sharif and B. Hassibi, "On the capacity of MIMO broadcast channels with partial side information," *IEEE Trans. Inf. Theory*, 51, no. 2, pp. 506–522, 2005.

[19] W. Li, T.-Q. Wu, and H.-C. Yang, "Enhancing RF energy harvesting performance with cooperative beam selection for wireless sensors," *Proc. IEEE Pacific Rim Conf. Commun., Comput. Signal Process. (PacRim2015)*, Victoria, Canada, August 2015, pp. 1–5.

[20] H.-C. Yang, P. Lu, H.-K. Sung, and Y.-C. Ko, "Exact sum-rate analysis of MIMO broadcast channel with random unitary beamforming," *IEEE Trans. Commun.*, 59, no. 11, pp. 2982–2986, 2011.

[21] T.-Q. Wu and H.-C. Yang, "RF energy harvesting with cooperative beam selection for wireless sensors," *IEEE Wireless Commun. Lett.*, 3, no. 6, pp. 585–588, 2014.

[22] S. Bi, C. K. Ho, and R. Zhang, "Wireless powered communication: opportunities and challenges," *IEEE Commun. Mag.*, 53, no. 4, pp. 117–125, 2015.

[23] Z. Ding, C. Zhong, D. W. K. Ng, et al., "Application of smart antenna technologies in simultaneous wireless information and power transfer," *IEEE Commun. Mag.*, 53, no. 4, pp. 86–93, 2015.

[24] O. Ozel, K. Tutuncuoglu, S. Ulukus, and A. Yener, "Fundamental limits of energy harvesting communications," *IEEE Commun. Mag.*, 53, no. 4, pp. 126–132, 2015.

[25] J. Yang and S. Ulukus, "Optimal packet scheduling in an energy harvesting communication system," *IEEE Trans. Commun.*, 60, no. 1, pp. 220–230, 2012.

[26] K. Tutuncuoglu and A. Yener, "Optimum transmission policies for battery limited energy harvesting nodes," *IEEE Trans. Wireless Commun.*, 11, no. 3, pp. 1180–1189, 2012.

[27] C. Huang R. Zhang, and S. Cui, "Throughput maximization for the Gaussian relay channel with energy harvesting constraints," *IEEE J. Sel. Areas Commun.*, 31, no. 8, pp. 1469–1479, 2013.

[28] O. Ozel, K. Tutuncuoglu, J. Yang, S. Ulukus, and A. Yener, "Transmission with energy harvesting nodes in fading wireless channels: Optimal policies," *IEEE J. Sel. Areas Commun.*, 29, no. 8, pp. 1732–1743, 2011.

[29] M. Antepli, E. Uysal-Biyikoglu, and H. Erkal, "Optimal packet scheduling on an energy harvesting broadcast link," *IEEE J. Sel. Areas Commun.*, 29, no. 8, pp. 1721–1731, 2011.

[30] L. R. Varshney, "Transporting information and energy simultaneously," in *Proc. IEEE Int. Symp. Inf. Theory (ISIT)*, Toronto, July 2008, pp. 1612–1616.

[31] P. Grover and A. Sahai, "Shannon meets Tesla: Wireless information and power transfer," in *Proc. IEEE Int. Symp. Inf. Theory (ISIT)*, Austin, TX, June 2010, pp. 2363–2367.

[32] L. Liu, R. Zhang, and K. Chua, "Wireless information transfer with opportunistic energy harvesting," *IEEE Trans. Wireless Commun.*, 12, no. 1, pp. 288–300, 2013.

[33] X. Di, K. Xiong, P. Fan, and H.-C. Yang, "Simultaneous wireless information and power transfer in rateless coded cooperative relay networks," *IEEE Trans. Vehic. Technol.*, 66, no. 4, pp. 2981–2996, 2017.

12 Application: Massive MIMO Transmission

Massive MIMO is a key enabling technology of enhanced mobile broadband services in advanced wireless communication systems [1, 2]. In massive MIMO systems (also referred to as very large MIMO and large-scale antenna systems [LSAS]), the base station (BS) is equipped with a large number of antennas and serves a relatively small number of users simultaneously. Such systems can significantly improve the spectral and power efficiencies [3]. Another attractive feature of massive MIMO systems is that linear precoding schemes can achieve near-optimal performance [4]. However, massive MIMO systems face several critical implementation challenges. High hardware complexity is the most serious one. Conventional implementations require an independent radio frequency (RF) chain for each antenna. When the number of antennas becomes very large, the hardware complexity becomes prohibitively high, even for BSs. At the same time, operating a large number of RF chains will necessarily consume a large amount of power. Furthermore, baseband power consumption also increases significantly due to the required high-dimension computationally intensive signal processing operations.

In this chapter, we present and study low-complexity solutions for massive MIMO systems as another application of advanced transmission technologies presented in previous chapters. We consider two popular approaches for complexity reduction, namely antenna subset selection [5, 6] and hybrid analog and digital precoding [7]. For each approach, we first present the general idea and then discuss sample designs in detail. In particular, trace-based antenna subset selection and equal-gain transmission-based hybrid precoding solutions are respectively investigated. The performance and complexity trade-offs of different design options are illustrated through selected numerical examples.

12.1 Antenna Subset Selection for Massive MIMO

Antenna subset selection can effectively cope with the high hardware complexity and power consumption of massive MIMO systems by reducing the number of active RF chains [8, 9]. Since massive MIMO systems serve a relatively small number of users simultaneously, the number of active RF chains can be reduced without significantly affecting transmission performance. Reducing the number of active RF chains not only decreases the power consumption of power-hungry RF components but also limits the size of the precoding matrices, which in turn reduces the baseband computational complexity and power consumption.

The optimal antenna subset selection, with the exception of certain special cases [6], requires an exhaustive search over all antenna subsets, which renders it too computationally intensive to be implemented even in conventional MIMO systems, not to mention in a massive MIMO setting. Therefore, researchers have proposed different suboptimal schemes for antenna subset selection [10–12]. [10] proposed two suboptimal schemes, namely symbol error rate (SER) based and norm-based schemes, targeting the SER and norm of the effective channel between the BS and scheduled users, respectively. The key idea behind both schemes is to greedily reduce the number of active antennas until only the desired number of antennas remains active. The authors in [11] proposed three suboptimal schemes to further reduce the computational complexity. Among them, the revised signal-to-noise ratio (SNR) based scheme finds an antenna subset by removing the antenna that contributes the least to the individual user SNR in each step. The other two schemes (single-QR and max-QR schemes), on the other hand, add an antenna to the antenna subset in each step based on Gram–Schmidt orthogonalization of the channel vectors between BS antennas and scheduled users.

Interestingly, those schemes that sequentially add antennas to the antenna subset (such as single-QR and max-QR schemes) perform better than the schemes which reduce the size of the antenna subset to a desired size in a massive MIMO setting, despite having less computational complexity. With this observation in mind, we develop sequential antenna-subset selection schemes for a single-cell massive MIMO system [13]. The goal is to sequentially select an antenna to maximize the system sum rate for zero-forcing beamforming (ZFBF) transmission. To this end, we develop a novel closed-form expression of the trace of a matrix inverse in terms of Gram–Schmidt orthogonalization coefficients calculated with the QR decomposition of the matrix. Based on the derived expression, we present two schemes, namely the trace-based and the min-trace-based selection schemes. Unlike the single-QR and max-QR schemes, which use a lower bound on the user SNR, we use the exact expression of the SNR to achieve better performance without increasing the computational complexity.

12.1.1 System Model

We consider a cellular system employing massive MIMO technology, where the BS of the target cell, equipped with M antennas, simultaneously serves K scheduled users, where $M \gg K$. For the purpose of clarity, we presume that the users have only one antenna, noting that the extension to multiple-antenna users is straightforward. We focus on downlink transmission assuming that the BS obtains channel gain information by exploiting channel reciprocity with time division duplexing (TDD) implementation.

We use $\mathbf{h}_i$, $i = 1, 2, \ldots, M$, to denote the vector of complex channel gains between the ith BS antenna and all scheduled users. Matrices $\mathcal{H}_d$ denote the channel matrices between the selected BS antennas and the K scheduled users. The composite received signal vector at the users' side, denoted by $\mathbf{y}$, can be written as

12.1 Antenna Subset Selection for Massive MIMO

$$\mathbf{y} = \mathcal{H}_d \mathbf{x} + \mathbf{n}, \tag{12.1}$$

where $\mathbf{x}$ is the vector of symbols transmitted by the BS with a power of P_{tot}, $\mathbf{n}$ is a zero mean complex Gaussian noise vector with $\mathbf{E}[\mathbf{nn}^H] = \sigma^2 \mathbf{I}_{K+K_v}$, and $\mathbf{I}_n$ is an identity matrix of size $n \times n$. The BS precodes the data intended for its K selected users, $\mathbf{s}$, as

$$\mathbf{x} = \sqrt{\frac{P_{\text{tot}}}{\mathbf{Tr}(\mathbf{BB}^H)}} \mathbf{Bs}, \tag{12.2}$$

where $\mathbf{B}$ is the precoding matrix that maps K data symbols to N transmit symbols. Since ZFBF shows near-optimal performance when $M \gg K$, we will presume that the BS uses ZFBF to nullify the interuser interference. The exact expression of $\mathbf{B}$ in terms of $\mathcal{H}_d$ is given in the following sections.

12.1.2 Closed-Form Expression for Trace of Matrix Inverse

Presuming linear independence of channel vectors, the precoding matrix, $\mathbf{B}$, is given by [11, 14]

$$\mathbf{B} = \mathcal{H}_d^H \left(\mathcal{H}_d \mathcal{H}_d^H \right)^{-1}. \tag{12.3}$$

The resultant transmission leads to zero interuser interference, and the system sum rate is given by

$$R_{\text{sum}} = K \log_2 \left(1 + \frac{P_{\text{tot}}}{\sigma^2 \mathbf{Tr}(\mathcal{H}_d \mathcal{H}_d^H)^{-1}} \right). \tag{12.4}$$

As such, the system sum rate can be maximized by selecting those antennas that result in the smallest value of $\mathbf{Tr}(\mathcal{H}_d \mathcal{H}_d^H)^{-1}$. Unfortunately, finding the optimal antenna subset requires an exhaustive search, which becomes impractical in a massive MIMO setting. We first present an important result on $\mathbf{Tr}(\mathcal{H}_d \mathcal{H}_d^H)^{-1}$ in Theorem 12.1 that will be used in developing our proposed schemes.

THEOREM 12.1 *Consider a full-rank $m \times n$ matrix $\mathbf{A}$, $m \geq n$, with QR decomposition $\mathbf{A} = \mathbf{QR}$, where $\mathbf{Q}$ is an $m \times n$ matrix of orthonormal column vectors, and $\mathbf{R}$ is an upper triangular matrix. The trace of $(\mathbf{A}^H \mathbf{A})^{-1}$ is given by*

$$\mathbf{Tr}\left(\mathbf{A}^H \mathbf{A}\right)^{-1} = \sum_{k=1}^{n} \frac{1 + \sum_{t=1}^{k-1} |p_{k,t}|^2}{|r_{kk}|^2}, \tag{12.5}$$

where r_{ij} denotes the (i, j)th entry of $\mathbf{R}$, and

$$p_{k,t} = \frac{r_{tk} - \sum_{j=t+1}^{k-1} p_{k,j} r_{tj}}{r_{tt}}. \tag{12.6}$$

Proof. We now prove the theorem by mathematical induction. For $n = 1$, we can show that $\mathbf{Tr}(\mathbf{A}^H \mathbf{A})^{-1} = (\mathbf{A}^H \mathbf{A})^{-1} = 1/|r_{11}|^2$.

Application: Massive MIMO Transmission

Now, suppose that (12.5) holds for all positive integers less than n, where $n \geq 2$. Since the columns of $\mathbf{Q}$ are orthonormal, $\mathbf{A}^H\mathbf{A} = \mathbf{R}^H\mathbf{R}$, and $(\mathbf{A}^H\mathbf{A})^{-1} = \mathbf{R}^{-1}\mathbf{R}^{-H}$. Furthermore, $\mathbf{R}$ can be decomposed as

$$\mathbf{R} = \begin{bmatrix} \widetilde{\mathbf{R}} & \mathbf{v} \\ \mathbf{0} & r_{nn} \end{bmatrix}, \tag{12.7}$$

where $\mathbf{0}$ is a null vector, $\mathbf{v} = \begin{bmatrix} r_{1n} & r_{2n} & \cdots & r_{n-1n} \end{bmatrix}^T$, and

$$\widetilde{\mathbf{R}} = \begin{bmatrix} r_{11} & r_{12} & \cdots & r_{1n-1} \\ 0 & r_{22} & \cdots & r_{2n-1} \\ \vdots & \vdots & \ddots & \vdots \\ 0 & 0 & \cdots & r_{n-1n-1} \end{bmatrix}. \tag{12.8}$$

Since

$$\mathbf{R}^{-1} = \begin{bmatrix} \widetilde{\mathbf{R}}^{-1} & -\widetilde{\mathbf{R}}^{-1}\mathbf{v}/r_{nn} \\ \mathbf{0} & 1/r_{nn} \end{bmatrix}, \tag{12.9}$$

$\text{Tr}(\mathbf{R}^{-1}\mathbf{R}^{-H}) = \text{Tr}(\mathbf{A}^H\mathbf{A})^{-1}$ is given by

$$\text{Tr}(\mathbf{A}^H\mathbf{A})^{-1} = \text{Tr}(\widetilde{\mathbf{R}}^{-1}\widetilde{\mathbf{R}}^{-H}) + \frac{1 + \|\widetilde{\mathbf{R}}^{-1}\mathbf{v}\|^2}{|r_{nn}|^2}. \tag{12.10}$$

Since $\widetilde{\mathbf{R}}$ is an upper triangular matrix of size $n-1$, after applying (12.5) to the first term of (12.10), we have

$$\text{Tr}(\mathbf{A}^H\mathbf{A})^{-1} = \sum_{k=1}^{n-1} \frac{1 + \sum_{t=1}^{k-1}|p_{k,t}|^2}{|r_{kk}|^2} + \frac{1 + \|\widetilde{\mathbf{R}}^{-1}\mathbf{v}\|^2}{|r_{nn}|^2}. \tag{12.11}$$

Eq. (12.5) follows immediately by observing that $\widetilde{\mathbf{R}}\mathbf{p} = \mathbf{v}$, where $\mathbf{p} = \begin{bmatrix} p_{n,1} & p_{n,2} & \cdots & p_{n,n-1} \end{bmatrix}^T$. □

REMARK 12.1 Theorem 12.1 can be used to calculate the Frobenius norm of the pseudo-inverse of a matrix. Specifically, if $\mathbf{A}^\dagger = (\mathbf{A}^H\mathbf{A})^{-1}\mathbf{A}^H$ denotes the pseudo-inverse of $\mathbf{A}$, then $\|\mathbf{A}^\dagger\|_F^2$ can be shown to be given by

$$\|\mathbf{A}^\dagger\|_F^2 = \text{Tr}(\mathbf{A}^\dagger\mathbf{A}^{\dagger H}) = \text{Tr}(\mathbf{A}^H\mathbf{A})^{-1}, \tag{12.12}$$

which can be computed by (12.5).

REMARK 12.2 The kth summand in (12.5), T_k, depends only on the $k \times k$ top-left submatrix of $\mathbf{R}$. For example, the first entry, $T_1 = 1/|r_{11}|^2$ depends only on $[r_{11}]$. Table 12.1 lists some of the values of T_k in terms of r_{ij}. It is clear from Table 12.1 that T_k does not contain any r_{ij} term with either $i > k$ or $j > k$. This property of the summation given in (12.5) will be used in deriving our proposed schemes.

We now present our proposed schemes that sequentially add antennas to the antenna subset while minimizing the value of $\text{Tr}(\mathcal{H}_d\mathcal{H}_d^H)^{-1}$.

12.1 Antenna Subset Selection for Massive MIMO

Table 12.1 Trace terms, T_k, for different values of k.

k	T_k
1	$\dfrac{1}{|r_{11}|^2}$
2	$\dfrac{1}{|r_{22}|^2}\left(1+\dfrac{|r_{12}|^2}{|r_{11}|^2}\right)$
3	$\dfrac{1}{|r_{33}|^2}\left(1+\dfrac{|r_{23}|^2}{|r_{22}|^2}+\dfrac{|r_{13}-r_{12}\frac{r_{23}}{r_{22}}|^2}{|r_{11}|^2}\right)$
4	$\dfrac{1}{|r_{44}|^2}\left(1+\dfrac{|r_{34}|^2}{|r_{33}|^2}+\dfrac{\left|r_{24}-r_{23}\frac{r_{34}}{r_{33}}\right|^2}{|r_{22}|^2}\right.$ $\left.+\dfrac{\left|r_{14}-r_{13}\frac{r_{34}}{r_{33}}-r_{12}\frac{r_{24}-r_{23}\frac{r_{34}}{r_{33}}}{r_{22}}\right|^2}{|r_{11}|^2}\right)$

12.1.3 Trace-Based Selection Scheme

The basic objective of antenna subset selection schemes is to find a subset that results in small values of $\mathbf{Tr}(\mathcal{H}_d\mathcal{H}_d^H)^{-1}$. Since exhaustive search is highly computationally intensive, we use Theorem 12.1 to reduce the computational complexity. In particular, we utilize an important observation made in Remark 12.2 to find the antenna subset sequentially. More specifically, with QR-decomposition of $\mathcal{H}_d^H$, we have from Theorem 12.1

$$\mathbf{Tr}(\mathcal{H}_d\mathcal{H}_d^H)^{-1} = \sum_{k=1}^{K} T_k, \qquad (12.13)$$

where $T_k = \left(1+\sum_{t=1}^{k-1}|p_{k,t}|^2\right)/|r_{kk}|^2$, and $p_{k,t}$ is given in (12.6).

In the proposed trace-based antenna selection scheme, the antenna that results in the smallest value of T_k, denoted by k^*, is selected at the kth step. For example, the antenna with the highest value of corresponding $|r_{11}|^2$ is selected as the first antenna. Then the second antenna is chosen from the set of remaining antennas that results in the smallest value of $\frac{1}{|r_{2i}|^2}\left(1+\frac{|r_{1i}|^2}{|r_{11*}|^2}\right)$, where $i = 1, 2, \ldots, M$ and $i \neq 1^*$. This minimization can be easily carried out as T_2 depends only the second antenna after we have selected the first antenna in the first step. This process continues on until a total of K antennas are selected for transmission. Mathematically speaking,

$$k^* = \underset{i \in \mathcal{I}_k'}{\operatorname{argmin}} \frac{1}{|r_{ki}|^2}\left(1+\sum_{t=1}^{k-1}|p_{k,t,i}|^2\right), \qquad (12.14)$$

where $\mathcal{I}'_k$ denotes the index set of the antennas that have not been selected until the $(k-1)$th step, and

$$p_{k,t,i} = \frac{1}{r_{tt^*}}\left(r_{ti} - \sum_{j=t+1}^{k-1} p_{k,j,i} r_{tj^*}\right). \quad (12.15)$$

Since in multiuser MIMO systems, multiplexing gain is limited by min (N, K), selecting more than K antennas does not improve the multiplexing gain. As such, many schemes, such as single-QR scheme and max-QR scheme, select only $N = K$ antennas. To gain diversity benefits, we can select more than K antennas by repeating the aforementioned procedure on the channel matrix corresponding to unselected antennas, $\mathbf{H}_{\mathcal{I}'}$. Algorithm 12.1 shows a sample implementation of the proposed trace-based scheme.

Note that the minimization is carried out over all unselected *individual* antennas, unlike the exhaustive search, where the minimization of $\mathbf{Tr}(\mathcal{H}_d \mathcal{H}_d^H)^{-1}$ is carried out over $\binom{M}{N}$ subsets of N antennas. Compared with the exhaustive search, the trace-based scheme has significantly lower computational complexity.

REMARK 12.3 *Since the selection procedure ensures that r_{kk^*} is larger than r_{kj^*} for $j > k$, we can approximate $\mathbf{Tr}(\mathcal{H}_d \mathcal{H}_d^H)^{-1}$ by approximating $\widetilde{\mathbf{R}}$ in (12.8) by a diagonal matrix:*

$$\widetilde{\mathbf{R}} \approx \begin{bmatrix} r_{11} & 0 & \cdots & 0 \\ 0 & r_{22} & \cdots & 0 \\ \vdots & \vdots & \ddots & \vdots \\ 0 & 0 & \cdots & r_{n-1\,n-1} \end{bmatrix}. \quad (12.16)$$

By using (12.16) in (12.10), $\mathbf{Tr}(\mathcal{H}_d \mathcal{H}_d^H)^{-1}$ can be approximated as

$$\mathbf{Tr}(\mathcal{H}_d \mathcal{H}_d^H)^{-1} \approx \sum_{k=1}^{K} \frac{1}{|r_{kk^*}|^2}\left(1 + \sum_{t=1}^{k-1} \frac{|r_{tk^*}|^2}{|r_{tt^*}|^2}\right). \quad (12.17)$$

By using the trace approximation, the antenna selection at the kth step can be carried out:

$$k^* = \operatorname*{argmin}_{i \in \mathcal{I}'_k} \frac{1}{|r_{ki}|^2}\left(1 + \sum_{t=1}^{k-1} \frac{|r_{ti}|^2}{|r_{tt^*}|^2}\right), \quad (12.18)$$

which is equivalent to the heuristic scheme proposed in [15].

REMARK 12.4 *If we ignore the term $\widetilde{\mathbf{R}}^{-1}\mathbf{v}$ altogether in (12.10), we get*

$$\mathbf{Tr}(\mathcal{H}_d \mathcal{H}_d^H)^{-1} \approx \sum_{k=1}^{K} \frac{1}{|r_{kk^*}|^2}, \quad (12.19)$$

which results in the selection of the kth antenna as $k^* = \operatorname{argmax}_{i \in \mathcal{I}'_k} |r_{ki}|^2$. The resulting scheme becomes the single-QR scheme proposed in [11]. Quite intuitively, the more terms in $\mathbf{Tr}(\mathcal{H}_d \mathcal{H}_d^H)^{-1}$ we ignore, the worse the performance of antenna selection becomes. We will examine the accuracy of these approximations with the simulation results later.

Algorithm 12.1

1: $\mathcal{I}' \leftarrow \{1, 2, \cdots, M\}$
2: $\mathcal{I} \leftarrow \{\}$
3: **for** $i \in \mathcal{I}'$ **do**
4: $\quad \mathbf{v}_i \leftarrow \mathbf{h}_i$
5: $\quad r_{1i} \leftarrow |\mathbf{v}_i|$
6: $\quad T_i \leftarrow 1/|r_{1i}|^2$
7: **end for**
8: **for** $k = 1, 2, \cdots, K$ **do**
9: $\quad$ <u>Choose the best antenna</u>
10: $\quad k^* \leftarrow \text{argmin}_{i \in \mathcal{I}'} T_i$
11: $\quad \mathcal{I}' \leftarrow \mathcal{I}' - \{k^*\}$
12: $\quad \mathcal{I} \leftarrow \mathcal{I} \cup \{k^*\}$
13: $\quad$ <u>Perform Gram–Schmidt Orthogonalization</u>
14: $\quad \mathbf{u} \leftarrow \mathbf{v}_{k^*}/r_{kk^*}$
15: $\quad$ **for** $i \in \mathcal{I}'$ **do**
16: $\quad\quad r_{ki} \leftarrow \mathbf{u}^H \mathbf{v}_i$
17: $\quad\quad \mathbf{v}_i \leftarrow \mathbf{v}_i - r_{ki}\mathbf{u}$
18: $\quad\quad r_{k+1\,i} \leftarrow |\mathbf{v}_i|$
19: $\quad\quad$ **for** $t = k, k-1, \ldots, 1$ **do**
20: $\quad\quad\quad p_{i,t} = (r_{ti} - \sum_{j=t+1}^{k} p_{i,j} r_{tj^*})/r_{tt^*}$
21: $\quad\quad$ **end for**
22: $\quad\quad T_i = (1 + \sum_{t=1}^{k} |p_{i,t}|^2)/|r_{k+1\,i}|^2$
23: $\quad$ **end for**
24: **end for**
25: Repeat the algorithm on $\mathbf{H}_{\mathcal{I}'}$ if $|\mathcal{I}| < N$
26: **return** $\mathcal{I}$

12.1.4 Min-Trace-Based Selection Scheme

In the trace-based subset selection scheme, the minimization of $\mathbf{Tr}(\mathcal{H}_d \mathcal{H}_d^H)^{-1}$ is carried out by selecting an antenna that contributes the least to $\mathbf{Tr}(\mathcal{H}_d \mathcal{H}_d^H)^{-1}$ in each step. However, selecting the antenna with the highest corresponding channel vector norm in the first step may not always result in the minimum value of $\mathbf{Tr}(\mathcal{H}_d \mathcal{H}_d^H)^{-1}$. Inspired by the max-QR scheme [11], we arrive at an improved but slightly more computationally intensive suboptimal antenna subset scheme, named the min-trace-based antenna selection scheme. In this scheme, instead of always choosing the antenna with the highest corresponding channel vector norm as the first antenna, each antenna is chosen sequentially as the first antenna. For each choice, the remaining antennas are selected the same way as in the trace-based scheme, which leads to M selected antenna subsets. The antenna subset that results in the minimum value of $\sum_{k=1}^{K} \left(1 + \sum_{t=1}^{k-1} |p_{k,t}^*|^2\right)/|r_{kk^*}|^2 = \mathbf{Tr}(\mathcal{H}_d \mathcal{H}_d^H)^{-1}$ among all the subsets is then selected for the transmission. A possible implementation of the min-trace-based selection scheme is given in Algorithm 12.2.

Algorithm 12.2

1: $(|\mathbf{h}_{(i)}|^2, \mathtt{Indices}) \leftarrow \mathtt{sort}(|\mathbf{h}_i|^2, \texttt{'descend'})$
2: $minTrace \leftarrow \infty$
3: **for** $ind = 1, 2, \cdots, M$ **do**
4:    <u>Choose the first antenna</u>
5:    $1^\star = \mathtt{Indices}(ind)$
6:    $itrTrace \leftarrow 1/|\mathbf{h}_{1^\star}|^2$
7:    $\mathcal{I}' \leftarrow \{1, 2, \ldots, M\} - \{1^\star\}$
8:    $\mathcal{I} \leftarrow \{1^\star\}$
9:    <u>Perform Gram–Schmidt orthogonalization</u>
10:    $r_{11^\star} \leftarrow |\mathbf{h}_{1^\star}|$
11:    $\mathbf{u} \leftarrow \frac{\mathbf{h}_{1^\star}}{r_{11^\star}}$
12:    **for** $i \in \mathcal{I}'$ **do**
13:       $r_{1i} \leftarrow \mathbf{u}^H \mathbf{h}_i$
14:       $\mathbf{v}_i \leftarrow \mathbf{h}_i - r_{1i}\mathbf{u}$
15:       $r_{2i} \leftarrow |\mathbf{v}_i|$
16:       $T_i \leftarrow \frac{1}{|r_{2i}|^2}\left(1 + \frac{|r_{1i}|^2}{|r_{11^\star}|^2}\right)$
17:    **end for**
18:    **for** $k = 2, \ldots, K$ **do**
19:       <u>Choose the best antenna</u>
20:       $(minVal, k^*) \leftarrow \min_{i \in \mathcal{I}'} T_i$
21:       $itrTrace \leftarrow itrTrace + minVal$
22:       **if** $itrTrace > minTrace$ **then**
23:          **break**
24:       **end if**
25:       $\mathcal{I}' \leftarrow \mathcal{I}' - \{k^*\}$
26:       $\mathcal{I} \leftarrow \mathcal{I} \cup \{k^*\}$
27:       <u>Perform Gram–Schmidt Orthogonalization</u>
28:       $\mathbf{u} \leftarrow \mathbf{v}_{k^*}/r_{kk^*}$
29:       **for** $i \in \mathcal{I}'$ **do**
30:          $r_{ki} \leftarrow \mathbf{u}^H \mathbf{v}_i$
31:          $\mathbf{v}_i \leftarrow \mathbf{v}_i - r_{ki}\mathbf{u}$
32:          $r_{k+1\,i} \leftarrow |\mathbf{v}_i|$
33:          **for** $t = k, k-1, \ldots, 1$ **do**
34:             $p_{i,t} = (r_{ti} - \sum_{j=t+1}^{k} p_{i,j} r_{tj^*})/r_{tt^*}$
35:          **end for**
36:          $T_i = (1 + \sum_{t=1}^{k} |p_{i,t}|^2)/|r_{k+1\,i}|^2$
37:       **end for**
38: **end for**

```
39:     Choose the best subset
40:     if itrTrace < minTrace then
41:         minTrace ← itrTrace
42:         $\mathcal{I}_{\min}$ ← $\mathcal{I}$
43:     end if
44: end for
45: Repeat the algorithm on $\mathbf{H}_{\mathcal{I}'}$ if $|\mathcal{I}| < N$
46: return $\mathcal{I}_{\min}$
```

Table 12.2 Complexities of different antenna-subset selection schemes.

Selection scheme	Computational complexity
Trace-based	$\mathcal{O}(MNK)$
Single-QR [11]	$\mathcal{O}(MNK)$
Min-trace-based	$\mathcal{O}(M^2NK)$
Max-QR [11]	$\mathcal{O}(M^2NK)$
SNR-based [11, 14]	$\mathcal{O}((M^2 - N^2)K^2)$
SER-based [10]	$\mathcal{O}((M - N)M^3K)$
Norm-based [10]	$\mathcal{O}((M - N)M^3K)$
Optimal	$\mathcal{O}(M^N NK^2)$

12.1.5 Trade-off Analysis

We now compare the computational complexity and performance of the trace-based antenna selection schemes with existing schemes. Table 12.2 shows the computational complexities of different antenna-subset selection algorithms. The algorithms with the least computational complexity are tabulated above the algorithms with higher computational complexities. The optimal selection scheme has the highest computation complexity as it requires exhaustive search among $\binom{M}{N}$ subsets of N antennas, and $\mathbf{tr}(\mathbf{BB}^H)$ is computed for each antenna subset. We also consider the single-QR scheme [11], max-QR scheme [11], SNR-based [11, 14], and norm-based selection schemes [10]. The most computationally intensive operation involved in the trace-based antenna subset selection schemes is the Gram–Schmidt orthogonalization of the channel vectors to calculate the $\mathbf{R}$ matrix. Since orthogonalization has the computational complexity of $\mathcal{O}(MK)$ and is repeated N times, the computational complexity of the trace-based scheme is $\mathcal{O}(MNK)$. Moreover, in the min-trace-based scheme, M different subsets are computed using the trace-based scheme. Consequently, the computational complexity of the min-trace-based scheme is $\mathcal{O}(M^2NK)$. Note that the proposed trace-based and desired-user trace-based schemes are the least computationally intensive.

Fig. 12.1 shows the behavior of the sum rate, R_{sum}, of different suboptimal antenna subset selection schemes over i.i.d. Rayleigh fading channels with an average SNR

Application: Massive MIMO Transmission

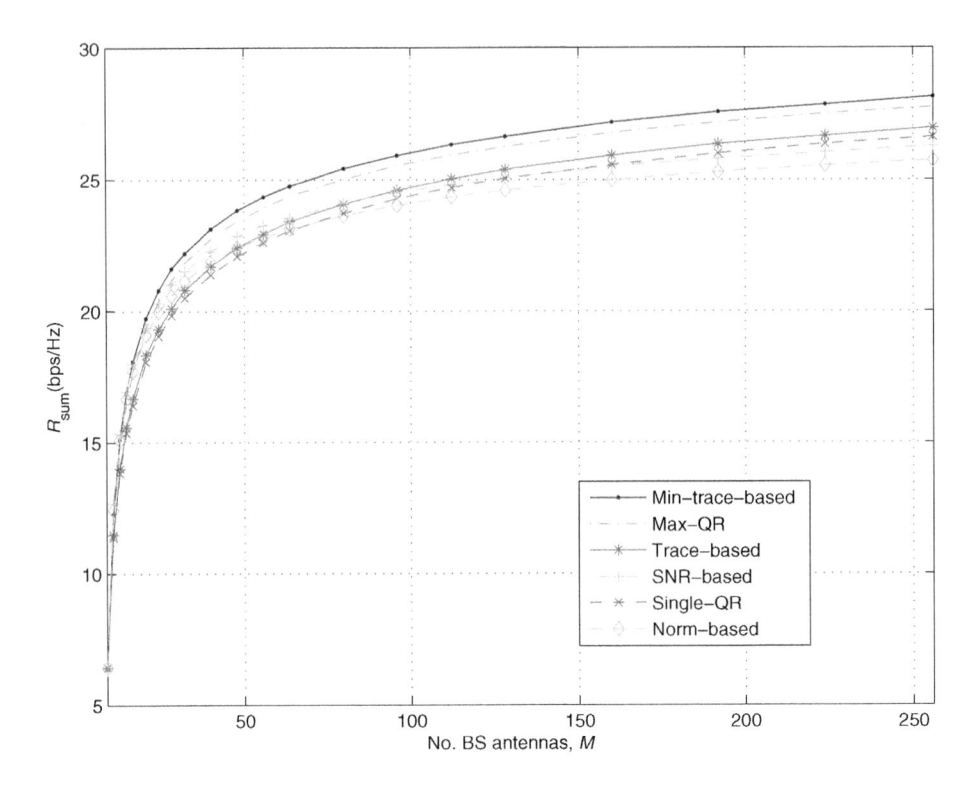

Figure 12.1 Comparison of R_{sum} for the proposed schemes against other antenna-subset selection schemes for $K = 8$ users served by 8 out of M BS antennas.

of 10 dB. We can see that the min-traced scheme enjoys the best performance and consistently outperforms the max-QR scheme which has similar complexity. There is mixed behavior among schemes with lower complexity. When the number of BS antennas M is relatively small, SNR-based and norm-based schemes have better performance. As M grows, the min-traced-based scheme starts to outperform all others. For high enough values of M, the performance of the trace-based scheme performance is only slightly poorer than that of the max-QR and min-trace-based schemes, which are much more computationally intensive, especially in a massive MIMO setting. The trace-based scheme performs better than the single-QR schemes for all values of M.

Fig. 12.2 shows R_{sum} of different subset selection schemes against the average SNR in an i.i.d. Rayleigh fading environment. Quite intuitively, the best performance is achieved by selecting all the antennas at the BS. Among the schemes, the optimal antenna selection scheme shows the best performance, while the performance gap between the optimal and other schemes is insignificant. Max-QR and fast global schemes show similar performance for all SNR ranges and perform slightly worse than the proposed min-trace-based scheme. Also, since M is small, the fast global scheme performs slightly better than the proposed trace-based scheme.

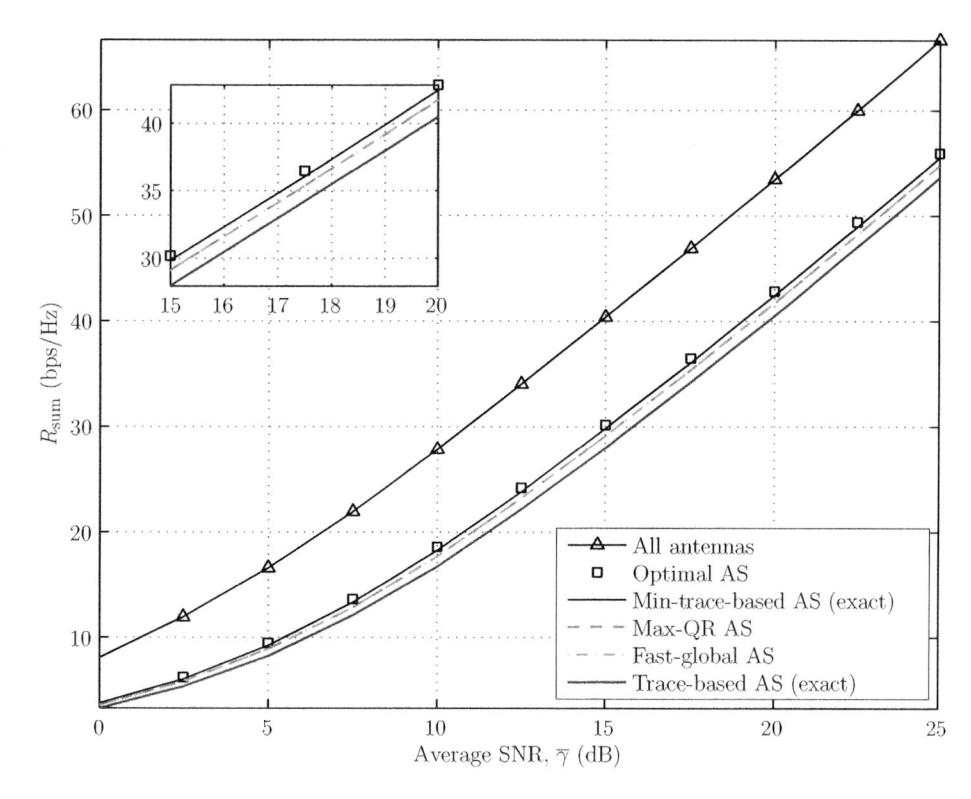

Figure 12.2 Comparison of R_{sum} for the proposed schemes against other antenna-subset selection schemes for $K = 8$ and $M = 16$. Reprint with permission from [13]. ©2018 KICS

12.2 Hybrid Precoding for Massive MIMO

Employing hybrid precoding at the BS can also reduce the hardware and processing complexity of massive MIMO systems. With hybrid precoding, the BS utilizes a limited number of RF chains to drive a large number of antennas using analog phased arrays. Although analog phased array has been used extensively for improving spatial resolution in radar systems, there has been only a limited application of analog beamforming in wireless communication systems. Some prominent applications of analog beamforming in communication systems include different WiFi and WPAN standards, such as IEEE 802.11ad and IEEE 802.15.13c [16, 17]. These systems rely on the beamforming gain of a phased array to improve system throughput with slight increase in system complexity, as compared with single-antenna systems. Unfortunately, conventional analog beamforming is suitable only for single-stream transmission as the transmitter has only one RF chain. Hybrid precoding, on the other hand, enables multi-stream/multi-user transmission by equipping the transmitter with more than one RF chain [18]. Specifically, the BS performs digital precoding at the baseband and applies analog beamforming using phase shifters to serve multiple scheduled users simultaneously.

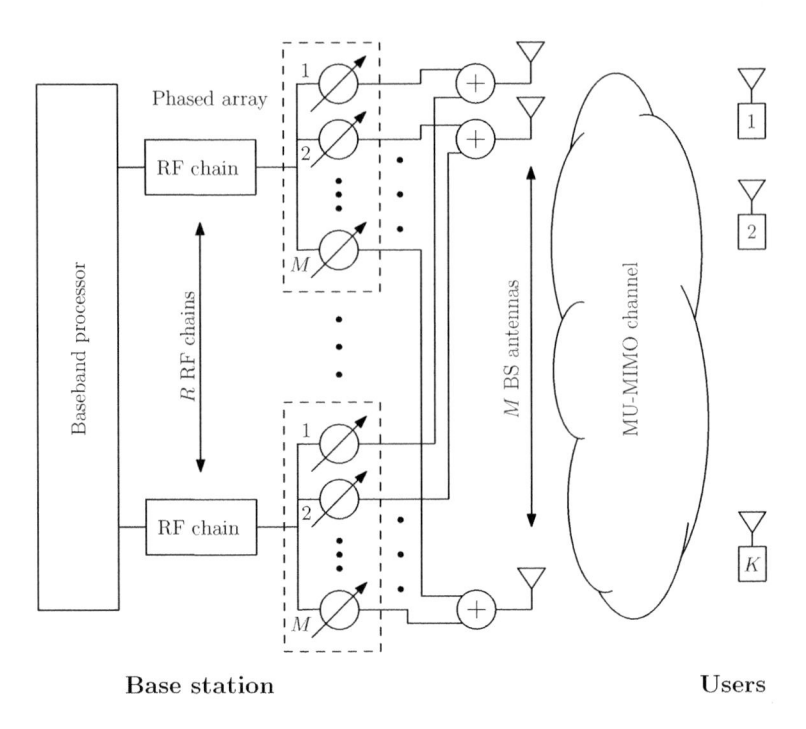

Figure 12.3 An illustration of a fully connected hybrid precoding architecture. Reprint with permission from [21]. ©2017 IET

Most work on hybrid precoding considers a fully connected architecture [7, 18–20]. Fig. 12.3 shows such an architecture where each RF chain is connected to *all* the antennas via phase shifters and RF adders. For a massive MIMO setting, a fully connected architecture has some limitations. Specifically, this architecture requires a large number of phase shifters and RF adders, resulting in a high hardware cost and power consumption. Note that extra components will necessarily introduce insertion loss. Second, as each RF chain feeds all the antennas, the operation becomes highly energy-intensive [18]. A subconnected architecture, in which each RF chain feeds only a subset of antennas as shown in Fig. 12.4, can be used to reduce hardware complexity and power consumption. Note that the subconnected architecture not only requires fewer phase shifters but also eliminates RF adders.

Here, we consider hybrid precoding design for a subconnected architecture to develop low-complexity solutions for multiuser massive MIMO systems [21]. Specifically, we adopt baseband ZFBF to nullify interuser interference and design the phased arrays sequentially using user channel information. Inspired by an important observation on the effect of phase shifting on user data rate, we develop an equal gain transmission (EGT)-based phase assignment strategy, namely sequential EGT. We further present two low-complexity phased array design strategies, called the best-user EGT and SVD (singular value decomposition) based EGT schemes, to improve the system performance.

12.2 Hybrid Precoding for Massive MIMO

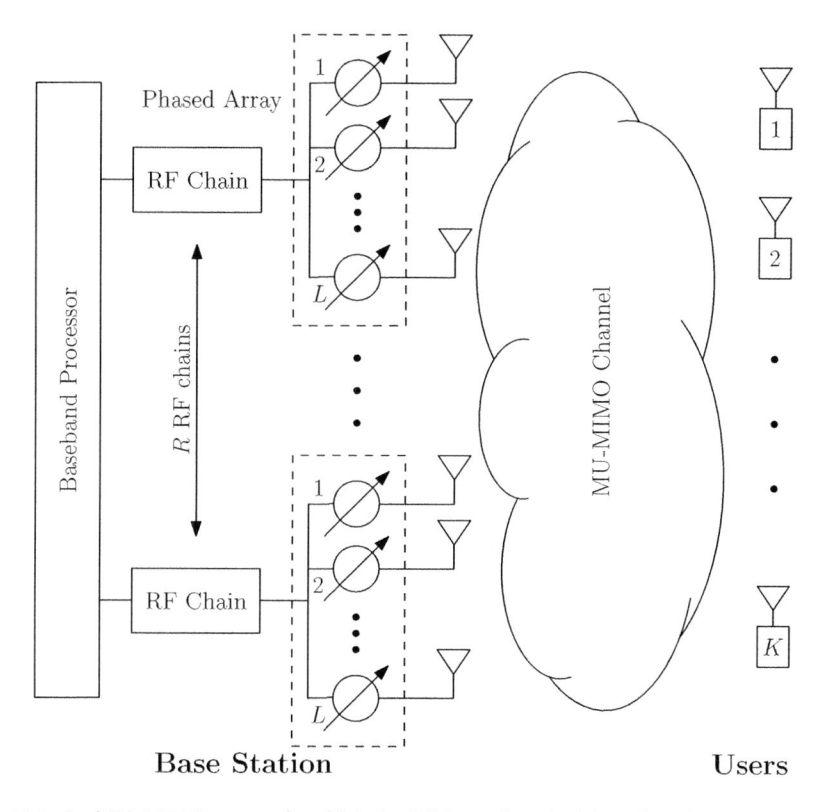

Figure 12.4 An MU-MIMO system in which the BS is equipped with R phased arrays. Reprint with permission from [21]. ©2017 IET

12.2.1 System Model

We consider a multiuser massive MIMO system in which the BS of a cell simultaneously serves K scheduled users. The BS is equipped with R RF chains, each of which feeds its information signal to a phased array with L phase shifters. Furthermore, each phase shifter drives an antenna, resulting in a total of $M = RL$ ($M \gg K$) antennas at the BS, as shown in Fig. 12.4. Compared with fully connected architecture, such subconnected architecture requires fewer phase shifters and no RF adder and, as such, is more suitable for practical implementation.

Let h_{ij} denote the complex channel gain between the ith BS antenna and the jth scheduled user, where $i = 1, 2, \ldots, M$, and $j = 1, 2, \ldots, K$. The composite received signal vector at the users' side, denoted by $\mathbf{y}$, can be written as

$$\mathbf{y} = \mathbf{H}\mathbf{x} + \mathbf{n}, \qquad (12.20)$$

where $\mathbf{x}$ is the vector of transmitted symbols from M antennas, $\mathbf{n}$ is the noise vector, and $\mathbf{H} = [h_{ij}]$ is the matrix of channel gains between the BS antennas and the users. The BS precodes the data intended for the scheduled users by applying baseband ZFBF and adjusting the phase of individual phase shifters in each analog-phased array.

Mathematically, the transmitted symbol vector, **x**, is given by

$$\mathbf{x} = \begin{bmatrix} \mathbf{w}_1 & \mathbf{0} & \cdots & \mathbf{0} \\ \mathbf{0} & \mathbf{w}_2 & \cdots & \mathbf{0} \\ \vdots & \vdots & \ddots & \vdots \\ \mathbf{0} & \mathbf{0} & \cdots & \mathbf{w}_R \end{bmatrix} \mathbf{Bs} = \mathbf{WBs}, \tag{12.21}$$

where $\mathbf{w}_r$ is the vector of complex phase shifts, $e^{j\theta_{ir}}$, $i = 1, 2, \ldots, L$, introduced by the rth phased array, where $r = 1, 2, \ldots, R$, **B** is the $R \times K$ ZF precoding matrix, and **s** is the vector of unity-power data symbols intended for K scheduled users.

The ZF precoding matrix, **B**, can be computed for a given phase-shift matrix **W** as

$$\mathbf{B} = \mathbf{W}^H \mathbf{H}^H \left(\mathbf{HWW}^H \mathbf{H}^H \right)^{-1} \mathbf{\Lambda}, \tag{12.22}$$

where $\mathbf{\Lambda}$ is a diagonal matrix introduced to adjust the power allocated to individual precoding vectors. We assume that the BS can perfectly estimate the downlink channel gains based on uplink transmission of pilot symbols from the users while exploring the channel reciprocity for TDD systems. This requires the transmission of at least L pilot symbols from each user. For the fully connected architecture, at least M orthogonal pilot symbols need to be transmitted to estimate the channel matrix. As such, the subconnected architecture reduces the channel estimation overhead cost by a factor of $M/L = R$ compared with the fully connected architecture.

The resulting hybrid beamforming transmission will result in zero interuser interference, and the composite received vector simplifies to

$$\mathbf{y} = \mathbf{\Lambda s} + \mathbf{n}. \tag{12.23}$$

Consequently, the system sum rate, R_{sum}, can be written as

$$R_{\text{sum}} = \sum_{k=1}^{K} \log_2 \left(1 + \frac{\lambda_k^2}{\sigma^2} \right), \tag{12.24}$$

where σ^2 is the noise variance at each receiver, and λ_k is the kth diagonal entry of the matrix $\mathbf{\Lambda}$. Denoting the BS total transmit power by P_{tot} and the kth column vector of $\mathbf{W}^H \mathbf{H}^H \left(\mathbf{HWW}^H \mathbf{H}^H \right)^{-1}$ by $\mathbf{a}_k$, the total power constraint implies that λ_k satisfy the following relationship:

$$\sum_{k=1}^{K} \lambda_k^2 \|\mathbf{a}_k\|^2 \leq P_{\text{tot}}. \tag{12.25}$$

Depending on the power allocation strategy, the constants λ_k take on different values. For example, with an equal-power allocation strategy, $\lambda_k^2 = P_{\text{tot}}/K\|\mathbf{a}_k\|^2$, and

$$R_{\text{sum}} = \sum_{k=1}^{K} \log_2 \left(1 + \frac{P_{\text{tot}}}{K\sigma^2 \|\mathbf{a}_k\|^2} \right). \tag{12.26}$$

The power can also be allocated in an optimal fashion to maximize the system sum rate [22]. Specifically, it can easily be verified that distributing the λ_k in a water-filling fashion as

$$\frac{\lambda_k^2}{\sigma^2} = \left(\frac{\mu}{\|\mathbf{a}_k\|^2} - 1\right)^+ \qquad (12.27)$$

results in the maximum sum rate of

$$R_{\text{sum}} = \sum_{\{k:\|\mathbf{a}_k\|^2 < \mu\}} \log_2\left(\frac{\mu}{\|\mathbf{a}_k\|^2}\right), \qquad (12.28)$$

where μ is found by

$$\sum_{k=1}^{K} \left(\mu - \|\mathbf{a}_k\|^2\right)^+ = \frac{P_{\text{tot}}}{\sigma^2}, \qquad (12.29)$$

and $(x)^+ = \max\{x, 0\}$. Unfortunately, neither strategy can ensure fairness among the users. In fact, depending on the channel conditions, the optimal-power allocation strategy may result in zero power allocation to a particular user. In a cellular system, maintaining fairness among the scheduled users can be essential for certain applications.

The equal-rate allocation strategy, on the other hand, ensures fairness among the users as each user gets the same portion of the total throughput. With the equal-rate allocation, we need $\lambda_k = \lambda_j$ for all k and j, where $k, j = 1, 2, \ldots, K$. Consequently, λ_k^2 can be shown to be given by $\lambda_k^2 = P_{\text{tot}} / \sum_{k=1}^{K} \|\mathbf{a}_k\|^2$, and the sum rate of the system for this strategy simplifies to

$$R_{\text{sum}} = K \log_2\left(1 + \frac{P_{\text{tot}}}{L\sigma^2 \text{Tr}(\mathbf{HWW}^H\mathbf{H}^H)^{-1}}\right). \qquad (12.30)$$

In this work, we consider the equal-rate allocation scheme and develop some hybrid precoding schemes based on the sum rate given in (12.30).

The sum rate of the resulting hybrid precoding system can be maximized by properly designing the analog beamforming vectors, $\mathbf{w}_r$. Specifically, the analog phase-shift values maximizing the sum rate of the system can be found by the following optimization problem:

$$\mathbf{W}^* = \underset{\mathbf{W}}{\operatorname{argmin}} \left\{\text{Tr}(\mathbf{HWW}^H\mathbf{H}^H)^{-1}\right\}, \qquad (12.31)$$

where $\mathbf{W}$ is given in (12.21). Unfortunately, the problem is nonconvex in nature and the solution of this type of problem is very difficult, if not impossible, for online implementation. In the following, we present three suboptimal beamforming schemes that design the analog beamforming vectors, $\mathbf{w}_r$ in a sequential manner based on the EGT scheme and study their performance in an i.i.d. Rayleigh fading environment.

Equal gain transmission is the optimal analog beamforming scheme for the single-user scenario. Consider a case in which the BS serves only one user. The system sum rate simplifies to

$$R_{\text{sum}} = \log_2\left(1 + \frac{P_{\text{tot}} \sum_{r=1}^{R} |\mathbf{h}_{1r}\mathbf{w}_r|^2}{L\sigma^2}\right), \qquad (12.32)$$

where R is the number of RF chains and $\mathbf{h}_{1r}$ is the (row) vector of the channel gains from the rth phased array antennas to the scheduled user. Here, the entries of $\mathbf{w}_r$ are of the form $e^{j\phi}$. Furthermore, the triangle inequality implies that $|\sum_{l=1}^{L} r_l e^{j\psi_l}| \leq \sum_{l=1}^{L} r_l$, where r_l and ψ_l are all reals, and $r_l \geq 0$. Since $\log_2(x)$ is a monotonically increasing function of x, R_{sum} can be shown to be upper bounded as

$$R_{\text{sum}} \leq \log_2 \left(1 + \frac{P_{\text{tot}} \sum_{r=1}^{R} \|\mathbf{h}_{1r}\|_1^2}{L\sigma^2}\right), \tag{12.33}$$

where $\|\mathbf{h}_{1r}\|_1$ denotes the l_1-norm of $\mathbf{h}_{1r}$. The equality holds when the jth entry of $\mathbf{w}_r$ is set to $e^{-j\phi_j}$, where ϕ_j is the phase angle of the jth entry of $\mathbf{h}_{1r}$, and the resulting EGT scheme achieves the maximum sum rate for the single-user system.

12.2.2 Sequential EGT Scheme

Before presenting our proposed analog precoding schemes, we first present an observation regarding the design of matrix $\mathbf{W}$ for ZF beamforming.

If the analog beamforming vectors, $\mathbf{w}_r$, are designed in a sequential fashion, the next $\mathbf{w}_r$ should be designed to maximize the magnitude of the diagonal entry of $\mathbf{HW}$ to effectively increase the sum rate given in (12.30).

Proof. This claim can be justified using mathematical induction as follows.

For the single-user case, $\mathbf{HW} = \mathbf{h}_{11}\mathbf{w}_1$. The optimal strategy is to perform EGT-based phase assignment, which maximizes $|\mathbf{h}_{11}\mathbf{w}_1|$.

For the two-users case, denoting the channel gain (row) vector from the ith phased array to the jth user by $\mathbf{h}_{ij}$, where $i = 1, 2$ and $j = 1, 2$, the matrix $\mathbf{HW}$ can be written as

$$\mathbf{HW} = \begin{bmatrix} \mathbf{h}_{11}\mathbf{w}_1 & \mathbf{h}_{12}\mathbf{w}_2 \\ \mathbf{h}_{21}\mathbf{w}_1 & \mathbf{h}_{22}\mathbf{w}_2 \end{bmatrix}. \tag{12.34}$$

We assume that the phase values of the first phased array are already designed to maximize $|\mathbf{h}_{11}\mathbf{w}_1|$ in the EGT fashion based on the channel gains. We now focus on the design of the second phased array beamforming vector, $\mathbf{w}_2$, such that $\text{Tr}(\mathbf{HWW}^H\mathbf{H}^H)^{-1}$ attains small values.

Since $\mathbf{w}_1$ is designed to maximize $|\mathbf{h}_{11}\mathbf{w}_1|$, $|\mathbf{h}_{21}\mathbf{w}_1|$ will be relatively very small, especially when L is large. Mathematically, the weak law of large numbers implies that

$$\lim_{L \to \infty} \frac{\mathbf{h}_{21}\mathbf{w}_1/L}{\mathbf{h}_{11}\mathbf{w}_1/L} = \frac{0}{\sqrt{\pi\overline{\gamma}/2}} = 0. \tag{12.35}$$

Here, we have used the fact that $\mathbf{h}_{21}\mathbf{w}_1$ is a sum of L i.i.d. zero mean Gaussian random variables, whereas $\mathbf{h}_{11}\mathbf{w}_1$ is a sum of L i.i.d Rayleigh random variables with the mean of $\sqrt{\pi\overline{\gamma}}/2$. As such, we can approximate $\mathbf{HW}$ as

$$\mathbf{HW} \approx \begin{bmatrix} \mathbf{h}_{11}\mathbf{w}_1 & \mathbf{h}_{12}\mathbf{w}_2 \\ 0 & \mathbf{h}_{22}\mathbf{w}_2 \end{bmatrix}. \tag{12.36}$$

As such, $(\mathbf{HWW}^H\mathbf{H}^H)^{-1}$ is approximately given by

$$(\mathbf{HWW}^H\mathbf{H}^H)^{-1} \approx \begin{bmatrix} \frac{1}{|\mathbf{h}_{11}\mathbf{w}_1|^2} & -\frac{\mathbf{h}_{12}\mathbf{w}_2}{\mathbf{h}_{22}\mathbf{w}_2|\mathbf{h}_{11}\mathbf{w}_1|^2} \\ -\frac{\mathbf{w}_2^H\mathbf{h}_{12}^H}{\mathbf{w}_2^H\mathbf{h}_{22}^H|\mathbf{h}_{11}\mathbf{w}_1|^2} & \frac{1}{|\mathbf{h}_{22}\mathbf{w}_2|^2} + \frac{|\mathbf{h}_{12}\mathbf{w}_2|^2}{|\mathbf{h}_{11}\mathbf{w}_1|^2|\mathbf{h}_{22}\mathbf{w}_2|^2} \end{bmatrix}. \tag{12.37}$$

12.2 Hybrid Precoding for Massive MIMO

Consequently, $\text{Tr}(\mathbf{HWW}^H\mathbf{H}^H)^{-1}$ can be approximately calculated as

$$\text{Tr}(\mathbf{HWW}^H\mathbf{H}^H)^{-1} \approx \frac{1}{|\mathbf{h}_{11}\mathbf{w}_1|^2} + \frac{1}{|\mathbf{h}_{22}\mathbf{w}_2|^2} + \frac{|\mathbf{h}_{12}\mathbf{w}_2|^2}{|\mathbf{h}_{11}\mathbf{w}_1|^2|\mathbf{h}_{22}\mathbf{w}_2|^2}. \quad (12.38)$$

Therefore, $\text{Tr}(\mathbf{HWW}^H\mathbf{H}^H)^{-1}$ can be effectively minimized by designing $\mathbf{w}_2$ such that $|\mathbf{h}_{22}\mathbf{w}_2|^2$ is maximized, and at the same time, $|\mathbf{h}_{12}\mathbf{w}_2|^2$ is minimized. Unfortunately, with the specific structure of $\mathbf{w}_2$ and the limited degrees of freedom, both objectives cannot be achieved simultaneously. An optimization problem to directly minimize (12.38) for given $\mathbf{w}_1$ can be formulated, but it is unfortunately a nonconvex problem and does not have a closed-form solution. On the other hand, noting that $|\mathbf{h}_{22}\mathbf{w}_2|^2$ appears in both the second and third terms of (12.38), and $\mathbf{w}_2$ that maximizes $|\mathbf{h}_{22}\mathbf{w}_2|^2$ will result in small values of $|\mathbf{h}_{21}\mathbf{w}_2|^2$ when L is large, we propose to design $\mathbf{w}_2$ to maximize $|\mathbf{h}_{22}\mathbf{w}_2|^2$ by computing $\mathbf{w}_2$ individual entries in an EGT fashion.

Fig. 12.5 compares the sum rates of the two-users scenario for different $\mathbf{w}_2$ design strategies against the average SNR, $\bar{\gamma} = P_{\text{tot}}\mathbb{E}[|h_{ij}|^2]/\sigma^2$, where $\mathbb{E}[\cdot]$ calculates the statistical average of a random variable. Observe that the EGT-based scheme (the proposed sequential EGT scheme) considerably outperforms the scheme in which $\mathbf{w}_2$ is designed to minimize $|\mathbf{h}_{12}\mathbf{w}_2|^2$. Furthermore, the EGT-based scheme performs nearly the same as

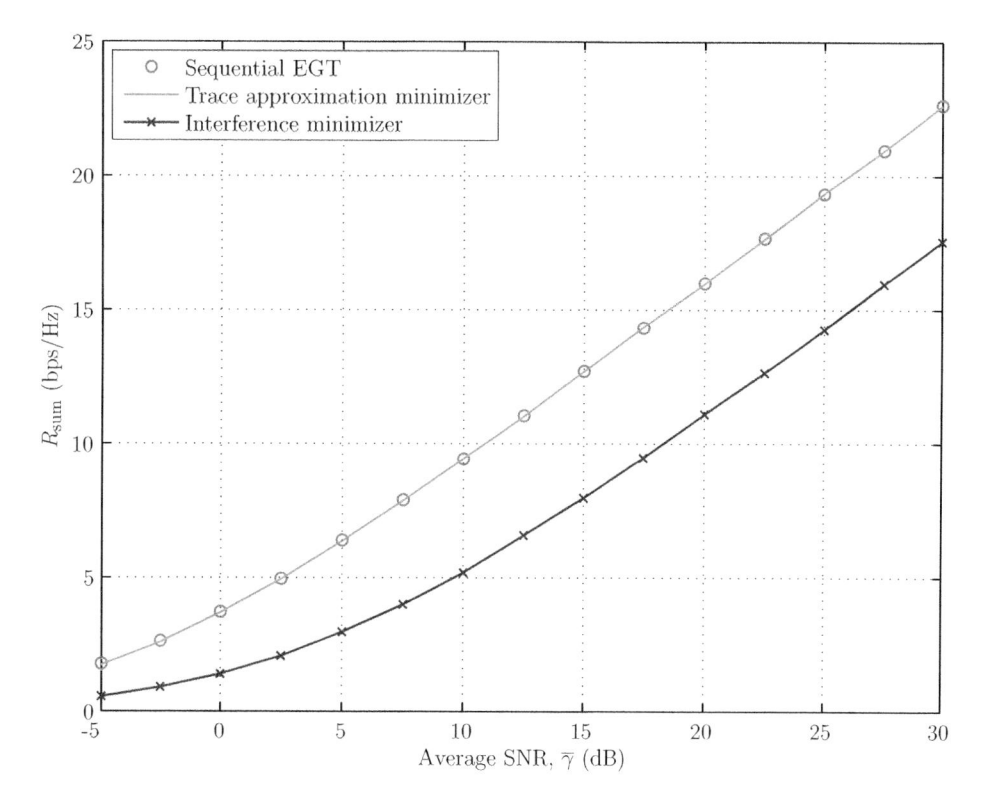

Figure 12.5 Comparison of the proposed sequential EGT with other trace minimization schemes. Reprint with permission from [21]. ©2017 IET

the scheme that directly minimizes (12.38), whose solution relies on numerical methods that may find a local minimum.

Now we consider a general scenario in which the BS serves K users simultaneously using $R = K$ RF chains. Building on the result of the two-users case, we presume that the first $K-1$ phased arrays are designed to maximize the first $K-1$ diagonal entries of $\mathbf{HW}$ in an EGT fashion. We design $\mathbf{w}_K$ for the Kth phased array such that $\text{Tr}(\mathbf{HWW}^H\mathbf{H}^H)^{-1}$ attains small values. For a massive MIMO setting, where L is large, $\mathbf{w}_k$ that maximizes $|\mathbf{h}_{kk}\mathbf{w}_k|$ will lead to small values of $|\mathbf{h}_{jk}\mathbf{w}_k|$ for $k, j = 1, 2, \ldots, K-1$, and $j \neq k$. As such, we can approximate $\mathbf{HW}$ as

$$\mathbf{HW} \approx \begin{bmatrix} \mathbf{D} & \mathbf{x} \\ \mathbf{0} & y \end{bmatrix}, \qquad (12.39)$$

where $\mathbf{D}$ is a diagonal matrix with diagonal entries, $\mathbf{h}_{kk}\mathbf{w}_k$, $k = 1, 2, \ldots, K-1$, $y = \mathbf{h}_{KK}\mathbf{w}_K$, and $\mathbf{x} = \begin{bmatrix} \mathbf{h}_{1K}\mathbf{w}_K & \mathbf{h}_{2K}\mathbf{w}_K & \ldots & \mathbf{h}_{K-1K}\mathbf{w}_K \end{bmatrix}^T$. Furthermore, $\text{Tr}(\mathbf{HWW}^H\mathbf{H}^H)^{-1}$ can be shown to be

$$\text{Tr}(\mathbf{HWW}^H\mathbf{H}^H)^{-1} \approx \text{Tr}(\mathbf{DD}^H)^{-1} + \frac{1}{|y|^2} + \frac{\|\mathbf{D}^{-1}\mathbf{x}\|^2}{|y|^2}. \qquad (12.40)$$

Following similar reasoning for the two-users case, maximizing $|y| = |\mathbf{h}_{KK}\mathbf{w}_K|$ by performing EGT-based phase-shift assignments for the Kth phased array will result in smaller values of $\text{Tr}(\mathbf{HWW}^H\mathbf{H}^H)^{-1}$. □

Based on the above justification, we propose a sequential design solution for the analog precoder. Specifically, the phased arrays are designed sequentially such that the diagonal entries of $\mathbf{HW}$ are maximized. With the proposed sequential EGT scheme, we assign the first phased array to the first user, the second array to the second user, and so on until each user is assigned to an array. If there are more phased arrays than the scheduled users, then this assignment procedure is repeated until all the phased arrays are assigned to a unique user.

12.2.3 Best-User EGT Scheme

In the sequential EGT scheme, the phased arrays are sequentially assigned to the scheduled users. This assignment strategy results in large diagonal entries and small off-diagonal entries of $\mathbf{HW}$. But we are not fully utilizing the diversity benefits offered by multiple users. In fact, the diagonal entries of $\mathbf{HW}$ can be further increased if we assign the phased array to the user that has the maximum l_1-norm of the corresponding channel vector. Particularly, when the number of RF chains, R, equals the number of scheduled users, K, the sequential EGT scheme minimizes the following approximation of $\text{Tr}(\mathbf{HWW}^H\mathbf{H}^H)^{-1}$ by maximizing $|\mathbf{h}_{kk}\mathbf{w}_k|$.

$$\text{Tr}(\mathbf{HWW}^H\mathbf{H}^H)^{-1} \approx \sum_{k=1}^{K} \frac{1}{|\mathbf{h}_{kk}\mathbf{w}_k|^2}. \qquad (12.41)$$

That is, the kth vector, $\mathbf{w}_k$, maximizes the kth entry of the approximation, i.e., $|\mathbf{h}_{kk}\mathbf{w}_k|$. Furthermore, different arrangements of users result in different values of

$\text{Tr}(\mathbf{HWW}^H\mathbf{H}^H)^{-1}$. If π denotes a permutation of $\{1, 2, \ldots, K\}$, then designing $\mathbf{w}_r$ by the sequential EGT scheme for the permutation π of the users results in

$$\text{Tr}(\mathbf{HWW}^H\mathbf{H}^H)^{-1} \approx \sum_{k=1}^{K} \frac{1}{\|\mathbf{h}_{\pi(k)k}\|_1^2}. \tag{12.42}$$

The optimal permutation that results in the minimum value of $\text{Tr}(\mathbf{HWW}^H\mathbf{H}^H)^{-1}$ requires an exhaustive search, which renders it too computationally intensive for online implementation. As such, we propose an alternative scheme, the best-user EGT scheme, that sequentially finds a permutation of $\{1, 2, \ldots, K\}$ that results in a smaller value of $\text{Tr}(\mathbf{HWW}^H\mathbf{H}^H)^{-1}$. In particular, the first element of the permutation is determined as

$$\pi(1) = \underset{i \in \mathcal{I}_1}{\arg\max} \, \|\mathbf{h}_{i1}\|_1. \tag{12.43}$$

The phase shifts are determined as $\mathbf{w}_1 = \exp(-j\angle \mathbf{h}_{\pi(1)1})$, where $\mathcal{I}_1 = \{1, 2, \ldots, K\}$. Similarly, the second element, $\pi(2)$ can be determined by

$$\pi(2) = \underset{i \in \mathcal{I}_2}{\arg\max} \, \|\mathbf{h}_{i2}\|_1, \tag{12.44}$$

where $\mathcal{I}_2 = \mathcal{I}_1 - \{\pi(1)\}$. In general,

$$\pi(r) = \underset{i \in \mathcal{I}_r}{\arg\max} \, \|\mathbf{h}_{ir}\|_1, \tag{12.45}$$

and $\mathbf{w}_r = \exp(-j\angle \mathbf{h}_{\pi(r)r})$, where $\mathcal{I}_r = \mathcal{I}_1 - \{\pi(1), \pi(2), \ldots, \pi(r-1)\}$. The assignment procedure is repeated if the number of phased arrays is greater than the number of scheduled users.

12.2.4 SVD-Based EGT Scheme

Both the sequential EGT and best-user EGT schemes aim to reduce the value of $\text{Tr}(\mathbf{HWW}^H\mathbf{H}^H)^{-1}$ by designing the analog precoder based on the EGT. Particularly, the phase-shift values of a phased array are determined based on the phase of channel vectors. This phase assignment strategy effectively uses the phased array to enable in-phase addition of different multipath components at the receiver side. But both the sequential EGT and best-user EGT schemes rely *only* on baseband ZFBF to eliminate the interuser interference. On the other hand, if the analog beamforming also contributes to the elimination of the interuser interference, the performance of the hybrid precoding scheme can be further improved.

In the following, we present a third scheme, the SVD-based EGT scheme, which improves the performance of the best-user EGT scheme by performing the orthogonalization of the channel matrix first. To this end, we consider the SVD of the matrix $\mathbf{H}$ as

$$\mathbf{H} = \mathbf{U\Sigma V}^H, \tag{12.46}$$

where both $\mathbf{U}$ and $\mathbf{\Sigma}$ are the matrices of size $K \times K$, whereas $\mathbf{V}$ is a matrix of size $M \times K$. $\mathbf{\Sigma}$ is a diagonal matrix with nonnegative diagonal entries sorted in descending order, while matrices $\mathbf{U}$ and $\mathbf{V}$ satisfy

$$\mathbf{UU}^H = \mathbf{U}^H\mathbf{U} = \mathbf{V}^H\mathbf{V} = \mathbf{I}_k, \tag{12.47}$$

where $\mathbf{I}_k$ is the identity matrix of dimension K. As with other proposed schemes, we try to design $\mathbf{W}$ such that the value of $\text{Tr}(\mathbf{HWW}^H\mathbf{H}^H)^{-1}$ is reduced. Using (12.46) and (12.47) and after some simplification, we get

$$\text{Tr}(\mathbf{HWW}^H\mathbf{H}^H)^{-1} = \text{Tr}\left(\boldsymbol{\Sigma}^{-2}(\mathbf{V}^H\mathbf{WW}^H\mathbf{V})^{-1}\right). \quad (12.48)$$

In the SVD-based EGT scheme, instead of setting the phase-shift values based on $\mathbf{H}$, we determine them based on $\mathbf{V}^H$. More specifically, the SVD-based EGT scheme first computes the SVD of $\mathbf{H}$ and then adjusts the phase-shift values in the same way as in the best-user EGT scheme applied to the matrix $\mathbf{V}^H$. Mathematically speaking, $\mathbf{V}^H$ can be rewritten as

$$\mathbf{V}^H = \begin{bmatrix} \mathbf{v}_{11} & \mathbf{v}_{12} & \cdots & \mathbf{v}_{1R} \\ \mathbf{v}_{21} & \mathbf{v}_{22} & \cdots & \mathbf{v}_{2R} \\ \vdots & \vdots & \ddots & \vdots \\ \mathbf{v}_{K1} & \mathbf{v}_{K2} & \cdots & \mathbf{v}_{KR} \end{bmatrix}. \quad (12.49)$$

Starting from the first column, we determine the phase values of $\mathbf{w}_r$ in the same way as in the best-user EGT scheme. In particular, $\mathbf{w}_1$ is determined by first computing l_1-norms of $\mathbf{v}_{k1}$, $k = 1, 2, \ldots, K$ and finding the maximum among them. Mathematically, we have $\mathbf{w}_1 = \exp\left(-j\angle \mathbf{v}_{\pi(1)1}\right)$, where

$$\pi(1) = \underset{i \in \mathcal{I}_1}{\text{argmax}}(\|\mathbf{v}_{i1}\|_1). \quad (12.50)$$

Similarly, $\mathbf{w}_r = \exp\left(-j\angle \mathbf{v}_{\pi(r)r}\right)$, where

$$\pi(r) = \underset{i \in \mathcal{I}_r}{\text{argmax}}(\|\mathbf{v}_{ir}\|_1), \quad (12.51)$$

and $\mathcal{I}_r = \{1, 2, \ldots, K\} - \{\pi(1), \pi(2), \ldots, \pi(r-1)\}$.

12.2.5 Complexity and Performance Comparison

The main computation in the sequential EGT scheme is the calculation and assignment of the phase-shift values of the phase shifters. Since, we have a total of M phase shifters, the computational complexity of the sequential EGT scheme is $\mathcal{O}(M)$. For the best-user EGT scheme, the most computationally intensive operation is the l_1-norm computation of K channel vectors for each phased array. Therefore, the computational complexity of the best-user EGT scheme is $\mathcal{O}(RLK) = \mathcal{O}(MK)$. For the SVD-based EGT scheme, we first compute the matrix $\mathbf{V}$ and calculate the phase values in the same way as in the best-user EGT scheme. Since the SVD of $\mathbf{H}$ is $\mathcal{O}(MK^2)$ and the best-user EGT scheme is $\mathcal{O}(MK)$, the computational complexity of the SVD-based EGT scheme turns out to be $\mathcal{O}(MK^2)$. With the exhaustive-search scheme based on EGT, the phased arrays are first sequentially assigned to every possible permutation of the users. Then, for each assignment, the sum rate is computed. Finally, the array-user assignment that results in the highest value of the sum rate is selected for the transmission. The most computationally intensive operation of the exhaustive-search-based EGT scheme is the computation of $\text{Tr}(\mathbf{HWW}^H\mathbf{H}^H)^{-1}$, which is $\mathcal{O}(MK + RK^2)$ and computed for each permutation of the

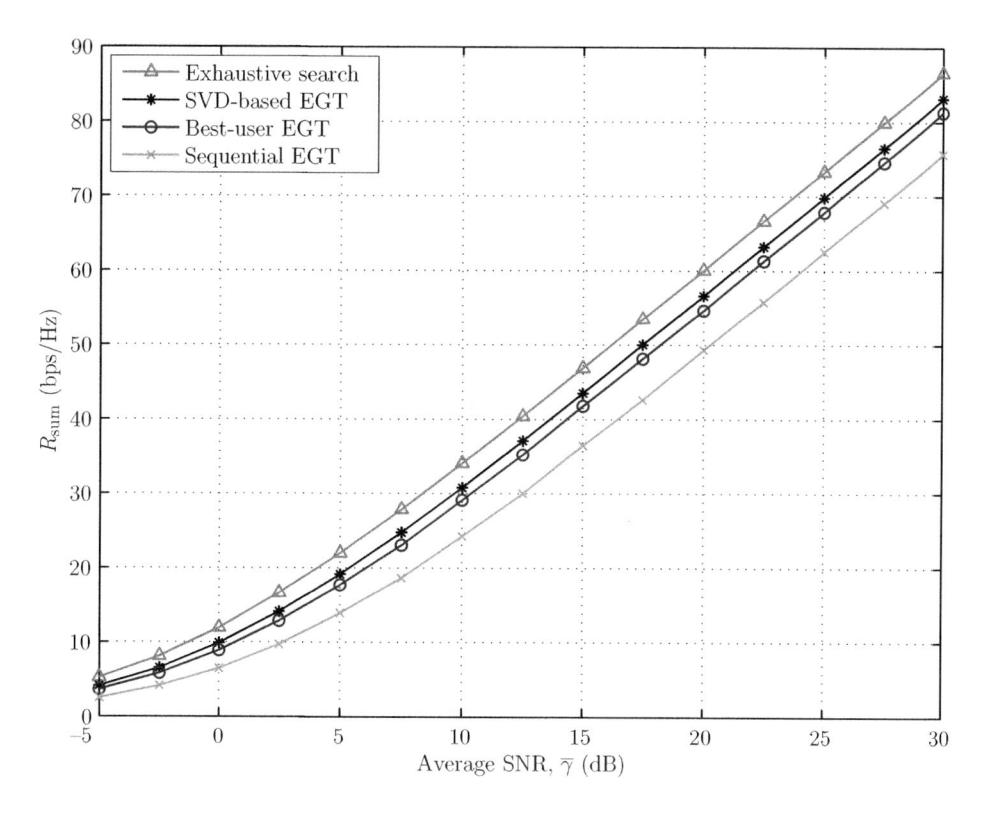

Figure 12.6 Sum rate of the proposed schemes in an i.i.d. Rayleigh fading environment against the average SNR of the individual channel. The number of phased arrays is $R = 8$, and each phased array has $L = 16$ phase shifters. Reprint with permission from [21]. ©2017 IET.

phase assignment. Therefore, the computational complexity of the exhaustive-search-based EGT scheme is $\mathcal{O}(K^R(MK + RK^2))$. As such, our proposed schemes have much lower complexity than the exhaustive-search-based scheme.

We study the sum rate performance on the proposed EGT-based schemes using Monte-Carlo simulation performed over 3000 trials in an i.i.d. Rayleigh fading environment. Fig. 12.6 shows the behavior of the sum rates for the proposed schemes against the individual link average SNR, $\bar{\gamma} = P_{\text{tot}}\mathbb{E}[|h_{ij}|^2]/\sigma^2$. For comparison purposes, we also plotted the sum rate of the exhaustive-search-based EGT scheme. As expected, the scheme with higher computational complexity outperforms the schemes with lower computational complexity. Moreover, the sum rates of all schemes improve by increasing the link average SNR. Lastly, among the proposed schemes, the SVD-based EGT scheme outperforms others for the whole SNR range, whereas the sequential EGT shows the worst performance as it has the lowest computational complexity.

Fig. 12.7 shows the sum rate of the EGT-based schemes against the number of scheduled users. We can see that the SVD-based and the best-user EGT schemes show similar performances, and both of them outperform the sequential EGT scheme for all the values of K and for both SNR values. The gap between the sum rates of the SVD-based EGT and the best-user EGT scheme grows as K increases. We also notice in Fig. 12.7

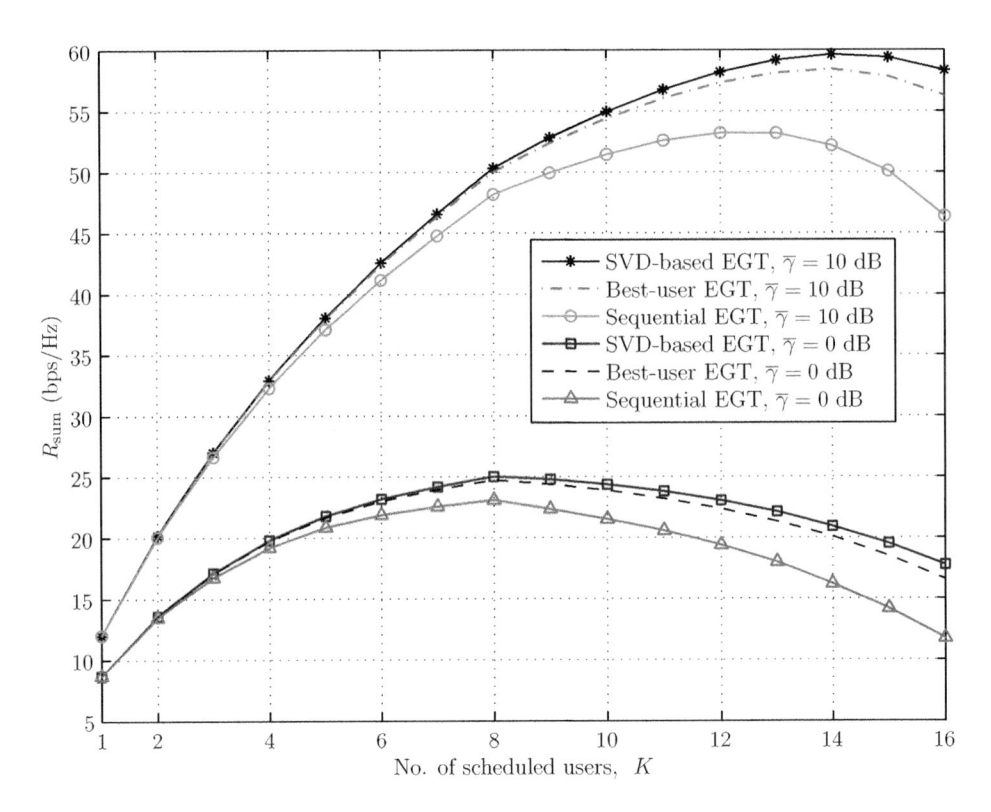

Figure 12.7 Sum rate of the proposed schemes against the number of scheduled users, K, in an i.i.d. Rayleigh fading environment with different average SNRs. The number of phased arrays is $R = 16$, and each phased array has $L = 32$ phase shifters. Reprint with permission from [21]. ©2017 IET.

that the sum rate of all the schemes shows a peak, and the optimal number of users, K^*, that maximizes the sum rate is not necessarily equal to the number of RF chains, R. Furthermore, the value of K^* varies by changing the individual link SNR. Specifically, K^* increases by increasing $\overline{\gamma}$ for all the schemes.

12.3 Summary

In this chapter, we studied two general approaches for low-complexity design of massive MIMO systems. In particular, we presented several suboptimal antenna-subset selection schemes based on matrix trace minimization. Both the trace-based and the min-trace-based schemes show better performance as compared with other existing schemes with similar computational complexity. We also investigated several hybrid precoding schemes for massive MIMO systems through joint-design EGT and ZFBF. We targeted subconnected architecture and sequential implementation for complexity reduction purposes. Various numerical results are presented and discussed to illustrate the performance versus complexity trade-offs involved.

12.4 Further Reading

Readers can refer to the book [23] for a complete survey of massive MIMO. Additional surveys on this promising topic include [24, 25]. The effect of nonideal hardware on massive MIMO transmission is studied in [26]. [27, 28] present another two antenna subsection selection schemes, particularly for massive MIMO systems. [19, 29] present a hybrid precoding scheme for a single receive scenario, whereas [20] studies zero-forcing-based hybrid precoding algorithms for a fully connected architecture.

References

[1] T. L. Marzetta, "Noncooperative cellular wireless with unlimited numbers of base station antennas," *IEEE Trans. Wireless Commun.*, 9, no. 11, pp. 3590–3600, 2010.

[2] E. Bjornson, E. G. Larsson, and T. L. Marzetta, "Massive MIMO: Ten myths and one critical question," *IEEE Commun. Mag.*, 54, no. 2, pp. 114–123, 2016.

[3] H. Q. Ngo, E. G. Larsson, and T. L. Marzetta, "Energy and spectral efficiency of very large multiuser MIMO systems," *IEEE Trans. Commun.*, 61, no. 4, pp. 1436–1449, 2013.

[4] F. Rusek, D. Persson, B. K. Lau, et al., "Scaling up MIMO: Opportunities and challenges with very large arrays," *IEEE Signal Process. Mag.*, 30, no. 1, pp. 40–60, 2013.

[5] A. F. Molisch, M. Z. Win, and J. H. Winters, "Reduced-complexity transmit/receive-diversity systems," *IEEE Trans. Signal Process.*, 51, no. 11, pp. 2729–2738, 2003.

[6] M. Gkizeli and G. N. Karystinos, "Maximum-SNR antenna selection among a large number of transmit antennas," *IEEE J. Sel. Topics Signal Process.*, 8, no. 5, pp. 891–901, 2014.

[7] S. Han, C.-L. I. Z. Xu, and C. Rowell, "Large-scale antenna systems with hybrid analog and digital beamforming for millimeter wave 5G," *IEEE Commun. Mag.*, 53, no. 1, pp. 186–194, 2015.

[8] A. Goldsmith, *Wireless Communications*, Cambridge University Press, 2005.

[9] M. K. Simon and M.-S. Alouini, *Digital Communication over Fading Channels*, 2nd ed., Wiley, 2005.

[10] R. Chen, J. Andrews, and R. Heath, "Efficient transmit antenna selection for multiuser MIMO systems with block diagonalization," in *IEEE Global Telecommun. Conf.*, Washington, DC, November 2007, pp. 3499–3503.

[11] M. Mohaisen and K. Chang, "On transmit antenna selection for multiuser MIMO systems with dirty paper coding," in *IEEE Int. Symp. Personal, Indoor Mobile Radio Commun.*, Tokyo September 2009, pp. 3074–3078.

[12] M. Benmimoune, E. Driouch, W. Ajib, and D. Massicotte, "Joint transmit antenna selection and user scheduling for massive MIMO systems," in *IEEE Wireless Commun. Netw. Conf.*, New Orleans, LA March 2015, pp. 381–386.

[13] M. Hanif, H.-C. Yang, G. Boudreau, E. Sich, and H. Seyedmehdi, "Antenna subset selection for massive MIMO systems: a trace-based sequential approach for sum rate maximization," *J. Commun. Netw.*, 20, no. 2, pp. 144–155, 2018.

[14] P. H. Lin and S. H. Tsai, "Performance analysis and algorithm designs for transmit antenna selection in linearly precoded multiuser MIMO systems," *IEEE Trans. Veh. Technol.*, 61, no. 4, pp. 1698–1708, 2012.

[15] M. Hanif, H.-C. Yang, G. Boudreau, E. Sich, and H. Seyedmehdi, "Low complexity antenna subset selection for massive MIMO systems with multi-cell cooperation," in *IEEE Globecom Workshops*, December 2015, pp. 1–5.

[16] J. Wang Z. Lan, C. Pyo et al., "Beam codebook based beamforming protocol for multi-Gbps millimeter-wave WPAN systems," *IEEE J. Sel. Areas Commun.*, 27, no. 8, pp. 1390–1399, 2009.

[17] S. Hur, T. Kim, D. J. Love, et al., "Millimeter wave beamforming for wireless backhaul and access in small cell networks," in *IEEE Trans. Commun.*, 61, no. 10, pp. 4391–4403, 2013.

[18] L. Liang, W. Xu, and X. Dong, "Low-complexity hybrid precoding in massive multiuser MIMO systems," *IEEE Wireless Commun. Lett.*, 3, no. 6, pp. 653–656, 2014.

[19] O. E. Ayach, S. Rajagopal, S. Abu-Surra, Z. Pi, and R. W. Heath, "Spatially sparse precoding in millimeter wave MIMO systems," *IEEE Trans. Wireless Commun.*, 13, no. 3, pp. 1499–1513, 2014.

[20] A. Alkhateeb, G. Leus, and R. W. Heath, "Limited feedback hybrid precoding for multi-user millimeter wave systems," in *IEEE Trans. Wireless Commun.*, 14, no. 11, pp. 6481–6494, 2015.

[21] M. Hanif, H.-C. Yang, G. Boudreau, E. Sich, and H. Seyedmehdi, "Low-complexity hybrid precoding for multi-user massive MIMO systems: A hybrid EGT/ZF approach," *IET Commun.*, 11, no. 5, pp. 765–771, 2017.

[22] T. Yoo and A. Goldsmith, "On the optimality of multiantenna broadcast scheduling using zero-forcing beamforming," *IEEE J. Sel. Areas Commun.*, 24, no. 3, pp. 528–541, 2006.

[23] T. Marzetta, E. G. Larsson, H. Yang, and H. Q. Ngo, *Fundamentals of Massive MIMO* Cambridge University Press, 2016.

[24] K. Zheng, L. Zhao, J. Mei, et al., "Survey of large-scale MIMO systems," *IEEE Commun. Surv. Tut.*, 17, no. 3, pp. 1738–1760, 2015.

[25] L. Lu, G. Y. Li, A. L. Swindlehurst, A. Ashikhmin, and R. Zhang, "An overview of massive MIMO: Benefits and challenges," *IEEE J. Sel. Topics Signal Process.*, 8, no. 5, pp. 742–758, 2014.

[26] E. Bjrnson, J. Hoydis, M. Kountouris, and M. Debbah, "Massive MIMO systems with non-ideal hardware: Energy efficiency, estimation, and capacity limits," *IEEE Trans. Inform. Theor.*, 60, no. 11, pp. 7112–7139, 2014.

[27] X. Gao, O. Edfors, F. Tufvesson, and E. G. Larsson, "Massive MIMO in real propagation environments: Do all antennas contribute equally?" *IEEE Trans. Commun.*, 63, no. 11, pp. 3917–3928, 2015.

[28] K. Elkhalil, A. Kammoun, T. Y. Al-Naffouri, and M. S. Alouini, "A blind antenna selection scheme for single-cell uplink massive MIMO," in *IEEE Globecom Workshops*, December 2016, pp. 1–6.

[29] J. Brady, N. Behdad, and A. M. Sayeed, "Beamspace MIMO for millimeter-wave communications: System architecture, modeling, analysis, and measurements," *IEEE Trans. Antennas Propagat.*, 61, no. 7, pp. 3814–3827, 2013.

Appendix: Order Statistics

The performance analysis of wireless communication systems requires the statistics of the signal-to-noise ratio (SNR) at the receiver. In the analysis of several advanced wireless communication techniques, we make use of order statistical results to obtain SNR statistics. This Appendix summarizes these results and their derivations for easy reference. Specifically, we first review the basic distribution functions of ordered random variables. After that, we derive several new order statistics results, including the joint distribution functions of partial sums of ordered random variables. The chapter is concluded with a discussion on the limiting distributions of extremes. Whenever appropriate, we use the exponential random variable special case as an illustrative example. Note that we focus on those order statistics results that are employed in the performance and complexity analysis of different wireless transmission technologies. For more thorough treatment of *order statistics*, the readers are referred to [1, 2].

A.1 Basic Distribution Functions

Order statistics deals with the distributions and statistical properties of the new random variables obtained after ordering the realizations of some random variables. Let γ_j, $j = 1, 2, \ldots, L$ denote L i.i.d. nonnegative random variables with common probability distribution function (PDF) $p_\gamma(\cdot)$ and cumulative density function (CDF) $F_\gamma(\cdot)$. Let $\gamma_{l:L}$ denote the random variable corresponding to the lth largest observation of the L original random variables, such that $\gamma_{1:L} \geq \gamma_{2:L} \geq \ldots \geq \gamma_{L:L}$. $\gamma_{l:L}$ is also called lth order statistics. The ordering process is illustrated in Fig. A.1.

A.1.1 Marginal and Joint Distributions

As an immediate result of the ordering process, these order statistics are no longer identically distributed [1]. The PDF of $\gamma_{l:L}$, $l = 1, 2, \ldots, L$, can be shown to be given by

$$p_{\gamma_{l:L}}(x) = \frac{L!}{(L-l)!(l-1)!}[F_\gamma(x)]^{L-l}[1 - F_\gamma(x)]^{l-1} p_\gamma(x). \tag{A.1}$$

The PDF of the largest random variable, $\gamma_{1:L}$, and the smallest random variable, $\gamma_{L:L}$, can be obtained as

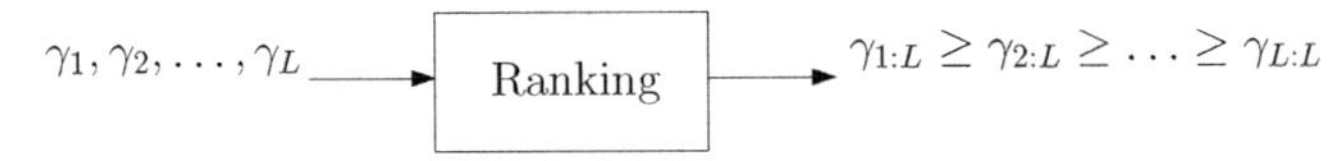

$$\gamma_1, \gamma_2, \ldots, \gamma_L \longrightarrow \boxed{\text{Ranking}} \longrightarrow \gamma_{1:L} \geq \gamma_{2:L} \geq \ldots \geq \gamma_{L:L}$$

Figure A.1 Ordering random variables.

$$p_{\gamma_{1:L}}(x) = L[F_\gamma(x)]^{L-1} p_\gamma(x), \qquad (A.2)$$

and

$$p_{\gamma_{L:L}}(x) = L[1 - F_\gamma(x)]^{L-1} p_\gamma(x), \qquad (A.3)$$

respectively.

Another consequence of the ordering operation is that the ordered random variables are independent of one another. Specifically, the joint PDF of arbitrary two order statistics, $\gamma_{l:L}$ and $\gamma_{k:L}$, $l < k$, can be shown to be given by [1]

$$p_{\gamma_{l:L}, \gamma_{k:L}}(x, y) = \frac{L!}{(l-1)!(k-l-1)!(L-k)!} \left(1 - F_\gamma(x)\right)^{l-1} p_\gamma(x) \qquad (A.4)$$
$$\times \left(F_\gamma(x) - F_\gamma(y)\right)^{k-l-1} p_\gamma(y) \left(F_\gamma(y)\right)^{L-k},$$

which is clearly not equal to the product of the marginal PDFs, $p_{\gamma_{l:L}}(x)$ and $p_{\gamma_{k:L}}(x)$. Furthermore, while the joint PDF of the L original random variables can be written as the product of individual PDFs, because of the independence property, as

$$p_{\gamma_1, \gamma_2, \ldots, \gamma_L}(x_1, x_2, \ldots, x_L) = \prod_{i=1}^{L} p_\gamma(x_i), \qquad (A.5)$$

the joint PDF of the L ordered random variables becomes

$$p_{\gamma_{1:L}, \gamma_{2:L}, \ldots, \gamma_{L:L}}(x_1, x_2, \ldots, x_L) = L! \prod_{i=1}^{L} p_\gamma(x_i), \quad x_1 \geq x_2 \geq \ldots \geq x_L. \qquad (A.6)$$

A.1.2 Conditional Distributions

Now let us consider the distribution of an ordered random variable, given the realization of another ordered random variable [1]. These results are summarized in the following two theorems.

THEOREM A.1 *The conditional distribution of the jth order statistics, $\gamma_{j:L}$, given that the lth order statistics $\gamma_{l:L}$ is equal to y, where $j > l$, is the distribution of the $L - j$th order statistics of $L - l$ i.i.d. random variables γ_j^-, whose PDF is the truncated PDF of the original random variable γ on the right at y, i.e.,*

$$p_{\gamma_{j:L} | \gamma_{l:L} = y}(x) = p_{\gamma_{L-j:L-l}^-}(x), \qquad (A.7)$$

where the PDF of γ_j^- is given by

$$p_{\gamma_j^-}(x) = \frac{p_\gamma(x)}{F_\gamma(y)}, x \leq y. \qquad (A.8)$$

Proof. By definition, the conditional PDF of $\gamma_{j:L}$ given $\gamma_{l:L} = y$ is

$$p_{\gamma_{j:L}|\gamma_{l:L}=y}(x) = \frac{p_{\gamma_{j:L},\gamma_{l:L}}(x,y)}{p_{\gamma_{l:L}}(y)}. \quad (A.9)$$

After substituting (A.4) and (A.1) into (A.9) and some manipulations, we can show that

$$p_{\gamma_{j:L}|\gamma_{l:L}=y}(x) = \frac{(L-l)!}{(j-l-1)!(L-j)!}\left(1 - \frac{F_\gamma(x)}{F_\gamma(y)}\right)^{j-l-1} \frac{p_\gamma(x)}{F_\gamma(y)}\left(\frac{F_\gamma(x)}{F_\gamma(y)}\right)^{L-j} \quad (A.10)$$

which is the PDF of the $L - j$th order statistics of $L - l$ i.i.d. random variables with distribution function $p_\gamma(x)/F_\gamma(y)$ ($x \leq y$). □

THEOREM A.2 *The conditional distribution of the jth order statistics, $\gamma_{j:L}$, given that the lth order statistics $\gamma_{l:L}$ is equal to y, where $j < l$, is the same as the distribution of the jth order statistics of $l - 1$ i.i.d. random variables γ_j^+, whose PDF is the truncated PDF of the original random variable γ on the left at y, i.e.,*

$$p_{\gamma_{j:L}|\gamma_{l:L}=y}(x) = p_{\gamma_{j:l-1}^+}(x), \quad (A.11)$$

where the PDF of γ_j^+ is given by

$$p_{\gamma_j^+}(x) = \frac{p_\gamma(x)}{1 - F_\gamma(y)}, \quad x \geq y. \quad (A.12)$$

The theorem can be similarly proved as Theorem A.1 [3]. These two theorems establish that order statistics of samples from a continuous distribution form a reversible Markov chain. Specifically, the distribution of $\gamma_{j:L}$ given that $\gamma_{l:L} = y$ with $j < l$ is independent of $\gamma_{l+1:L}, \gamma_{l+2:L}, \ldots, \gamma_{L:L}$ and the distribution of $\gamma_{j:L}$ given that $\gamma_{l:L} = y$ with $j > l$ is independent of $\gamma_{l-1:L}, \gamma_{l-2:L}, \ldots, \gamma_{1:L}$. These results become very useful in the derivation of the following sections.

A.2 Distribution of Partial Sum of the Largest Order Statistics

In the analysis of certain advance diversity combining schemes, we require the statistics of the sum of the largest order statistics, i.e., $\sum_{i=1}^{L_s} \gamma_{i:L}$, where $L_s < L$ [4–7]. The direct calculation of such a distribution function can be very tedious [8], as the largest L_s order statistics $\gamma_{i:L}, i = 1, 2, \ldots, L_s$ are correlated random variables with distribution function

$$p_{\gamma_{1:L},\gamma_{2:L},\ldots,\gamma_{L_s:L}}(x_1, x_2, \ldots, x_{L_s}) = \frac{L!}{(L-L_s)!}[F_\gamma(x_{L_s})]^{L-L_s} \prod_{i=1}^{L_s} p_\gamma(x_i),$$

$$x_1 \geq x_2 \geq \ldots \geq x_{L_s}. \quad (A.13)$$

A.2.1 Exponential special case

When γ_i are i.i.d. exponential random variables, i.e., the PDF of γ_i is commonly given by

$$p_\gamma(x) = \frac{1}{\overline{\gamma}} e^{-\frac{x}{\overline{\gamma}}}, \quad x \geq 0, \tag{A.14}$$

where $\overline{\gamma}$ is the common mean, we can invoke the classical result by Sukhatme to convert the sum of correlated random variables to the sum of independent random variables [4, 9].

THEOREM A.3 *The spacings between ordered exponential random variables $x_l = \gamma_{l:L} - \gamma_{l+1:L}$, $l = 1, 2, \ldots, L$, are independently exponential random variables with distribution functions given by*

$$p_{x_l}(x) = \frac{l}{\overline{\gamma}} e^{-lx/\overline{\gamma}}, \; x \geq 0. \tag{A.15}$$

Proof. Omitted. □

The sum of the largest order statistics can be rewritten as

$$\sum_{i=1}^{L_s} \gamma_{i:L} = \sum_{i=1}^{L_s} \sum_{l=i}^{L} x_l \tag{A.16}$$
$$= x_1 + 2x_2 + \ldots + L_s x_{L_s} + L_s x_{L_s+1} + \ldots + L_s x_L,$$

which becomes the sum of independent random variables. We can then follow the moment generating function (MGF) approach to derive the statistics of the sum. Specifically, the MGF of lx_k can be shown to be given by

$$\mathcal{M}_{lx_k}(s) = \int_0^\infty p_{lx_k}(x) e^{sx} dx = \left(1 - \frac{s\overline{\gamma}l}{k}\right)^{-1}. \tag{A.17}$$

Consequently, the MGF of the partial sum $\sum_{i=1}^{L_s} \gamma_{i:L}$ can be obtained as the product of the individual MGF as

$$\mathcal{M}_{\sum_{i=1}^{L_s} \gamma_{i:L}}(s) = (1 - s\overline{\gamma})^{-L_s} \prod_{l=L_s+1}^{L} \left(1 - \frac{s\overline{\gamma}L_s}{l}\right)^{-1}. \tag{A.18}$$

Finally, the PDF and CDF of $\sum_{i=1}^{L_s} \gamma_{i:L}$ can be routinely obtained after applying proper inverse Laplace transform, for which closed-form results can be obtained. For example, the PDF of $\sum_{i=1}^{L_s} \gamma_{i:L}$ is given by

$$p_{\sum_{i=1}^{L_s} \gamma_{i:L}}(x) = \frac{L!}{(L-L_s)! L_s!} e^{-\frac{x}{\overline{\gamma}}} \left[\frac{x^{L_s-1}}{\overline{\gamma}^{L_s}(L_s-1)!} \right.$$
$$+ \frac{1}{\overline{\gamma}} \sum_{l=1}^{L-L_s} (-1)^{L_s+l-1} \frac{(L-L_s)!}{(L-L_s-l)! l!} \left(\frac{L_s}{l}\right)^{L_s-1}$$
$$\left. \times \left(e^{-\frac{lx}{L_s \overline{\gamma}}} - \sum_{m=0}^{L_s-2} \frac{1}{m!} \left(-\frac{lx}{L_s \overline{\gamma}}\right)^m \right) \right]. \tag{A.19}$$

A.2.2 General Case

The statistics of $\sum_{i=1}^{L_s} \gamma_{i:L}$ becomes more challenging to obtain when γ_i are not exponentially distributed. Existing solutions usually involve several folds of integration, which does not lead to convenient mathematical evaluation [5]. Based on the results in the previous section, we propose an alternative approach to obtain the statistics of $\sum_{i=1}^{L_s} \gamma_{i:L}$ for the general case. The basic idea is to treat the partial sum as the sum of two correlated random variables as

$$\sum_{i=1}^{L_s} \gamma_{i:L} = \sum_{i=1}^{L_s-1} \gamma_{i:L} + \gamma_{L_s:L}. \quad (A.20)$$

Then, the PDF of $\sum_{i=1}^{L_s} \gamma_{i:L}$ can be obtained from the joint PDF of $\gamma_{L_s:L}$ and $\sum_{i=1}^{L_s-1} \gamma_{i:L} \equiv y_{L_s}$, denoted by $p_{\gamma_{L_s:L}, y_{L_s}}(\cdot, \cdot)$ as

$$p_{\sum_{i=1}^{L_s} \gamma_{i:L}}(x) = \int_0^\infty p_{\gamma_{L_s:L}, y_{L_s}}(x-y, y) dy. \quad (A.21)$$

We now derive the general result of the joint PDF of the lth order statistics, $\gamma_{l:L}$, and the partial sum of the first $l - 1$ order statistics, $y_l = \sum_{j=1}^{l-1} \gamma_{j:L}$, $p_{\gamma_{l:L}, y_l}(\cdot, \cdot)$. Note that while the γ_j are independent random variables, y_l and $\gamma_{l:L}$ are correlated random variables. Applying the Bayesian formula, we can write $p_{\gamma_{l:L}, y_l}(\cdot, \cdot)$ as

$$p_{\gamma_{l:L}, y_l}(x, y) = p_{\gamma_{l:L}}(x) \times p_{y_l | \gamma_{l:L}=x}(y), \quad (A.22)$$

where $p_{\gamma_{l:L}}(\cdot)$ is the PDF of the lth order statistics, given in (A.1), and $p_{y_l | \gamma_{l:L}=x}(\cdot)$ is the conditional PDF of the sum of the first $l - 1$ order statistics given that $\gamma_{l:L}$ is equal to x. The conditional PDF $p_{y_l | \gamma_{l:L}=x}(\cdot)$ can be determined with the help of the following corollary of Theorem A.2 [10].

COROLLARY A.1 *The conditional PDF of the sum of the first $l - 1$ order statistics of L i.i.d. random variables ($l \leq L$), given that the lth order statistics is equal to y, is the same as the distribution of the sum of $l - 1$ different i.i.d. random variables whose PDF is the truncated PDF of the original random variable on the left at y, i.e.,*

$$p_{y_l | \gamma_{l:L}=y}(x) = p_{\sum_{j=1}^{L=l-1} \gamma_j^+}(x), \quad (A.23)$$

where the PDF of γ_j^+ was given in (A.12).

Proof. Theorem A.2 states that given that the lth order statistics $\gamma_{l:L} = y$, the conditional distribution of the jth order statistics with $j < l$ is the same as the distribution of the jth order statistics of $l - 1$ different i.i.d. random variables whose PDF is the truncated PDF of the original random variable on the left at y, as shown in (A.12). Therefore, the distribution of the sum of the first $l - 1$ order statistics of L i.i.d. random variables given that the lth order statistics $\gamma_{l:L} = y$ is the same as the distribution of the sum of the $l - 1$ order statistics of $l - 1$ i.i.d. random variables with the truncated distribution. The proof of this corollary is completed by noting that the ordering operation does not affect the statistics of the sum of the random variables. $\square$

Appendix: Order Statistics

Based on the above corollary, the conditional PDF in (A.22) can be obtained as the PDF of the sum of $l-1$ i.i.d. random variables. The MGF approach can also apply, which is demonstrated for the exponential r. v. case in the following example.

Example A.1 When γ_i are i.i.d. exponentially distributed with CDF given by

$$F_\gamma(x) = 1 - e^{-\frac{x}{\bar{\gamma}}}, \quad \gamma \geq 0, \tag{A.24}$$

the truncated PDF in (A.12) specializes to

$$p_{\gamma_j^+}(y) = \frac{1}{\bar{\gamma}} e^{-\frac{x-y}{\bar{\gamma}}}, \quad x \geq y. \tag{A.25}$$

The corresponding MGF of γ_j^+ can be shown to be given by

$$M_{\gamma_j^+}(t) = \int_0^{+\infty} p_{\gamma_j^+}(y) e^{sx} dx = \frac{1}{1 - s\bar{\gamma}} e^{sy}. \tag{A.26}$$

Consequently, the MGF of the sum of $l-1$ i.i.d. random variable γ_j^+ can be shown to be given by

$$M_{\sum_{j=1}^{l-1} \gamma_j^+}(t) = [M_{\gamma_j^+}(t)]^{l-1} = \frac{1}{(1 - s\bar{\gamma})^{l-1}} e^{(l-1)sx}, \tag{A.27}$$

which is also the MGF of the conditional random variable. After carrying out proper inverse Laplace transform, we can obtain the conditional PDF of the sum of the first $l-1$ order statistics $p_{y_l|\gamma_{l:L}=y}(\cdot)$ as

$$\begin{aligned}
p_{y_l|\gamma_{l:L}=y}(x) &= p_{\sum_{j=1}^{l-1} \gamma_j^+}(x) \\
&= \frac{1}{(l-2)!\bar{\gamma}^{l-1}} [x - (l-1)y]^{(l-2)} e^{-\frac{x-(l-1)y}{\bar{\gamma}}}, \quad x \geq (l-1)y.
\end{aligned} \tag{A.28}$$

Finally, we can obtain a generic expression of the joint PDF $p_{\gamma_{l:L},y_l}(\cdot,\cdot)$ after appropriate substitution into (A.22) as

$$p_{\gamma_{l:L},y_l}(x,y) = \frac{L!}{(L-l)!(l-1)!} [F_\gamma(x)]^{L-l} [1 - F_\gamma(x)]^{l-1} p_\gamma(x) p_{\sum_{j=1}^{l-1} \gamma_j^+}(y). \tag{A.29}$$

Note that since the random variable γ_j are nonnegative and $\gamma_{j:L} \geq \gamma_{l:L}$ for all $j = 1, 2, \ldots, l-1$, we have $y_l = \sum_{j=1}^{l-1} \gamma_{j:L} \geq (l-1)\gamma_{l:L}$. Therefore, the support of the joint PDF $p_{\gamma_{l:L},y_l}(x,y)$ is the region $x > 0, y > (l-1)x$.

For the exponential r. v. special case, noting that the PDF of the lth order statistics of L i.i.d. exponential random variables specializes to

$$p_{\gamma_{l:L}}(x) = \frac{L!}{\bar{\gamma}(l-1)!} \sum_{j=0}^{L-l} \frac{(-1)^j}{(L-l-j)!j!} e^{-\frac{(l+j)x}{\bar{\gamma}}}, \tag{A.30}$$

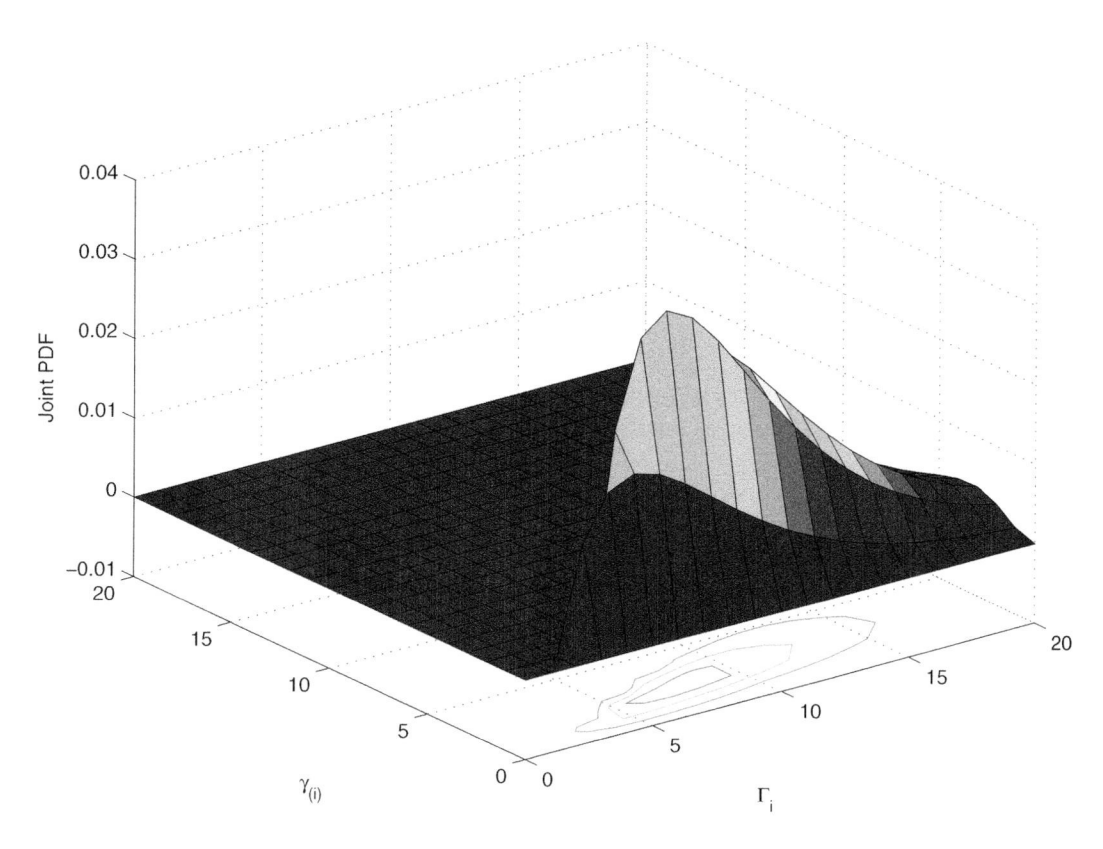

Figure A.2 Joint PDF of Γ_{i-1} and $\gamma_{(i)}$ for the exponential random variable case ($L = 6$, $i = 3$, and $\overline{\gamma} = 6$ dB). Reprint with permission from [10]. ©2006 IEEE.

we obtain the joint PDF of $\gamma_{l:L}$ and y_l, by substituting (A.30) and (A.28) into (A.22), as

$$p_{\gamma_{l:L}, y_l}(x, y) = \sum_{j=0}^{L-l} \frac{(-1)^j L!}{(L-l-j)!(l-1)!(l-2)!j!\overline{\gamma}^l} [y - (l-1)x]^{(l-2)} e^{-\frac{y+(j+1)x}{\overline{\gamma}}},$$
$$x \geq 0, \ y \geq (l-1)x. \tag{A.31}$$

As an illustration, the joint PDF of $\gamma_{l:L}$ and y_l for the exponential special case, as given in (A.31), is plotted in Fig. A.2. Note that since $l = 3$ here, the joint PDF is equal to zero when $y_l < 2\gamma_{l:L}$.

A.3   Joint Distributions of Partial Sums

In this section, we generalize the result in the previous section by deriving the joint distributions of partial sums of order statistics [3]. We consider two scenarios depending on whether all ordered random variables are involved or not.

A.3.1 Cases Involving All Random Variables

We first investigate the three-dimensional joint PDF $p_{y_l,\gamma_{l:L},z_l}(y,\gamma,z)$, where y_l is the sum of the first $l-1$ order statistics, i.e., $y_l = \sum_{j=1}^{l-1} \gamma_{j:L}$, and z_l is the sum of the last $L-l$ order statistics, i.e., $z_l = \sum_{j=l+1}^{L} \gamma_{j:L}$. The definition of y_l and z_l is illustrated as.

$$y_l = \underbrace{\sum_{i=1}^{l-1} \gamma_{i:L}}_{\gamma_{1:L},\ldots,\gamma_{l-1:L}}, \gamma_{l:L}, \underbrace{z_l = \sum_{i=l+1}^{k-1} \gamma_{i:L}}_{\gamma_{l+1:L},\ldots,\gamma_{L:L}}. \qquad (A.32)$$

Applying the Bayesian rule twice, we can write this joint PDF $p_{y_l,\gamma_{l:L},z_l}(y,\gamma,z)$ as the product of the PDF of $\gamma_{l:L}$ and two conditional PDFs as

$$p_{y_l,\gamma_{l:L},z_l}(y,\gamma,z) = p_{\gamma_{l:L}}(\gamma) \times p_{z_l|\gamma_{l:L}=\gamma}(z) \times p_{y_l|\gamma_{l:L}=\gamma,z_l=z}(y). \qquad (A.33)$$

The generic expression of $p_{\gamma_{l:L}}(\cdot)$ was given in (A.1) in terms of the common PDF and CDF of γ_i, $p_\gamma(\gamma)$, and $P_\gamma(\gamma)$. As noted earlier, the order statistics of samples from a continuous distribution form a Markov chain, the conditional distribution of the jth order statistics of L i.i.d. random samples, given that the lth order statistics is equal to γ ($j < l$) and that the sum of the last $L - l$ order statistics is equal to z, is the same as the conditional distribution of the jth order statistics given only that the lth order statistics is equal to γ, i.e.,

$$p_{\gamma_{j:L}|\gamma_{l:L}=\gamma,\sum_{k=l+1}^{L}\gamma_{k:L}=z}(y) = p_{\gamma_{j:L}|\gamma_{l:L}=\gamma}(y). \qquad (A.34)$$

With the application of Corollary A.1, we have

$$p_{y_l|\gamma_{l:L}=\gamma,z_l=z}(y) = p_{\sum_{j=1}^{l-1}\gamma_j^+}(y), \; y > (l-1)\gamma, \qquad (A.35)$$

where the PDF of γ_j^+ was given in (A.12)

The conditional PDF $p_{z_l|\gamma_{l:L}=\gamma}(z)$ can be determined with the help of the following corollary [3].

COROLLARY A.2 *The conditional distribution of the sum of the last $L - l$ order statistics of L i.i.d. random samples ($l \leq L$), given that the lth order statistics is equal to γ, is the same as the distribution of the sum of $L - l$ different i.i.d. random variables whose PDF is the PDF of the original (unordered) random variable truncated on the right of γ, i.e.,*

$$p_{z_l|\gamma_{l:L}=\gamma}(z) = p_{\sum_{j=1}^{L-l}\gamma_j^-}(z), \; z < (L-l)\gamma, \qquad (A.36)$$

where the PDF of γ_j^- is given by

$$p_{\gamma_j^-}(x) = \frac{p_\gamma(x)}{F_\gamma(\gamma)}, \quad 0 < x < \gamma. \qquad (A.37)$$

Proof. Based on Theorem A.1, the distribution of the sum of the last $L - l$ order statistics of L i.i.d. random variables given that the lth order statistics $\gamma_{l:L} = \gamma$ is the same as the distribution of the sum of $L - l$ order statistics of $L - l$ i.i.d. random variables

A.3 Joint Distributions of Partial Sums

with truncated distribution. The proof is completed by noting that the ordering operation does not affect the statistics of the sum of random variables. □

Finally, we obtain the joint PDF $p_{y_l, \gamma_{l:L}, z_l}(y, \gamma, z)$ as

$$p_{y_l, \gamma_{l:L}, z_l}(y, \gamma, z) = p_{\gamma_{l:L}}(\gamma) p_{\sum_{j=1}^{L-l} \gamma_j^-}(z) p_{\sum_{j=1}^{l-1} \gamma_j^+}(y), \quad \text{(A.38)}$$
$$\gamma > 0, y > (l-1)\gamma, z < (L-l)\gamma.$$

Example A.2 When γ_i are i.i.d. exponentially distributed, $p_{\gamma_j^-}(x)$ defined in (A.37) specializes to

$$p_{\gamma_j^-}(x) = \frac{\frac{1}{\overline{\gamma}} e^{-\frac{x}{\overline{\gamma}}}}{1 - e^{-\frac{\gamma}{\overline{\gamma}}}}, \quad x < \gamma. \quad \text{(A.39)}$$

Following again the MGF approach, we can obtain the closed-form expression for $p_{z_l | \gamma_{l:L} = \gamma}(\cdot)$ as

$$p_{\sum_{j=1}^{L-l} \gamma_j^-}(z) = \sum_{i=0}^{L-l} \binom{L-l}{i} \frac{e^{-z/\overline{\gamma}}}{(1 - e^{-\gamma/\overline{\gamma}})^{L-l}} \frac{(-1)^i (z - i\gamma)^{L-l-1}}{\overline{\gamma}^{L-l}(L-l-1)!} \mathcal{U}(z - i\gamma), \quad \text{(A.40)}$$

where $\mathcal{U}(\cdot)$ is the unit step function.

It follows, after properly substituting (A.14), (A.24), (A.28), and (A.36) into (A.38), we can obtain a closed-form expression for the joint PDF $p_{y_l, \gamma_{l:L}, z_l}(y, \gamma, z)$ as

$$p_{y_l, \gamma_{l:L}, z_l}(y, \gamma, z) = \frac{L!}{(L-l)!(l-1)!\overline{\gamma}^L} \frac{[y - (l-1)\gamma]^{l-2}}{(l-2)!(L-l-1)!} e^{-\frac{y+\gamma+z}{\overline{\gamma}}} \mathcal{U}(y - (l-1)\gamma)$$
$$\times \sum_{i=0}^{L-l} \binom{L-l}{i} (-1)^i (z - i\gamma)^{L-l-1} \mathcal{U}(z - i\gamma),$$
$$\gamma > 0, y > (l-1)\gamma, z < (L-l)\gamma. \quad \text{(A.41)}$$

We can apply the three-dimensional joint PDF of partial sums derived above to obtain the joint PDF of the lth order statistics and the sum of the remaining $L - 1$ order statistics. Specifically, setting $l = 1$, (A.38) reduces to the joint PDF of the largest random variable and the sum of the remaining ones, denoted by $p_{\gamma_{1:L}, z_1}(\gamma, z)$, which is given by

$$p_{\gamma_{1:L}, z_1}(\gamma, z) = p_{\gamma_{1:L}}(\gamma) p_{\sum_{j=1}^{L-1} \gamma_j^-}(z), \quad \gamma > 0, z < (L-l)\gamma. \quad \text{(A.42)}$$

For the general case where $1 < l < L$, we can obtain the joint PDF of $\gamma_{l:L}$ and $\sum_{i \neq l} \gamma_{i:L}$, while noting that the latter is equal to $y_l + z_l$, as

$$p_{\gamma_{l:L}, y_l + z_l}(\gamma, w) = \int_0^{(L-l)\gamma} p_{y_l, \gamma_{l:L}, z_l}(w - z, \gamma, z) \, dz, \quad y > (l-1)\gamma. \quad \text{(A.43)}$$

These joint PDFs are useful in the performance analysis of different wireless transmission technologies in later chapters.

A.3.2 Cases Only Involving the Largest Random Variables

We now consider the joint PDFs involving only the first k ordered random variables, i.e., $\gamma_{l:L}$, $l = 1, 2, \ldots, k$. More specifically, we are interested in obtaining the joint PDF of the lth largest random variable ($l < k$) and the sum of the remaining $k - 1$ ones, which is denoted by $\sum_{j=1, j \neq l}^{k} \gamma_{j:L}$ [11]. Note that the case of $l = k$ has been addressed in the previous section.

We first consider the general case where $1 < l < k - 1$. In this case, we start with the joint PDF of the four random variables y_l, $\gamma_{l:L}$, z_l^k, $\gamma_{k:L}$, where y_l and z_l^k are partial sums defined based on the first k order statistics, as

$$\underbrace{y_l = \sum_{i=1}^{l-1} \gamma_{i:L}}_{\gamma_{1:L}, \ldots, \gamma_{l-1:L}}, \gamma_{l:L}, \underbrace{z_l^k = \sum_{i=l+1}^{k-1} \gamma_{i:L}}_{\gamma_{l+1:L}, \ldots, \gamma_{k-1:L}}, \gamma_{k:L}, \gamma_{k+1:L}, \ldots, \gamma_{L:L}. \tag{A.44}$$

Applying the Bayesian rule twice, the four-dimensional joint PDF under consideration, $p_{y_l, \gamma_{l:L}, z_l^k, \gamma_{k:L}}(y, \gamma, z, \beta)$, can be written as

$$p_{y_l, \gamma_{l:L}, z_l^k, \gamma_{k:L}}(y, \gamma, z, \beta) = p_{\gamma_{l:L}, \gamma_{k:L}}(\gamma, \beta) \cdot p_{y_l | \gamma_{l:L} = \gamma, \gamma_{k:L} = \beta}(y)$$
$$\times p_{z_l^k | \gamma_{l:L} = \gamma, \gamma_{k:L} = \beta, y_l = y}(z), \tag{A.45}$$

where $p_{\gamma_{l:L}, \gamma_{k:L}}(\cdot, \cdot)$ is the joint PDF of $\gamma_{l:L}$ and $\gamma_{k:L}$, the generic expression of which was given in (A.4). As shown earlier, ordered random variables satisfy the Markovian property. Therefore, the conditional PDFs $p_{y_l | \gamma_{l:L} = \gamma, \gamma_{k:L} = \beta}(y)$ and $p_{z_l^k | \gamma_{l:L} = \gamma, \gamma_{k:L} = \beta, y_l = y}(z)$ are equivalent to $p_{y_l | \gamma_{l:L} = \gamma}(y)$ and $p_{z_l^k | \gamma_{l:L} = \gamma, \gamma_{k:L} = \beta}(z)$, respectively. With the application of Corollary A.1, we obtain the conditional PDF $p_{y_l | \gamma_{l:L} = \gamma, \gamma_{k:L} = \beta}(y)$ as

$$p_{y_l | \gamma_{l:L} = \gamma, \gamma_{k:L} = \beta}(y) = p_{y_l | \gamma_{l:L} = \gamma}(y)$$
$$= p_{\sum_{i=1}^{l-1} \gamma_i^+}(y), \quad 0 < (l-1)\gamma < y, \tag{A.46}$$

where γ_i^+ is the random variable with truncated distribution given in (A.12).

The conditional PDF $p_{z_l^k | \gamma_{l:L} = \gamma, \gamma_{k:L} = \beta, y_l = y}(z)$, or equivalently $p_{z_l^k | \gamma_{l:L} = \gamma, \gamma_{k:L} = \beta}(z)$, can be determined with the help of the following corollary [11].

COROLLARY A.3 *The conditional distribution of the sum of $k - l + 1$ order statistics, i.e., $\sum_{i=l+1}^{k-1} \gamma_{i:L}$, given that the lth order statistics is equal to γ and the kth order statistics is equal to β, is the same as the distribution of the sum of $k - l + 1$ different i.i.d. random variables whose PDF is the truncated PDF of the original random variable on the right at γ and on the left at β, i.e.,*

$$p_{z_l^k | \gamma_{l:L} = \gamma, \gamma_{k:L} = \beta}(z) = p_{\sum_{i=1}^{k-l-1} \gamma_i^{\pm}}(z),$$
$$0 < \beta < \frac{z}{k - l - 1} < \gamma, \tag{A.47}$$

where $\gamma_i^{\pm}$ denotes the random variable with PDF

$$p_{\gamma_i^{\pm}}(x) = \frac{p_\gamma(x)}{F_\gamma(\gamma) - F_\gamma(\beta)}, \quad \beta \leq x \leq \gamma.$$

A.3 Joint Distributions of Partial Sums

The results can be easily proved after sequentially applying Corollary A.1 and Corollary A.2.

Finally, after substituting (A.4), (A.46), and (A.47) into (A.45), we obtain the generic expression of the joint PDF $p_{y_l, \gamma_{1:L}, z_1^k, \gamma_{k:L}}(y, \gamma, z, \beta)$ as

$$p_{y_l, \gamma_{1:L}, z_1^k, \gamma_{k:L}}(y, \gamma, z, \beta) = p_{\gamma_{1:L}, \gamma_{k:L}}(\gamma, \beta) \cdot p_{\sum_{i=1}^{l-1} \gamma_i^+}(y) \cdot p_{\sum_{i=1}^{k-l-1} \gamma_i^\pm}(z), \quad (A.48)$$

$$0 < \beta < \gamma, y > (l-1)\gamma, \beta < \frac{z}{k-l-1} < \gamma.$$

In the case of $k = 1$, it is sufficient to consider the three-dimensional joint PDF of $p_{\gamma_{1:L}, z_1^k, \gamma_{k:L}}(\gamma, z, \beta)$, which can be written as

$$p_{\gamma_{1:L}, z_1^k, \gamma_{k:L}}(\gamma, z, \beta) = p_{\gamma_{1:L}, \gamma_{k:L}}(\gamma, \beta) \cdot p_{z_1^k | \gamma_{1:L} = \gamma, \gamma_{k:L} = \beta}(z). \quad (A.49)$$

Applying Corollary A.3, the 3D joint PDF can be obtained as

$$p_{\gamma_{1:L}, z_1^k, \gamma_{k:L}}(\gamma, z, \beta) = p_{\gamma_{1:L}, \gamma_{k:L}}(\gamma, \beta) \cdot p_{\sum_{i=1}^{k-2} \gamma_i^\pm}(z), \quad (A.50)$$

$$0 < \beta < \gamma, (k-2)\beta < z < (k-2)\gamma.$$

When $l = k - 1$, the three-dimensional joint PDF of interest becomes $p_{y_{k-1}, \gamma_{k-1:L}, \gamma_{k:L}}(y, \gamma, \beta)$, which can be written, after applying the Bayesian rule, as

$$p_{y_{k-1}, \gamma_{k-1:L}, \gamma_{k:L}}(y, \gamma, \beta) = p_{\gamma_{k-1:L}, \gamma_{k:L}}(\gamma, \beta) \cdot p_{y_{k-1} | \gamma_{k-1:L} = \gamma, \gamma_{k:L} = \beta}(y). \quad (A.51)$$

It follows, after proper substitution, that $p_{y_{k-1}, \gamma_{k-1:L}, \gamma_{k:L}}(y, \gamma, \beta)$ can be shown to be given by

$$p_{y_{k-1}, \gamma_{k-1:L} \gamma_{k:L}}(y, \gamma, \beta) = p_{\gamma_{k-1:L}, \gamma_{k:L}}(\gamma, \beta) \cdot p_{\sum_{i=1}^{k-2} \gamma_i^+}(y), \quad (A.52)$$

$$0 < \beta < \gamma, y > (k-2)\gamma.$$

The three-dimensional joint PDFs $p_{\gamma_{1:L}, z_1^k, \gamma_{k:L}}(\gamma, z, \beta)$ and $p_{y_{k-1}, \gamma_{k-1:L} \gamma_{k:L}}(y, \gamma, \beta)$ for the exponential random variable special case can be similarly obtained after proper substitutions into (A.50) and (A.52).

Example A.3 When γ_i are i.i.d. exponential random variables, it can also be shown that $p_{\gamma_i^\pm}(x)$ specializes to

$$p_{\gamma_i^\pm}(x) = \frac{e^{-x/\overline{\gamma}}}{\overline{\gamma}\left(e^{-\beta/\overline{\gamma}} - e^{-\gamma/\overline{\gamma}}\right)}, \quad \beta \leq x \leq \gamma. \quad (A.53)$$

The corresponding MGF is given by

$$\mathcal{M}_{\gamma_i^\pm}(s) = \int_\beta^\gamma p_{\gamma_i^\pm}(x) e^{sx} dx = \frac{e^{s\beta - \beta/\overline{\gamma}} - e^{s\gamma - \gamma/\overline{\gamma}}}{(e^{-\beta/\overline{\gamma}} - e^{-\gamma/\overline{\gamma}})(1 - s\overline{\gamma})}. \quad (A.54)$$

Noting that the MGF of $\sum_{i=1}^{k-l-1} \gamma_i^\pm$ is equal to $[\mathcal{M}_{\gamma_i^\pm}(s)]^{k-l-1}$, we can obtain the closed-form expression for (A.47) by taking the inverse Laplace transform as

$$p_{z_l^k|\gamma_{l:L}=\gamma,\gamma_{k:L}=\beta,y_l=y}(z) = \mathcal{L}^{-1}\{[\mathcal{M}_{\gamma_i^\pm}(s)]^{k-l-1}\} \quad (A.55)$$

$$= \sum_{j=0}^{k-l-1}\binom{k-l-1}{j}\frac{(-1)^j e^{-z/\overline{\gamma}}[z-\beta(k-l-j-1)-\gamma j]^{k-l-2}}{[(e^{-\beta/\overline{\gamma}}-e^{-\gamma/\overline{\gamma}})\overline{\gamma}]^{k-l-1}(k-l-2)!}$$

$$\times \mathcal{U}(z-\beta(k-l-j-1)-\gamma j), (k-l-1)\beta < z < (k-l-1)\gamma.$$

Meanwhile, the joint PDF of $\gamma_{l:L}$ and $\gamma_{k:L}$ given in (A.4) becomes

$$p_{\gamma_{l:L},\gamma_{k:L}}(x,y) = \frac{L!}{(l-1)!(k-l-1)!(L-k)!\overline{\gamma}^2}e^{-\frac{lx+y}{\overline{\gamma}}} \quad (A.56)$$

$$\times \left(e^{-\frac{y}{\overline{\gamma}}}-e^{-\frac{x}{\overline{\gamma}}}\right)^{k-l-1}\cdot\left(1-e^{-\frac{y}{\overline{\gamma}}}\right)^{L-k}.$$

After substituting (A.28), (A.55), and (A.56) into (A.48), we can obtain a closed-form expression for the four-dimensional joint PDF, $p_{y_l,\gamma_{l:L},z_l^k,\gamma_{k:L}}(y,\gamma,z,\beta)$, for the exponential random variable case as

$$p_{y_l,\gamma_{l:L},z_l^k,\gamma_{k:L}}(y,\gamma,z,\beta) \quad (A.57)$$

$$= \frac{L! e^{-(y+\gamma+z+\beta)/\overline{\gamma}}(1-e^{-\beta/\overline{\gamma}})^{L-k}[y-(l-1)\gamma]^{l-2}}{(L-k)!(k-l-1)!(k-l-2)!(l-1)!(l-2)!\overline{\gamma}^k}$$

$$\times \sum_{j=0}^{k-l-1}\binom{k-l-1}{j}(-1)^j[b-\beta(k-l-j-1)-\gamma j]^{k-l-2}$$

$$\times \mathcal{U}(\gamma)\mathcal{U}(\gamma-\beta)\mathcal{U}(y-(l-1)\gamma)\mathcal{U}(z-\beta(k-l-j-1)-\gamma j),$$

$$(k-l-1)\beta < z < (k-l-1)\gamma.$$

Remark: We notice from the above three examples that the derivation of the joint PDFs often involves the statistics of the sum of some truncated random variables. For convenience, we summarize the PDF and MGF of the truncated random variables corresponding to the received SNR of the three most popular fading channel models in Table A.1.

With these multiple-dimensional joint PDFs available, we can readily derive the joint PDF of any partial sums involving the largest k ordered random variables. For example, we can obtain the joint PDF of $Y = y_l + \gamma_{l:L}$ and $Z = z_l^k + \gamma_{k:L}$, $p_{Y,Z}(y,z)$ as

$$p_{Y,Z}(y,z) = \int_0^{\frac{z}{k-l}}\int_{\frac{z}{k-l}}^{\frac{y}{l}} p_{y_l,\gamma_{l:L},z_l^k,\gamma_{k:L}}(y-\gamma,\gamma,z-\beta,\beta)d\gamma d\beta, \quad y > \frac{l}{k-l}z. \quad (A.58)$$

In addition, the joint PDF $\gamma_{l:L}$ and $W = y_l + z_l^k + \gamma_{k:L}$, denoted by $p_{\gamma_{l:L},W}(\gamma,w)$, can be calculated as

A.3 Joint Distributions of Partial Sums

Table A.1 Truncated MGFs for the three fading models under consideration.

	$M_{\gamma^+}(s)$ where γ^+ has PDF $p_{\gamma_j^+}(x) = \frac{p_\gamma(x)}{1-P_\gamma(\gamma_T)}$, $x \geq \gamma_T$.
Rayleigh	$\dfrac{1}{1-s\bar{\gamma}} e^{s\gamma_T}$
Rice	$\dfrac{1+K}{1+K-s\bar{\gamma}} e^{\frac{s\bar{\gamma}K}{1+K-s\bar{\gamma}}} \dfrac{Q_1\left(\sqrt{\frac{2K(1+K)}{1+K-s\bar{\gamma}}}, \sqrt{2(1+K-s\bar{\gamma})\frac{\gamma_T}{\bar{\gamma}}}\right)}{Q_1\left(\sqrt{2K}, \sqrt{2(1+K)\frac{\gamma_T}{\bar{\gamma}}}\right)}$
Nakagami-m	$\left(1-\dfrac{s\bar{\gamma}}{m}\right)^{-m} \dfrac{\Gamma\left(m, \frac{m\gamma_T}{\bar{\gamma}} - s\gamma_T\right)}{\Gamma\left(m, \frac{m\gamma_T}{\bar{\gamma}}\right)}$

	$M_{\gamma^-}(s)$ where γ^- has PDF $p_{\gamma_j^-}(x) = \frac{p_\gamma(x)}{P_\gamma(\gamma_T)}$, $0 < x < \gamma_T$.
Rayleigh	$\dfrac{1}{1-s\bar{\gamma}} \dfrac{1 - e^{s\gamma_T - \frac{\gamma_T}{\bar{\gamma}}}}{1 - e^{-\frac{\gamma_T}{\bar{\gamma}}}}$
Rice	$\dfrac{1+K}{1+K-s\bar{\gamma}} e^{\frac{s\bar{\gamma}K}{1+K-s\bar{\gamma}}} \dfrac{1 - Q_1\left(\sqrt{\frac{2K(1+K)}{1+K-s\bar{\gamma}}}, \sqrt{2(1+K-s\bar{\gamma})\frac{\gamma_T}{\bar{\gamma}}}\right)}{1 - Q_1\left(\sqrt{2K}, \sqrt{2(1+K)\frac{\gamma_T}{\bar{\gamma}}}\right)}$
Nakagami-m	$\left(1-\dfrac{s\bar{\gamma}}{m}\right)^{-m} \dfrac{1 - \Gamma\left(m, \frac{m\gamma_T}{\bar{\gamma}} - s\gamma_T\right)/\Gamma(m)}{1 - \Gamma\left(m, \frac{m\gamma_T}{\bar{\gamma}}\right)/\Gamma(m)}$

	$M_{\gamma^\pm}(s)$ where $\gamma^\pm$ has $p_{\gamma_i^\pm}(x) = \frac{p_\gamma(x)}{P_\gamma(\gamma_{T_1}) - P_\gamma(\gamma_{T_2})}$, $\gamma_{T_2} \leq x \leq \gamma_{T_1}$.
Rayleigh	$\dfrac{1}{1-s\bar{\gamma}} \dfrac{e^{s\gamma_{T_2} - \frac{\gamma_{T_2}}{\bar{\gamma}}} - e^{s\gamma_{T_1} - \frac{\gamma_{T_1}}{\bar{\gamma}}}}{e^{-\frac{\gamma_{T_2}}{\bar{\gamma}}} - e^{-\frac{\gamma_{T_1}}{\bar{\gamma}}}}$
Rice	$\dfrac{\frac{1+K}{1+K-s\bar{\gamma}} e^{\frac{s\bar{\gamma}K}{1+K-s\bar{\gamma}}} Q_1\left(\sqrt{\frac{2K(1+K)}{1+K-s\bar{\gamma}}}, \sqrt{2(1+K-s\bar{\gamma})\frac{\gamma_{T_2}}{\bar{\gamma}}}\right)}{Q_1\left(\sqrt{2K}, \sqrt{2(1+K)\frac{\gamma_{T_2}}{\bar{\gamma}}}\right) - Q_1\left(\sqrt{2K}, \sqrt{2(1+K)\frac{\gamma_{T_1}}{\bar{\gamma}}}\right)}$ $- \dfrac{\frac{1+K}{1+K-s\bar{\gamma}} e^{\frac{s\bar{\gamma}K}{1+K-s\bar{\gamma}}} Q_1\left(\sqrt{\frac{2K(1+K)}{1+K-s\bar{\gamma}}}, \sqrt{2(1+K-s\bar{\gamma})\frac{\gamma_{T_1}}{\bar{\gamma}}}\right)}{Q_1\left(\sqrt{2K}, \sqrt{2(1+K)\frac{\gamma_{T_2}}{\bar{\gamma}}}\right) - Q_1\left(\sqrt{2K}, \sqrt{2(1+K)\frac{\gamma_{T_1}}{\bar{\gamma}}}\right)}$
Nakagami-m	$\left(1-\dfrac{s\bar{\gamma}}{m}\right)^{-m} \dfrac{\Gamma\left(m, \frac{m\gamma_{T_2}}{\bar{\gamma}} - s\gamma_{T_2}\right) - \Gamma\left(m, \frac{m\gamma_{T_1}}{\bar{\gamma}} - s\gamma_{T_1}\right)}{\Gamma\left(m, \frac{m\gamma_{T_2}}{\bar{\gamma}}\right) - \Gamma\left(m, \frac{m\gamma_{T_1}}{\bar{\gamma}}\right)}$

$$p_{\gamma_{l:L},W}(\gamma,w) = \begin{cases} \displaystyle\int_{(k-2)w/(k-1)}^{(k-2)\gamma} p_{\gamma_{1:L},z_1^k,\gamma_{k:L}}(\gamma,z,w-z)dz, & l=1; \\ \displaystyle\int_0^\gamma \int_{(k-l-1)\beta}^{(k-l-1)\gamma} p_{\gamma_l,\gamma_{l:L},z_l^k,\gamma_{k:L}}(w-z-\beta,\gamma,z,\beta)dzd\beta, & 1 < l < k-1; \\ \displaystyle\int_0^\gamma p_{\gamma_{k-1},\gamma_{k-1:L}\gamma_{k:L}}(w-\beta,\gamma,\beta)d\beta, & l=k-1. \end{cases} \quad (A.59)$$

Note that only finite integrations of the joint PDFs are involved in these calculations. Therefore, even though a closed-form expression is tedious to obtain, this joint probability can be easily calculated with mathematical software, such as Mathematica or Maple.

A.4 Limiting Distributions of Extreme Order Statistics

The limiting distribution of the largest order statistics from L i.i.d. samples can be useful in the asymptotic analysis of multiuser wireless systems. It has been established that the limiting distribution of $\gamma_{1:L}$, i.e., $\lim_{L\to+\infty} F_{\gamma_{1:L}}(x)$, if it exists, must be one of the following three types: [1, 12]

- Fréchet distribution with CDF

$$F^{(1)}(x) = \exp(-x^{-\alpha}), x > 0, \alpha > 0; \quad (A.60)$$

- Weibull distribution with CDF

$$F^{(2)}(x) = \begin{cases} \exp[-(-x)^{\alpha}], & x \leq 0, \alpha > 0, \\ 1, & x > 0; \end{cases} \quad (A.61)$$

- Gumbel distribution with CDF

$$F^{(3)}(x) = \exp(-e^{-x}). \quad (A.62)$$

Specifically, there exist constants $a_L > 0$ and b_L such that

$$\lim_{L\to+\infty} P_{\gamma_{1:L}}(a_L x + b_L) = F^{(i)}(x), \ i = 1, 2, \text{ or } 3. \quad (A.63)$$

Note that these three distributions are members of the family of generalized extreme-value (GEV) distribution with CDF given by

$$F^{GEV}(x) = \exp\left\{-\left[1 + \xi\left(\frac{x-\mu}{\sigma}\right)\right]^{-1/\xi}\right\}, \quad (A.64)$$

where ξ, μ, and σ are constant parameters. Specifically, when $\xi = 0$, the CDF of GEV distribution simplifies to

$$F^{GEV}(x) = \exp\left(-e^{-\frac{x-\mu}{\sigma}}\right), \quad (A.65)$$

which is of the Gumbel type. Furthermore, setting $\xi > 0$ and $\xi < 0$ will lead to Fréchet and Weibull types, respectively.

The type of the limiting distribution depends on the properties of the distribution functions of the original unordered random variables. It can be shown that if the distribution functions of γ_i satisfy

$$\lim_{x\to+\infty} \frac{x p_\gamma(x)}{1 - F_\gamma(x)} = \alpha, \quad (A.66)$$

for some constant $\alpha > 0$, then the limiting distribution of $\gamma_{1:L}$ will be of the Fréchet type. This indicates that for some constant $a_L > 0$, we have

$$\lim_{L\to+\infty} F_{\gamma_{1:L}}(a_L x) = F^{(1)}(x), \quad (A.67)$$

where a_L can be computed by solving $1 - P_\gamma(a_L) = 1/L$ based on the characteristic of extremes. If, instead, the following condition is satisfied by the distribution of γ_i

$$\lim_{x\to+\infty} \frac{1 - F_\gamma(x)}{p_\gamma(x)} = \alpha, \quad (A.68)$$

where the constant $\alpha > 0$, then the limiting distribution of $\gamma_{1:L}$ will be of the Gumbel type. More specifically, we have

$$\lim_{L \to +\infty} F_{\gamma_{1:L}}(x) = F^{(3)}\left(\frac{x - b_L}{a_L}\right) = \exp\left(-e^{-\frac{x-b_L}{a_L}}\right), \quad \text{(A.69)}$$

where the constants b_L and a_L can be computed by sequentially solving equations $1 - F_\gamma(b_L) = 1/L$ and $1 - F_\gamma(a_L + b_L) = 1/(eL)$.

Example A.4 When γ_i follow an i.i.d. exponential distribution with PDF $p_\gamma(x) = e^{-x}$, we can easily verify the condition in (A.68) holds and, as such, the limiting distribution of $\gamma_{1:L}$ is of the Gumbel type. Applying the results in (A.69), we can show that $a_L = 1$ and $b_L = \log L$. Finally, we have

$$\lim_{L \to +\infty} F_{\gamma_{1:L} - \log L}(x) = \lim_{L \to +\infty} \Pr[\gamma_{1:L} < \log L + x] = \exp\left(-e^{-x}\right). \quad \text{(A.70)}$$

A.5 Summary

In this Appendix, the relevant order statistics results used in the book were presented in a systematic fashion. Starting from the basic distribution functions of order random variables, we derived the distribution functions of the partial sum and the joint distribution function of several partial sums. The exponential random variable special case is used for illustration whenever feasible. The results of limiting distribution of ordered random variables are also summarized.

A.6 Further Reading

For more thorough coverage of order statistics, the reader may refer to [1]. Section 9.11.2 of [13] addresses the distribution function of the sum of the largest order statistics for nonidentically distributed original random variables. [14] presents a novel MGF based-analytical framework for the joint distribution function of several partial sums.

References

[1] H. A. David, *Order Statistics*. Wiley, 1981.
[2] N. Balakrishnan and C. R. Rao, *Handbook of Statistics 17: Order Statistics: Applications*, 2nd ed., Elsevier, 1998.
[3] Y.-C. Ko, H.-C. Yang, S.-S. Eom, and M.-S. Alouini, "Adaptive modulation and diversity combining based on output-threshold MRC," *IEEE Trans. Wireless Commun.*, 6, no. 10, pp. 3728–3737, 2007.

[4] M.-S. Alouini and M. K. Simon, "An MGF-based performance analysis of generalized selective combining over Rayleigh fading channels," *IEEE Trans. Commun.*, 48, no. 3, pp. 401–415, 2000.

[5] Y. Ma and C. C. Chai, "Unified error probability analysis for generalized selection combining in Nakagami fading channels," *IEEE J. Select. Areas Commun.*, 18, no. 11, pp. 2198–2210, 2000.

[6] M. Z. Win and J. H. Winters, "Virtual branch analysis of symbol error probability for hybrid selection/maximal-ratio combining Rayleigh fading," *IEEE Trans. Commun.*, 49, no. 11, pp. 1926–1934, 2001.

[7] A. Annamalai and C. Tellambura, "Analysis of hybrid selection/maximal-ratio diversity combiners with Gaussian errors," *IEEE Trans. Wireless Commun.*, 1, no. 3, pp. 498–511, 2002.

[8] Y. Roy, J.-Y. Chouinard, and S. A. Mahmoud, "Selection diversity combining with multiple antennas for MM-wave indoor wireless channels," *IEEE J. Select. Areas Commun.*, 14, no. 4, pp. 674–682, 1998.

[9] P. V. Sukhatme, "Tests of significance for samples of the population with two degrees of freedom," *Ann. Eugenics*, 8, pp. 52–56, 1937.

[10] H.-C. Yang, "New results on ordered statistics and analysis of minimum-selection generalized selection combining (GSC)," *IEEE Trans. Wireless Commun.*, 5, no. 7, pp. 1876–1885, 2006.

[11] S. Choi, M.-S. Alouini, K. A. Qaraqe, and H.-C. Yang, "Finger replacement method for RAKE receivers in the soft handover region," *IEEE Trans. Wireless Commun.*, 7, no. 4, pp. 1152–1156, 2008.

[12] N. T. Uzgoren, "The asymptotic development of the distribution of the extreme values of a sample," in *Studies in Mathematics and Mechanics Presented to Richard von Mises*, Academic Press, 1954, pp. 346–353.

[13] M. K. Simon and M.-S. Alouini, *Digital Communications over Generalized Fading Channels*, 2nd ed., Wiley, 2004.

[14] S. Nam, M.-S. Alouini, and H.-C. Yang, "An MGF-based unified framework to determine the joint statistics of partial sums of ordered random variables," *IEEE Trans. Inform. Theor.*, 56, no. 11, pp. 5655–5672, 2010.

Index

absolute SNR-based scheduling, 149
absolute threshold GSC, 61
adaptive beam activation based on conditional best beam index feedback, 203
adaptive combining, 84, 319
adaptive diversity, 84
adaptive equalization, 43
adaptive modulation and coding, 33
adaptive transmission, 31
Alamouti's scheme, 30
all beam feedback strategy, 214
amplify and forward, 227
antenna subset selection, 114, 367
average access rate, 149
average access time, 149
average error rate, 13
average feedback load, 116, 156
average service time, 263, 274
average spectrum efficiency, 158
average waiting time, 262

bandwidth efficient AMDC, 131, 134
bandwidth efficient and power greedy AMDC, 131, 135
best beam feedback strategy, 216
best beam index strategy, 193
best beam SINR and index strategy, 193
best-user EGT, 385
block change scheme, 123

centralized OR, 232
channel-blind transmission, 343
channel coherence bandwidth, 5
channel coherence time, 9
channel reciprocity, 368
channel-aware transmission, 348
closed-loop transmit diversity, 113
code division multiple access, 48
cognitive radio, 256
continuous sensing, 264
cooperative beam selection, 352
cooperative relay, 240
cyclic prefix, 46

decode and forward, 228
delay-insensitive, 336
delay-sensitive, 336
direct-sequence spread spectrum, 48
dirty paper coding, 180
distributed OR, 233
diversity combining, 20, 55
diversity gain, 36
diversity order, 37

energy harvesting, 335
equal gain combining, 24
equal gain transmission, 378
equal SNR power reallocation, 171
equalization, 41
extended delivery time, 264

fading, 1
fairness issue, 148
false alarm, 277
free-space optical, 303
frequency selective fading, 16, 41
frequency-hopping spread spectrum, 52
full channel inversion, 32
full GSC scheme, 123
fully connected architecture, 378

generalized selection combining, 55
generalized selection multiuser scheduling, 160
generalized switch and examine combining, 66
generalized switch and examine combining with post-examine selection, 72
Gram–Schmidt orthogonalization, 368
greedy weight clique ZFBF, 184
GSC with threshold test per branch, 60

hard-switching, 303
hybrid analog and digital precoding, 367
hybrid FSO/RF, 303
hybrid precoding, 377

imperfect sensing, 277
incremental relaying, 246

Index

interblock interference, 46
intersymbol interference, 18
interuser interference, 179
interweave, 256

joint adaptive modulation and adaptive combining, 323
joint adaptive modulation and diversity combining, 129

linear bandpass modulation, 10
linear precoding, 180

massive MIMO, 367
max–min relay selection, 247
maximum ratio combining, 22
millimeter wave, 303
MIMO broadcast channel, 179
MIMO transmission, 36
min-trace-based antenna selection, 373
minimum selection GSC, 91
misdetection, 277
multicarrier transmission, 44
multicarrier/OFDM, 6, 45
multipath fading, 4
multiple-antenna transmission and reception, 36
multiuser diversity, 146
multiuser MIMO system, 179
multiuser parallel scheduling, 160
multiuser selection diversity, 147

Nakagami fading, 7
nonregenerative relaying, 227
non–work-preserving strategy, 266
nonlinear precoding, 180
normalized SNR-based user scheduling, 151
normalized threshold GSC, 61

on–off-based scheduling, 162
open-loop transmit diversity, 113
opportunistic regenerative relaying, 240
opportunistic relaying, 230
opportunistic spectrum access, 256
optimum combining, 212
orthogonal frequency division multiplexing (OFDM), 41
outage probability, 8, 13
output threshold GSC, 84
output threshold MRC, 102
overlaid sensor transmission, 335

packet service time, 274
path diversity, 51
path loss, 2
peak-to-average power ratio, 47
periodic sensing, 264
power adaptation, 32, 290

power allocation, 169
power delay profile, 4
power-efficient AMDC, 131
precoding, 179
proportional fair scheduling, 148

quantized best beam SINR and index strategy, 193

RAKE finger management, 121
RAKE receiver, 51, 120
random beamforming, 189
random unitary beamforming, 179, 352
ranking maintained power reallocation, 170
rate adaptation, 33
rate-maintained power reallocation, 171
Rayleigh fading, 7
regenerative relay, 228
relay selection, 230
relay transmission, 226
RF energy harvesting, 335
Rician fading, 7
RMS delay spread, 5

scheduling outage, 155
secondary transmission, 256
selection combining, 21, 212
selective fading, 5
selective multiuser diversity, 155
sequential EGT, 384
shadowing, 3
SHO overhead, 125
signal to interference plus noise ratio, 18, 290
slow fading, 9
soft handover, 120
soft-switching, 303
space-time block coding, 30
spatial multiplexing gain, 36
spectrum opportunities, 257
spectrum sharing, 257, 281
spread spectrum transmission, 48
subconnected architecture, 378
successive projection ZFBF, 184
sum-rate capacity, 180
SVD-based EGT, 385
switch and examine combining, 25
switch and examine combining with post-examining selection, 26
switch and stay combining, 24
switched multiuser diversity, 157
switched-based multiuser scheduling, 164

tapped delay line, 5
temporal spectrum opportunity, 258
threshold combining, 24
trace of matrix inverse, 369
trace-based antenna selection, 371
transmit antenna replacement, 115